Remote Sensing of Night-Time Light

Remote Sensing of Night-Time Light

Editors

Ran Goldblatt
Steven Louis Rubinyi
Hogeun Park

MDPI • Basel • Beijing • Wuhan • Barcelona • Belgrade • Manchester • Tokyo • Cluj • Tianjin

Editors
Ran Goldblatt
New Light Technologies Inc.
Washington DC, USA

Steven Louis Rubinyi
The World Bank
Washington, DC, USA

Hogeun Park
The World Bank
Washington, DC, USA

Editorial Office
MDPI
St. Alban-Anlage 66
4052 Basel, Switzerland

This is a reprint of articles from the Special Issue published online in the open access journal *Remote Sensing* (ISSN 2072-4292) (available at: https://www.mdpi.com/journal/remotesensing/special_issues/RS_Night_Time_Light).

For citation purposes, cite each article independently as indicated on the article page online and as indicated below:

LastName, A.A.; LastName, B.B.; LastName, C.C. Article Title. *Journal Name* **Year**, *Volume Number*, Page Range.

ISBN 978-3-0365-7178-2 (Hbk)
ISBN 978-3-0365-7179-9 (PDF)

Contents

About the Editors

Ran Goldblatt

Dr. Ran Goldblatt is the Chief Scientist of New Light Technologies Inc. in Washington DC. He is a GIS and Remote Sensing analyst with 25+ years of experience in geospatial data management, modeling, analysis and visualization. He is an expert in remote sensing analysis methods of satellite and aerial imagery, including processing of multispectral satellite data, image classification, feature extraction, remotely sensed impact evaluation and time-series analysis of land cover / land use. He has led multiple ground-breaking projects related to the use of daytime and night-time remotely sensed imagery to detect long-term environmental and social processes, including the development of machine learning methodologies and approaches to map urbanization and track economic trends across space and time. He has authored and co-authored close to 40 peer-reviewed publications and serves as an Editor and Guest Editor of several Special Issues and Research Topics in leading journals.

Steven Louis Rubinyi

Dr. Steven Louis Rubinyi is a Disaster Risk Management Specialist at the World Bank, where he has worked for ten years at a thematic level on communities of practice aimed at leveraging geospatial solutions across the World Bank investment portfolio in developing countries. In his current role, he is based in Hanoi, Vietnam, where he works across the East Asia region to bring emerging digital technologies to cities to support their resilience planning efforts. In previous roles, he led the resilience planning thematic area of the World Bank's City Resilience Program, contributed towards urban and coastal resilience projects in Bangladesh, and helped to coordinate the World Bank's support for housing reconstruction following the 2015 Nepal Earthquake. He holds a PhD from the Environmental Change Institute at the University of Oxford, where his research focused on leveraging earth observation data for insights into disaster risk management as well as geostatistical disaggregation of socioeconomic population characteristics.

Hogeun Park

Dr. Hogeun Park is an Urban Specialist with the Middle East and North Africa (MENA) region of the World Bank's Urban, Disaster Risk Management, Resilience, and Land Global Practice. His work focuses on the intersection of urban planning and spatial analytics, which enabled us to deliver the projects with empirical and rigorous evidence. Prior to joining the Bank, Hogeun was a Big Pixel postdoctoral fellow in UC San Diego. He has published widely in leading peer-reviewed journals on the topic of urban development, spatial analytics, and remote sensing. He completed his PhD in Planning from Michigan State University with a fellowship from the Korean government. His previous research and work experience also include the Korea International Cooperation Agency (KOICA), International Rice Research Institute (IRRI), the Asia Foundation (TAF), and the National Socio-Environmental Synthesis Center (SESYNC) .

Preface to "Remote Sensing of Night-Time Light"

Satellite measurements of night-time lights are one of the most widely recognized remotely-sensed indicators of human activity on Earth. The amount of light that is emitted from Earth at night and that is captured by sensors on board satellites that orbit Earth can provide meaningful information related to the distribution and characteristics of urban areas, urbanization processes, socio-economic dynamics, local and regional GDP, the economic impacts of conflicts, the impacts of epidemics on economic activity, the impacts of shocks such as natural or man-made disasters, and more.

With the increased availability of remotely-sensed measurements of night-time lights, which are captured, for example, by the Defense Meteorological Satellite Program's Operational Linescan System (DMSP-OLS) and the Visible Infrared Imaging Radiometer Suite (VIIRS) sensors, new possibilities emerge in understanding how Earth is changing, allowing improvements in decision-making, guiding policy and improving governance. Today, remotely sensed observations of night-time lights provide a key instrument for understanding almost every aspect of human activity on Earth across space and time. Advances in the availability and the quality of night-time light data, together with improvements in data storage capabilities and the development of new analytical methods and workflows for analyzing the data, are accompanied with an ongoing increase in the number of scientific applications that exploit remotely sensed night-time lights in almost every academic discipline.

This book includes 16 articles that were published in the Remote Sensing Special Issue "Remote Sensing of Night-Time Light" and that highlight novel research on the remote sensing of night-time lights. The book aims to stimulate progress in the remote sensing research domain related to the utilization of night-time lights in a wide range of multidisciplinary and interdisciplinary domains, including urban studies, ecology, economics, engineering, oceanography, sociology, epidemiology, and more.

As the Guest Editors of this Special Issue, we appreciate the professionalism and support from all contributors, editors, and reviewers for their dedication and efforts toward the publication of this book. We would like to thank the Managing Editor of this Special Issue, Ms. Darcy Zhang, for her continuous support and assistance which helped make this SI a success.

Ran Goldblatt, Steven Louis Rubinyi, and Hogeun Park
Editors

remote sensing

MDPI

Article

The Dimming of Lights in China during the COVID-19 Pandemic

Christopher D. Elvidge [1,*], Tilottama Ghosh [1], Feng-Chi Hsu [1], Mikhail Zhizhin [1,2] and Morgan Bazilian [3]

1. Earth Observation Group, Payne Institute for Public Policy, Colorado School of Mines, Golden, CO 80401, USA; tghosh@mines.edu (T.G.); fengchihsu@mines.edu (F.-C.H.); mzhizhin@mines.edu (M.Z.)
2. Russian Space Research Institute, 117997 Moscow, Russia
3. Payne Institute for Public Policy, Colorado School of Mines, Golden, CO 80401, USA; mbazilian@mines.edu

* Correspondence: celvidge@mines.edu

Received: 26 July 2020; Accepted: 28 August 2020; Published: 2 September 2020

Abstract: A satellite survey of the cumulative radiant emissions from electric lighting across China reveals a large radiance decline in lighting from December 2019 to February 2020—the peak of the lockdown established to suppress the spread of COVID-19 infections. To illustrate the changes, an analysis was also conducted on a reference set from a year prior to the pandemic. In the reference period, the majority (62%) of China's population lived in administrative units that became brighter in March 2019 relative to December 2018. The situation reversed in February 2020, when 82% of the population lived in administrative units where lighting dimmed as a result of the pandemic. The dimming has also been demonstrated with difference images for the reference and pandemic image pairs, scattergrams, and a nightly temporal profile. The results indicate that it should be feasible to monitor declines and recovery in economic activity levels using nighttime lighting as a proxy.

Keywords: VIIRS; Day-night band (DNB); Nighttime lights; COVID-19; Pandemic

1. Introduction

During the first half of 2020, the world experienced economic collapse due to the COVID-19 pandemic since the Great Depression in the late 1920s [1]. It is well established that low-light imaging satellite sensors are able to detect the dimming and loss of lighting after disaster events [2–4], wars [5,6], and economic collapse [7]. In order to explore the possibility that satellite observation can be used to track the decline and recovery of economic activity worldwide associated with the COVID-19 pandemic, we conducted an analysis of lighting changes in China, the first country that experienced the pandemic. Large portions of the Chinese economy shut down, as the government imposed a national lockdown forcing most people to remain at home [8]. Previous satellite studies indicate that the nighttime lights of China have been growing for several decades [9]. For this study, we use specialized low-light imaging satellite data collected by the NASA/NOAA Visible Infrared Imaging Radiometer Suite (VIIRS) day-night band (DNB). The DNB light intensification was designed to detect moonlit clouds in the visible band [10]. With a million-fold amplification in the signal, the DNB also detects electric lighting present at the Earth's surface.

Because clouds can block the detection of surface lighting, many scientists interested in using satellites to observe nighttime lights choose to work with cloud-free temporal composites. Annual cloud-free composites have sufficient temporal leverage to fill in seasonal data outages from sunlight and to filter out ephemeral features unrelated to electric lighting [11]. These include biomass burning and energetic particle hits from the aurora and the South Atlantic Anomaly. In some cloud-prone areas, such as Java Island, it takes a full year to get a sufficient number of cloud-free observations to generate a sharp and clear image of nighttime lights. However, in cases where the user wishes to study short–term changes in lighting, the phenomena of interest are averaged away and cannot be seen clearly in an annual average. In such a case, monthly and nightly data products are available and can be considered for use. Studies conducted with nightly images are well suited for small geographic extents, where an analyst can select sets before and after images that are cloud-free. Another way to work with nightly data is to assemble temporal profiles for individual locations and then filter out clouds or cloud plus moonlight. The monthly composites differ from the annual VIIRS nighttime lights products, in that they are not filtered to remove ephemeral events, like biomass burning. In addition, background areas having no detectable lighting are not zeroed out in the monthly composites. It is left to the user to perform filtering based on their use case.

This analysis is based primarily on two pairs of monthly cloud-free composites and the sum of the light brightness for 300+ level 2 administrative units [12] in each of four months. For the subject pair, we used the composites of December 2019 and February 2020 and calculated the percentage change in brightness in February relative to that in December 2019. The composites in December 2019 were selected, because December 2019 is the last full month prior to the public recognition of the outbreak in January. We intentionally skipped over January 2020 to avoid lighting changes associated with the Chinese New Year (CNY), which fell on 25 January 2020. During the CNY, many institutions close for two weeks centered on the New Year date, and many people return to their provinces. To avoid any possible CNY effects, we selected the February 2020 composite set for the analysis, corresponding to the peak lockdown period. To provide a reference, the same analysis was performed on a pair of months from the previous year: December 2018 and March 2019. The February 2019 composite was skipped in favor of March 2019, because the CNY in 2019 was on February 5.

The monthly composites include a pair of images—the average radiance and the tally of cloud-free coverages. Radiance difference images were calculated and used to produce colorized maps for specific cities. In addition, we calculated the percentage change in brightness for more than 300 level 2 administrative units [12]. The grid cell set used for the four months in the % change calculation was standardized by blocking grid cells having low cloud-free coverages or extremely low radiances. Then, the radiances were summed for each administrative unit. These sums were then differenced, and a percent difference was calculated.

In addition, we examined the evidence of dimming in a nightly temporal profile to show how the exact start dates of dimming and the level of recovery can be derived as well as an assessment of the lighting recovery level.

2. Materials and Methods

We presented results from three different styles of DNB data products: filtered monthly composites, temporal profiles from monthly composites, and a nightly temporal profile. The one we spent the most time on were the two monthly pairs (December 2018 to March 2019 and December 2019 to February 2020). Here, we filtered the composites to extract a consistent set of lit grid cells for the comparison of the pre-pandemic brightness levels and the dimming of lights during the pandemic. The percent of dimming during the pandemic was then checked against the report confirming COVID-19 infections published by the Harvard Dataverse [13]. We then presented results from the entire time series of cloud-free composites to check whether the pandemic dimming was unique. Finally, we presented a nightly temporal profile to identify dates for the onset of dimming and recovery, plus the recovery level relative to the pre-pandemic brightness level.

Preparation of the two monthly pairs: The methods used to generate monthly cloud-free average radiances were described by Elvidge et al. [11]. These were rough composites; in that they were not filtered to remove background or biomass burning. The key to using the monthly products to analyze the dimming of lights across administrative units was to mask out background and poor-quality data, focusing the analysis on a consistent set of lit grid cells from each of the four focus months. In this case, we filtered the composites based on low cloud-free coverages, low radiance levels, and snow cover.

Filtering on low cloud-free coverages: We visually examined the cloud-free coverage grids for China and found there were data gaps with zero cloud-free coverages in every month (Figure 1). In addition, we decided to filter out grid cells having single cloud-free coverages and focused the study on areas having larger numbers of usable observations. For each month, a binary on-off mask was generated for grid cells having less than two cloud-free coverages. The final mask for low cloud-free coverages was set, so that a grid cell was excluded from the study if it dipped below two cloud-free coverages in any of the four months.

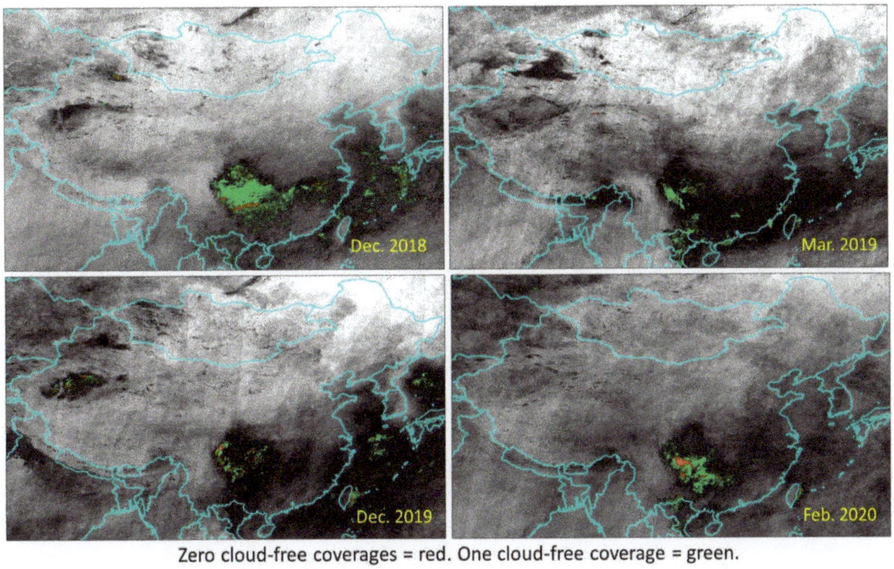

Figure 1. Marking of grid cells having less than two cloud-free coverages in each of the four study months.

Filtering based on low average radiance: We filtered out grid cells that dipped below one $nW/cm^2/sr$ in any of the four months. This filtering had two primary purposes. The first was to zero out background areas, where no lighting was detected. Although the average radiance in background areas was quite low, it varied both spatially and temporally (Figure 2). In many of China's western provinces, the areas of lighting are dwarfed by the spatial extent of the background. If the background radiances are summed with the lights, the result will be dominated by the temporal variations in the background. The other reason to filter based on low average radiance was to eliminate biomass burning. Fires are quite bright in the DNB; however, they typically occur in background areas lacking electric lighting detectable by VIIRS. Biomass burning was filtered from the four study months by filtering out grid cells that dipped below one nW in any of the four months.

Filtering of snow cover: The high visible wavelength reflectivity of snow increased the brightness of lighting observed by satellite. Figure 3 shows an example of snow–induced brightening along a transect through Harbin, China, where snow was absent in December 2018 but consistently present during the December 2019 compositing period. The December 2018 radiance was consistently dimmer than the December 2019 levels. At present, there is no technique for removing snow effects. Therefore, grid cells having snow cover in any of the four months were excluded from the change detection analysis. A snow cover mask was produced for each of the four months using NOAA Advanced Microwave Sounding Unit-A (AMSU-A) daily snow cover product [14], tallying the number of times snow cover was detected for the same set of days used in the DNB cloud-free composites (Figure 4). Monthly snow cover tallies of one and two were filtered out to remove false detections. This removed a narrow rim of AMSU-A detections surrounding the consistently snow-covered areas.

Then, the four monthly snow products were added together to generate a mask used to zero out lighting affected by snow cover in any of the four months.

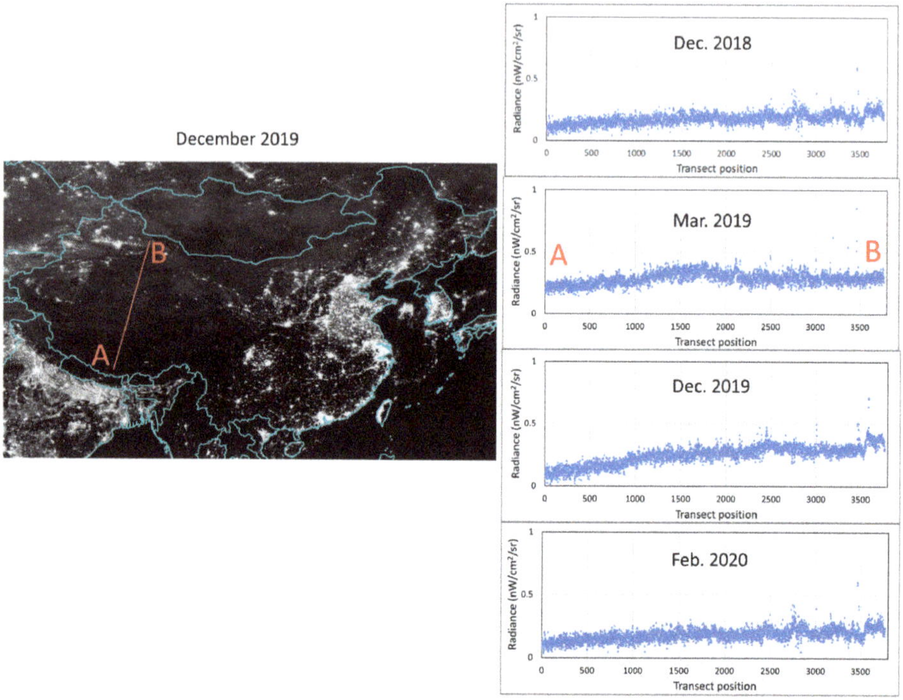

Figure 2. Average radiance transect across background areas in Western China.

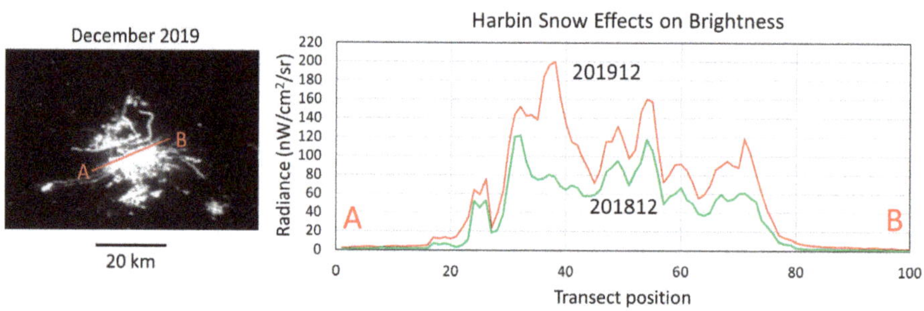

Figure 3. Radiance transect across Harbin, China showing the brightening effect of snow cover present in December 2019. The December 2018 radiances were consistently dimmer than the December 2019 ones.

Sum-of-lights (SOL) calculation: After masking, the radiances of the remaining grid cells were multiplied by the surface area of 15 arc-second grid cells prior to summing inside the level 2 administrative units. This "latitude adjustment" was necessary, because the ground footprint of the grid cells declined with latitude (Figure 5). The grid cells at the equator covered 0.857 km^2 and decreased with latitude, following a cosine function [15]. China spans latitudes from 18 to 53°N. Therefore, in the south, the grid cells cover 0.204 km^2, and in the far north they only cover 0.129 km^2. The SOL was calculated by summing the grid cell radiance inside the level 2 administrative units after the latitude adjustment. These sums were then differenced, and a percent difference was calculated. The population of the administrative units was tallied with the Landscan population grid from 2018 [16] along with a latitude adjustment.

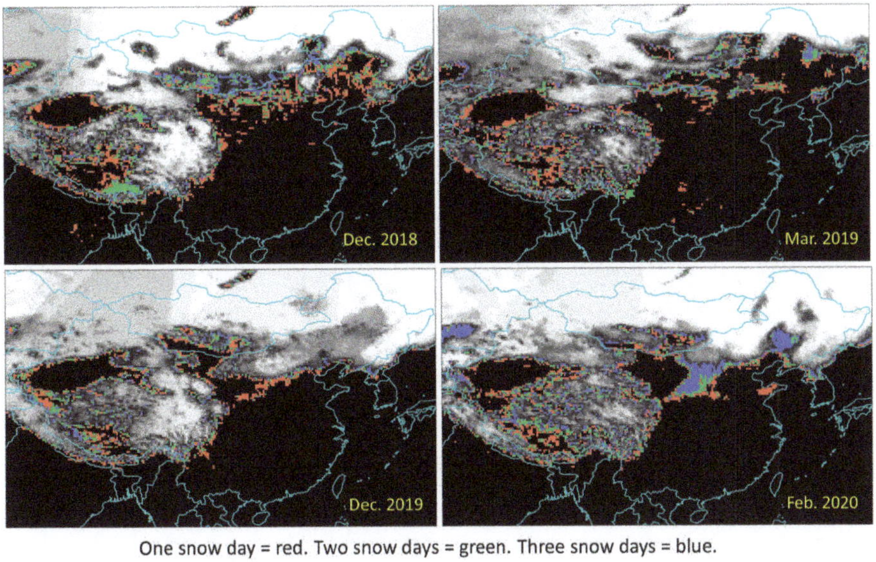

Figure 4. Advanced Microwave Sounding Unit-A (AMSU-A) snow cover tallies matched to the dates used in the day-night band (DNB) monthly cloud-free composites.

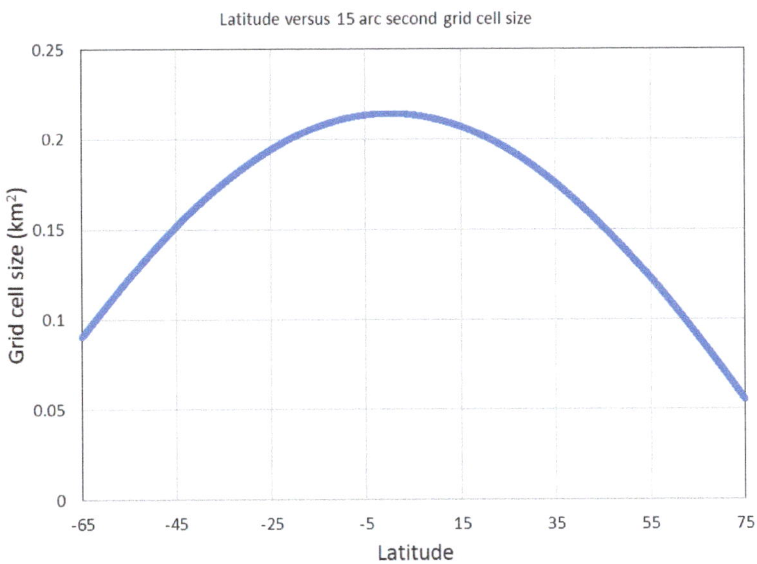

Figure 5. Dependence of grid cell size on latitude. Ground footprint sizes for 15 arc-second grid cells were largest at the equator and declined moving in either direction away from the equator.

3. Results

Colorized difference images: To consider initial COVID-19 effects, we examined the change in brightness between the two monthly pairs for several major urban areas in China. By coloring gains in brightness as cyan and declines in brightness as red, it was possible to compare the color patterns for the pre-pandemic pair (December 2018 and March 2019) and the pair that straddled the pandemic lockdown (December 2019 and February 2020). Figure 6 shows results for five major urban areas: Wuhan, Beijing, Guangzhou, Shanghai, and Xi'an. It can be seen in all five areas that there was generally a mix of dimming and brightening in each monthly pair. Brightening dominates in the pre-pandemic pair. In contrast, the lockdown pairs showed extensive areas experienced a decline in brightness in February 2020. For Wuhan, there was a relatively even mix of increase and decline in brightness for February 2020. This qualitative observation was confirmed by the quantitative analysis results in Table 1, which shows Wuhan lighting increased by 8% in the pre-pandemic pair and declined by 3% during the lockdown period. The decline in lockdown lighting appeared much more substantial in Beijing (Figure 6), and this was confirmed by the results in Table 1, where the city's lighting increased by 3% in the reference pair but declined by 24% during the lockdown. Shanghai and Guangzhou fell between Wuhan and Beijing in terms of the diming of lights, both witnessing a 12% decline during the lockdown. Xi'an also showed extensive areas of dimming during the lockdown period. The quantitative results in Table 1 showed Xi'an's light brightness grew by 23% during the reference period and declined by 19% during the lockdown.

Table 1. Percent difference in brightness for the pre-pandemic and pandemic monthly pairs in seven of China's major urban areas.

Name	% Difference (The Brightness in March 2019 Minus the Brightness in December 2018)	% Difference (The Brightness in February 2020 Minus the Brightness in December 2019)	Population
Beijing	3.13	−24.08	20,389,641
Chongqing	3.57	−16.41	29,924,960
Guangzhou	12.24	-8.78	13,146,601
Shanghai	−4.77	−12.96	24,058,784
Tianjin	−0.89	−12.95	13,569,917
Wuhan	8.47	−3.03	10,076,555
Xi'an	23.52	−19.42	8,856,884

Percent change maps with population: A pair of maps were created to spatially visualize the change in brightness across the entire country in reference to population numbers (Figures 7 and 8). The maps show the outlines of the subnational units and have circles with different sizes based on population and color coded to five classes of change in brightness. There were three color-coded classes for increased brightness: yellow, light green, and dark green. There were two color-coded classes for declines in brightness: orange and red. The pre-pandemic reference (Figure 7) set showed that the heavily populated areas in Central China had yellow and green circles predominating, indicating that growth in lighting was widespread. In contrast, the pandemic map (Figure 8) shows the predominance of orange and red circles, indicating the prevalence of declines in lighting.

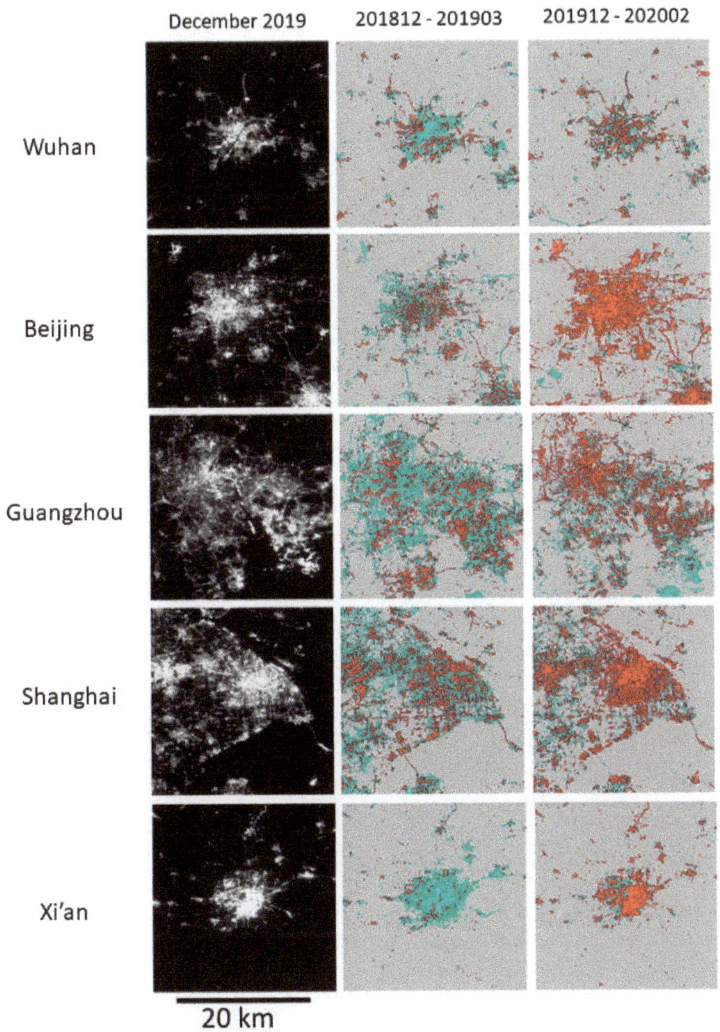

Figure 6. Examples of February 2020 dimming in five of China's major cities.

Histograms of percent differences versus population: Figure 9 shows population tallies versus percent difference in brightness for the two pairs of months. In the pre-pandemic reference set, 61% of the population was located in administrative units, where the brightness of lighting increased. In contrast, 82% of China's population lived in administrative units in February 2020, where the brightness of lighting decreased.

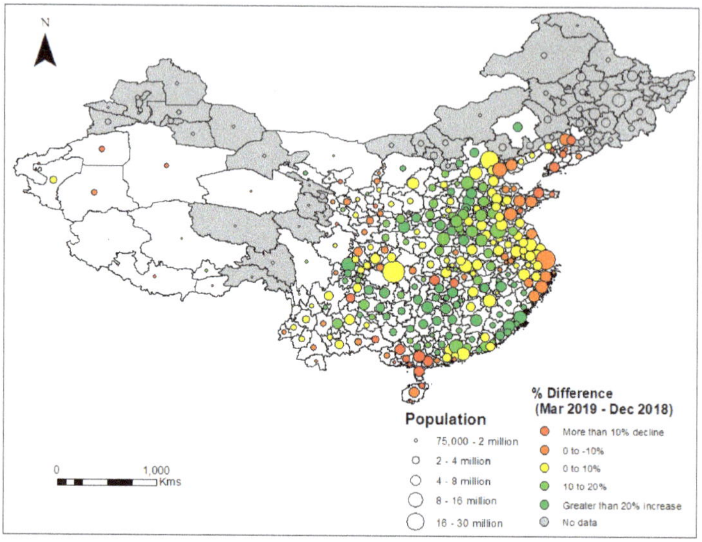

Figure 7. Map of the % changes in brightness for the reference period (brightness in December 2018 minus brightness in March 2019). The administrative unit boundaries are shown in black, and population size classes of each are indicated with circles with five diameters placed at the administrative unit's centroid. Administrative units, where lighting was completely masked out, are gray. There were three classes for increased brightness in March 2019, indicated by yellow, light green, and dark green. The orange and red classes indicated declines in brightness in February 2020. Yellow and green classes predominated in the central core of China's densely populated heartland. (The tabular data used to generate the maps see Supplementary Materials).

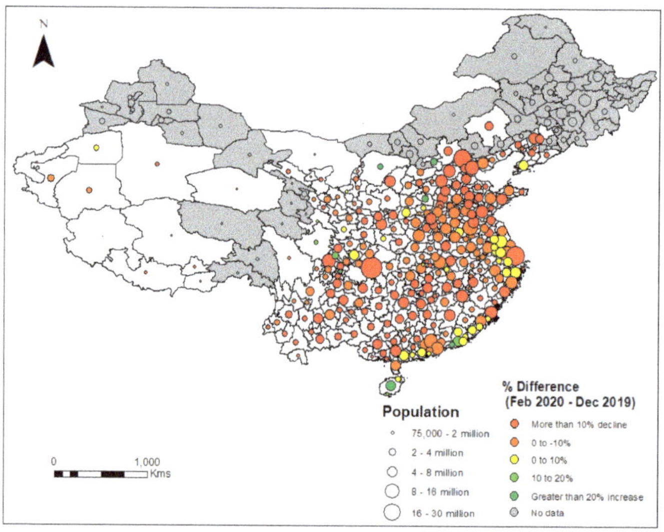

Figure 8. Map of the % changes in brightness for the study pair spanning the pandemic (the brightness in December 2019 minus the brightness in February 2020). Decline in brightness, indicated by orange and red circles, predominated in the central core of China's densely populated heartland. (The tabular data used to generate the maps see Supplementary Materials).

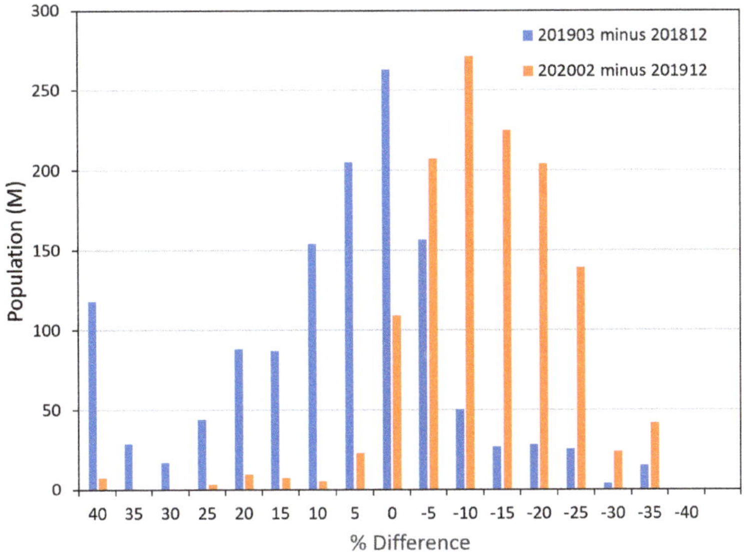

Figure 9. Population histograms for a gradation of brightening and dimming levels for the reference period (blue) and study period (orange). The population numbers shift between the two periods. In the reference period, the majority (62%) lived in administrative units that became brighter in March 2019 relative to in December 2018. The situation reversed in February 2020, when 82% of the population lived in administrative units where lighting dimmed as a result of the pandemic.

Beijing scattergrams: Another way to see the magnitude of the dimming was using the comparison between pre-pandemic and pandemic scattergrams. Figure 10 shows scattergrams made with all of the grid cells in Beijing. The pre-pandemic scattergram was on the left, and the pandemic scattergram was on the right. In both cases, December is on the X axis. The slope of the regression line for December 2018 and March 2019 was 1.002, indicating no significant change in radiance levels between these two months. The slope dropped to 0.713 in the pandemic scattergram due to substantial dimming in February 2020.

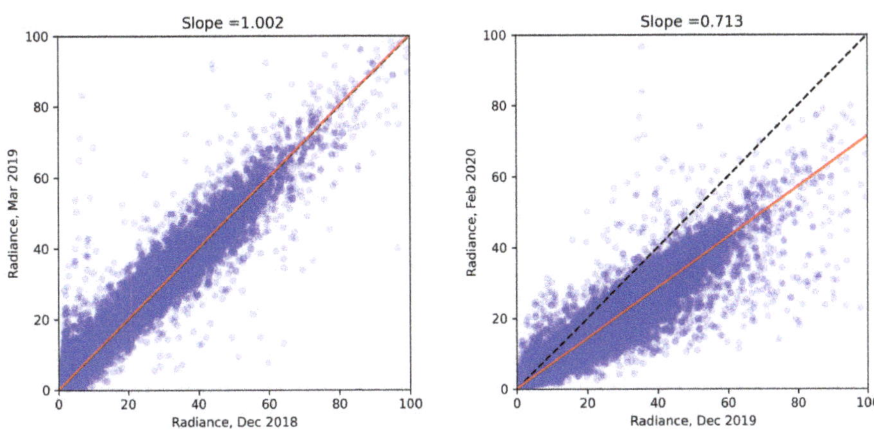

Figure 10. Pre-pandemic and pandemic scattergrams of the monthly average radiances for grid cells in Beijing.

Correlation with confirmed cases: The percent change in SOL from the pandemic monthly pair was matched to the number of confirmed COVID-19 cases per million people from January 15 to the end of February 2020 [13]. The results (Figure 11) indicated that dimming occurred in many cities with few confirmed cases.

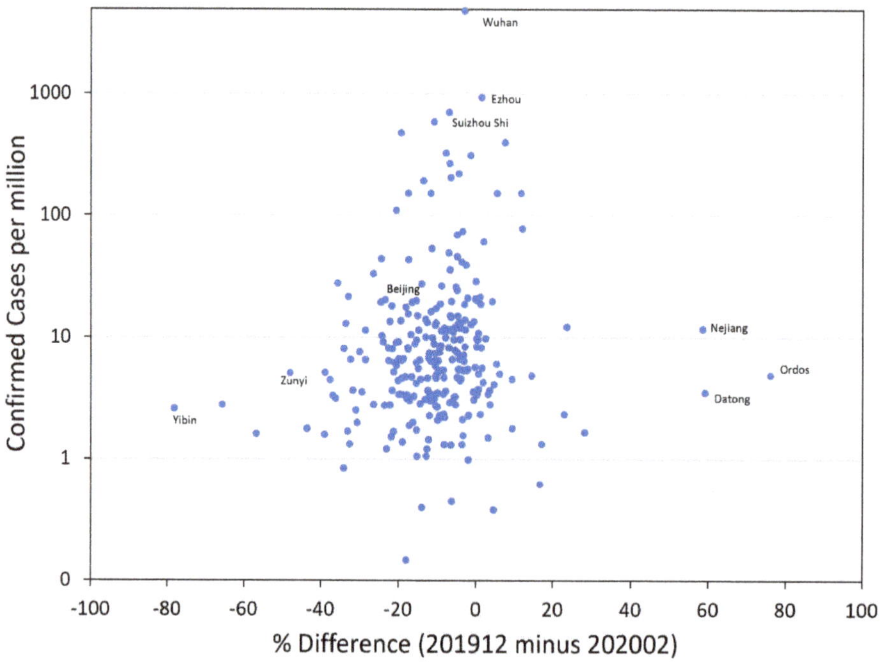

Figure 11. Percent change in monthly sum-of-lights (SOL) from the pandemic study pair versus confirmed COVID-19 cases per million people.

Is the February 2020 dimming unique? To examine whether the February 2020 dimming is unique, we constructed a data cube with 96 monthly cloud-free composites of Beijing, spanning from April 2012 to March 2020. In examining the monthly temporal profiles for individual grid cells, it was possible to see a radiance dip in February 2020 relative to those in the previous two months and March 2020 (Figure 12). However, there were other dips along the profiles in magnitude comparable to that in February 2020. To determine if the additional dimming events were isolated geographically or spatially extensive, we constructed a temporal slice along a north-south transect across the data cube (Figure 13). Spatially extensive dimming showed up as dark vertical lines. There were two black vertical lines during the early part of the record and three month-wide data gaps centered on the summer solstice caused by the exclusion of stray light contamination prior to the advent of the "stray light correction" [17]. However, there were additional dark vertical lines indicating widescale dimming of lights. Regardless of where the transect was taken, the same set of months appeared as dark vertical lines. The most prominent of these were February 2014, July and November 2015, July 2018, and February 2020. December 2019 was bright with sharply defined features (Figure 14). To investigate the cause of the dimming, average radiance images in each of these months were compared to that in December 2019 (Figure 14). The December 2019 contrast enhancement was applied to each of the other monthly images. It can be seen that December 2019 has the brightest features and the details were sharp. The February 2020 geographic feature details were comparable to those in December 2019, but the radiance levels were dimmed. The other months exhibited dimming with the blurring of spatial details. The examination of the cloud-free coverages revealed that the pre-pandemic dimmed months were short with cloud-free coverages and had multiple data gaps with zero cloud-free coverages (Figure 15).

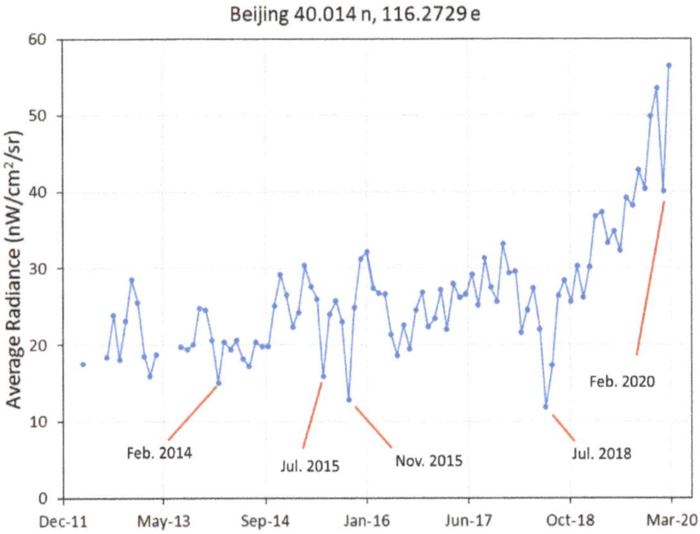

Figure 12. Temporal profile of monthly average radiances from the DNB cloud-free composite time series for a grid cell in Beijing. Dimming during the pandemic is expressed by a shard dip in radiance in February 2020. Similar dips were found in other months.

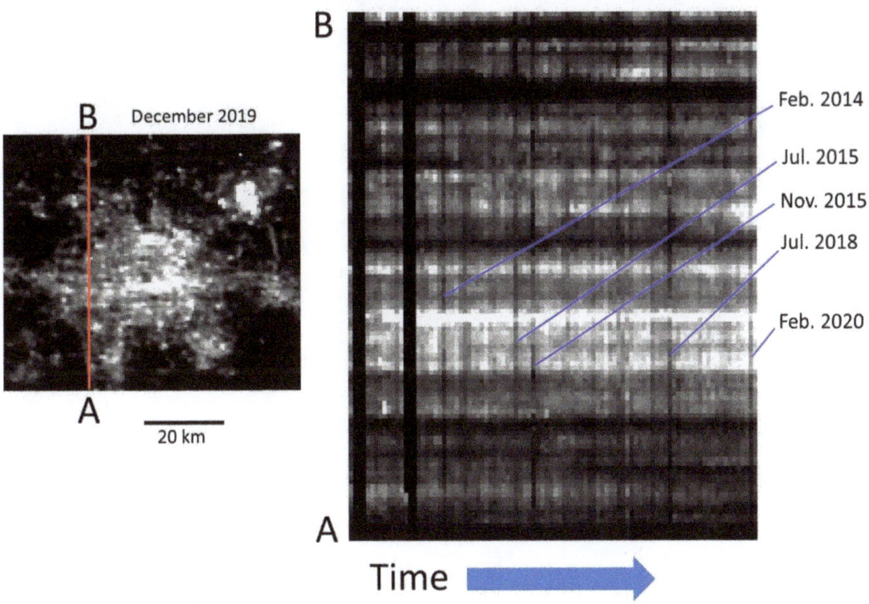

Figure 13. A temporal slice from a north-south transect across Beijing showing dark vertical stripes in months, where dimming was spatially extensive.

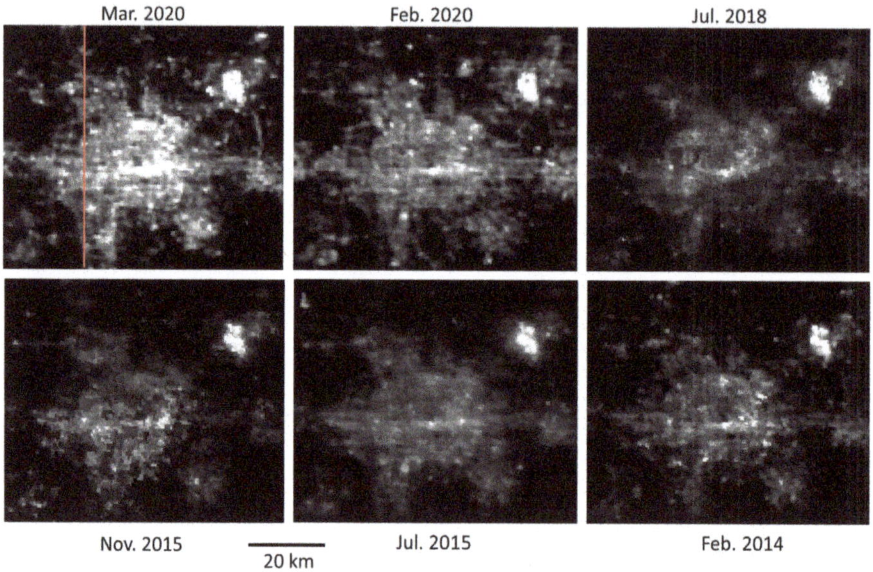

Figure 14. Comparison of Beijing DNB monthly composite images having spatially extensive dimming relative to that in December 2019.

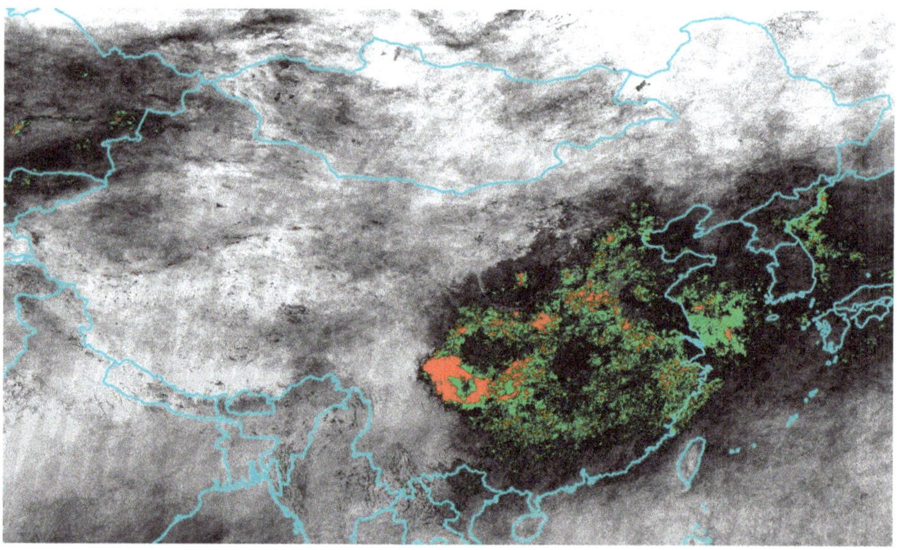

Figure 15. November 2015 cloud-free coverages. This is one of the months showing spatially extensive DNB dimming. A large part of Eastern China had low numbers of cloud-free coverages. Zero cloud-free coverages are in red, and one cloud-free coverage is in green.

***Examination of dimming and recovery with a nightly temporal profile*:** One of the grid cells with the sharpest decline in lighting during February 2020 covered the entrance to Shanghai Disneyland. The site closed on 25 January 2020 and partially reopened on 9 March 2020 [18]. A nightly temporal profile of the entrance is shown in Figure 16, only filtered to remove cloud observations. The start dates of dimming and recovery exactly matched the reported closure dates. The dimmed state was 42% of the pre-closure radiance level. However, the recovery state brightness through 6 June 2020 was 81% of the pre-closure level.

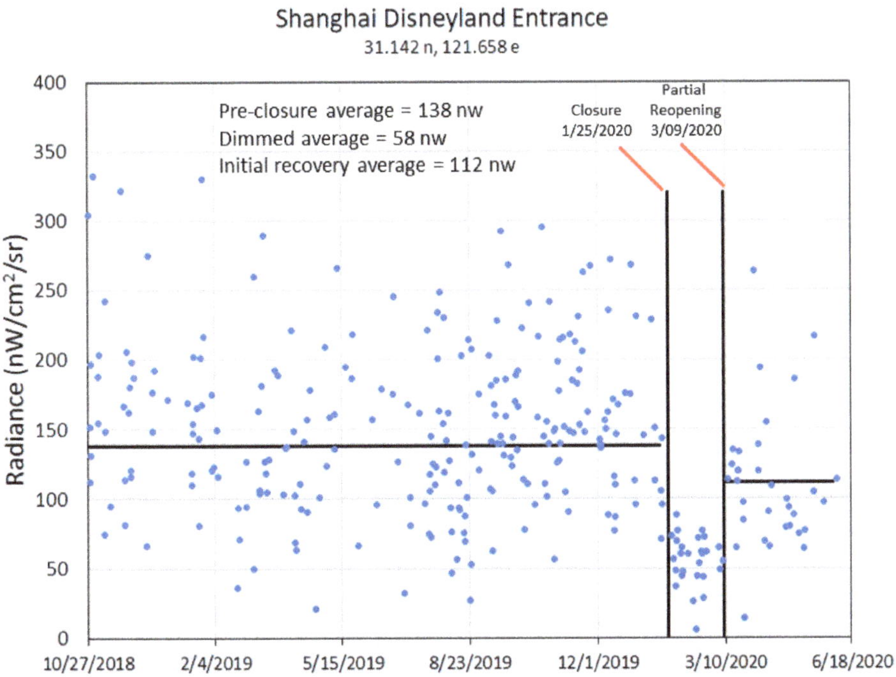

Figure 16. Nightly temporal profile of DNB radiances for the grid cell covering Shanghai Disneyland.

4. Discussion

The decline in the brightness of lighting in China during the COVID-19 lockdown was in the 3–24% range and was observed in four different ways. The simplest way to see the dimming was by comparing radiance difference images for two selected pairs of monthly composite images, matching seasonally and avoiding the Chinese New Year. We used the brightness in December 2018 minus the brightness in March 2019 as the pre-pandemic pair (reference set) and the brightness in December 2019 minus the brightness in February 2020 as the pair impacted by the lockdown (pandemic set). The reference and pandemic difference images were density-sliced to enhance the visual interpretation of dimming and brightening. Major cities in China readily showed dimming in the difference images.

Another way to see the dimming is by using full province SOL tallies for more than 344 level 2 administrative units. The difference between the reference and pandemic set tallies made it possible to see the prevalence of dimming in February 2020. We conducted a t-test of unequal variances to ascertain whether the difference of the mean between the pre-pandemic and pandemic pair was significant or not. The t-test of unequal variances at a 0.05 significance level was associated with a significant effect t (649) of −9.60 and p of 1.77×10^{-20}. Thus, the pandemic difference pair had a larger statistically significant mean than the pre-pandemic pair (Table 2).

Table 2. *t*-test for the difference pairs.

	Difference (201903–201812)	*Difference (202002–201912)*
Mean	−460.85	835.54
Variance	2,393,271.83	3,885,643.08
Observations	344.00	344
Hypothesized Mean Difference	0.00	
df	649.00	
t Stat	−9.60	
P(T<=t) one-tail	8.83×10^{-21}	
t Critical one-tail	1.65	
P(T<=t) two-tail	1.77×10^{-20}	
t Critical two-tail	1.96	

Radiance scattergrams offer another way to visualize the dimming. In the reference set, the scattergram data cloud was aligned with the diagonal, indicating no systematic dimming or brightening between the two months of the reference set. In contrast, the scattergram of the pandemic set shows the entire data cloud shifted to lower radiance levels in February 2020 relative to in December 2019.

The fourth approach used to see the dimming is with nightly temporal profiles. Here, it was possible to see the precise date when dimming started, the radiance level of the lighting during the dimming period, the date range of recovery, and the recovery level.

An important aspect of this study is the filtering to exclude grid cells impacted by snow cover, low numbers of cloud-free coverages, and background areas, where lighting was detected.

5. Conclusions

During the first half of 2020, nearly the entire population of the world lived under some form of human contact restrictions aimed at reducing the transmission of COVID-19. The restrictions in China began with a lockdown in Wuhan on 24 January 2020, and these were quickly extended across China. The lockdown featured the closures of schools, restaurants, travel, and many factories and commercial centers. People were required to stay at home. The associated economic impacts are significant and are still being tallied.

China's lighting and electricity demand have long been on an upward trajectory, tracking the expansion in construction and economic activity levels. This pattern went in reverse during the COVID-19 lockdown, resulting in dramatic declines in the brightness of electric lighting observed by satellite. The declines can be seen clearly in multiple cities with the difference image calculated by the brightness in December 2019 minus the brightness in February 2020. While brightness declines were prevalent, it was common to find patches with lighting increase mixed with the areas having decline. The situation was reversed in the pre-pandemic difference images, where brightening prevailed amid a patchwork pattern of dimming and brightening. The brightness decline in February 2020 lighting levels can also be seen in scattergrams formed with the December 2019 radiances.

With nightly DNB radiance profiles, it is possible to identify the start dates of dimming and recovery, plus the recovery level relative to the pre-pandemic condition. This was demonstrated for the grid cell containing the Shanghai Disneyland entrance. This site was chosen, because it is one of a few facilities having precise closure and partial reopening dates published. The dimming and onset dates of recovery observed by VIIRS exactly matched the closure and reopening dates. The level of lighting recovery following the reopening was 81% of the pre-pandemic lighting level.

In exploring the DNB monthly composite time series, we found several issues that other researchers should be aware of. First, the background radiance levels varied both spatially and temporally. If the focus of an analysis is electric lighting, it is important to zero out the background using a radiance threshold. In nearly every monthly composite, there were data gaps, where zero cloud-free observations were available. These should be excluded from multitemporal analyses in a way to avoid corrupting change detection results. Snow cover resulted in increased brightness of light, which can also lead to erroneous conclusions on brightness changes. Finally, we found the evidence that there were cases, where undetected clouds resulted in the blurring and dimming of lighting features. This suggested that a lighting sharpness index could be used in conjunction with cloud detections to improve both the annual and monthly DNB composites.

This study confirms that satellites are able to detect the dimming of lighting associated with the current pandemic lockdowns. We found no correlation between the percent dimming in February 2020 and COVID-19 confirmed case numbers. This suggested that the dimming of lights was primarily associated with "stay-at-home" lockdowns designed to slow the virus transmission. It is possible that the recovery pattern of lighting can be used as an indicator of the economic recovery. Disaster and conflict lighting outages are typically due to damage to generation or distribution systems. In the case of COVID-19, the power generation and distribution systems remain intact. Therefore, in theory, if the lights dim, they could return to normal shortly after government lockdowns are lifted. If the lighting fails to return to pre-lockdown levels, this is an indication that the economic damage continues to affect lighting. The results from this study indicate that the dimming and recovery of lights can be effectively monitored from space and that recovery levels can be rated relative to pre-pandemic radiance levels.

Supplementary Materials: The tabular data used to generate the maps shown in Figures 7 and 8 are available online at http://www.mdpi.com/2072-4292/12/17/2851/s1.

Author Contributions: The author contributions are as follows: C.D.E. developed the concept for the paper, supervised the analysis and wrote the first draft of the paper. T.G. processed the monthly cloud-free composites, conducted the analysis and generated the figures. F.-C.H. coordinated the data ingest and processing. M.Z. developed the geolocation algorithm. M.B. edited the paper and contributed to the conceptualization. All authors have read and agreed to the published version of the manuscript.

Funding: The algorithm development for the production of the cloud-free DNB composites was funded under a NASA research grant. The algorithm development for the production of nightly DNB profiles was funded by the Rockefeller Foundation.

Acknowledgments: The authors sincerely appreciate the NASA/NOAA Joint Polar Satellite System (JPSS) for providing the VIIRS data used in this study. The algorithm development for the production of the cloud-free DNB composites is funded under an NASA research grant. The algorithm development for the production of nightly DNB profiles is funded by the Rockefeller Foundation.

Conflicts of Interest: The authors declare no conflict of interest.

References

1. Craven, M.; Mysore, M.; Singhal, S.; Smit, S.; Wilson, M. COVID-19: Implications for Business. Available online: https://www.mckinsey.com/business-functions/risk/our-insights/covid-19-implications-for-business (accessed on 20 August 2020).
2. Elvidge, C.D.; Baugh, K.E.; Hobson, V.R.; Kihn, E.A.; Kroehl, H.W. Detection of fires and power outages using DMSP-OLS data. In *Remote Sensing Change Detection: Environmental Monitoring Methods and Applications*; Lunetta, R.S., Elvidge, C.D., Eds.; Taylor and Francis: London, UK, 1999; pp. 123–135.
3. Kohiyama, M.; Hayashi, H.; Maki, N.; Higashida, M.; Kroehl, H.W.; Elvidge, C.D.; Hobson, V.R. Early damaged area estimation system using DMSP-OLS night-time imagery. *Int. J. Remote Sens.* **2004**, *25*, 2015–2036. [CrossRef]
4. Román, M.O.; Stokes, E.C.; Shrestha, R.; Wang, Z.; Schultz, L.; Carlo, E.A.S.; Sun, Q.; Bell, J.; Molthan, A.; Kalb, V.; et al. Satellite-based assessment of electricity restoration efforts in Puerto Rico after Hurricane Maria. *PLoS ONE* **2019**, *14*, e0218883. [CrossRef] [PubMed]
5. Li, X.; Li, D. Can night-time light images play a role in evaluating the Syrian Crisis? *Int. J. Remote Sens.* **2014**, *35*, 6648–6661. [CrossRef]

6. Coscieme, L.; Sutton, P.C.; Anderson, S.; Liu, Q.; Elvidge, C.D. Dark Times: Nighttime satellite imagery as a detector of regional disparity and the geography of conflict. *Gisci. Remote Sens.* **2017**, *54*, 118–139. [CrossRef]
7. Elvidge, C.D.; Hsu, F.-C.; Baugh, K.E.; Ghosh, T. Lighting Tracks Transition in Eastern Europe. In *Land-Cover and Land-Use Changes in Eastern Europe after the Collapse of the Soviet Union in 1991*; Gutman, G., Radeloff, V., Eds.; Springer International Publishing: Cham, Switzerland, 2017; pp. 35–56. ISBN 9783319426389.
8. Kupferschmidt, K.; Cohen, J. Can China's COVID-19 strategy work elsewhere? *Science* **2020**, *367*, 1061–1062. [CrossRef] [PubMed]
9. Elvidge, C.; Hsu, F.-C.; Baugh, K.; Ghosh, T. National Trends in Satellite-Observed Lighting: 1992–2012. *Remote Sens. Appl. Ser.* **2014**, *23*, 97–120.
10. Miller, S.; Straka, W.; Mills, S.; Elvidge, C.; Lee, T.; Solbrig, J.; Walther, A.; Heidinger, A.; Weiss, S. Illuminating the Capabilities of the Suomi National Polar-Orbiting Partnership (NPP) Visible Infrared Imaging Radiometer Suite (VIIRS) Day/Night Band. *Remote Sens.* **2013**, *5*, 6717–6766. [CrossRef]
11. Elvidge, C.D.; Baugh, K.; Zhizhin, M.; Hsu, F.C.; Ghosh, T. VIIRS night-time lights. *Int. J. Remote Sens.* **2017**, *38*, 5860–5879. [CrossRef]
12. Database of Global Administrative Areas. Available online: https://gadm.org/ (accessed on 20 August 2020).
13. China Data Lab. China COVID-19 Daily Cases with Basemap. Harvard Dataverse, V23. 2020. Available online: https://doi.org/10.7910/DVN/MR5IJN (accessed on 20 August 2020).
14. Grody, N.; Weng, F.; Ferraro, R. Application of AMSU for obtaining Water Vapor, Cloud Liquid Water, Precipitation, Snow Cover and Sea Ice Concentration. In Proceedings of the Tenth International TOVS Study Conference, Boulder, Colorado, USA, 27 January–2 February 1999; pp. 230–240.
15. Karney, C.F.F. Algorithms for geodesics. *J. Geod.* **2013**, *87*, 43–55. [CrossRef]
16. Bhaduri, B.; Bright, E.; Coleman, P.; Dobson, J. LandScan: Locating people is what matters. *Geoinformatics* **2002**, *5*, 34–37.
17. Mills, S.; Weiss, S.; Liang, C. VIIRS day/night band (DNB) stray light characterization and correction. *Proc. Spie Int. Soc. Opt. Eng.* **2013**, *8866*. [CrossRef]
18. Davis, R. Shanghai Disney Resort, Closed in January Due to Coronavirus, Set to Partially Reopen Monday. Available online: https://variety.com/2020/biz/news/shanghai-disney-resort-disneyland-reopen-china-coronavirus-1203527299/ (accessed on 20 August 2020).

remote sensing

Article

Examining the Economic and Environmental Impacts of COVID-19 Using Earth Observation Data

William Straka III [1,*], Shobha Kondragunta [2], Zigang Wei [3], Hai Zhang [3], Steven D. Miller [4] and Alexander Watts [5]

1 Cooperative Institute for Meteorological Satellite Studies, University of Wisconsin-Madison, Madison, WI 53706, USA
2 National Oceanic and Atmospheric Administration, 5200 Auth Rd., Camp Springs, MD 20746, USA; shobha.kondragunta@noaa.gov
3 I. M. Systems Group, 5830 University Research Court, College Park, MD 20740, USA; zigang.wei@noaa.gov (Z.W.); hai.zhang@noaa.gov (H.Z.)
4 Cooperative Institute for Research in the Atmosphere, Colorado State University, Fort Collins, CO 80523, USA; Steven.Miller@colostate.edu
5 BlueDot Inc., Toronto, ON M5J 1A7, Canada; alexander@bluedot.global
* Correspondence: william.straka@ssec.wisc.edu

Abstract: The COVID-19 pandemic has infected almost 73 million people and is responsible for over 1.63 million fatalities worldwide since early December 2019, when it was first reported in Wuhan, China. In the early stages of the pandemic, social distancing measures, such as lockdown restrictions, were applied in a non-uniform way across the world to reduce the spread of the virus. While such restrictions contributed to flattening the curve in places like Italy, Germany, and South Korea, it plunged the economy in the United States to a level of recession not seen since WWII, while also improving air quality due to the reduced mobility. Using daily Earth observation data (Day/Night Band (DNB) from the National Oceanic and Atmospheric Administration Suomi-NPP and NO_2 measurements from the TROPOspheric Monitoring Instrument TROPOMI) along with monthly averaged cell phone derived mobility data, we examined the economic and environmental impacts of lockdowns in Los Angeles, California; Chicago, Illinois; Washington DC from February to April 2020—encompassing the most profound shutdown measures taken in the U.S. The preliminary analysis revealed that the reduction in mobility involved two major observable impacts: (i) improved air quality (a reduction in NO_2 and $PM_{2.5}$ concentration), but (ii) reduced economic activity (a decrease in energy consumption as measured by the radiance from the DNB data) that impacted on gross domestic product, poverty levels, and the unemployment rate. With the continuing rise of COVID-19 cases and declining economic conditions, such knowledge can be combined with unemployment and demographic data to develop policies and strategies for the safe reopening of the economy while preserving our environment and protecting vulnerable populations susceptible to COVID-19 infection.

Keywords: COVID-19; earth observation data; nitrogen dioxide (NO_2); night light imagery (VIIRS); mobility; environmental impacts

Citation: Straka, W., III; Kondragunta, S.; Wei, Z.; Zhang, H.; Miller, S.D.; Watts, A. Examining the Economic and Environmental Impacts of COVID-19 Using Earth Observation Data. *Remote Sens.* **2021**, *13*, 5. https://dx.doi.org/10.3390/rs13010005

Received: 28 September 2020
Accepted: 17 December 2020
Published: 22 December 2020

1. Introduction

The spread of severe acute respiratory syndrome coronavirus 2 (SARS-CoV-2, COVID-19) was first reported in Wuhan, China, in early 2020 by human-to-human transmission [1–4], which became a global pandemic by early March 2020 and affected more than 150 countries within weeks. The World Health Organization declared a global health emergency related to COVID-19 on 30 January 2020 and a worldwide pandemic on 11 March 2020 [5]. As of November 2020, COVID-19 has infected over 73 million people worldwide and over 16 million people in the US, causing more than 1.63 million and almost 301,000 mortalities worldwide and in the US, respectively, as of 15 December 2020 [6].

Because the health effects of COVID-19 are unknown and the virus has a high mortality rate, precautionary measures like wearing face coverings, social distancing, and handwashing have been suggested, while clinical studies are underway to develop vaccines and potential cures. Social distancing, which includes full lock down (stay at home), partial lock down (travel if there is a need), quarantine (isolation for certain duration), and maintaining a minimum separation distance while in public, is considered an effective measure to curb the spread of a pandemic, as was seen during the 1918 flu pandemic [7]. Lockdown restrictions (full to partial) within the US were first implemented in early March and were lifted by end of May. During this time frame, only essential businesses (e.g., hospitals, gas stations, grocery stores) were operational.

An immediate impact of lockdown restrictions was on mobility and its reduction, which not only curtailed the infection rate of COVID-19 but also reduced pollution level [8] and resulted in a 4.8% drop in GDP [9]. Since the beginning of the lockdown, several studies have examined the economic effects of the lockdown measures. Bonaccorsi et al. [10], in their study of the economic impacts of the lockdown in Italy, reported that the lockdown not only reduced national and state level fiscal revenues but also caused a segregation effect in municipalities with high inequality and high concentration of low-income population due to the mobility reduction, such that the municipalities were at a higher risk of experiencing poverty and slower recovery. Chen et al. [11] also reported that mobility restrictions in areas experiencing high COVID-19 deaths had an increase in unemployment insurance claims and reduction in economic activity both in the US and Europe; a similar trend was also reported in China [12]. Although the extent of the economic disruption due to the COVID-19 pandemic is still unknown, it is evident that GDP growth is declining because of the pandemic.

Not surprisingly, the most polluted regions of the globe saw improved air quality during the lockdown. Several studies have reported that lockdown measures contributed to emission reduction as measured by the concentration of nitrogen dioxide (NO_2), carbon monoxide (CO), sulphur dioxide (SO_2), PM_{10}, and $PM_{2.5}$ (particulate matter in ug/m^3 for particles smaller than 10 and 2.5 µm in median diameter, respectively). It is important to note that there is a direct relationship between NO_2 and population density [13], as it is an indication of economic activity. This can be a byproduct of industrial activity (e.g., power plants, industrial plants) as well as from combustion due to the operation of vehicles (cars, trucks, buses, trains, etc.).

During and following the lockdown in China, many cities experienced a drop in NO_2 concentration by 40–60%, an increase in mean ozone concentration by a factor of 1.5–2, and a 35% drop in $PM_{2.5}$ [14–16]. Similar trends regarding concentrations of NO_2, CO, PM_{10}, and $PM_{2.5}$ were reported for Delhi and Mumbai in India and several cities in Europe [17–19]. Even NASA and the European Space Agency (ESA) have been using satellite imagery to monitor the environmental conditions resulting from COVID-19 response measures due to the reduced fuel emission (https://eodashboard.org/). Evidently, lockdown measures have contributed to a reduction in different gasses which contribute to the creation of photochemical smog along with other pollutants hazardous to human health. Long-term exposure to NO_2 and $PM_{2.5}$ have been linked to respiratory diseases that have been identified as a contributing factor to fatality from COVID-19 [20]. The linkages between prior exposure to $PM_{2.5}$ and mortality due to infection caused by COVID-19 were presented by a Harvard study (Wu et al., 2020 [21]).

The purpose of this study was to determine if the locations of NO_2 as well as the decrease in power consumption, as measured by the radiance from the Visible Infrared Imaging Radiometer Suite (VIIRS) Day/Night Band, aligned with economic activity centers and to understand how the underlying population density related to these locations. Using three case study locations (Los Angeles, CA; Chicago, IL; Washington, DC), we examined how NO_2 concentration and economic activity (VIIRS Day/Night Band imagery was used as a proxy) changed due to change the reduction in mobility patterns during February, March, and April of 2020. This preliminary study demonstrates how satellite-derived information (e.g., NO_2 concentration and light intensity—a proxy for energy consumption) can be used to explore variation in economic activities in near real-time resulting from a global pandemic or an isolated, local incident (weather or attack on infrastructure) that causes a reduction or shutdown of the power grid, etc.

2. Data and Methodology

This study focused on three major metropolitan areas in the United States: Chicago, IL; Los Angeles (LA) County, CA; Washington DC, where similar "stay at home" measures were implemented on 12th, 19th, and 20th of March 2020, respectively. The lockdown measures ensured that only essential businesses and services were open in the three cities. The other reason for choosing these cities was the availability of mobility data at a high confidence level.

Each metropolis has a unique economy. LA County is the third largest metropolitan area in the United States with an estimated GDP of $1.05 trillion in 2018 [9], and an estimated population of about 10 million as of July 2019 (US Census Bureau 2020) (Figure 1a). Its economy includes a variety of industries ranging from entertainment to shipping to software, which require long commutes over the vast highway system. The Chicago metropolitan area has an estimated 2018 GDP of $689 billion [9] and an estimated population of 2.7 million in 2019 (US Census Bureau 2020) (Figure 1b). While the majority of Chicago's economy includes financial and professional service sectors concentrated in the downtown part of the city requiring lengthy commutes, there is a heavy manufacturing presence in the Chicago metropolitan area. The Washington DC metropolis is primarily focused on businesses and the support of the federal government and had an estimated 2018 GDP of over $540 billion [9] and an estimated population of 9.81 million in 2019 (US Census Bureau 2020) (Figure 1c).

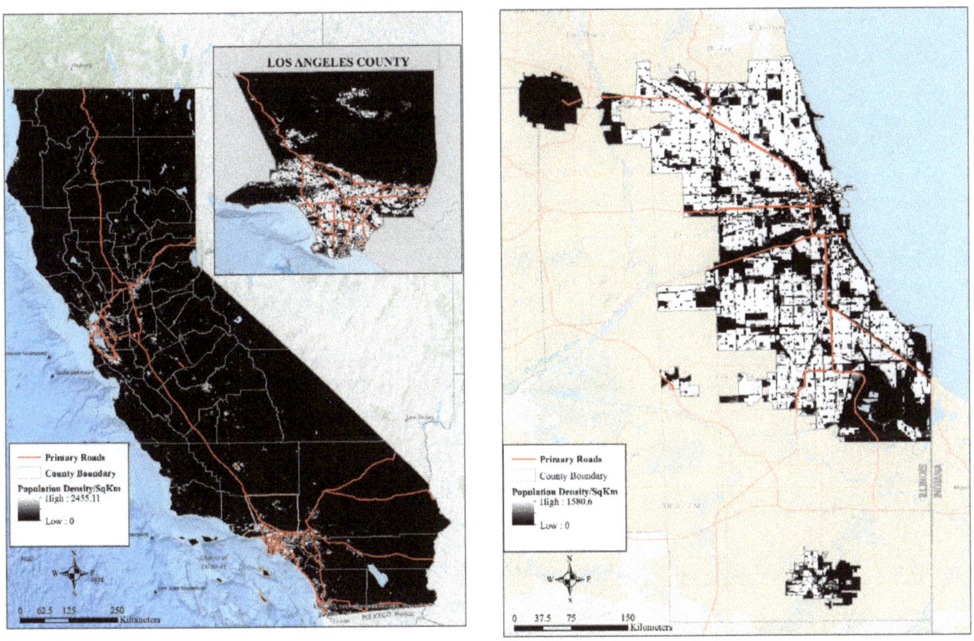

(a) (b)

Figure 1. *Cont.*

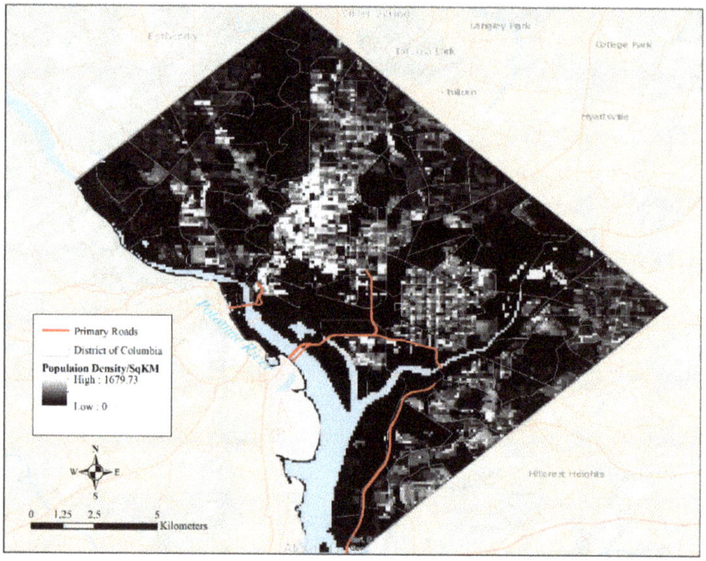

(c)

Figure 1. Population density distribution in 2020 for California and LA County (**a**), Chicago, IL (**b**), and Washington, District of Columbia (**c**).

In the three metropolises, there is a heavy reliance on long home-to-work commutes, which is a factor for heavy pollution. According to the American Lung Association's 2020 "State of the Air" report, LA, Chicago, and Washington DC (are ranked 1, 16th, and 20th among the 25 most ozone-polluted cities in the US [22]. While each of these regions contain numerous small-to-medium coal/gas power-generating facilities, the largest power-generating facilities in Chicago and Washington DC are nuclear power plants, which do not emit any NO_2 or greenhouse gases. The city of LA has four natural gas-fired generating stations within city boundaries, which are medium in size. The city primarily receives power from Utah, Arizona, and Nevada.

In the three metropolises, there is heavy commercial traffic (trucks), particularly in the port and heavy industry regions of the city as well as a wide range of age and type of vehicles. In addition, all three cities have large mass transportation infrastructure (buses, subways, trains) which are relied upon by the population, particularly in the downtown areas. The largest difference between the three study cities is that the type of fuel utilized in the Los Angeles area. For many decades, California has had a unique blend of gasoline which was developed to help reduce the amount of smog, particularly in the LA basin.

In terms of climatology, as observed from the National Oceanic and Atmospheric Administration's (NOAA) National Climatic Data Center (NCDC), the three sites do have varying climatology. Being in the northcentral part of the United States, Chicago is by far the coldest and snowiest location of each of the study regions. It has an average monthly snowfall of 9.1" in February to an average of 1" of snow in April, with an average temperature ranging from −1.6 °C to roughly 5 °C over the three months of study. In general, this would translate to an increase in activity. Similarly, in the Washington, DC metropolitan area, the average temperature is roughly from 4 °C to 15 °C over the three months of study. As one would expect, the LA basin is much warmer than both Chicago and Washington, DC. However, all of the areas generally have precipitation including both rain and snow during the study period. While the temperature and primary precipitation are different, all three study locations observe an increase temperatures and increased precipitation, typically in the form of rain, during the three-month study period. This would generally lead to an overall increase in movement.

Materials and Methods

For the study, we used tropospheric column NO_2 data from Sentinel 5P TROPOspheric Monitoring Instrument (TROPOMI) from the ESA datahub (https://scihub.copernicus.eu/) to examine the environmental effects of the lockdowns, VIIRS Day/Night Band (DNB) data to capture changes in energy consumption using light intensity (a proxy for economic activity), surface $PM_{2.5}$ data from the AirNow monitoring network (airnow.gov), cell phone derived mobility data from BlueDot software company under academic agreement to use the data for research purposes, and 2020 population distribution data from WorldPop (https://www.worldpop.org/). To maintain consistency among datasets, monthly averaged data were used for the time period studied. A discussion of the datasets and analytics used to examine the impact of the lockdowns on economic activities and environmental conditions is presented below.

Sentinel-5P TROPOMI

The Sentinel-5 Precursor (Sentinel-5P) satellite is a low-earth orbiting satellite developed by the ESA as part of the Copernicus Programme. It flies in a sun-synchronous ascending node orbit at roughly 824 km in altitude. The primary instrument onboard the satellite is TROPOspheric Monitoring Instrument (TROPOMI), a spectrometer designed to sense ultraviolet (UV, 270–320 nm), visible (VIS, 310–500 nm), near (NIR, 675–775 nm). and short-wavelength infrared (SWIR, 2305–2385 nm) radiation [23], and monitor trace gases (O_3, CH_4, CO, NO_2, SO_2) as well as aerosol index and layer height. The S5P TROPOMI is an air quality mission that observes air quality related to trace gases and aerosols at high spatial resolution. The NO_2 product used in this study was available at a 3.5 km $\times$ 5 km spatial resolution. For our analysis, we remapped the pixel level TROPOMI data to $0.25^\circ \times 0.25^\circ$ resolution.

Suomi NPP VIIRS

The Suomi National Polar-Orbiting Partnership (S-NPP) and NOAA-20 satellites are polar orbiting environmental satellites launched in a sun-synchronous 1330 local time ascending node orbit. The S-NPP and NOAA-20 are spaced one-half orbit apart (~50 min) from each other. Each satellite orbits the Earth at a roughly 834 km altitude and completes a single orbit in ~101 min. Both satellites carry the VIIRS instrument, which collects both visible and infrared imagery spanning from 0.4–12 μm, combining key capabilities of several legacy instruments. The VIIRS includes a DNB capable of sensing visible/near-infrared (500–900 nm) during both day- and nighttime (low-light) conditions. At night, it is sensitive to small amounts (~7 orders of magnitude fainter than daylight) of light present in its band pass and is capable of detecting from its orbital altitude the light emitted from a single isolated streetlamp [24–27]. The DNB data have been used for a wide range of applications such as fire detection, meteorological phenomena, observations of anthropogenic light sources, like ship tracking and fishery monitoring [28,29], as well as to estimate electrical usage and power outage, industrial output, and economic activities [30–33].

VIIRS DNB Radiance Data Creation

To discern trends in human activity and mobility, monthly composites of city light intensity from S-NPP were created. The measured DNB radiances from the nights where there was little to no moonlight (roughly the day after last quarter to the day after the first quarter phase of the moon, or approximately 14 consecutive nights of each lunar cycle) were cloud-cleared using the NOAA's Interface Data Processing Segment (IDPS) VIIRS Cloud Mask to create a nightly screened image [34]. These images were then remapped to a common 15 arc second grid and combined into a single monthly composite of the brightest radiance for each pixel in the grid.

A key point to note is the need to account for the stray light which occurs in the DNB. Stray light arises from flaws in the light shielding of the satellite, where non-earth-scene light enters from either the VIIRS scan cavity or through the nadir door and solar diffuser openings. This results in a "gray" haze in the data, which can extend as far south as the southern United States in the summer months due to the tilt of the earth and exists in the DNB for both S-NPP and NOAA-20. A post-processing correction is applied to the data to remove this stray light [35], allowing for a more consistent radiance across as the stray light region is traversed.

Mobility Data

Human mobility was approximated using anonymized, population-aggregated, near real-time, mobile device GPS location data provided by Veraset (Veraset, San Francisco, CA, USA, 2020), a data-as-a-service vendor. The location data were used to calculate median maximum distance, i.e., the distance between farthest check-in and home location in km (mean daily median by census tract). "Home" was defined as the location (~0.6 square km grid cell) where a device was primarily located between 12 am and 5 am local time. Maximum distances from home were calculated daily for each device, and a daily median value was assigned at the census tract level. This data were averaged over the course of a given month to correlate with the analysis of the DNB as well as to filter the day-to-day noise in mobility patterns and discern over all trends in human activity. Census tracts with less than 5 daily devices were excluded from the analysis. This type of data has been utilized to study the effectiveness of "stay-at-home" orders during COVID-19 [36] as well as by emergency management agencies to help determine the likely areas for COVID-19 to spread. Due to the fact of privacy laws in other countries, this study used the data available for cities in the United States.

Population Distribution Data

Population data sets were obtained from WorldPop [37], which provides high-resolution, open, and contemporary data about human population distribution across the world. We obtained the 2020 population data for the three study sites as GeoTiff files at a spatial resolution of 100 m × 100 m (3 arc seconds). The gridded population data were created using a top-down approach [37], which was adjusted to match the United Nations Population estimate. Using the census tract boundaries (the spatial scale at which mobility data were generated) and the gridded population data, we determined the population density at the tract level for the implementation of the spatial data mining approaches.

Spatial Data Mining

The following spatial statistics methods were used to examine the spatial variation of mobility, energy usage, the NO_2 concentration over the three months, and also to understand the relationship among the variables, while the underlying population density did not change during February–April.

Hotspot Analysis, also known as the Getis–Ord (Gi*) spatial statistic, was used to identify the spatial clusters of hotspots (i.e., features of high values) and coldspots (i.e., features of low values). The Gi* spatial statistic estimates the spatial dependency effect of an attribute based on a specified spatial relationship among the features. The spatial relationship can be based on identifying features within a certain distance from a feature or assigning a weight to features based on their distance from a feature. This statistic identifies statistically significant clusters of high and low values represented by a *z*-score and *p*-values (more details about the statistical method can be found in References [38,39]).

To identify the variation in mobility and energy consumption (using DNB radiance values), an optimized hotspot analysis was implemented. The Optimized Hot Spot Analysis tool in ArcGIS Pro [40] implements incremental spatial autocorrelation that performs Global Moran's I for a series of distances to measure the intensity of clustering at each distance. The intensity of clustering represented by a *z*-score identifies the optimal distance that allows for pronounced clustering. This peak optimal distance was used as the threshold distance for identifying clusters in the hotspot analysis. This tool automatically aggregates data and identifies the significant distance threshold to aid with cluster identification.

Spatial Autocorrelation: Evidently, mobility and energy consumption dropped as a result of lockdown across the study sites. Because of the spatial variation of high and low population density clusters in the study site, Moran's I was used to identify the spatial autocorrelation in the parameters (mobility, DNB intensity and NO_2 concentration) using Moran's I. While Global Moran's I describes the spatial dependency and association across the study site, Local Moran's I (Local Indicator Spatial Association (LISA)) identifies the degree of association between a census tract and its neighbors (more details about these statistics can be found in References [41–43]).

Visualization: The visualization of GIS data throughout this paper were created using ArcGIS® software by Esri. ArcGIS® and ArcMap™ are the intellectual property of Esri and are used herein under license. Copyright Esri©. All rights reserved. For more information about Esri® software, please visit www.esri.com.

3. Results—Effects of Lockdown

3.1. Population Density

Before exploring the effects of the lockdowns in our three case study cities, we explored the population density distribution at the census tract level (the administrative boundary that was used to generate mobility data). The main purpose was to identify the high–low density clusters to understand the effects of lockdown on subsequent changes in mobility patterns, nitrogen dioxide concentrations as well as power usage due to the change (i.e., drop) in economic activities. Figure 2a,b depict the population density distribution for LA County at the census tract level. These data show that the highest density tracts were clustered in the south-central part of LA County (specifically, surrounding LA city, highlighted by the square in Figure 2a), and this area was surrounded by high to moderate density tracts (hotspots in Figure 2b). The northern half, southwestern part, and southernmost part of the county were occupied by very low-density census tracts (coldspots in Figure 2b). Essentially, the southern half of the county is densely populated with high spatial variability, while the northern half of the county is sparsely populated.

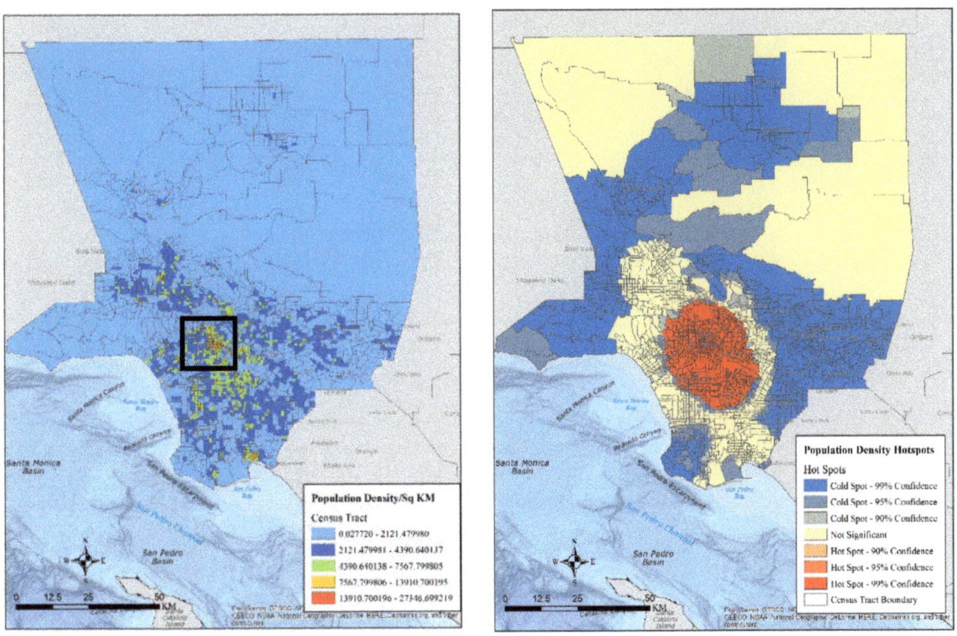

(**a**) (**b**)

Figure 2. Population density distribution at the census tract in LA County (population/square KM) (**a**) and high- and low-density census tract clusters in LA County (**b**).

In the city of Chicago, moderate to heavily populated tracts are clustered in the northern and northeastern part of the city (Figure 3a). While the sparsely populated tracts were clusters in the southern and northwestern part of the city; the low population density tracts were spread out across the city. From the hotspot analysis output, it was clear that the heavy density tracts were clustered along the north-eastern part of the city (hotspots with high significance in Figure 3b), along the lake shore, while the low population density clusters were located in the southern and northwestern parts of the city (coldspots in Figure 3b). There was a random distribution of moderate to low density tracts across the entire city (Figure 3b).

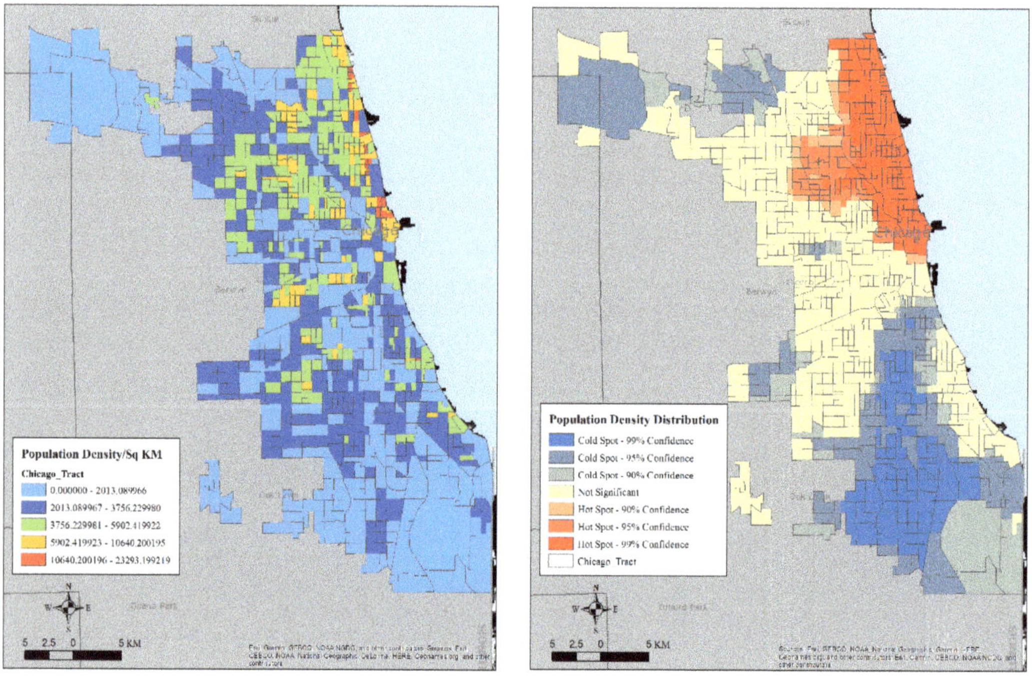

(**a**) (**b**)

Figure 3. Population density distribution at census tract in Chicago (population/square KM) (**a**) and high- and low-density census tract clusters in Chicago (**b**).

The majority of Washington, DC population is concentrated in the central part of the city, with a few clusters of low and very low-density tracts are spread out across the entire city (Figure 4a,b). It was clear from the hotspot analysis that the densely population tracts were clustered in the center of Washington, DC with few clusters of low-density population. The remainders of the tracts were less densely populated.

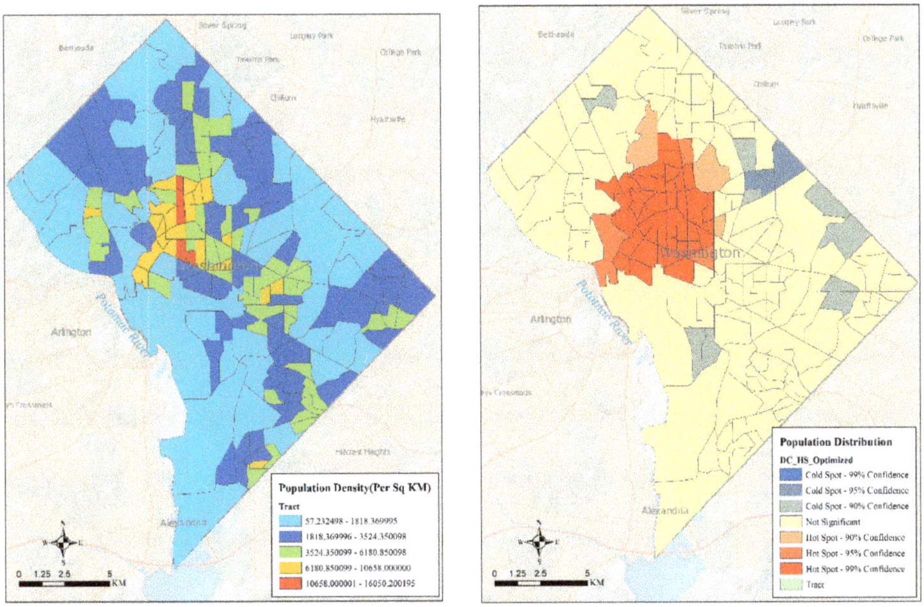

(a) (b)

Figure 4. Population density distribution at census tract in DC (population/square KM) (**a**) and high- and low-density census tract clusters in Washington, DC. (**b**).

3.2. *Variations in Mobility Patterns*

The immediate impact of lockdown measures in the cities was a reduction in mobility due to the tele-working and shutdown of all businesses except essential businesses and services. Because stay-at-home measures were in place by March 2020 in all three cities, we explored the change in mobility (distance and pattern) at the census tract level. Figures 5 and 6 show the mobility distribution in LA County during February through April. The data indicate that mean travel distance dropped from 268 km in February to 50 km in March and 42.6 km in April. In February, the high mobility areas were concentrated in the downtown area of LA City as well as in the northeast and northwest part of LA County, which has a very sparse population (Figures 5a and 2a).

Evidently, the highest distance traveled dropped by more than 75% in the entire county. While more than 50 km distance on average was traveled in the northwestern part of the county, mobility was no more than 20 km in the downtown area of the county (central part of the county closer to LA City (red box in Figure 5a)). Essentially, the lock-down-induced telecommuting appears to have impacted on the reduction of mobility in LA city by more than 90%. Moderate mobility was still observed in March near Malibu, Long Beach, and Santa Monica (southwestern part of the county represented by black boxes in Figure 5b). Although there were some pockets of high mobility in the high-density areas of LA County, mobility appeared to have dropped in areas surrounding the downtown LA (central part of the County) but was still higher in sparsely populated counties of LA along the northeast and northwestern part of the county. The maximum distanced traveled in LA County by April was approximately 42 km (Figure 5c). Nevertheless, the areas experiencing moderate mobility remained the same as they were in March, and these areas included the low-density tracts of the county as well as Santa Monica, Long Beach, and Malibu. Preliminary analysis of the income data from the US Census (2018 American

Community Survey) revealed that the mobility reduced significantly in the impoverished part of LA County.

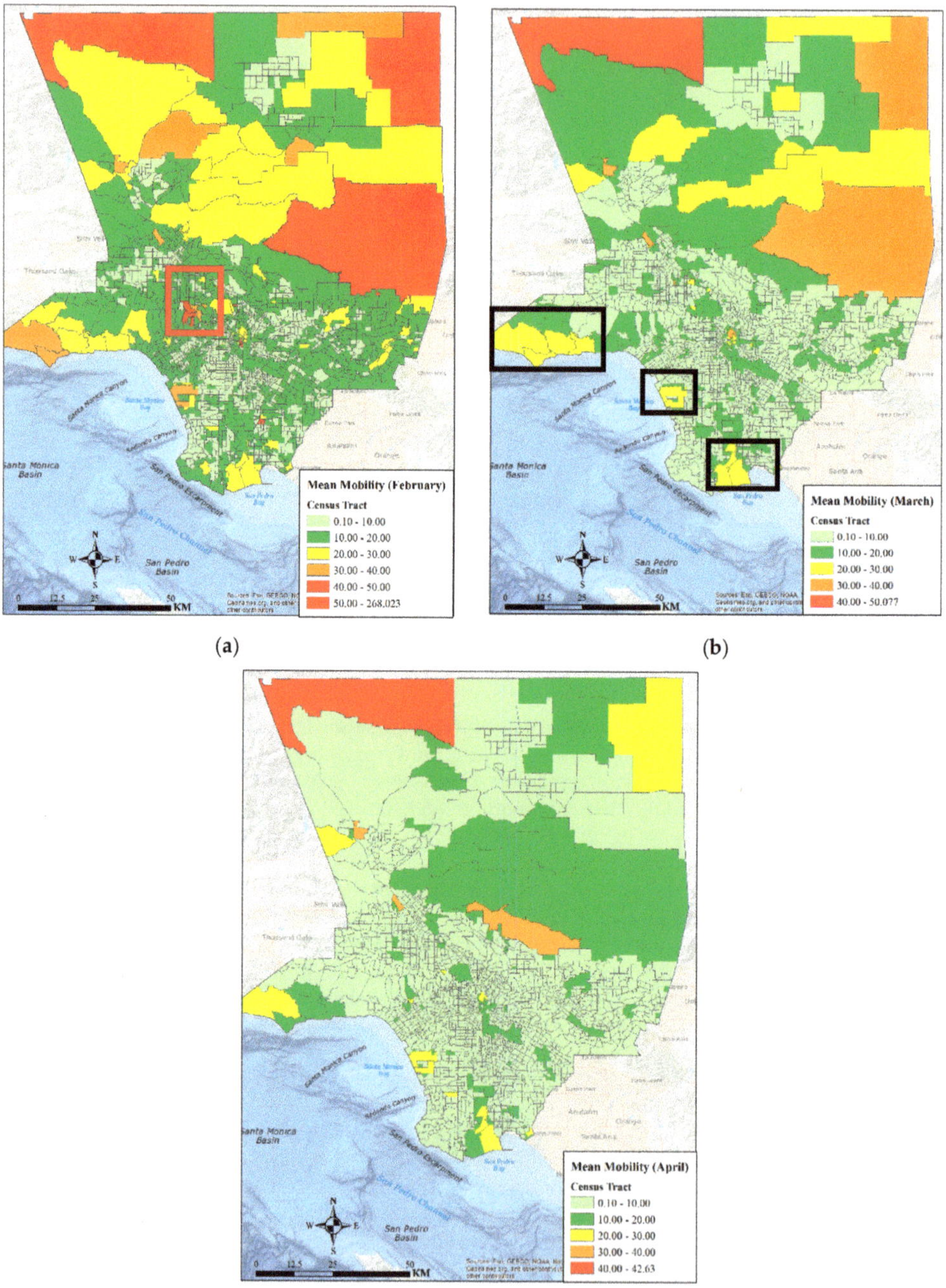

(**a**) (**b**) (**c**)

Figure 5. Mean mobility distribution in LA in February (**a**), March (**b**), and April (**c**).

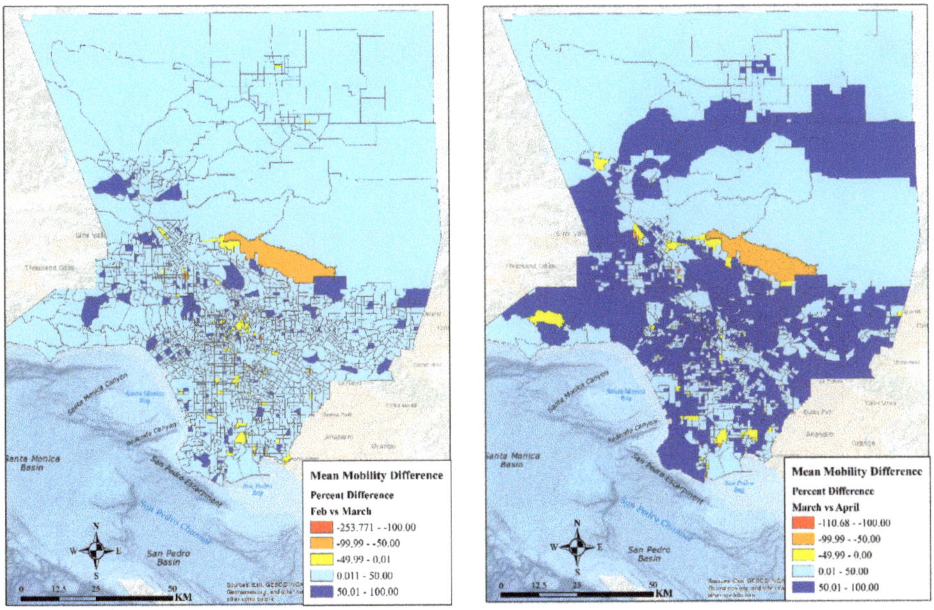

(a) (b)

Figure 6. Mean mobility percentage difference in LA between February and March (**a**) and between March and April (**b**).

Although mobility was reduced by 50–100% by March in many parts of LA County, including closer to LA City (Figure 6a), majority of LA experienced a drop in mobility by 50–100% by April (Figure 6b). During February–April, mobility increased significantly only in few census tracts of LA. A future analysis of the tracts experiencing mobility increase will be conducted to explore the effects of the underlying socio-economic characteristics as well as businesses that might have contributed to the mobility increase.

Before the lockdown, travel in Chicago in February appears to have been concentrated near the downtown area (black box in Figure 7a), near the Chicago Midway International Airport (northwestern part of the City), Whiting (southern part), and near the Chicago Midway International Airport (red box in Figure 7a). It also appears that travel in Chicago was not concentrated in the high-density tracts that are located in the northeastern part of the metropolis (Figure 3a).

Following the lockdown, by March, the maximum distance traveled dropped by 50% (81 km to 44.6 km) (Figure 7b). However, the moderate to very high mobility clusters were still located near the downtown area of Chicago, near the airports, and Whiting (Figure 7b, red and black boxes). Mobility dropped by another 50% in April in Chicago (Figure 7c), but like February and March, high to moderate mobility areas were present in the central Southern and Southwestern part of Chicago (Figure 7c). Evidently, mobility was still higher near O'Hare International Airport (northwestern part of Chicago. Figure 7a–c).

Figure 8a,b depict the percent change in mobility in Chicago during February through April. Immediately after the lockdown, mobility dropped by more than 50% in many tracts across Chicago, while it also increased by more than 100% in the northeastern part of the City (black box in Figure 8a) near Uptown—a residential neighborhood. By April, mobility reduced by more than 75% (Figure 8b) across the entire city. The mobility reduction was evident in the central, northeastern, northwestern (near O'Hare International Airport), and southern (near Whiting) part of the City. Although the mobility reduction near the airports was approximately 25%, the reduction in residential neighborhoods of the city (high density tracts, Figure 3a) appeared to be due to the fact of tele-commuting.

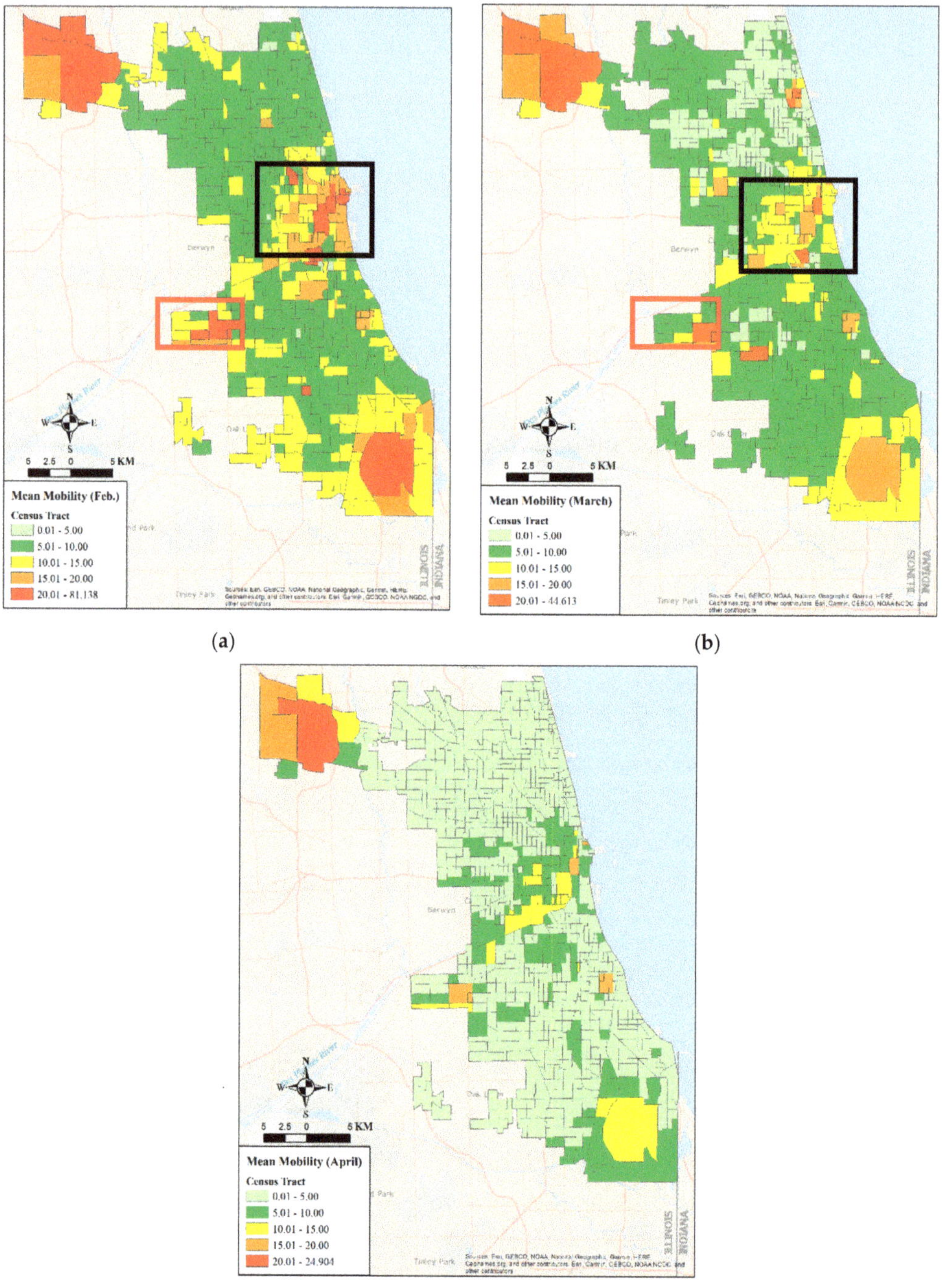

(a) (b) (c)

Figure 7. Mean mobility distribution in Chicago in February (**a**), in March (**b**), and April (**c**).

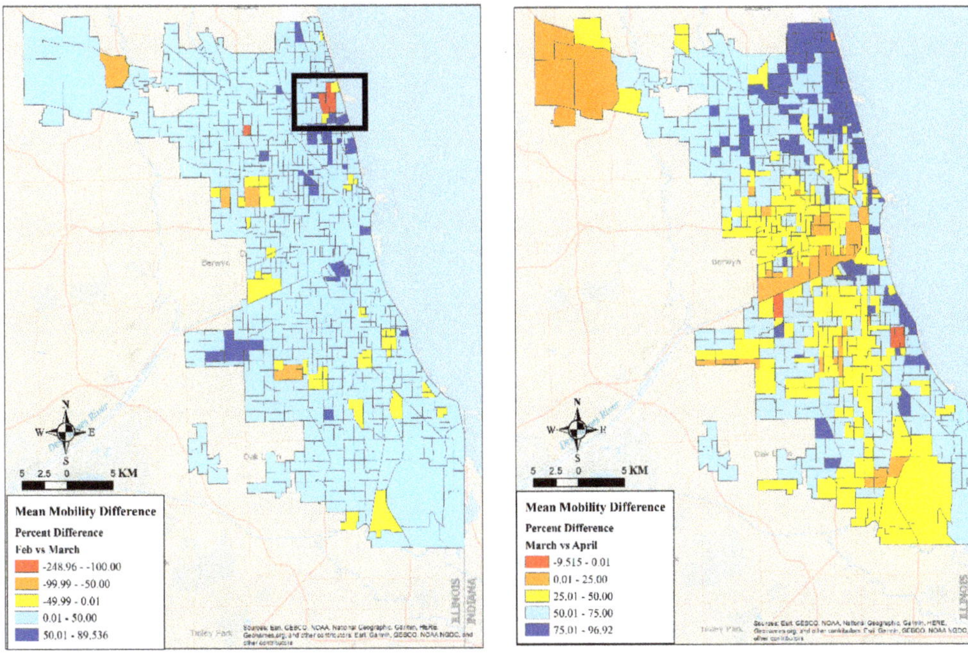

(a) (b)

Figure 8. Mean mobility percent difference in Chicago between February and March (**a**) and between March and April (**b**).

The densely populated tracts in Washington DC were clustered near the central and southeastern part (Figure 4a). The central part of DC was also where White House is located. Therefore, it is no surprise that mobility was higher in February in the central part of DC (near the White House and the downtown area) (Figure 9a). In March, after the beginning of the lockdown, mobility dropped in Washington DC by almost 50% from about 73 km to 38 km (Figure 9a,b), but the highest mobility was reported to be near the White House, Capital Hills, and the Washington, DC downtown. Residential neighborhoods surrounding the central part of DC exhibited low mobility. In comparison to March, mobility dropped by almost 50% from in April. However, the clusters of high to moderate mobility were still concentrated near White House, downtown DC, and Capital Hills (central part of DC, Figure 9c) where most of the policy makers were meeting regularly to address the spread of the pandemic. Mobility appeared to have dropped significantly in the residential areas of DC (surrounding areas of White House and downtown), which could be attributed to tele-working.

Between February and March, mobility dropped by more than 50% in few places across DC, but mobility was higher near the White House in March (Figure 10a). By April, however, mobility dropped by at least 16% percent across DC, and it was higher than 90% in a few locations (Figure 10b). Even the central part of DC (near White House and downtown areas) experienced a 25–50% reduction in mobility by April. While mobility reduced in high-density tracts (nearer to downtown area) immediately after the lockdown in March, by April, all across DC significant reductions in mobility were observed. However, unlike LA and Chicago, DC did not experience an increase in mobility in March or April. This could be attributed to the fact that LA County has sparsely populated census tracts as opposed to DC and Chicago, which are mainly occupied urban tracts.

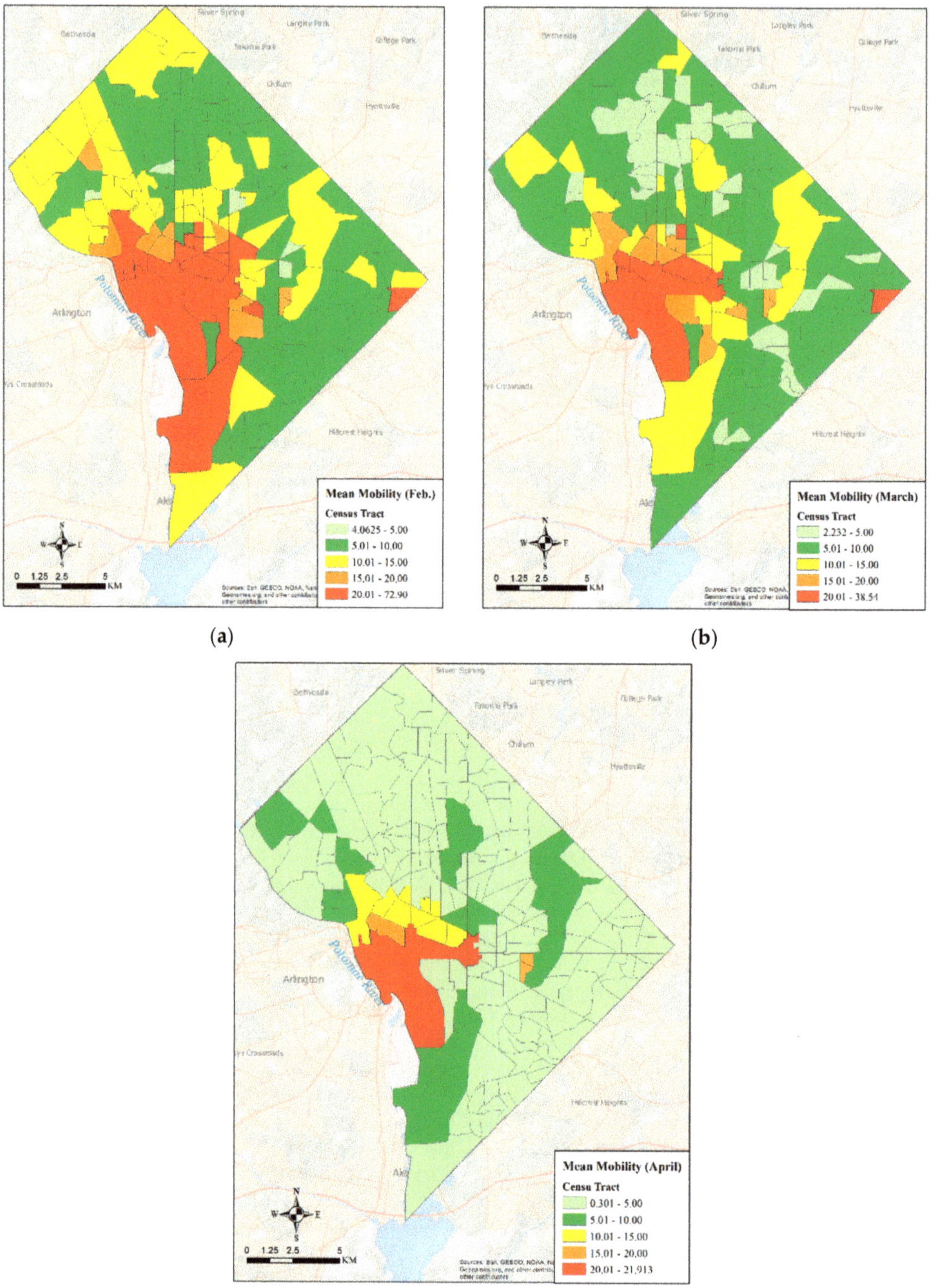

(**a**) (**b**) (**c**)

Figure 9. Mean mobility distribution in Washington, DC in February (**a**), in March (**b**), and April (**c**).

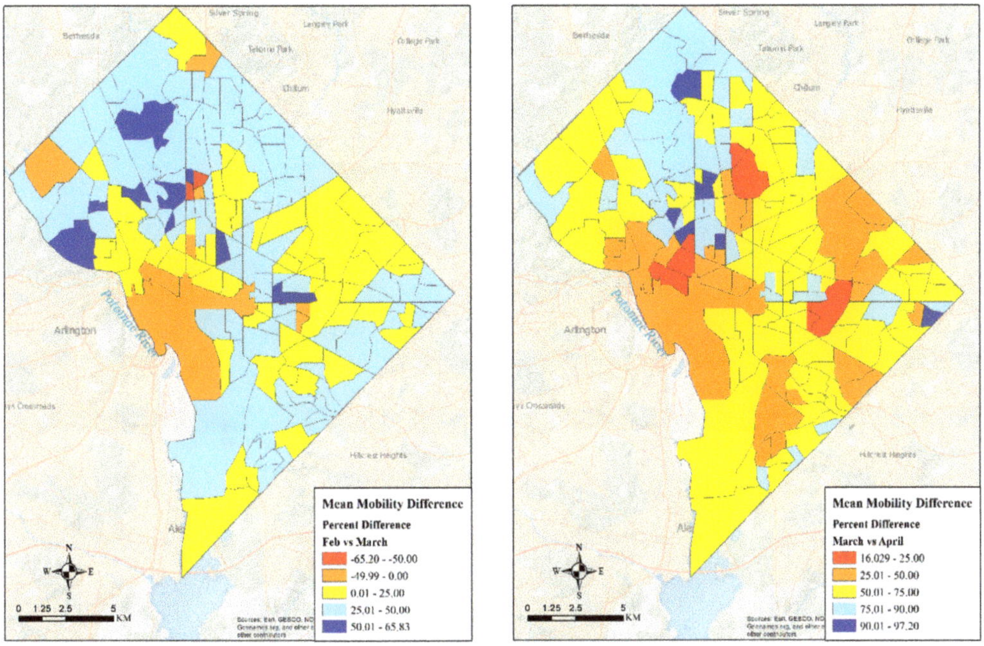

(**a**) (**b**)

Figure 10. Mean mobility percentage difference in Washington DC between February and March (**a**) and between March and April (**b**).

3.3. Analysis of NO_2

NO_2 is one of many byproducts of industrial processes that are considered hazardous to the health of humans and the environment (EPA, 2015). From the NO_2 trends observed by TROPOMI over Los Angeles (Figure 11), it can be seen that there was a large reduction in the monthly average total tropospheric column NO_2 from February to March, and the reduction continued through April.

More than half of the signal was due to the lockdown, with only the highest concentrations of NO_2 (due to industrial activity) remaining near the LA Port region, where most of the refinery and industrial capability in Los Angeles is located, as well as the Inland Empire (San Bernardino Valley), which is a major shipping hub. A recent study [44] showed that while total NO_2 reduction in Los Angeles during 15 March 15 to 30 April 2020 compared to the same time period in 2019 (Business as Usual, BAU) was about 66%, NO_2 reduction due to the lockdown measures was 35% and with the remaining 31% being due to the fact of seasonality. Even during the lockdown, it would be expected that there would be some industrial activity to support essential services. The trend in NO_2 is correlated to the mobility pattern observed during the lockdown. In Figure 12, the left image (a) is a histogram of the total column NO_2 for February, March, and April 2020, and the right image (b) is the distribution of distance a given cellphone travelled during the daytime period (i.e., when most movement occurs) for each corresponding month. As can be seen from Figure 12, February and March exhibited a wider range of mobility compared to April. The curve shifted to the left with high concentrations (tails of the curves), nearly 50% lower than the values observed in February and May.

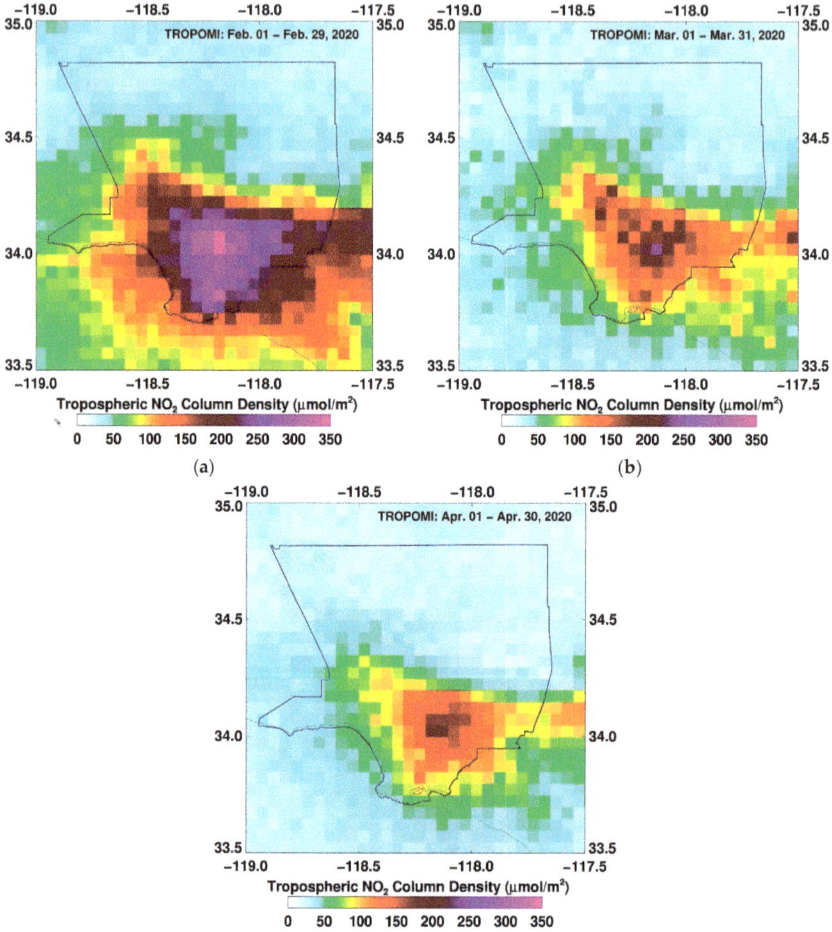

Figure 11. Distribution of NO_2 concentrations in Los Angeles, CA during February (**a**), March (**b**), and April (**c**).

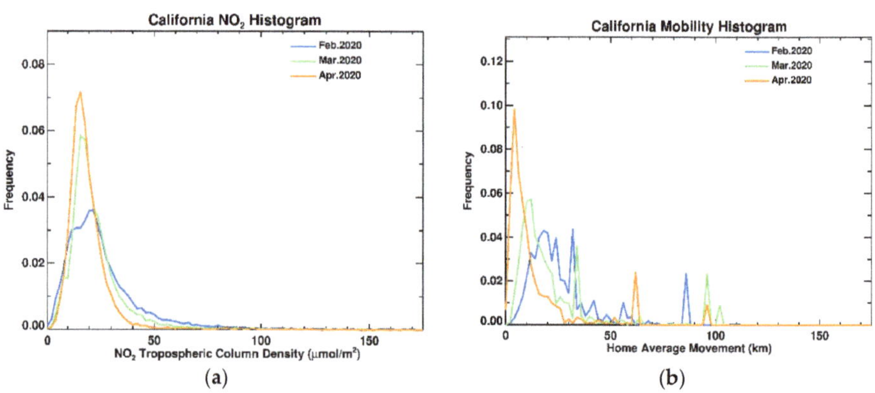

Figure 12. Histogram of NO_2 concentration in California (**a**) and mobility distribution (**b**).

Chicago, Illinois had an earlier lockdown than Los Angeles. As discussed previously, the economy is focused around the downtown region by Lake Michigan in the financial and professional services sectors. However, there is a heavy manufacturing presence in the Chicago metropolitan area, particularly close to the southeastern part of Lake Michigan and into the western part of Indiana.

As one might expect, the areas where people commute to on a regular basis showed a dramatic decrease in NO_2 in the downtown region, while the areas of heavy industry, such as powerplants and refineries, remained at elevated (though reduced) NO_2 levels (Figure 13). A recent study [44] reported that reductions in NO_2 as observed by TROPOMI due to the lockdown were ~14% for 15 March to 30 April 2020 compared to the same time in 2019.

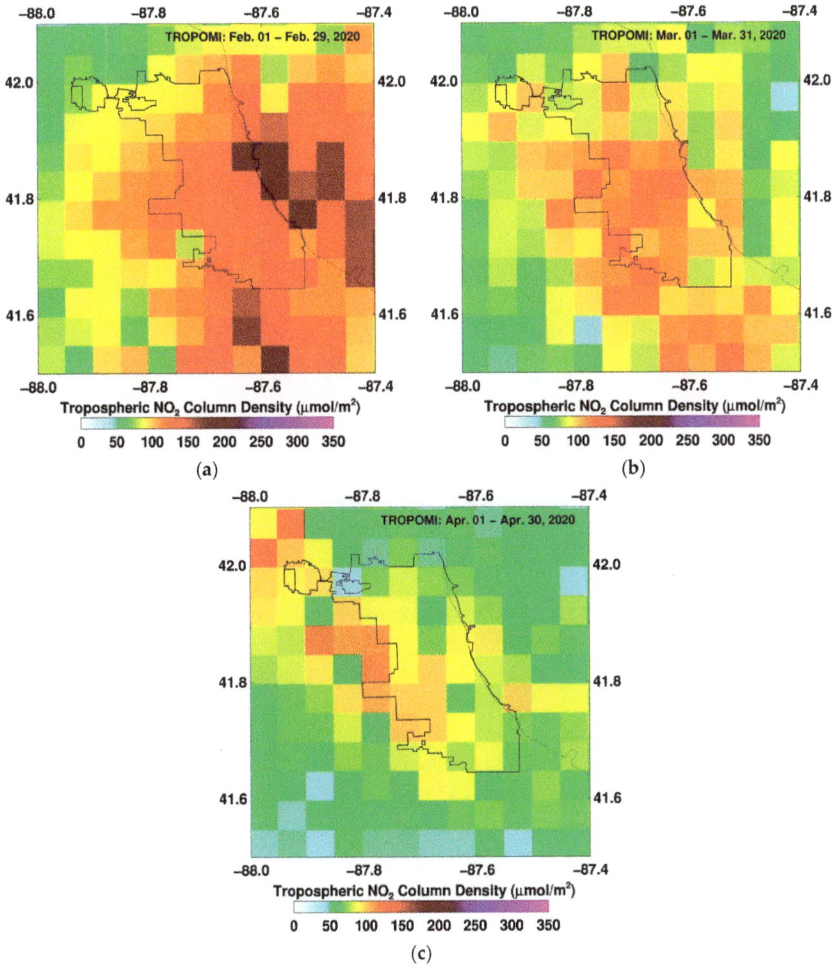

Figure 13. Distribution of NO_2 concentration in Chicago in February (**a**), March (**b**) and April (**c**).

While there was a similar trend in decreased mobility with Los Angeles, the spread of the total column of NO_2 was much narrower in Chicago (Figure 14). This is partially due to the fact that the area observed was much smaller than the Los Angeles basin. The other notable difference was that that the average distance for commuting was much shorter for Chicago than Los Angeles.

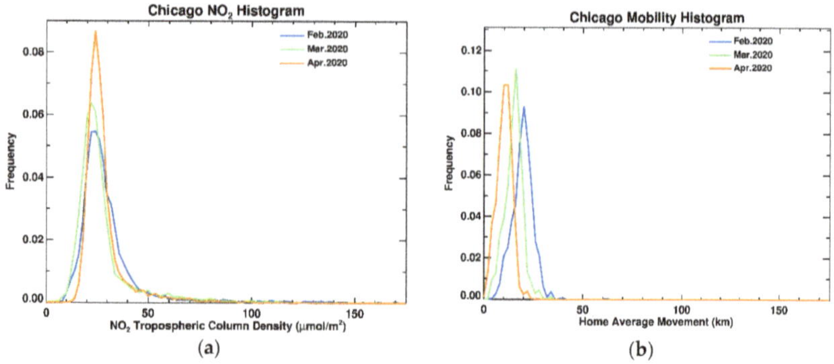

Figure 14. Histogram of NO_2 concentration in Chicago (**a**) and mobility distribution (**b**).

The Washington, DC metropolitan area is heavily driven by businesses and federal agencies. However, unlike Los Angeles and Chicago, there is no heavy industry. Maryland and the DC area implemented their lockdown on 17 March. The primarily I-95 travel corridor in Maryland, Delaware, and southern New Jersey can easily be seen in Figure 15, which is the total tropospheric NO_2 column density for February.

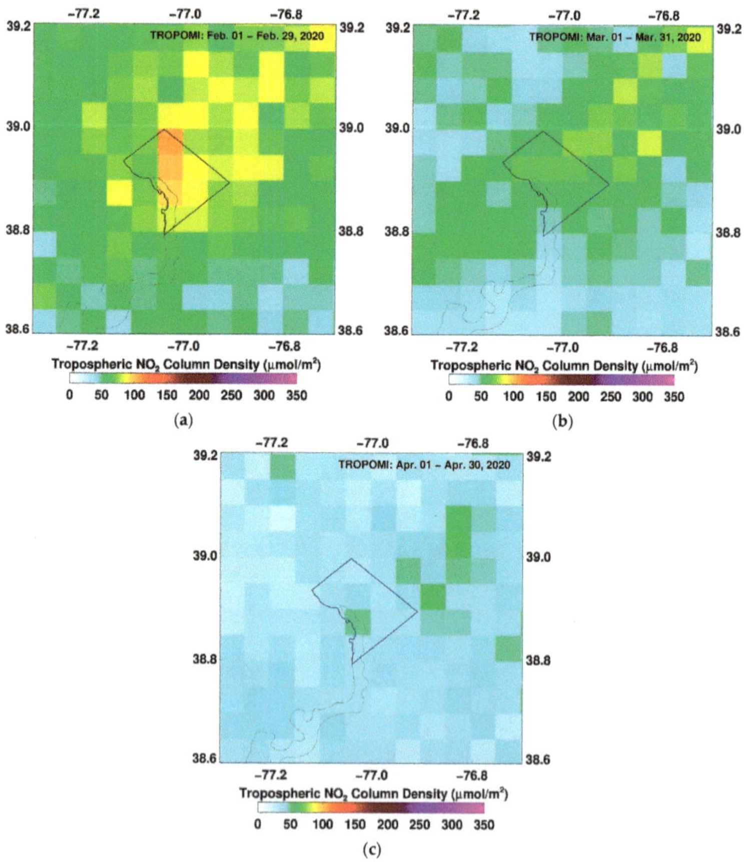

Figure 15. Distribution of NO_2 concentration in Washington DC in February (**a**), March (**b**) and April (**c**).

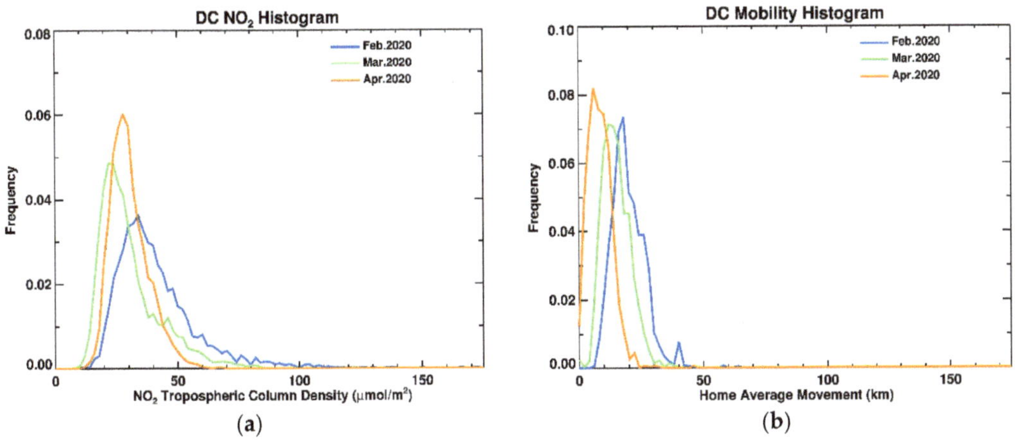

Figure 16. Histogram of NO_2 concentration in Washington, DC (**a**) and mobility distribution (**b**).

Table 1 shows the comparison between the monthly averaged NO_2 concentrations and mobility information distance over the entire region of each study site. As can be seen, the average mobility distance was at its lowest in the April 2020 peak distance, which corresponds with the lowest average NO_2 concentrations in all three sites.

Table 1. Monthly average distribution of NO_2 (μmoles/m^2) concentration and mobility (km).

	NO_2 (μmoles/m^2)			Mobility (km)		
City	**February**	**March**	**April**	**February**	**March**	**April**
Washington, DC	89.6	62.7	36.1	13.3	10.1	5.4
Chicago	123.9	102.7	82.1	12.4	9.2	5.4
Los Angeles	108.8	65.6	51.3	24.1	17.6	11.5

There were some variations between each of the sites that can be seen in Table 1. Los Angeles, for example, had a significant drop in movement between February and March, which was accompanied by a significant drop in NO_2 density. The leveling off in mobility did not directly correlate to the steady decrease in the rate of NO_2 density. There are a number of factors which could be a reason for this, including the time it takes to turn off various industrial processes or the number of cars on the highway.

In the case of Chicago, their lockdown was not as abrupt (only 17% drop in NO_2 for 26% drop in mobility between February and March), but even so, the decrease in movement resulted in a decrease in NO_2 density. The Washington DC metropolitan area is notable in that the shutdown did not occur until late March, meaning that the largest drop in mobility would have occurred in late-March. Even then, the NO_2 density decreased by 30% between February and March while the drop in NO_2 between March and April is quite significant, at 42%. The mandatory telework is continuing in the Washington, DC area and the trend in NO_2 for the whole year (2020) will shed light on how policy makers can introduce work schedules to the federal employees in the area to minimize air pollution.

The photochemical smog that leads to poor air quality is a chemical soup of noxious gases (NO_x = NO+ NO_2) among other volatile organic compounds (VOCs) that lead to ozone and $PM_{2.5}$ formation. Ozone is harmful to humans as well as plants, whereas $PM_{2.5}$ is harmful to humans. Both are pollutants that were declared as criteria pollutants by the United States Environmental Protection Agency (EPA). While NO_2 and VOCs are precursors for ozone and secondary aerosol formation, $PM_{2.5}$ can also be directly emitted (soot from cars) or photochemically formed from NO_2, SO_2, and VOCs which are precursors.

Because of the extreme reductions in the SO2 emissions beginning in the 1970s to curb acid rain, SO_2 is no longer a main source for secondary aerosol formation. NO_2 and VOCs remain the main precursors leading to the formation of secondary nitrate and organic aerosols. Figure 17 shows $PM_{2.5}$ concentrations in February, March, and April of 2020 decreased compared to the same months in 2019 with the exception of February 2020 in Los Angeles which was higher than the values observed in February 2019. Note that the lockdowns did not start until March and the differences could be due to the unique seasonal differences between the two years. Of the three cities, Chicago saw the largest reduction in $PM_{2.5}$.

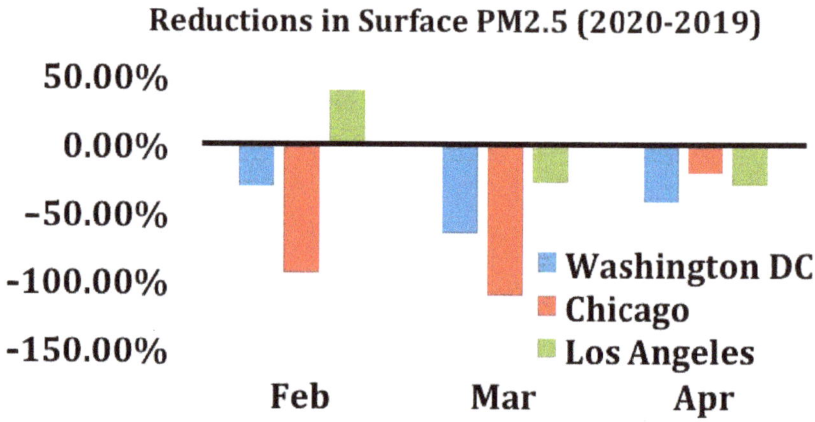

Figure 17. Distribution of surface $PM_{2.5}$ in Washington, DC, Chicago, and LA during February–March.

3.4. Distribution of Nighttime Lights (NTL)

The Day/Night Band (DNB) on S-NPP (and NOAA-20) has the ability to detect visible/near-infrared (500–900 nm spectral response) imagery for both day and nighttime conditions. The instrument is sensitive enough to detect not just the light emitted from a single isolated streetlamp but also emitted light from the mesosphere reflected off cloud features as well as density perturbations within the mesosphere itself. While it has a wide range of applications, the DNB was used here as a proxy of power usage (economic activities). Because the DNB is only able to observe electrical light at night, generally around 1–2 a.m. local time, this means that it is only a measurement of nighttime/early morning activity. Despite this limitation, there is significant activity at night like traffic movement that is captured by the DNB imagery.

Immediately after the lockdown measures were in place, businesses were shut down and the majority of the economic activities stopped in LA as is evident from Figure 18b, except for some activities that were still ongoing near downtown LA and the Long Beach area. The reduction in activities in March aligns with the reduction in mobility seen in LA County, except for the downtown area and near Long Beach where probably the port activities were still underway to some extent (Figure 5). The limited traffic movement in March (Figure 18b) could be due to the travel to essential businesses, such as grocery stores and hospitals. Although the lockdown measures were still active in April, economic activities appear to have resumed in LA in April (Figure 18c). The highest NTL intensity values (nWatts$\cdot$cm$^{-2}\cdot$sr^{-1}) in February, March and April were 7610, 5730, and 3569 respectively. Evidently, the April NTL intensity was 53% lower than February radiance and ~38% lower than March radiance. However, it is clear that the economic activities and associated mobilities in April were concentrated in the downtown LA area, near the port in Long Beach, and along the LA-San Bernardino and LA–San Fernando corridors as evident from Figure 18d,e, where the blue indicates a measured increase in light intensity, while red indicates a measured decrease in intensity (radiance). This color scheme has been used in other studies regarding DNB radiance differences [29–32,45].

(a)

(b)

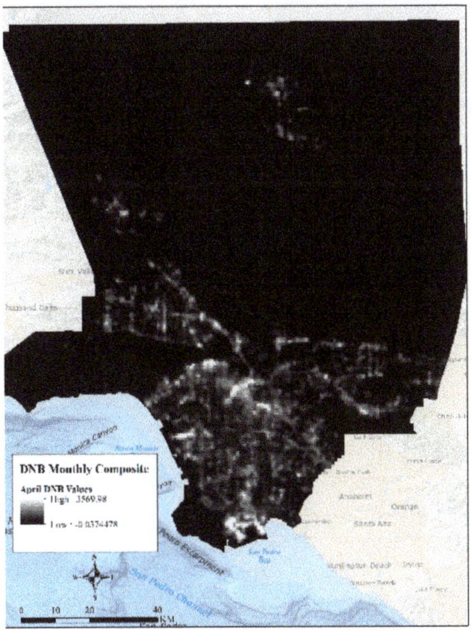

(c)

Figure 18. *Cont.*

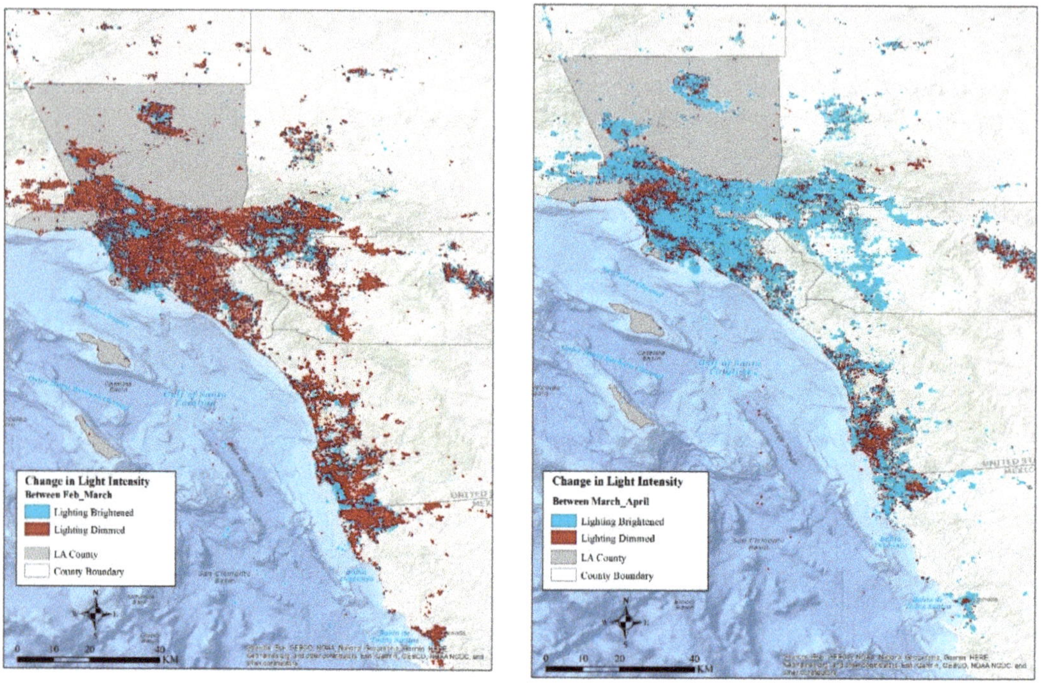

(d) (e)

Figure 18. Nighttime light (NTL) intensity distribution from VIIRS for Los Angeles in February (**a**), March (**b**), April (**c**); change in NTL intensity in LA between February and March (**d**) and between March and April (**e**).

Not surprisingly a similar pattern was observed in Chicago in March after the lockdown measures were in place in Illinois, particularly in the Chicago Metropolitan area. Economic activities dropped in Chicago region except for the downtown Chicago area near Millennium Park (Figure 19). Some activities are also observed near Rosemont area (O'Hare International Airport) and Clearing (Chicago Midway International Airport). Comparison of the light intensity between February and March (892 and 839 nWatts·cm^{-2}·sr^{-1}, respectively) indicates that the reduction in economic activities was not that drastic as was the case in LA (Figure 19d). By April, economic activities and traffic movement started in the Chicago region (Figure 19c), but the comparison of light intensity between March and April (839 and 1207 nWatts·cm^{-2}·sr^{-1}, respectively) (Figure 19e) indicates that economic activities and mobility increased way more than what was observed in February in certain parts of Chicago rather than the broader region. In fact, the radiance dropped by 6% in March, but increased by ~44% in April. This is corroborated by $PM_{2.5}$ observations in April, which showed very limited reduction in April 2020 as compared to April 2019, whereas March 2020 showed substantial reductions in $PM_{2.5}$ compared to March 2019 (Figure 17). The activities are concentrated in the Chicago metropolis rather than beyond the metropolis in the suburbs as was the case in February.

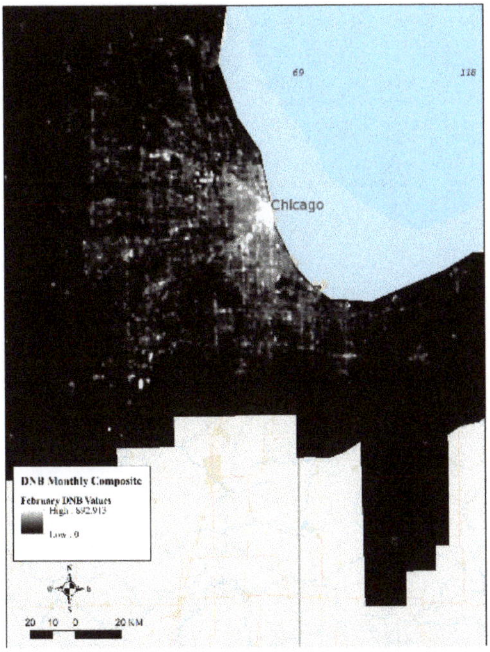

(**a**)

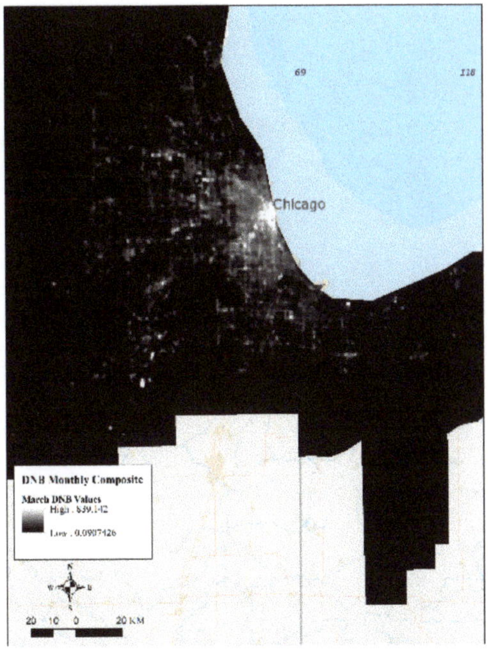

(**b**)

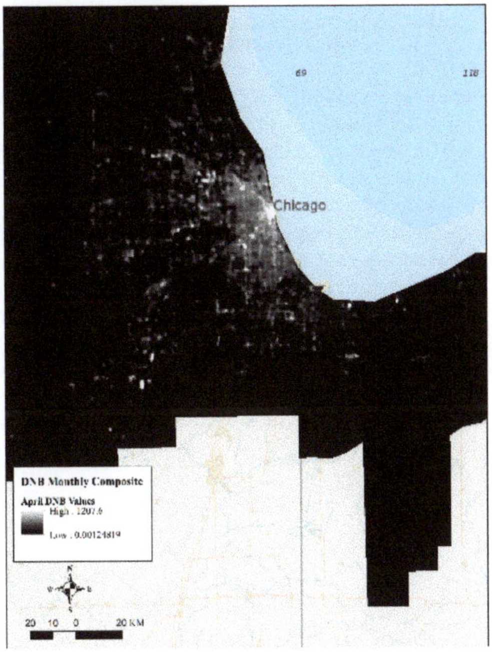

(**c**)

Figure 19. *Cont.*

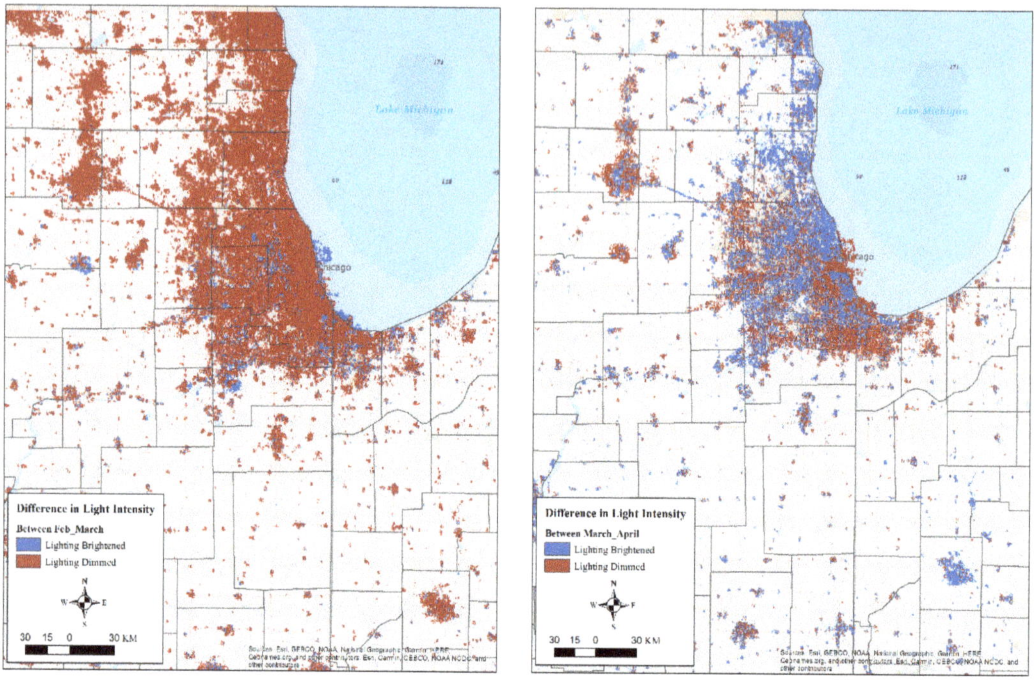

(d) **(e)**

Figure 19. Nighttime light (NTL) intensity distribution from VIIRS for Chicago in February (**a**), March (**b**), April (**c**); change in NTL intensity in Chicago between February and March (**d**) and between March and April (**e**).

The pattern of reduced economic activities and mobility in March followed by an increase in activities in April was observed in DC and its surrounding urban areas of Pittsburgh, Maryland, etc. Although the radiance values dropped by 22% in DC (Figure 20d, from 490 to 383 nWatts·cm^{-2}·sr^{-1}), there was an increase in activity (lighting was bright) in the neighboring areas of DC in Arlington and Bethesda. An obvious decline in activity in College Park was also observed due to the closure of the university campus and federal buildings near the campus. The drop in central DC in March was concentrated in the downtown area where businesses closed down immediately following the lockdown orders. By April, though economic activities increased to some extent (radiance increased by 7% than what was observed in March, Figure 20e), economic activities in the DC area beyond the downtown and White house area reduced in April as most employees started working remotely. This drop in radiance aligns with what was observed with regard to mobility change during March and April in the broader DC region. By contrast to DC metro area, economic activities and traffic movement appear to have resumed by April in the Baltimore, Columbia, MD areas.

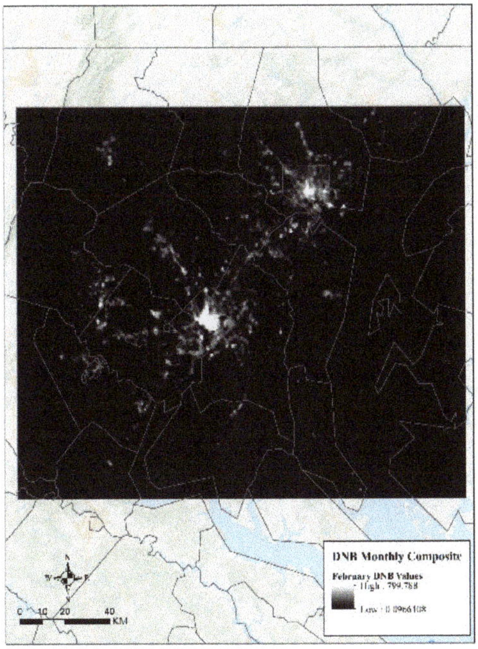

(**a**)

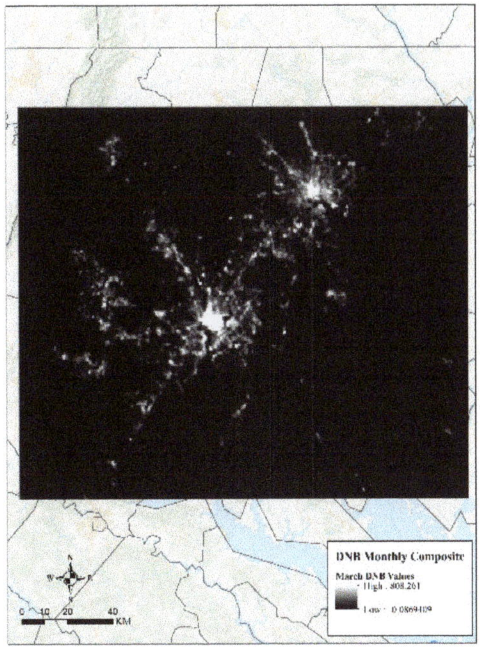

(**b**)

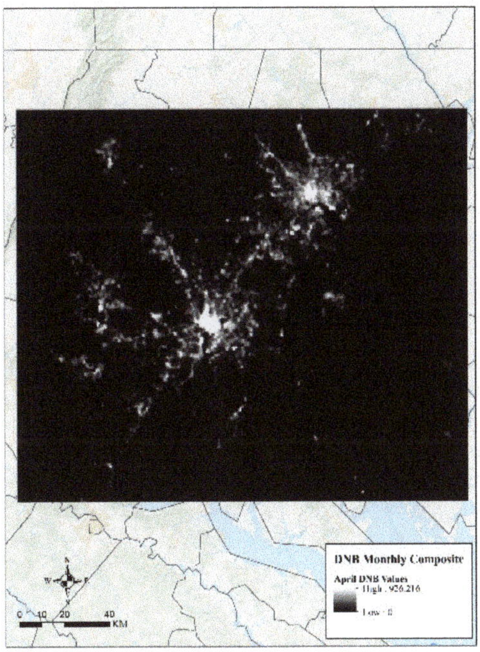

(**c**)

Figure 20. *Cont.*

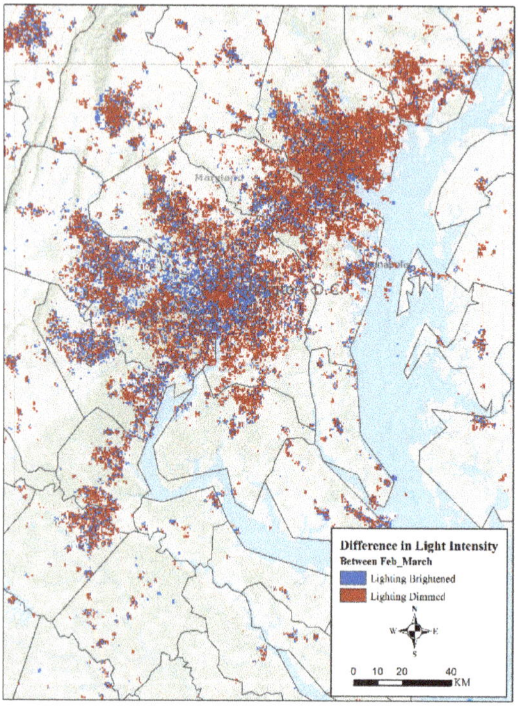

(d)

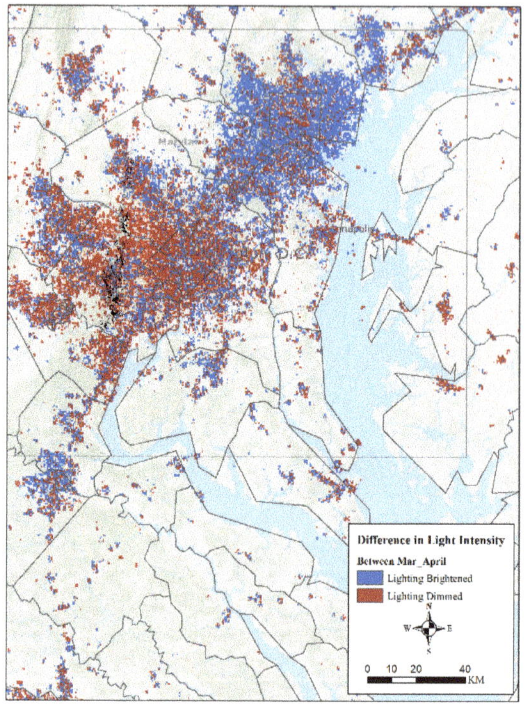

(e)

Figure 20. Nighttime light (NTL) intensity distribution from VIIRS for Washington DC in February (**a**), March (**b**), April (**c**); change in NTL intensity in DC between February and March (**d**) and between March and April (**e**).

4. Discussion and Conclusions

This paper presented an initial study of the impacts of lockdown measures in response to the novel coronavirus disease 2019 (i.e., COVID-19) on economic activities and environmental conditions in three cities across the US. The study revealed that the lockdown measures have two conflicting effects: (i) first, a reduction in mobility contributed to a decrease in economic activity that subsequently impacts rise of poverty, and (ii) second, a mobility reduction also reduced pollution, specifically, the concentration of NO_2, which is a key precursor for photochemical smog production. Long-term exposure to NO_2 and $PM_{2.5}$ have been linked to respiratory diseases that have been identified as a contributing factor in fatality from COVID-19 (Ogen 2020, [20]). The linkages between prior exposure to $PM_{2.5}$ and mortality due to infection from COVID-19 was presented by a Harvard study (Wu et al. [21]). A recent study [46] observed the effects of lockdown measures as a result from COVID-19 in four major metropolis areas, including Los Angeles. Connerton et al. [46] stated that they utilized air quality data from local air monitoring agencies for the study areas. For LA, Connerton et al. [45] NO_2 concentration was taken from the South Coast Air Quality Monitoring District dataset for a single month of data (March) from 2 locations: Central Los Angeles and North Orange County. While this provides a high temporal dataset, it is not necessarily representative of the entire LA basin. This study utilized satellite-based observation from the TROPOMI instrument on Sentinel-5P, which covers the entire LA basin at the expense of finer scale temporal changes. Both Connerton et al. [46] and this study showed a net decrease in NO_2 concentration over both Los Angeles and New York city areas. However, the rate of decrease was different for LA, which was 38%

(Table 2 in [46]) verses 25% for Wu et al. [21], which utilized satellite-based observations. Furthermore, while Connerton et al. [46] only looked at the change in a single month (i.e., March), as previously noted, while Wu et al. [21] explored the changing trend for three months. There is a natural variability of NO_2 that occurs with the changing of temperatures through the various months. However, the rate of change shown in this study as well as others [21,46] is larger than what naturally would occur.

Following the lockdown in February, mobility, economic activities, and NO_2 concentration dropped in LA, Chicago, and DC. While the reduction in economic activities (as seen from DNB data) in LA in March appears to be a result of the complete shutdown of all businesses; this was not the case in Chicago and DC. In fact, unlike LA, where some activities were underway in the LA downtown area and Long Beach, in Chicago and DC, economic activities and mobility were continuing in March in the broader metro areas surrounding the downtown as well as seaports and airport facilities. This supports the findings by Elvidge et al. (2020) [45], namely, that the reduction in nighttime lights was not due to normal variability. Rather they were the result of the "stay at home" orders implemented by the various cities in this study.

By April, economic activities and travel to essential businesses had resumed in LA, Chicago, and DC. It appears that economic activities resumed in downtown areas with high population densities, and mobility (observed from mobility data and VIIRS data) resumed along corridors connecting to high-density areas in LA and broader DC region as well. Preliminary analyses of median household incomes in the study sites from the US Bureau of Economic Analysis dataset appears to reveal that the mobility reduction was more pronounced in low-income and poor neighborhoods of LA and Chicago rather than in the affluent areas, which probably were occupied by service sector employees. Given that businesses, specifically, those catering to the service sector were not fully functional and employees started tele-commuting, it is not surprising that despite an increase in mobility and economic activity (related to travel to points of interest like grocery stores, hospitals etc.) in April, both NO_2 and $PM_{2.5}$ concentration were reduced in LA, Chicago, and DC during March–April. A major takeaway from this study is that while the lockdown measures helped reduce the transmission of the disease, they also reduced mobility, thereby disrupting economic activities and improving air quality. This study demonstrates how satellite imagery can be used to examine the change in economic activities and resulting air quality in near real-time as well as to examine the change in energy usage (a proxy for economic activities). It is also obvious that reduction in NO_2 concentration is a result of reduction in overall mobility irrespective of the economic activity centers and underlying population density distribution. The study also revealed that lockdown measures ensured limited mobility and economic activity in the business districts as well as in high density neighborhoods with essential businesses and corridors connecting major activity centers and high-density areas. This study also revealed that low-income neighborhoods experienced the brunt of the lockdown where mobility was lower as opposed to affluent areas. This information could be combined with demographic data, COVID-19 cases and energy consumption information to identify potential areas for disease spread and test site locations and explore the varying patterns of energy consumption such that strategies can be developed to address energy supply–demand relationships as well as reduce disease spread and NO_2 concentration while resuming economic activity. Research is currently underway to further explore the correlation between information provided by the Day\Night Band, as shown in this study, and various economic indicators, such as unemployment numbers, GDP statistics, and mobility patterns. Future research will also explore the relationships from different cities with varying population densities and across time.

Author Contributions: W.S.III developed the concept and methodology of this analysis; mobility data were provided by A.W.; Day\Night Band data analysis and background was provided by S.D.M. and W.S.III The analyses of the NO_2 and $PM_{2.5}$ data were performed by Z.W., H.Z. and S.K. Analysis and visualization were performed by Z.W., H.Z. and S.K. The writing and reviewing was performed by W.S.III and S.K. All authors have read and agreed to the published version of the manuscript.

Funding: This research received no external funding.

Data Availability Statement: No new data were created or analyzed in this study. Data sharing is not applicable to this article.

Acknowledgments: NOAA's JPSS Proving Ground and Risk Reduction (PGRR) Program helped provide funding for this effort. The authors sincerely appreciate the NASA/NOAA Joint Polar Satellite System (JPSS) Program for providing the VIIRS data used in this study. The authors wish to acknowledge Chris Elvidge and Earth Observation Group, Payne Institute for Public Policy, Colorado School of Mines for the development for the production of the cloud-free DNB composites utilized in this analysis. The authors also wish to acknowledge the extraordinary effort of Bandana Kar from Oak Ridge National Laboratory in helping with conceptualization, methodology development, and analysis, and we also wish to acknowledge Jaimie Johns and Jack Forsythe from BlueDot Inc. for their contribution in cleaning and filtering the anonymized mobile decide location data. The manuscript contents are solely the opinions of the authors and do not constitute a statement of policy, decision, or position on behalf of NOAA or the US Government.

Conflicts of Interest: The authors declare no conflict of interest.

References

1. Hui, D.S.; Azhar, E.I.; Madani, T.A.; Ntoumi, F.; Kock, R.; Dar, O.; Ippolito, G.; Mchugh, T.D.; Memish, Z.A.; Drosten, C.; et al. The continuing 2019-nCoV epidemic threat of novel coronaviruses to global health—The latest 2019 novel coronavirus outbreak in Wuhan, China. *Int. J. Infect. Dis.* **2020**, *91*, 264–652. [CrossRef] [PubMed]
2. Paules, C.I.; Marston, H.D.; Fauci, A.S. Coronavirus Infections—More Than Just the Common Cold. *JAMA* **2020**, *323*, 707. [CrossRef]
3. Li, Q.; Guan, X.; Wu, P.; Wang, X.; Zhou, L.; Tong, Y.; Ren, R.; Leung, K.S.M.; Lau, E.H.Y.; Wong, J.Y.; et al. Early Transmission Dynamics in Wuhan, China, of Novel Coronavirus–Infected Pneumonia. *N. Engl. J. Med.* **2020**, *382*, 1199–1207. [CrossRef] [PubMed]
4. Wong, J.; Goh, Q.Y.; Tan, Z.; Lie, S.A.; Tay, Y.C.; Ng, S.Y.; Soh, C.R. Preparing for a COVID-19 pandemic: A review of operating room outbreak response measures in a large tertiary hospital in Singapore. *Can. J. Anesth.* **2020**, *67*, 732–745. [CrossRef] [PubMed]
5. World Health Organization (WHO). WHO Announces COVID-19 Outbreak a Pandemic. Available online: https://www.euro.who.int/en/health-topics/health-emergencies/coronavirus-CoVID-19/news/news/2020/3662/who-announces-CoVID-19-outbreak-a-pandemic (accessed on 19 September 2020).
6. Johns Hopkins University of Medicine (JHU), COVID-19 Data in Motion. Available online: https://coronavirus.jhu.edu/ (accessed on 19 September 2020).
7. Strochlic, N.; Champine, R.D. How Some Cities 'Flattened the Curve' during the 1918 Flu Pandemic. National Geographic. Available online: https://www.nationalgeographic.com/history/2020/03/how-cities-flattened-curve-1918-spanish-flu-pande668mic-coronavirus/ (accessed on 19 September 2020).
8. Sharma, S.; Zhang, M.; Anshika; Gao, J.; Zhang, H.; Kota, S.H. Effect of restricted emissions during COVID-19 on air quality in India. *Sci. Total Environ.* **2020**, *728*, 138878. [CrossRef]
9. U.S. Bureau of Economic Analysis. CAGDP2 Gross Domestic Product (GDP) by County and Metropolitan Area. Available online: https://apps.bea.gov/itable/iTable.cfm?ReqID=70&step=1 (accessed on 19 September 2020).
10. Bonaccorsi, G.; Pierri, F.; Cinelli, M.; Flori, A.; Galeazzi, A.; Porcelli, F.; Schmidt, A.L.; Valensise, C.M.; Scala, A.; Quattrociocchi, W.; et al. Economic and social consequences of human mobility restrictions under COVID-19. *Proc. Natl. Acad. Sci. USA* **2020**, *117*, 15530–15535. [CrossRef]
11. Chen, S.; Igan, D.; Pierri, N.; Presbitero, A. The Economic Impact of COVID-19 in Europe and the US: Outbreaks and Individual Behaviour Matter a Great Deal, Non-Pharmaceutical Interventions Matter Less. VOXEU-CEPR. Available online: https://voxeu.org/article/economic-impact-CoVID-19-europe-and-us (accessed on 19 September 2020).
12. Huang, J.; Wang, H.; Xiong, H.; Fan, M.; Zhou, A.; Li, Y.; Dou, D. Quantifying the Economic Impact of 681 COVID-19 in Mainland China Using Human Mobility Data. Available online: https://arxiv.org/abs/2005.03010 (accessed on 19 September 2020).
13. Lamsal, L.N.; Martin, R.V.; Parrish, D.D.; Krotkov, N.A. Scaling Relationship for NO_2 Pollution and Urban Population Size: A Satellite Perspective. *Environ. Sci. Technol.* **2013**, *47*, 7855–7861. [CrossRef]
14. Bauwens, M.; Compernolle, S.; StavrakouiD, T.; Müllerid, J.-F.; Van Gent, J.; Eskes, H.; Levelt, P.F.; Van Der, A.R.; Veefkind, J.P.; Vlietinck, J.; et al. Impact of Coronavirus Outbreak on NO 2 Pollution Assessed Using TROPOMI and OMI Observations. *Geophys. Res. Lett.* **2020**, *47*, 688. [CrossRef]
15. Shi, X.; BrasseuriD, G.P. The Response in Air Quality to the Reduction of Chinese Economic Activities during the COVID-19 Outbreak. *Geophys. Res. Lett.* **2020**, *47*, 691. [CrossRef]

16. He, G.; Pan, Y.; Tanaka, T. The short-term impacts of COVID-19 lockdown on urban air pollution in China. *Nat. Sustain.* **2020**, *3*, 1005–1011. [CrossRef]
17. Mahato, S.; Pal, S.; Ghosh, K.G. Effect of lockdown amid COVID-19 pandemic on air quality of the megacity Delhi, India. *Sci. Total Environ.* **2020**, *730*, 139086. [CrossRef] [PubMed]
18. Kumari, P.; Toshniwal, D. Impact of lockdown measures during COVID-19 on air quality– A case study of India. *Int. J. Environ. Health Res.* **2020**, 1–8. [CrossRef] [PubMed]
19. UNECE. Declines in air Pollution Due to COVID-19 Lockdown Show Need for Comprehensive Emission Reduction Strategies. UNECE Sustainable Development GOALS. Available online: https://www.unece.org/info/media/news/environment/2020/declines-in-air-pollution-due-to-CoVID-19-lolockdown-show-need-for-comprehensive-emission-reduction-strategies/doc.html (accessed on 19 September 2020).
20. Ogen, Y. Assessing nitrogen dioxide (NO_2) levels as a contributing factor to coronavirus (COVID-19) fatality. *Sci. Total Environ.* **2020**, *726*, 138605. [CrossRef] [PubMed]
21. Wu, X.; Nethery, R.C.; Sabath, M.B.; Braun, D.; Dominici, F. Air pollution and COVID-19 mortality in the United States: Strengths and limitations of an ecological regression analysis. *Sci. Adv.* **2020**, *6*, eabd4049. [CrossRef]
22. American Lung Association. State of the Air 2020. American Lung Association. Available online: https://wtop.com/wp-content/uploads/2020/04/SOTA-2020-Proof-8-Embargoed.pdf (accessed on 19 September 2020).
23. Veefkind, J.; Aben, E.; McMullan, K.; Förster, H.; De Vries, J.; Otter, G.; Claas, J.; Eskes, H.; De Haan, J.; Kleipool, Q.; et al. TROPOMI on the ESA Sentinel-5 Precursor: A GMES mission for global observations of the atmospheric composition for climate, air quality and ozone layer applications. *Remote Sens. Environ.* **2012**, *120*, 70–83. [CrossRef]
24. Lee, T.E.; Miller, S.D.; Turk, F.J.; Schueler, C.; Julian, R.; Deyo, S.; Dills, P.; Wang, S. The NPOESS VIIRS Day/Night Visible Sensor. *Bull. Am. Meteorol. Soc.* **2006**, *87*, 191–200. [CrossRef]
25. Liao, L.B.; Weiss, S.; Mills, S.; Hauss, B. Suomi NPP VIIRS Day and Night Band (DNB) on-orbit performance. *J. Geophys. Res. Atmos.* **2013**, *118*, 12705–12718. [CrossRef]
26. Mills, S.; Jacobson, E.; Jaron, J.; McCarthy, J.; Ohnuki, T.; Plonski, M.; Searcy, D.; Weiss, S. Calibration of the VIIRS Day/Night Band (DNB). *Remote Sens.* **2015**, *7*, 718–988.
27. Jacobson, E.; Ibara, A.; Lucas, M.; Menzel, R.; Murphey, H.; Yin, F.; Yokoyama, K. Operation and characterization of the Day/Night Band (DNB) for the NPP Visible/Infrared Imager Radiometer Suite (VIIRS). In Proceedings of the 6th Annual Symposium on Future National Operational Environmental Satellite Systems-NPOESS and GOES-R, Boston, MA, USA, 20 January 2010; p. 349.
28. Straka, W.; Seaman, C.J.; Baugh, K.E.; Cole, K.; Stevens, E.; Miller, S.D. Utilization of the Suomi National Polar-Orbiting Partnership (NPP) Visible Infrared Imaging Radiometer Suite (VIIRS) Day/Night Band for Arctic Ship Tracking and Fisheries Management. *Remote Sens.* **2015**, *7*, 971–989. [CrossRef]
29. Miller, S.D.; Straka, W.; Mills, S.P.; Elvidge, C.D.; Lee, T.F.; Solbrig, J.; Walther, A.; Heidinger, A.K.; Weiss, S.C. Illuminating the Capabilities of the Suomi National Polar-Orbiting Partnership (NPP) Visible Infrared Imaging Radiometer Suite (VIIRS) Day/Night Band. *Remote Sens.* **2013**, *5*, 6717–6766. [CrossRef]
30. Bennett, M.M.; Smith, L.C. Advances in using multitemporal night-time lights satellite imagery to detect, estimate, and monitor socioeconomic dynamics. *Remote Sens. Environ.* **2017**, *192*, 176–197. [CrossRef]
31. Elvidge, C.D.; Baugh, K.E.; Kihn, E.A.; Kroehl, H.W.; Davis, E.R.; Davis, C.W. Relation between satellite observed visible-near infrared emissions, population, economic activity and electric power consumption. *Int. J. Remote Sens.* **1997**, *18*, 1373–1379. [CrossRef]
32. Jing, X.; Shao, X.; Cao, C.; Fu, X.; Yan, L. Comparison between the Suomi-NPP Day-Night Band and DMSP-OLS for Correlating Socio-Economic Variables at the Provincial Level in China. *Remote Sens.* **2016**, *8*, 17. [CrossRef]
33. Yeh, C.; Perez, A.; Driscoll, A.; Azzari, G.; Tang, Z.; Lobell, D.; Ermon, S.; Burke, M. Using publicly available satellite imagery and deep learning to understand economic well-being in Africa. *Nat. Commun.* **2020**, *11*, 1–11. [CrossRef] [PubMed]
34. Kopp, T.J.; Thomas, W.; Heidinger, A.K.; Botambekov, D.; Frey, R.A.; Hutchison, K.D.; Iisager, B.D.; Brueske, K.; Reed, B. The VIIRS Cloud Mask: Progress in the first year of S-NPP toward a common 741 cloud detection scheme. *J. Geophys. Res. Atmos.* **2014**, *119*, 2441–2456. [CrossRef]
35. Mills, S.; Weiss, S.; Liang, C. VIIRS day/night band (DNB) stray light characterization and correction. In Proceedings of the Earth Observing Systems XVIII. *SPIE Intl. Soc. Opt. Eng.* **2013**, *8866*, 88661P.
36. Lasry, A.; Kidder, D.; Hast, M.; Poovey, J.; Sunshine, G.; Zviedrite, N.; Ahmed, F.; Ethier, K.A. Timing of Community Mitigation and Changes in Reported COVID-19 and Community Mobility—Four US 746 Metropolitan Areas. *MMWR Morb. Mortal. Wkly. Rep.* **2020**, *69*, 451–457. [CrossRef]
37. WorldPop. Mapping Populations—WorldPop Gridded Population Estimate Datasets and Tools. 2020. Available online: https://www.worldpop.org/methods/populations (accessed on 20 September 2020).
38. Getis, A.; Ord, J.K. The Analysis of Spatial Association by Use of Distance Statistics. *Geogr. Anal.* **2010**, *24*, 189–206. [CrossRef]
39. Ord, J.K.; Getis, A. Local Spatial Autocorrelation Statistics: Distributional Issues and an Application. *Geogr. Anal.* **1995**, *27*, 286–306. [CrossRef]

40. ESRI. How Optimized Hot Spot Analysis Works. ESRI. 2020. Available online: https://pro.arcgis.com/en/pro-app/tool-reference/spatial-statistics/how-optimized-hot-spot-analysis-works.htm#:~{}:text=The%20Optimized%20Hot%20Spot%20Analysis%20tool%20identifies%20peak%20distances%20using,becomes%20the%20scale%20of%20analysis.&text=This%20distance%20corresponds%20to%20758the,Getis%2DOrd%20Gi*)%20tool (accessed on 20 September 2020).
41. Anselin, L. Local Indicators of Spatial Association-LISA. *Geogr. Anal.* **1995**, *27*, 93–115. [CrossRef]
42. Anselin, L.; Syabri, I.; Kho, Y. GeoDa: An Introduction to Spatial Data Analysis. *Handb. Appl. Spat. Anal.* **2010**, *38*, 73–89. [CrossRef]
43. Anselin, L. The Moran Scatterplot as an ESDA Tool to Assess Local Instability in Spatial Association. In *Spatial Analytical Perspectives on GIS*; Informa UK Limited: London, UK, 2019; pp. 111–126.
44. Goldberg, D.L.; Anenberg, S.C.; Griffin, D.; McLinden, C.A.; Lu, Z.; Streets, D.G. Disentangling the impact of the COVID-19 lockdowns on urban NO 2 from natural variability. *Geophys. Res. Lett.* **2020**, *47*, e2020GL089269. [CrossRef] [PubMed]
45. Elvidge, C.D.; Ghosh, T.; Hsu, F.-C.; Zhizhin, M.; Bazilian, M. The Dimming of Lights in China during the COVID-19 Pandemic. *Remote Sens.* **2020**, *12*, 2851. [CrossRef]
46. Connerton, P.; Vicente de Assunção, J.; Maura de Miranda, R.; Dorothée Slovic, A.; José Pérez-Martínez, P.; Ribeiro, H. Air quality during COVID-19 in four megacities: Lessons and challenges for public health. *Int. J. Environ. Res. Public Health* **2020**, *17*, 775. [CrossRef] [PubMed]

remote sensing

MDPI

Technical Note

Pandemic Induced Changes in Economic Activity around African Protected Areas Captured through Night-Time Light Data

Anupam Anand [1,*] and Do-Hyung Kim [2]

[1] Independent Evaluation Office, Global Environment Facility, Washington, DC 20006, USA
[2] Office of Innovation, UNICEF, New York, NY 10017, USA; dokim@unicef.org
* Correspondence: aanand2@thegef.org

Abstract: The importance of tourism for development is widely recognized. Travel restrictions imposed to contain the spread of COVID-19 have brought tourism to a halt. Tourism is one of the key sectors driving change in Africa and is based exclusively on natural assets, with wildlife being the main attraction. Economic activities, therefore, are clustered around conservation and protected areas. We used night-time light data as a proxy measure for economic activity to assess change due to the pandemic. Our analysis shows that overall, 75 percent of the 8427 protected areas saw a decrease in light intensity in varying degrees in all countries and across IUCN protected area categories, including in popular protected area destinations, indicating a reduction in tourism-related economic activities. As countries discuss COVID-19 recovery, the methods using spatially explicit data illustrated in this paper can assess the extent of change, inform decision-making, and prioritize recovery efforts.

Keywords: VIIRS; night-time lights; COVID-19; pandemic; protected areas; wildlife tourism

Citation: Anand, A.; Kim, D.-H. Pandemic Induced Changes in Economic Activity around African Protected Areas Captured through Night-Time Light Data. *Remote Sens.* **2021**, *13*, 314. https://doi.org/10.3390/rs13020314

Received: 1 December 2020
Accepted: 11 January 2021
Published: 18 January 2021

Publisher's Note: MDPI stays neutral with regard to jurisdictional claims in published maps and institutional affiliations.

1. Introduction

Travel and tourism are vital economic sectors closely tied with socio-economic progress and recognized for their potential to contribute to poverty reduction and development. This sector accounts for 10.3 percent of global GDP, making it larger than agriculture. In 2019 alone, it created one in four new jobs. In Africa, the tourism sector employs around 24.6 million people and contributes $169 billion to the continent's economy combined, representing 7.1% of Africa's GDP [1]. Over 30 African countries have identified tourism as a national priority within the Enhanced Integrated Framework (EIF), a multilateral partnership dedicated to providing trade-related assistance to less developed countries [2]. Tourism in the majority of African countries is centered around their protected area (PA) systems. Africa's 8400 protected areas (PAs) produce approximately US$48 billion in revenue [3]. Tourism revenue is also the primary source of funding for PA agencies on the continent. The impact of COVID-19 on conserved and protected areas have been estimated to range from economic, social, and ecological to management and enforcement [4–6].

Protected areas, initially envisioned to protect iconic landscapes, seascapes, and wildlife, now cover 15% of the earth's land surface [7]. Besides the traditional functions of habitat and biodiversity protection, PAs today are directly linked to supporting human livelihoods and well-being, providing ecosystem services, and contributing to climate change mitigation and adaptation [8]. Protected areas are hotspots of biodiversity conservation and often overlap with areas of poverty and underdevelopment. Tourism revenues provide the vital means to address both conservation and socio-economic development objectives of communities living in and around PAs and beyond. A recent study found how households around PAs with tourism also had higher wealth levels (by 17%) and a lower likelihood of poverty (by 16%) compared to similar households living far from protected areas [9]. Empirical evidence also suggests that nature-based tourism can be positive for people and wildlife [10–12].

COVID-19, which started around December of 2019, drew significant impacts on the travel industry with travel restrictions that began in January 2020. The World Health Organization (WHO) declared it a pandemic on 11 March 2020. Immediately countries across the world introduced movement restrictions and confinement policies ranging from voluntary compliance to complete ban on non-essential travel for effective control of the pandemic. The tourism sector (non-essential travel) is the most directly impacted by lockdown measures but also offers the opportunity to examine the effectiveness of confinement measures including controlling the spread of COVID-19 and its broader impacts.

By the end of April 2020, all tourist destinations worldwide have imposed travel restrictions, out of which 72% have completely closed their borders for international tourism [13]. The World Tourism Organization (WTO) has stated that international tourism could decline by 60–80% in 2020. This decline could mean a total collapse of the tourism sector in Africa, jeopardizing the progress made in recent years and threatening the survival of millions.

The impact of infectious diseases on the travel and tourism industry has been recognized during the 2014 Ebola outbreak in West Africa. The effect of that outbreak extended beyond the Ebola zone and was felt across the continent [14]. Travel and tourism in Africa took a massive hit, with broader macroeconomic and social impacts, as per studies conducted a few years later after the outbreak ended [15].

At the date of this paper, we cannot estimate the socio-economic impacts of the COVID-19 pandemic due to the scarcity of socio-economic data, travel restrictions, and persisting uncertainty about the pandemic duration and about how recovery will take place. In this scenario, remote sensing approaches can be a useful first step to assess the extent of change induced by this pandemic. Few studies recently have used remote sensing data to demonstrate the impact of the pandemic on economic activities and the environment [16–18].

Night-time light data have been successfully used as a proxy measure for economic activity and combined with other data to assess economic growth and development in various contexts and scales [19–21]. Using night-time light data as a proxy indicator, studies have estimated economic activities on global, national, and subnational levels in developing economies [22–24], the spatial distribution of GDP [20,25], and mapped urbanization dynamics [24,26,27]. Recent studies in China [17,28], Germany [29] and India [18] have shown how nighttime lights can be used as a proxy to monitor change and recovery in economic activity levels due to lockdown imposed by the pandemic. Studies have shown that night-time lights do provide a reliable indicator to capture economic activity. However, the relationship seems to be statistically more robust for developing economies than developed ones [19]. Developed economies are less dependent on physical infrastructure, and the services sector contributes more to the overall economic output.

In this paper, we use the night-time lights as a proxy indicator to compare the changes in economic activities before and during the pandemic around PAs in Africa to assess change patterns.

This study has two main objectives:

1. Demonstrate the application of NTL to capture the changes in economic activities around PAs before and during the COVID-19 pandemic.
2. Assess the changes and patterns and how are they related to other contextual conditions.

The study looks at the economic activities around a five-kilometer buffer of protected areas in Africa because most economic and tourist activities occur within that zone. This study includes 8427 PAs, of which 7763 were terrestrial protected areas, whereas 443 were coastal and marine, and 221 only marine PAs. These protected areas have different protection levels and include international designations such as UNESCO World Heritage sites, Ramsar sites, and UNESCO-MAB reserves.

We compared the night-time lights data for April 2020 with the data from April 2019. We choose April 2020 to observe the change in economic activities due to COVID-19 because all tourist destinations worldwide have imposed travel restrictions by then [13], and mobility around national parks in Africa was also at its lowest (Figure 1).

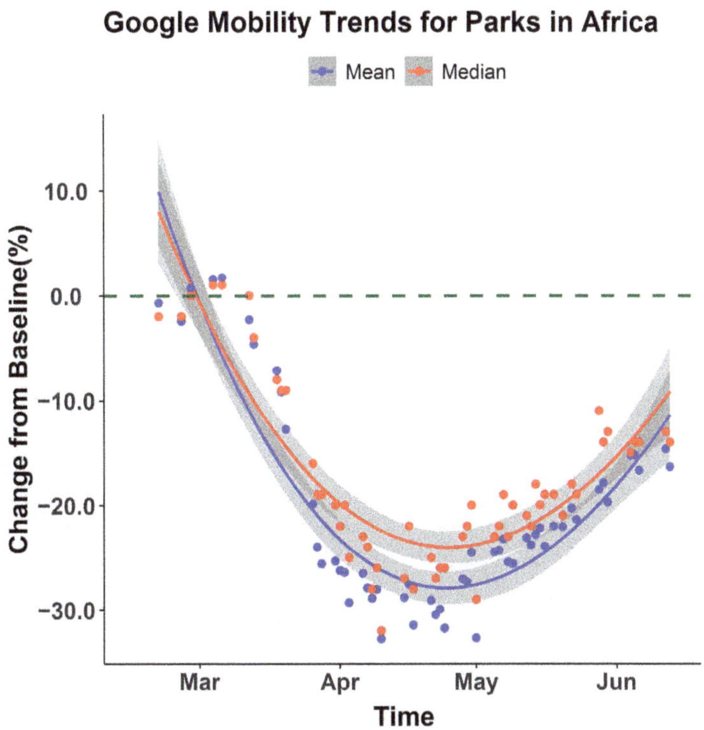

Figure 1. Google mobility trends for parks in Africa. As per the Google mobility data, parks typically refer to official national parks and not the general outdoors found in rural areas.

The results are presented to highlight COVID-19 pandemic-induced travel restrictions on economic activities around PAs in Africa. Future research could include large socio-economic datasets and site-specific data gathered through surveys and interviews with selected PA and tourism stakeholders.

2. Materials and Methods

We used two globally available data sets for this study, the World Database on Protected Areas (WDPA) downloaded from the protected planet website [30], and daily VIIRS VNP46A1 product, the source of night-time light data.

World Database on Protected Areas: We used the April 2020 version of the WDPA, which provides boundaries, designation, and other information. The WDPA is the most comprehensive database on protected areas and provides unified standards and unique opportunities for large-scale conservation studies [31].

The WDPA contains both the boundary and point locations for the sites that do not have precise boundaries. We did not include Libya and Somalia data as they do not have boundary information in the WDPA. We also did not include any regional or country-level PA database because the WDPA is the only authoritative dataset following globally consistent standards and is regularly validated and updated to maintain the highest data qualities.

Following the WDPA best practice guidelines [32], we used the higher and stricter IUCN designation for overlapping areas.

The WDPA also classifies PAs by IUCN management categories and governance types, as reported by the data provider. Sixty-six percent of protected areas in the WDPA had an IUCN Management Category, and 88 percent had a governance type as of May 2019 [33].

Night-time lights data: The Day/Night Band (DNB) sensor of the Visible Infrared Imaging Radiometer Suite (VIIRS) provides global daily measurements of nocturnal visible and near-infrared (NIR) light that is sensitive in lowlight conditions. It also allows for the generation of new science-quality night-time products. We used the daily VIIRS VNP46A1 product as the source for night-time light change observation [34]. DNB radiance band (DNB_At_Sensor_Radian ce_500 m) was used to create average night-time observation composition for April 2019 and April 2020. We used quality flag bands QF_Clound, QF_Cirrus, and QF_DNB to filter low-quality pixels. Moon illumination fractions were also adjusted using moon illumination information from the Moon_Illumination_Fraction band. We used the Moon_Illumination_Fraction layer included in the level-3 VNP46A1 data and subtracted moon illuminated fraction from radiance values to reduce the lunar effect. See Figure 2a,b below for a visual comparison between processed vs. unprocessed VIIRS night-time data.

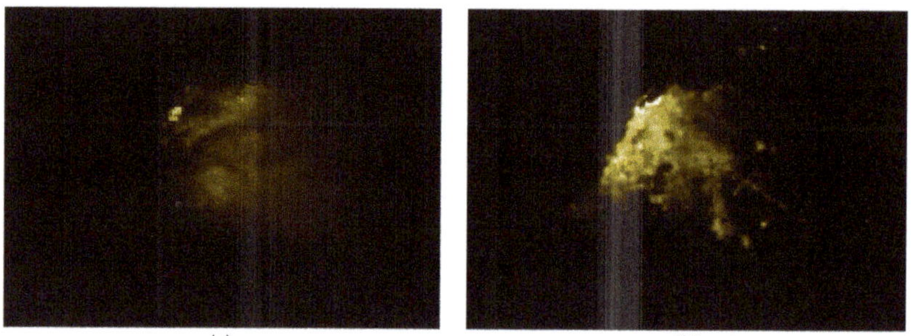

(a) (b)

Figure 2. (**a**) Luanda city area in Angola is seen in unprocessed VIIRS night-time data acquired on 16 April 2020; (**b**) the image shows the same location in the processed VIIRS night-time data corrected for moon illumination and bad pixels using quality flags and then compositing for the entirety of April 2020.

Limitations with Night-Time Light: We only have 30 days of observations for each year—so ephemeral lights are not corrected.

We used level-3 VNP46A1 data available at LAADS DAAC [35]. Products such as black marble are not available; that is why level 3 data were used to demonstrate its utility despite some limitations. Future studies may consider using more advanced data products as they become available.

Median or mean: In theory, mean should be used to capture light values by sporadic activities in camps. However, mean values can be significantly affected by ephemeral lights. Median, however, can remove the effects of outliers and remove the signals from temporal activities. In light of these limitations, we report both the statistics.

Night-time light trend comparison: Changes in the NTL trend for the PAs and their surrounding areas were identified and then validated using high-resolution satellite imagery to identify the tourist camps, parking lots, and lodges. These are the most common locations where tourism-related activities take place captured through night-time lights data. We also performed the paired Wilcox test between the pair of median night-time light values between April 2019 and April 2020 at the African continent's protected areas to confirm the changes in the night-time lights that we observed as statistically significant.

Research [36,37] shows that transition to LEDs may produce a reduction of the radiance on VIIRS, and therefore reduction in NTL intensity may not be solely due to the lockdown. However, studies [37] also suggest that transition to LED is least in Africa and that kerosine to dry-cell battery LED usage is more prevalent [38]. The Global Energy Review 2020 [39] has also indicated an overall reduction in electricity demand due to the pandemic. Therefore, it is unlikely that the observed changes in nighttime light within a span of one year (April 2019–April 2020) in and around the protected areas were due to LED transition. However future studies could explore the role of LED transition and observed NTL changes during the pandemic around protected areas in Africa.

3. Results

3.1. Change in Light Intensity in African PAs

The results show an overall decrease in the mean and median light intensity between the two years. The mean intensity for April 2019 was 0.37, whereas it decreased to 0.34 in April 2020, about an 8 percent decrease. Similarly, the median values also reduced from 0.32 to 0.28, a decline of about 12.5 percent. The paired Wilcox test reveals that the test's p-value is $< 2.2 \times 10^{-16}$, which is less than the significance level alpha = 0.05. We can conclude that the median light intensity for 2020 is significantly less compared to 2019 (Figure 3).

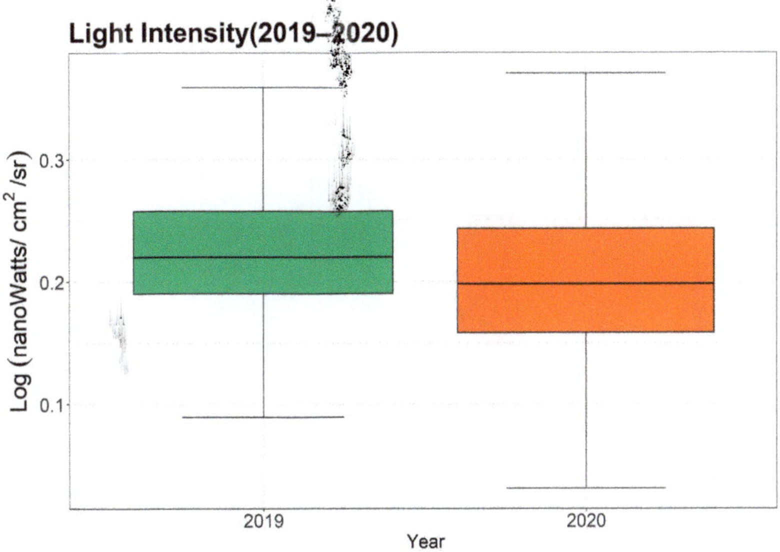

Figure 3. Comparison of night-time lights intensity between 2019 and 2020.

Overall, 75 percent of the 8427 protected areas saw a decrease in light intensity in varying degrees. There is considerable heterogeneity in the changes captured in percentage values for the overall PAs, and the PAs showing the highest reduction in light intensity. The difference could be due to various factors, including differences in funding allocations for PAs, the enforcement of curfews, types of businesses around PAs, access to electricity, and the kind of mass tourism or proximity to urban areas.

Results indicate that fifty percent of PAs have about a 12 percent decrease in the mean light intensity, and 25 percent have seen a reduction of 23 percent or more in the studied period. The median decline in all the PAs was 13 percent overall, 13 percent in terrestrial, 4 percent in coastal and marine, and 8 percent decrease in the marine-only PAs.

An analysis of the proportion of PAs with decreased light intensity for each country reveals that almost all the PAs in all the countries saw a decrease in the light intensity by varying degrees (Figure 4). The map in Figure 4 below, based on an equal interval, reveals that 80–100 percent of PAs saw a reduction in night-time light intensity in most African countries. These include PAs in Botswana, Kenya, Namibia, and South Africa, heavily dependent on tourism revenues.

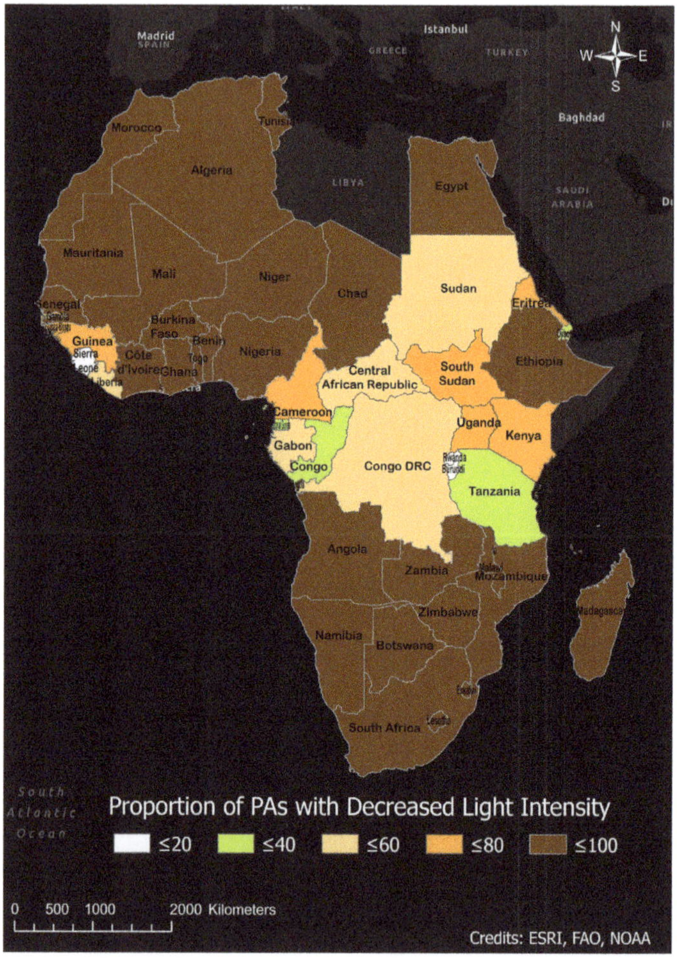

Figure 4. Map showing the proportion of protected areas (PAs) with decreased light intensity.

Less than 20% of PAs in Sierra Leone, Rwanda, and Burundi show a reduction in light intensity. It should be noted that rural access to electricity (% of rural population) in Burundi and Sierra Leone is less than 8% [40], which may explain why those countries see low change rates in light intensity.

Except for a few places, the light intensity has decreased overall for the same countries; for example, in Kenya and South Africa (Figure A1a,b in Appendix C). Minor increase in light intensity was observed in and around some of the coastal and marine protected areas in South Africa likely due to businesses catering to domestic customers to stay afloat.

3.2. *Protected Areas with Maximum Decrease in Light Intensity*

For the analysis, we have two subgroups—terrestrial and the other two, which included coastal and marine, and marine only PAs. The top 20 PAs within each of these two categories (Figure A2a,b). Seven of the top 20 terrestrial PAs that have seen the highest decrease in light intensity are from South Africa, followed by PAs in Nigeria and Kenya (Figure A2a). Similarly, eleven of the top 20 coastal and marine PAs that have seen the highest decrease in light intensity are also in South Africa (Figure A2b).

The Merja Zerga Biological Reserve in Morocco (Figure 5a), which has experienced the maximum drop in the light intensity, is also a Ramsar Site. It is a permanent hunting reserve known for intensive tourism activities, bird watching, fishing, grazing, and rush harvesting. High-resolution images of the area show decreased movement and economic activity, as demonstrated through the number of boats in the harbor and vehicles in the area (Figure 5a,b).

(a) (b)

Figure 5. Decrease in tourism activity can be seen in the decreased numbers of vehicles in the parking area at Merja Zerga, a Ramsar Site. (**a**) Before the pandemic in April 2019; (**b**) during the pandemic in April 2020. (Image© 2021 Digital Globe, NextView License).

3.3. Change in Night-Time Light Intensity by IUCN PA Management Categories

We looked at the light intensity changes in the protected areas for the different IUCN PA Management Categories (Table A1 in Appendix B). The IUCN categories help the countries classify protected areas based on their primary management objectives [33,41].

All the protected areas across the IUCN management categories show a decrease in the median value of light intensity in April 2020 (Figure A3a).

In terms of the proportion of total PAs showing a decrease irrespective of the intensity, more than 65 percent PAs across all management categories showed a decline in light intensity where 72 percent PAs under high protection saw a reduction in light intensity (Figure A3b).

The WDPA database also consists of protected areas that are Ramsar sites or World Heritage Sites (WHS). In total, there were 337 sites with international designations, which included 245 Ramsar Sites and Wetland of International Importance, 39 UNESCO-MAB Biosphere Reserves, and 53 World Heritage Sites (natural or mixed). All the WHS sites saw a decrease in the median and mean values (Table A2 in Appendix D). A similar trend was also seen in the Ramsar Sites.

In Kenya, Lake Nakuru saw the highest decrease by 56 percent, followed by Mejen Djebel in Tunisia by 50 percent. Among the MABs, the Zembra Zembretta Iles biosphere reserve in Tunisia saw the most decline in light intensity, by 49 percent. Among WHS, the Kenyan rift valley and the Okavango Delta in Botswana saw the maximum decrease by 45 and 38 percent, respectively.

In terms of the total proportion, 87 percent of Ramsar sites saw a decrease in light intensity, followed by 77 percent of the UNESCO MABs and 67 percent of the WHS (Figure 6).

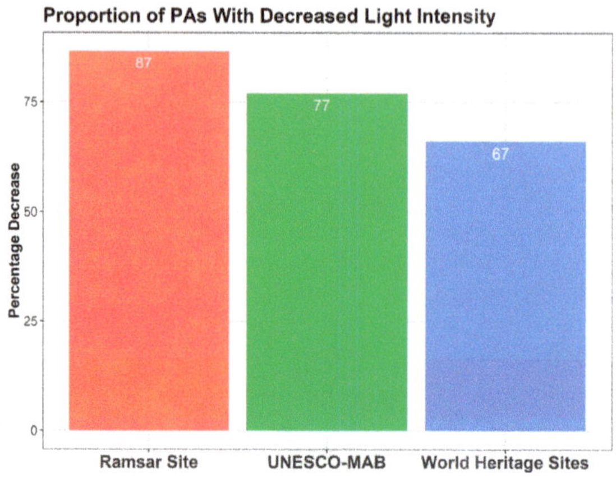

Figure 6. The proportion of PAs with decreased light intensity across international designations.

3.4. Change in Night-Time Light Intensity by Governance Type

Protected areas exist under the authority of diverse actors, including governments, indigenous peoples and local communities, private actors, and combinations of these. The IUCN governance categories help classify PAs according to who holds authority, responsibility, and accountability for them [33,42]. Besides assessing change in light intensity in PAs by the IUCN management categories, we also looked at the changes in the light intensity in PAs by IUCN governance types.

All PAs across all governance categories show decreased light intensity (median value) in 2020 (Figure 7a). In terms of the proportion of total PAs showing a reduction in light intensity, more than 70 percent of PAs across all governance categories showed a decrease (Figure 7b). PAs managed by non-profits showed a smaller decline in light intensity.

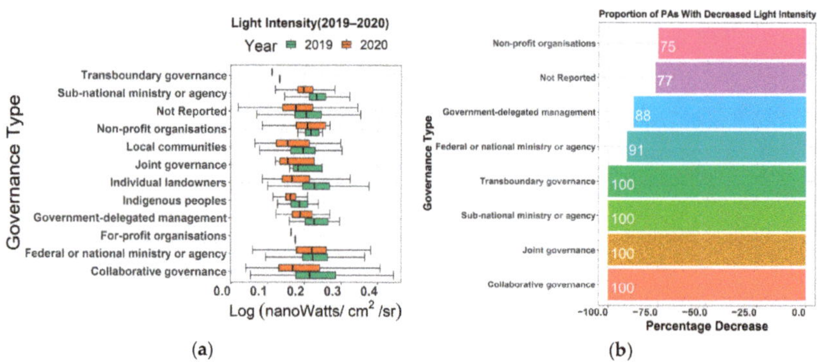

Figure 7. (**a**) Comparing light intensity across governance types; (**b**) proportions of PAs with decreased light intensity across governance types.

The difference across these categories could be due to differences in funding sources, availability of emergency resources to maintain essential services, sustaining minimum operational requirements, and closing the protected areas.

3.5. Status in Popular PA Destinations

We chose three popular, biodiverse, and often-visited PA destinations in Africa—Serengeti, Masai Mara, and Hell's Gate National Park to examine the decrease in light intensity. Overall, night-time light intensity decreased around all three national parks—

Serengeti (−11 percent), Masai Mara (−23 percent), and HGNP (−23 percent), indicating a decrease in economic activity sparked by the pandemic induced lockdown and travel restrictions (Figure 8a–c).

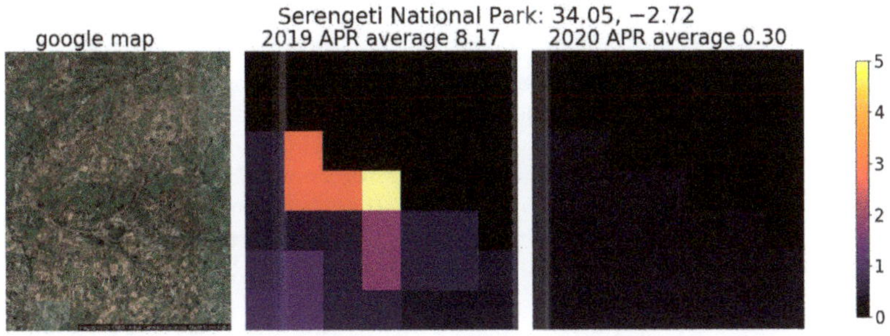

(**a**)

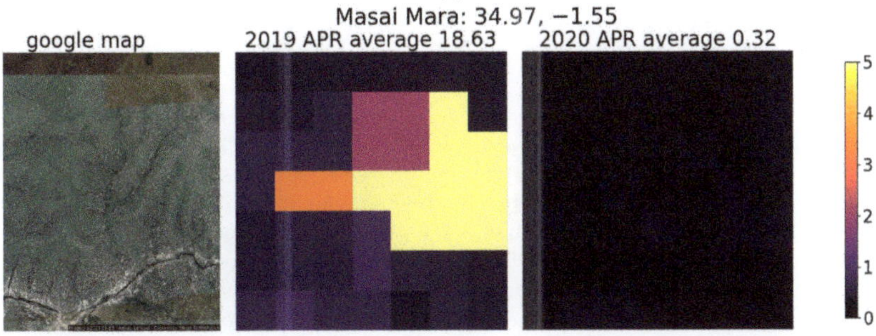

(**b**)

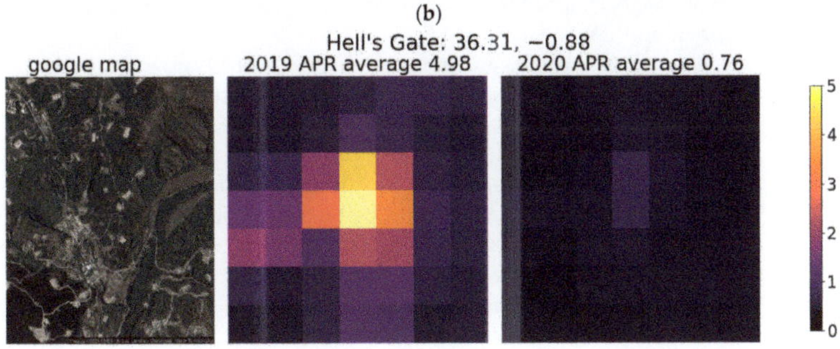

(**c**)

Figure 8. On the left part of each panel, satellite images show the popular tourist lodges, camp settlements, and markets around the three parks. The NTL data for these same sites, showing before (in the middle part of each panel) and after (right part of each panel), indicate that the locations have undergone a decrease in the light intensity. (**a**) Serengeti National Park; (**b**) Masai Mara; (**c**) Hell's Gate National Park.

Appendix A provides contextual information about these three parks and travel restrictions in the countries where they are located. In the absence of ground verification, the additional information helps explain the changes observed through the analysis of NTL data.

4. Discussion

The study demonstrates one of the many applications of NTL used as a proxy indicator for economic activities and highlights how the level of economic activity in its entirety has gone down around protected areas during the pandemic. This study also illustrates the utility of NTL as a low cost and efficient way of assessing impact when the traditional approaches could be challenging, posed by movement restrictions and time constraints. The findings reinforce what we are beginning to understand as the impact of lockdown and stay-at-home order on many economic sectors. The findings also indicate lockdown measures have been largely effective even in remote areas while highlighting the trade-off between controlling the pandemic and economic losses. Our study is the first to apply NTL to assess the change in economic activities around PAs in the context of COVID-19. The main challenge in using NTL for such rapid assessment is the lack of real-time analysis-ready data. We overcame this challenge by creating a monthly composite based on daily data, which was time-consuming. Future studies that focus on creating such a dataset could concentrate on providing analysis-ready datasets with lunar illumination correction. Such datasets would pave the way for rapid multitemporal analysis to trace change and the impact of economic recovery over time. As socio-economic data on the impacts of the lockdown become available, future studies can use these data sources for validation of the findings.

5. Conclusions

Protected areas can positively impact poverty reduction in developing countries, and nature-based tourism has been the critical strategy of conservation and development interventions to create a win–win situation for both nature and people. Long-term conservation and poverty reduction efforts around the protected and conserved areas depend on the tourism sector. Electrification in Africa has grown over the decades and has been used to assess economic activities. The method presented here shows how earth observation can be used to understand the extent of change in economic activities and provided a quick way to assess the reduction in economic impact and its spatial distribution. Future studies could also use NTL and traditional statistics to understand the pandemic's impact in specific areas and PAs.

As governments and agencies discuss post COVID-19 recovery and focus on rebuilding their economies, spatially explicit data illustrated in this paper can be used to target alternative livelihoods, enterprise development, and conservation. The method can be further applied to identify the areas most affected in terms of economic loss, conservation value, and loss of livelihoods. Furthermore, remote sensing methods with selections of relevant indicators, as travel and mobility data become available at various scales, can be used to assess the efficacy of compliance measures, including restrictions on non-essential travel to control the pandemic.

Resuming tourism activities will be gradual, and reopening parks and facilities will be challenging. Most settlements around PAs are small and lack the necessary health care infrastructure, and could be overwhelmed by an influx of cases after reopening. Spatially explicit data can be used to monitor and plan in these areas, including around tourism facilities, and targeted for further research, and collection of microdata and perception data as part of the medium to long-term recovery efforts.

Author Contributions: Conceptualization, A.A. and D.-H.K.; methodology, A.A. and D.-H.K.; writing—original draft preparation, A.A. and D.-H.K.; writing—review and editing, A.A. and D.-H.K. All authors have read and agreed to the published version of the manuscript.

Funding: This research received no external funding.

Institutional Review Board Statement: Not Applicable.

Informed Consent Statement: Not Applicable.

Data Availability Statement: Publicly available datasets were analyzed in this study. This data can be found here: https://ladsweb.modaps.eosdis.nasa.gov/archive/allData/5000/VNP46A1/.

Acknowledgments: We thank the five anonymous reviewers for their feedback and comments that helped improve this paper's quality.

Conflicts of Interest: The authors declare no conflict of interest. The views expressed in the manuscript are those of the authors and do not necessarily reflect the opinion or position of their employers.

Appendix A. Additional Information on Three National Parks and Travel Restrictions

Appendix A.1. Serengeti National Park, Tanzania

In Tanzania, tourism is one of the fastest-growing sectors and has contributed 17.5% of the GDP and 30% of the foreign exchange earnings in 2014/15. It is the second-highest employer after agriculture. Tanzania's Serengeti National Park is well-known for its wildlife migration tourism and as its oldest protected area. It is both a World Heritage Site and a Biosphere Reserve. The Serengeti contributes substantially to the country's revenue. It is a significant employment source centered on nature-based tourism activities, such as game viewing, safaris, bush meals, and visits to cultural and historical sites. The park accepts visitors throughout the year. However, the best season is between mid-May and mid-October. Available visitor records show that domestic and foreign tourists' arrival has steadily increased with the highest number of visitors (1,196,284) in 2018/2019 in the past ten years. Tourist Facilities park includes nine public campsites, 140 seasonal campsites operated by private tour operators, two rest houses, six lodges, one youth hostel, bandas (thatched-roof huts), and tented lodges. COVID-19 containment measures in Tanzania have been markedly less strict than many neighboring states, with the government expressing the need to protect the economy rather locking the country down.

Appendix A.2. Maasai Mara Nature Reserve, Kenya

The Maasai Mara Nature Reserve, established as a world heritage site in 1989, is north of the Serengeti National Park. It is one of the world's most famous tourist attractions and contributes significantly to Kenya's economy. The Maasai Mara ecosystem is home to approximately 25% of Kenya's wildlife. It is estimated that there are 200,000 visitors per year to the Maasai Mara [43]. Since the establishment of the first tourist lodge, Kekorok, in 1963, the Maasai Mara has seen more than 60 lodges open. Kenya is enforcing strict lockdown measures. Reports from Mara conservancies indicate no tourists in March–April 2020—the likelihood is that there will be next to no tourists in the ensuing year [44].

Appendix A.3. Hell's Gate National Park, Kenya

Established in 1984, Hell's Gate National Park is small and covers 68.25 square km. It is located in the northwestern part of Kenya around Lake Naivasha, approximately 100 km from Nairobi, 56 km from Nakuru, and 10 km from Naivasha town [45]. It is a popular destination for weekend getaways and retreats because of its proximity to urban areas. Explorers Fisher and Thomson in 1883 named it after a narrow opening in the cliffs, once a tributary of a prehistoric lake that fed early humans in the Rift Valley. The park also has a geothermal spa nearby, which is the only one of its kind in East and Central Africa [46]. It is home to a variety of savanna mammals and birds, including zebras and gazelles. The diverse topography and geological setting with towers and cliffs, trails, hot springs, and gorges offer a range of outdoor activities inside the Park. There are three campsites in the National Park—Endachata Campsite, Naiburta Campsite, and Oldubai Campsite. Park visitation dropped by 99 percent in national parks in Kenya at the height of the Covid-19 pandemic [47].

Appendix B

Table A1. Protected area management categories' definitions from IUCN [1].

	IUCN Category	Description
Ia	Strict Nature Reserve	Strictly protected: human visitation, use, and impacts are strictly controlled and limited to ensure the protection of the conservation values Protect biodiversity and geological/geomorphic features Can serve as reference areas for scientific research and monitoring
Ib	Wilderness Area	Unmodified or slightly modified areas without permanent or significant human habitation Protected and managed to preserve their natural condition
II	National Park	Large natural or near natural areas Protect large-scale ecological processes, along with the complement of species and ecosystems characteristic of the area Provide a foundation for environmentally and culturally compatible, spiritual, scientific, educational, recreational, and visitor opportunities
III	Natural Monument or Feature	Protect a specific natural monument, which can be a landform, sea mount, submarine cavern, geological features such as a cave, or even a living feature such as an ancient grove Generally, quite small, protected areas and often have high visitor value
IV	Habitat Species Management Area	Protect particular species or habitats and management reflects this priority Can need regular, active interventions to address the requirements of specific species or to maintain habitats
V	Protected Landscape/Seascape	Interaction of people and nature over time has produced an area of distinct character with significant, ecological, biological, cultural, and scenic value Safeguarding the integrity of this interaction is vital to protecting and sustaining the area and its associated nature conservation and other values
VI	Protected area with sustainable use of natural resources	Conserve ecosystems and habitats together with associated cultural values and traditional natural resource management systems Most of the area in a natural condition, where a proportion is under sustainable natural resource management and where low-level non-industrial use of natural resources compatible with nature conservation

[1] Source: https://www.iucn.org.

Appendix C

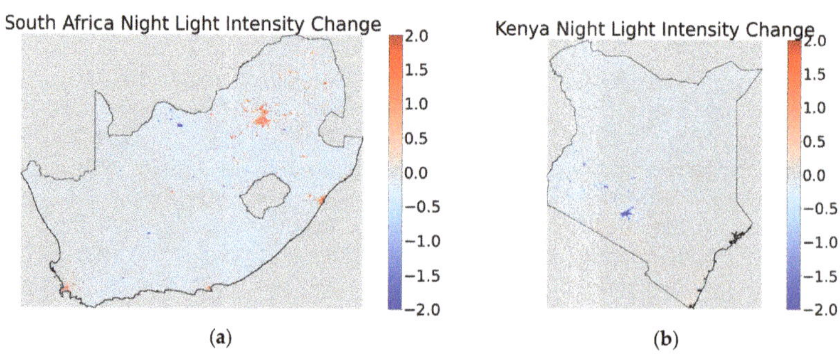

Figure A1. (**a**) Change in night light intensity in South Africa; (**b**) change in night light intensity in Kenya.

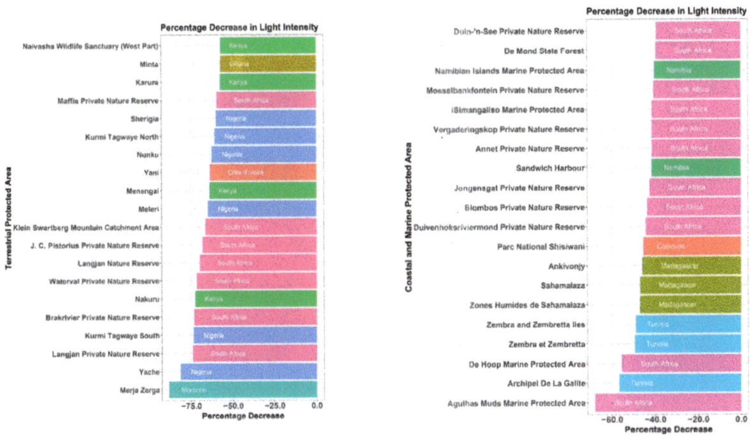

(a) (b)

Figure A2. (**a**) Percentage decrease in light intensity in Terrestrial PAs with the highest decrease in the light intensity; (**b**) percentage decrease in light intensity in coastal and marine PAs with the highest decrease in the light intensity.

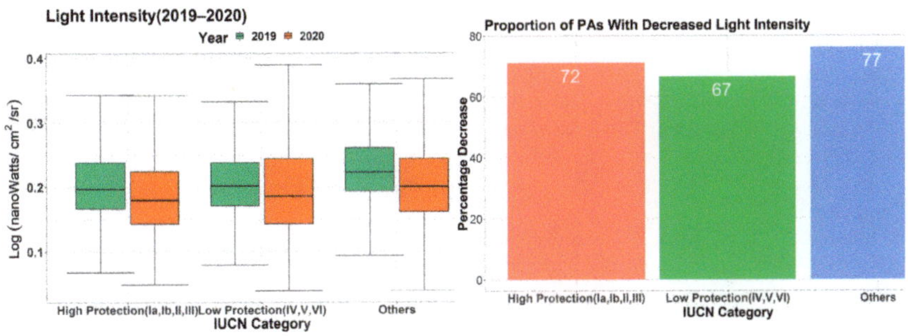

(a) (b)

Figure A3. (**a**) Comparison of light intensity across IUCN categories; (**b**) proportion of PAs with decreased light intensity across IUCN categories.

Appendix D

Table A2. The decrease in mean and median (light intensity) for PAs with international designations.

Sites	Mean Percentage Decrease	Median Percentage Decrease
Ramsar	−17.925	−15.236
UNESCO-MAB	−10.694	−10.207
World Heritage	−14.93	−12.12

References

1. WTTC. Urgent Appeal to International Community to Support African Travel and Tourism Sector. Available online: https://wttc.org/News-Article/Urgent-Appeal-to-International-Community-to-Support-African-Travel-and-Tourism-Sector (accessed on 14 June 2020).
2. *Towards Measuring the Economic Value of Wildlife Watching Tourism in Africa—Briefing Paper*; World Tourism Organization (UNWTO) (Ed.) World Tourism Organization (UNWTO): Madrid, Spain, 2015; ISBN 978-92-844-1675-2.

3. United Nations Environment Programme. Emerging Zoonotic Diseases and Links to Ecosystem Health—UNEP Frontiers 2016 Chapter. Available online: http://www.unenvironment.org/resources/emerging-zoonotic-diseases-and-links-ecosystem-health-unep-frontiers-2016-chapter (accessed on 27 April 2020).
4. PROTECTED AREA EXECUTIVE DIRECTORS CONFERENCE The Impact of COVID-19 on Protected Areas Operations and Programmes Proceedings of the First Conference of Africa's Heads of Protected Areas. 2020. Available online: https://www.iucn.org/sites/dev/files/content/documents/2020/report_on_the_impact_of_covid_19_doc_july_10.pdf (accessed on 15 July 2020).
5. Hockings, M.; Dudley, N.; Ellio, W.; Napolitano, M.; MacKinnon, K.; Pasha, M.; Phillips, A.; Woodley, S.; Appleton, M.; Chassot, O.; et al. Editorial Essay: Covid-19 and Protected and Conserved Areas. *Parks* **2020**, *26*. Available online: https://parksjournal.com/wp-content/uploads/2020/06/Hockings-et-al-10.2305-IUCN.CH_.2020.PARKS-26-1MH.en_-1.pdf (accessed on 3 July 2020). [CrossRef]
6. Lindsey, P.; Allan, J.; Brehony, P.; Dickman, A.; Robson, A.; Begg, C.; Bhammar, H.; Blanken, L.; Breuer, T.; Fitzgerald, K.; et al. Conserving Africa's Wildlife and Wildlands through the COVID-19 Crisis and Beyond. *Nat. Ecol. Evol.* **2020**, 1–11. [CrossRef]
7. UNEP-WCMC; IUCN; NGS. *Protected Planet Live Report 2021*; UNEP-WCMC; IUCN; NGS: Cambridge UK; Gland, Switzerland; Washington, DC, USA, 2021; Available online: https://livereport.protectedplanet.net/chapter-2 (accessed on 16 January 2021).
8. Watson, J.E.M.; Dudley, N.; Segan, D.B.; Hockings, M. The Performance and Potential of Protected Areas. *Nature* **2014**, *515*, 67–73. [CrossRef]
9. Naidoo, R.; Gerkey, D.; Hole, D.; Pfaff, A.; Ellis, A.M.; Golden, C.D.; Herrera, D.; Johnson, K.; Mulligan, M.; Ricketts, T.H.; et al. Evaluating the Impacts of Protected Areas on Human Well-Being across the Developing World. *Sci. Adv.* **2019**, *5*, eaav3006. [CrossRef]
10. Ferraro, P.J.; Hanauer, M.M. Quantifying Causal Mechanisms to Determine How Protected Areas Affect Poverty through Changes in Ecosystem Services and Infrastructure | PNAS. Available online: https://www.pnas.org/content/111/11/4332 (accessed on 16 June 2020).
11. Balmford, A.; Green, J.M.H.; Anderson, M.; Beresford, J.; Huang, C.; Naidoo, R.; Walpole, M.; Manica, A. Walk on the Wild Side: Estimating the Global Magnitude of Visits to Protected Areas. *PLOS Biol.* **2015**, *13*, e1002074. [CrossRef]
12. den Braber, B.; Evans, K.L.; Oldekop, J.A. Impact of Protected Areas on Poverty, Extreme Poverty, and Inequality in Nepal. *Conserv. Lett.* **2018**, *11*, e12576. [CrossRef]
13. UNWTO. *Covid-19 Related Travel Restrictions—A Global Review for Tourism*; UNWTO: Madrid, Spain, 2020.
14. Poletto, C.; Gomes, M.F.; Pastore y Piontti, A.; Rossi, L.; Bioglio, L.; Chao, D.L.; Longini, I.M., Jr.; Halloran, M.E.; Colizza, V.; Vespignani, A. Assessing the Impact of Travel Restrictions on International Spread of the 2014 West African Ebola Epidemic. *Eurosurveillance* **2014**, *19*. [CrossRef]
15. Huber, C.; Finelli, L.; Stevens, W. The Economic and Social Burden of the 2014 Ebola Outbreak in West Africa. *J. Infect. Dis.* **2018**, *218*, S698–S704. [CrossRef]
16. Sannino, A.; D'Emilio, M.; Castellano, P.; Amoruso, S.; Boselli, A. Analysis of Air Quality during the COVID-19 Pandemic Lockdown in Naples (Italy). *Aerosol Air Qual. Res.* **2021**, *21*, 200381. [CrossRef]
17. Liu, Q.; Sha, D.; Liu, W.; Houser, P.; Zhang, L.; Hou, R.; Lan, H.; Flynn, C.; Lu, M.; Hu, T.; et al. Spatiotemporal Patterns of COVID-19 Impact on Human Activities and Environment in Mainland China Using Nighttime Light and Air Quality Data. *Remote Sens.* **2020**, *12*, 1576. [CrossRef]
18. Ghosh, T.; Elvidge, C.D.; Hsu, F.-C.; Zhizhin, M.; Bazilian, M. The Dimming of Lights in India during the COVID-19 Pandemic. *Remote Sens.* **2020**, *12*, 3289. [CrossRef]
19. Mellander, C.; Lobo, J.; Stolarick, K.; Matheson, Z. Night-Time Light Data: A Good Proxy Measure for Economic Activity? *PLoS ONE* **2015**, *10*, e0139779. [CrossRef]
20. Aubrecht, C.; Jaiteh, M.; De Sherbinin, A. *Global Assesment of Light Pollution Impact on Protected Areas*; CIESIN: Palisades, NY, USA, 2010. Available online: http://www.ciesin.org/documents/light-pollution-Jan2010.pdf (accessed on 9 May 2020).
21. Xiang, W.; Tan, M. Changes in Light Pollution and the Causing Factors in China's Protected Areas, 1992–2012. *Remote Sens.* **2017**, *9*, 1026. [CrossRef]
22. Henderson, J.V.; Storeygard, A.; Weil, D.N. Measuring Economic Growth from Outer Space. *Am. Econ. Rev.* **2012**, *102*, 994–1028. [CrossRef] [PubMed]
23. Keola, S.; Andersson, M.; Hall, O. Monitoring Economic Development from Space: Using Nighttime Light and Land Cover Data to Measure Economic Growth. *World Dev.* **2015**, *66*, 322–334. [CrossRef]
24. Ma, T.; Zhou, C.; Pei, T.; Haynie, S.; Fan, J. Quantitative Estimation of Urbanization Dynamics Using Time Series of DMSP/OLS Nighttime Light Data: A Comparative Case Study from China's Cities. *Remote Sens. Environ.* **2012**, *124*, 99–107. [CrossRef]
25. Shi, K.; Yu, B.; Huang, Y.; Hu, Y.; Yin, B.; Chen, Z.; Chen, L.; Wu, J. Evaluating the Ability of NPP-VIIRS Nighttime Light Data to Estimate the Gross Domestic Product and the Electric Power Consumption of China at Multiple Scales: A Comparison with DMSP-OLS Data. *Remote Sens.* **2014**, *6*, 1705–1724. [CrossRef]
26. Zhang, Q.; Seto, K.C. Mapping Urbanization Dynamics at Regional and Global Scales Using Multi-Temporal DMSP/OLS Nighttime Light Data. *Remote Sens. Environ.* **2011**, *115*, 2320–2329. [CrossRef]
27. Liu, Z.; He, C.; Zhang, Q.; Huang, Q.; Yang, Y. Extracting the Dynamics of Urban Expansion in China Using DMSP-OLS Nighttime Light Data from 1992 to 2008. *Landsc. Urban Plan.* **2012**, *106*, 62–72. [CrossRef]

28. Elvidge, C.D.; Ghosh, T.; Hsu, F.-C.; Zhizhin, M.; Bazilian, M. The Dimming of Lights in China during the COVID-19 Pandemic. *Remote Sens.* **2020**, *12*, 2851. [CrossRef]
29. Jechow, A.; Hölker, F. Evidence That Reduced Air and Road Traffic Decreased Artificial Night-Time Skyglow during COVID-19 Lockdown in Berlin, Germany. *Remote Sens.* **2020**, *12*, 3412. [CrossRef]
30. UNEP-WCMC and IUCN Protected Planet: The World Database on Protected Areas (WDPA). Available online: https://www.protectedplanet.net/ (accessed on 14 June 2020).
31. Liu, X.; Blackburn, T.M.; Song, T.; Wang, X.; Huang, C.; Li, Y. Animal Invaders Threaten Protected Areas Worldwide. *Nat. Commun.* **2020**, *11*, 2892. [CrossRef] [PubMed]
32. UNEP-WCMC and IUCN Protected Planet—Technical Resources: Calculating Protected Area Coverage. Available online: https://www.protectedplanet.net/c/calculating-protected-area-coverage (accessed on 14 June 2020).
33. UNEP-WCMC. *User Manual for the World Database on Protected Areas and World Database on Other Effective Area-based Conservation Measures: 1.6*; UNEP-WCMC: Cambridge, UK, 2019; Available online: http://wcmc.io/WDPA_Manual (accessed on 9 May 2020).
34. Román, M.O.; Wang, Z.; Sun, Q.; Kalb, V.; Miller, S.D.; Molthan, A.; Schultz, L.; Bell, J.; Stokes, E.C.; Pandey, B.; et al. NASA's Black Marble Nighttime Lights Product Suite. *Remote Sens. Environ.* **2018**, *210*, 113–143. [CrossRef]
35. Data Processing Levels | Earthdata. Available online: https://earthdata.nasa.gov/collaborate/open-data-services-and-software/data-information-policy/data-levels/ (accessed on 29 December 2020).
36. Coesfeld, J.; Anderson, S.J.; Baugh, K.; Elvidge, C.D.; Schernthanner, H.; Kyba, C.C.M. Variation of Individual Location Radiance in VIIRS DNB Monthly Composite Images. *Remote Sens.* **2018**, *10*, 1964. [CrossRef]
37. Kyba, C.C.M.; Kuester, T.; de Miguel, A.S.; Baugh, K.; Jechow, A.; Hölker, F.; Bennie, J.; Elvidge, C.D.; Gaston, K.J.; Guanter, L. Artificially Lit Surface of Earth at Night Increasing in Radiance and Extent. *Sci. Adv.* **2017**, *3*, e1701528. [CrossRef]
38. Bensch, G.; Peters, J.; Sievert, M. The Lighting Transition in Rural Africa—From Kerosene to Battery-Powered LED and the Emerging Disposal Problem. *Energy Sustain. Dev.* **2017**, *39*, 13–20. [CrossRef]
39. Global Energy Review 2020—Analysis. Available online: https://www.iea.org/reports/global-energy-review-2020 (accessed on 29 December 2020).
40. World Bank Access to Electricity, Rural (% of Rural Population)—Sub-Saharan Africa | Data. Available online: https://data.worldbank.org/indicator/EG.ELC.ACCS.RU.ZS?locations=ZG (accessed on 14 June 2020).
41. Dudley, N. *Guidelines for Applying Protected Area Management Categories*; IUCN: Gland, Switzerland, 2008; ISBN 978-2-8317-1086-0.
42. Borrini-Feyerabend, G.; Dudley, N.; Jaeger, T.; Lassen, B.; Pathak Broome, N.; Phillips, A.; Sandwith, T. *Governance of Protected Areas: From understanding to action. Best Practice Protected Area Guidelines Series No. 20*; IUCN: Gland, Switzerland, 2013; ISBN 978-2-8317-1608-4.
43. Onchwati, J.; Sommerville, H.; Brockway, N. Sustainable tourism development in the Masai Mara National Reserve, Kenya, East Africa. *WIT Trans. Ecol. Environ.* **2010**, *139*, 319–330.
44. Mara Conservancy April 2020. Available online: https://www.maratriangle.org/november-2016-1/april-2020 (accessed on 17 June 2020).
45. KWS 2020 Hells Gate National Park | Kenya Wildlife Service. Available online: http://www.kws.go.ke/content/hells-gate-national-park (accessed on 5 January 2021).
46. Kariuki, P.G.; Onyiengo, D.Z.; Odhiambo, S.P.O. Influence of Critical Success Factors on Competitive Position of Hell's Gate National Park as a Tourist Destination. *IOSR-JBM* **2016**, *18*, 4.
47. The Standard. Available online: https://www.standardmedia.co.ke/adblock?u=https%3A%2F%2Fwww.standardmedia.co.ke%2Frift-valley%2Farticle%2F2001388437%2Fhells-gate-national-park-welcomes-more-visitors (accessed on 5 January 2021).

remote sensing

MDPI

Article

What Can We Learn from Nighttime Lights for Small Geographies? Measurement Errors and Heterogeneous Elasticities

Richard Bluhm [1,2,*] and Gordon C. McCord [3]

1 Institute of Macroeconomics, Leibniz University Hannover, 30167 Hannover, Germany
2 Department of Political Science, University of California San Diego, 9500 Gilman Drive #0521, La Jolla, CA 92093, USA
3 School of Global Policy and Strategy, University of California San Diego, 9500 Gilman Drive #0519, La Jolla, CA 92093, USA; gmccord@ucsd.edu
* Correspondence: bluhm@mak.uni-hannover.de; Tel.: +49-511-762-5655

Abstract: Nighttime lights are routinely used as a proxy for economic activity when official statistics are unavailable and are increasingly applied to study the effects of shocks or policy interventions at small geographic scales. The implicit assumption is that the ability of nighttime lights to pick up changes in GDP does not depend on local characteristics of the region under investigation or the scale of aggregation. This study uses panel data on regional GDP growth from six countries, and nighttime lights from the Defense Meteorological Satellite Program (DMSP) to investigate potential nonlinearities and measurement errors in the light production function. Our results for high statistical capacity countries (the United States and Germany) show that nightlights are significantly less responsive to changes in GDP at higher baseline level of GDP, higher population densities, and for agricultural GDP. We provide evidence that these nonlinearities are too large to be caused by differences in measurement errors across regions. We find similar but noisier relationships in other high-income countries (Italy and Spain) and emerging economies (Brazil and China). We also present results for different aggregation schemes and find that the overall relationship, including the nonlinearity, is stable across regions of different shapes and sizes but becomes noisier when regions become few and large. These findings have important implications for studies using nighttime lights to evaluate the economic effects of shocks or policy interventions. On average, nighttime lights pick up changes in GDP across many different levels of aggregation, down to relatively small geographies. However, the nonlinearity we document in this paper implies that some studies may fail to detect policy-relevant effects in places where lights react little to changes in economic activity or they may mistakenly attribute this heterogeneity to the treatment effect of their independent variable of interest.

Keywords: nighttime lights; GDP; nonlinearity; panel data; MAUP; aggregation

Citation: Bluhm, R.; McCord, G.C. What Can We Learn from Nighttime Lights for Small Geographies? Measurement Errors and Heterogeneous Elasticities. *Remote Sens.* **2022**, *14*, 1190. https://doi.org/10.3390/rs14051190

Academic Editors: Ran Goldblatt, Steven Louis Rubinyi and Hogeun Park

Received: 27 December 2021
Accepted: 21 February 2022
Published: 28 February 2022

1. Introduction

A growing literature in economics and other fields uses nighttime lights as a proxy for economic activity when data on gross domestic product (GDP or GDP per capita) are unavailable. Nightlights have been used to compare economic activity across geographic units at a variety of scales, from seminal work at the country level [1–3], over states/provinces [4], down to the level of cities [5], villages [6,7], and grid cells [2]. An increasingly common application of nightlights is to measure regional or local economic impacts of shocks (such as floods [8], sanctions [9], or transportation cost shocks [5]) or spatially-targeted policies (such as a rural employment scheme [10], federal transfers to a state [11], or regional favoritism by political leaders [4]). However, little is known about whether the nightlights-GDP relationship at smaller geographic scales is constant across locations. Note that while we focus on the relationship between nightlights and GDP, other studies have used survey data to study the correlation with income or income proxies when GDP data are unavailable [6,12,13]. Beyond GDP, nightlights have been used to as a proxy for population density [7,14], electrification [15], and infrastructure [16].

Interpreting how changes in nightlights reflect unmeasured changes in economic activity requires having a reliable estimate from settings where both nightlights and economic activity are measured. Recent work has explored how the relationship between growth in nightlights and economic activity varies across contexts, both at national scale [17] and at smaller geographies [7]. Some studies conclude that the subnational nightlights-GDP relationship is not stable in emerging and advanced economies [18]. Others find that luminosity is ineffective as a proxy for economic activity in less densely populated counties in the United States and may be a poor predictor of GDP in areas where agriculture plays a crucial role in the economy [19]. Given that physical features underlying economic activity usually vary by economic sector (across agriculture and industry, for example) and by population density, it seems plausible that the relationship between changes in nightlights and economic growth varies across contexts. However, it is often overlooked that the presence of measurement errors in nighttime lights and GDP complicates the interpretation of such findings and makes it difficult to separate the influence of measurement errors from heterogeneity in the data generating process. As a result, little is known about how stable this relationship is at finer levels of spatial disaggregation and what this potential heterogeneity might imply for studies using nightlights as a proxy for economic activity. Unfortunately, this raises the potential of incorrect inference regarding the economic effects of a policy, investment or shock. If, for example, nightlights are not associated with changes in economic activity over some range of GDP or population density, then a policy intervention or public investment in these kinds of areas might be incorrectly labeled as ineffective if researchers estimate its economic impacts solely through nightlights. Similarly, variation in the nightlights-GDP relationship might lead researchers to conclude that a policy had heterogeneous impacts on economic activity across locations, even when the true economic impact does not vary.

This paper makes three contributions. First, we compile subnational economic data from several high and middle income countries together with data on nighttime lights and discuss the sources of measurement errors in both. Second, we estimate the nightlights-GDP elasticity—the percentage change in nightlights associated with a one percent change in economic activity—in two different ways: (i) unconditionally for different income groups to gauge whether the relative attenuation factors consistent with a constant elasticity are plausible, and (ii) by conditioning the elasticity on population density. Our results illustrate significant nonlinearity at subnational scale. The estimated elasticities tend to fall with higher levels of GDP, higher population densities, and the size of the agricultural sector. We find that the required variation in measurement errors would have to be implausibly large and follow an improbable pattern to explain these findings, which implies that the structural elasticity likely varies. Finally, we study changes in the nightlights-output relationship at various configurations of geographic aggregation. We find that the average relationship and the evidence in favor of nonlinearity is remarkably stable across geographies of different shapes and sizes, even though the influence of measurement errors in lights and GDP should decrease as units become larger. Simultaneously, the variation of estimates around the average increases as the number of units falls. Given the previous evidence, this suggests that in addition to sampling variation, spatial correlation in industrial composition and population density contributes to this pattern.

2. Materials and Methods

Applied researchers typically use nighttime lights as a proxy for GDP when data on income or production are unavailable. In the canonical regression equation setup, GDP would be on the left hand side, if it were observed, and some policy intervention or other variable of interest on the right hand side. The researcher is interested in the treatment effect of the policy on GDP. Replacing GDP by nighttime lights (whether as a sum or density per unit area) implies that the policy parameter of interest is not statistically identified.

Instead, the researcher obtains a product of the policy effect on GDP, say τ, multiplied by the elasticity of nighttime lights with respect to GDP, say β. To see this, let the policy equation of interest be $y_i^G = \tau d_i + e_i$ while the structural relationship between lights and GDP is $y_i^L = \beta y_i^G + \epsilon_i$, where y^G is GDP, y^L is nighttime light output and d_i is the variable for the policy being evaluated. Both quantities are in logs and could be measured in per capita terms. Since GDP is not observed, the researcher actually estimates $y_i^L = \theta d_i + \epsilon_i$. In this case, it is straightforward to show that the probability limit of $\hat{\theta}$ is $\beta\tau$. This is precisely why the literature typically multiplies the policy parameter by an estimate of the *inverse elasticity* of lights with respect to GDP, in order to relate their estimates back to changes in aggregate income [3]. The elasticity of lights with respect to output is typically assumed to be constant. In fact, usually an estimate of 0.3 [3,4] is used in order to back out the effect of the policy on GDP. However, if there is heterogeneity in how nightlights react to GDP, then this has an important implication: any systematic variation in the elasticity of nightlights with respect to GDP will translate into differences in the estimated effect of the policy across locations, even when the true effect of the policy is constant.

Unfortunately, both nighttime lights and GDP data are subject to measurement error. The DMSP nighttime lights that are used in most of the literature suffer from four sources of error: (1) bottom-coding as a result of filtering and limited detection of low lights, (2) topcoding as a result of sensor saturation in bright areas, (3) blooming or overglow as a result of atmospheric scattering and "pollution" from adjacent light sources, exacerbated by geolocation errors, and (4) a lack of inter-annual calibration which makes it impossible to convert the recorded digital numbers into a physical quantity such as radiance [5,20–23]. Moreover, as we discuss below, measuring subnational GDP in all countries often involves assumptions about the location of certain economic activity, or interpolations of baseline year surveys for industrial or agricultural output conducted infrequently [24]. GDP data in developing countries are particularly error-prone and could be subject to outright manipulation [25]. The presence of measurement error on both sides of the equation has long been recognized in the economics literature focused on estimating the relationship between nighttime lights and GDP, or optimal combinations of both [2,3,17,26].

The newer VIIRS day-night band solves many of the legacy technology issues of the DMSP system. The sensor has a smaller ground footprint, better ability to detect lights at both ends of the spectrum, and is radiometrically calibrated. This paper focuses on DMSP for two reasons. First, most applications of nightlights in economics research have used DMSP given the longer time series (VIIRS data are only available from 2012 onward, resulting in short panels with GDP). Second, VIIRS satellites observe nighttime lights much closer to midnight when most production and consumption have ceased. Nevertheless, evidence suggests that the VIIRS data vastly outperforms the DMSP system when it comes to predicting GDP [19]. While the number of studies using VIIRS is undoubtedly on the rise, we use the DMSP data in this analysis to speak to the existing literature and leave a similar exercise with VIIRS for future work.

In our setting, the key implication of measurement error in both variables is that we cannot simply take estimates obtained from regressions on subsamples, or nonlinear regressions, at face value and definitively conclude that there is nonlinearity in the structural elasticity. Differences in the elasticity could be due to structural differences in how light reacts to GDP (e.g., higher densities in cities combined with economies of scale in light use or differential light consumption by different economic sectors) or due to differences in measurement error (e.g., due to greater informality or difficulty measuring the activity of some regions). To circumvent this issue, we calculate how large the measurement error would need to be in order to fully explain the observed variation in the elasticity. We can then ask if the implied measurement error is plausible in countries where we expect these errors to be small and constant across units, and then use these estimates to assess the variation in countries where measurement errors should play a larger role. We study what the literature calls the structural relationship (how lights react to GDP), as opposed the predictive relationship (how lights predict GDP), but make no attempt to estimate the structural relationship net of measurement error. Other work focuses on the country level and proposes a method of identifying the elasticity under the presence of very general forms of measurement error using proxies for the statistical capacity of each country [17]. We use subnational data where variation in statistical capacity is not helpful for identification.

Our article examines the nightlights-GDP relationship in Brazil, China, Germany, Italy, Spain and the United States, spanning a range of quality in national accounts (and by extension subnational accounts) in order to gauge variation in the structural relationship in both higher and lower measurement error contexts. In addition, these countries are large and diverse in terms of their geography and economic structure, have subnational GDP data available for a suitable number of years, and have a large number of second-level administrative units (county/municipality/district) to ensure sufficient statistical power for estimation. These countries also vary in terms of quality grades for their national accounts, which we take as a proxy for the quality of their subnational accounts. For example, if we take quality grades from the 1994 Penn World Table 5.6 and 2008 Penn World Table 6.1 in order to reflect data quality during years corresponding to our study period, then the United States and Germany have an A grade, Italy an A−, Spain a B+, Brazil a B and China a C [27,28].

2.1. Empirical Strategy

We study the potential heterogeneity in the income elasticity of lights across subnational units using two different but related approaches. Both take the constant elasticity model underlying most of the applied literature as the benchmark and then set up different conditions which would lead us to reject this model. We focus on GDP (as opposed to other measures such as GDP per capita) because the economics literature studying the relationship between nightlights and economic activity or employing nightlights as a proxy focuses on GDP instead of GDP per capita. The implicit assumption is that growth in nighttime lights increases equally in population growth and growth in per capita incomes. Secondly, GDP per capita is not always positively correlated to total economic activity.

First, we specify single variable regressions of the observed nighttime lights (y_{it}^L) on GDP per area (y_{it}^G), both in logarithms, for samples split according to quartiles of average GDP. More formally, for each subsample we specify

$$y_{it}^L = \beta y_{it}^G + \mu_i + \psi_t + \epsilon_{it} \quad (1)$$

where μ_i and ψ_t represent geographic unit and year fixed effects, respectively, and ϵ_{it} is an idiosyncratic error.

Together with the specification in logarithms, the inclusion of unit fixed effects implies that we are relating a region's growth in nighttime lights to growth in GDP. The inclusion of time fixed effects allows for country-specific factors that vary over time but influence every administrative unit in the same way (such as differences in the ability of sensors to pick up nighttime lights, due to satellite changes or orbital or sensor degradation). We are therefore estimating the relationship of changes in nightlights and economic activity within a geographic unit over time, and not how the level of nightlights is associated with economic activity across geographic units. This is the relevant estimation to inform empirical exercises studying the impact of a shock, policy or investment at subnational level.

Our parameter of interest, β, is the income elasticity of nighttime lights, a unitless measure indicating the percentage luminosity increase in response to a one percent change in GDP. We estimate this parameter separately for subsamples split according to quartiles of average GDP to obtain four coefficients ($\hat{\beta}^k$ for $k = 1,2,3,4$). These four groups of different average GDP are a proxy for subnational differences in statistical capacity. Under the conditional mean independence assumptions typically made in the related literature [2,3,26] the resulting estimates will converge to the true coefficient times an attenuation factor. If the true relationship were constant across all subsamples of the data and the variance of the measurement error in GDP were constant, then the estimated elasticities should be very similar in each partition of the data (with some expected sampling variation). More formally, $\hat{\beta}^k \xrightarrow{p} \beta\lambda$. If the true relationship is constant but the GDP error variance differs across subsamples, then the estimated coefficients will be attenuated differently, such that $\hat{\beta}^k \xrightarrow{p} \beta\lambda^k$. In this case, it is plausible that the highest GDP category would have the smallest measurement error in GDP, resulting in the least attenuation of the coefficient. However, our analysis does not presume any particular pattern of attenuation. Appendix A derives these results and provides details on the required assumptions.

We use this relationship to study whether the differences in elasticities and the implied attenuation factors are plausible. If the true structural elasticity is assumed constant, the ratio of any two point estimates indicates how different measurement errors must be in order explain the variation in the estimated elasticities across the two subsamples. For each country, we estimate and report the ratios $\theta^k = \beta^k/\beta^1$ comparing the quarter k to the data up to the first quartile. If, for example, the variance of measurement errors in GDP falls with higher incomes, then the sequence of coefficient estimates would be rising and we would interpret a ratio of, say, $\hat{\theta}^4 = 0.5/0.25$ as 'the income elasticity of nighttime lights has to be twice as attenuated in the lowest GDP quartile than the highest GDP quartile for the constant elasticity model to be true'. The pattern would be reversed if the variance of measurement errors in GDP increases with income. If we cannot reject the hypothesis that the ratio is different from one, then we conclude that their attenuation factors could have been the same.

Going one step further, we interpret values of θ^k statistically different from 1 in countries with high statistical capacity as evidence against the constant elasticity model typically assumed in the applied literature. Of course, it is plausible that the variation in measurement errors of subnational GDP is substantial in developing countries, where there may be more variation in informality across regions, or in the capacity of the statistical apparatus to collect economic data. However, in developed countries with uniformly high statistical capacity, we should not observe significant differences in the signal-to-noise ratio of GDP across regions, and, in fact, expect the estimated $\hat{\theta}$s to be close to one. Even if the signal-to-noise ratio in the best measured part of the data were, say, 0.8, then a doubling or halving of this ratio for some regions would imply implausibly large differences in measurement error in high capacity countries. This is why our sample of countries deliberately spans highest quality subnational accounts (the US or Germany) and countries where regional GDP is estimated with less precision (China or Brazil). We use the estimates from developed economies to better understand the relative roles of "structural" nonlinearities and measurement errors in the light-output relationship in developing economies, where nighttime lights are most often used as a proxy for local GDP.

Note that our assumptions measurement errors in GDP and lights also imply that the standard errors of the estimated coefficients are biased. While the sign is indeterminate, it can be shown that the resulting t-statistics are underestimated. The standard errors of the estimated relative attenuation factors are also affected by measurement errors. We leave potential solutions to these issues for future research and note that our estimates of the uncertainties are not free of large sample biases.

Our second approach to studying heterogeneity in the income elasticity models the variation along a third variable:

$$y^L_{it} = (\beta + \gamma z_i)y^G_{it} + \mu_i + \psi_t + e_{it} = \beta y^G_{it} + \gamma(y^G_{it} z_i) + \mu_i + \psi_t + \epsilon_{it} \tag{2}$$

where z_i is the logarithm of population density in the first year of the data and all other variables are defined as before. We focus on population density since the light-GDP relationship could vary along this dimension for several reasons. For example, fixed costs of light infrastructure can be large at low population densities, while at high densities, economies of scale and vertical city growth may decrease the responsiveness of lights to changes in GDP.

Given that both light and GDP per area are in logarithms, taking the derivative of Equation (2) with respect to y^G_{it} results in an elasticity that varies with initial population density. For this specification, we are no longer interested in ratios of these elasticities at different points in the distribution but instead focus on the sign and significance of $\hat{\gamma}$. It is important to note that all coefficient estimates in specifications with multiple independent variables, of which at least one is measured with error, are biased in unknown directions. However, if we are willing to assume that population density in the initial year is measured without error and make additional independence and linearity assumptions, then $\hat{\gamma}$ converges to zero in probability if the constant elasticity model is correct (see Appendix A for details on these results). Moreover, $\hat{\beta}$ will still be attenuated by the same signal-to-noise ratio as in Equation (1). Hence, we may compare the estimates of $\hat{\beta}$ across the long and short regressions and, with some caveats, interpret a significant result on $\hat{\gamma}$ as further evidence against the constant elasticity model.

Another likely source of structural differences in the nightlight-economic output relationship are differences in industrial composition. For example, if nighttime lights fail to pick up changes in agricultural GDP [29,30], then differences in sectoral composition across regions are sufficient to generate variation in the measured light-output relationship. Given that agricultural areas are typically less densely populated than regions with large manufacturing or service hubs, this would also imply some variation of the light-output elasticity with respect to population density. Moreover, it is an open question whether lights primarily respond to value-creation in industry or services and whether the elasticity is constant within each economic sector. Light in densely populated urban areas with a high concentration of services, for example, might not scale linearly in output. To explore this in our data, we run regressions of nightlights on GDP separately for agricultural, industry (including construction), and service sector GDP, with and without interactions with population density. Just as with aggregate output, we cannot simply compare the coefficients for different sectors to gauge whether structural elasticity varies because measurement errors are likely to vary across sectors. Instead, maintaining the same assumption from above, we again ask if the implied relative measurement errors are plausible and check whether the interaction with population density is significant.

Finally, we analyze whether aggregation to geographic units of different shapes and sizes changes the pattern of the resulting estimates. This may occur for two reasons related to our analytical framework. First, aggregating smaller units to larger units could reduce measurement errors in both GDP and lights. For example, county GDP errors due to downscaling state data or workplace versus place of residence mismatches are offset when small regions are grouped together. Similarly, overglow of nightlights into neighboring regions becomes internalized when analyzing larger areas, and the relative importance of topcoding decreases as the size of units increases. A second reason for why aggregation could affect the pattern of elasticities is the grouping of smaller units (with potentially different structural elasticities) into larger, more economically mixed units. Regardless, assessing how geographic scale affects the observed elasticities can help explain the disparate findings documented in the literature, which typically finds large variation in elasticities across countries and across different levels of aggregation within the same country [18,19].

Previous research on the Modifiable Areal Unit Problem (MAUP) in urban economics suggests that size and shape of administrative units matters little in comparison to other specification issues but also finds that aggregating to large units can distort the underlying relationship [31]. No study has systematically examined this problem in the context of nighttime lights and GDP.

We use methods from the literature on the MAUP [31,32] to study whether the design of regions creates variation in sectoral composition and population density, which, in turn, generates differences in the structural nightlights-economic output relationship. Specifically, we use disaggregated data on local GDP for the continental United States (3080 counties) and Brazil (5569 municipalities) to create many alternative administrative divisions of varying shapes and sizes. We construct simulated partitions for a given number of administrative units, k, using the following random-seed-and-grow algorithm [31]. We start out with the finest level of aggregation into n units. First, we randomly pick one seed unit. Second, we identify the unit's closest neighbor before merging these two units so that now there are $n-1$ units remaining. We repeat these steps until $n = k$. The resulting partition is geographically contiguous. We run the algorithm 1000 times for every 200th number of units from $\underline{k} = 50$ to some country-specific maximum $\bar{k}$. We take 50 as a lower bound since this is the number of US states and then simulate the result for $k \in \{50, 200, 400, 600, \ldots, \bar{k}\}$, where $\bar{k}$ is 3000 for the US and 5400 for Brazil. This results in thousands of alternative divisions of the Unites States and Brazil over which we can aggregate nighttime lights, GDP and population, and estimate the specifications given in Equations (1) and (2). A persistence of nonlinearity in simulations with high levels of aggregation (where measurement errors become less severe) would provide additional evidence that the structural elasticity varies.

2.2. Data

We compile data on subnational GDP, sectoral composition, and population from a variety of sources. For each country, Table 1 lists the smallest geography for which subnational GDP data are available, the number of units, years for which the data are available, the industrial classification used (if available), and the primary source. We deflate current local currency units by the national GDP deflator from the World Development Indicators if the data are not provided in real (constant price) quantities. In the case of the U.S. and European countries, we compute the GDP in each industrial sector by aggregating all NAICS or NACE sectors to a three-sector classification (agriculture, manufacturing and construction, and services). We also obtain high-quality vector geometries representing the geographical units within each country from national statistical offices or other public databases. Data for China is not widely available, which is why we use GDP (in USD) for 299 prefectures from the Economist Intelligence Unit and supplemented this data with information from annual yearbooks and the CEIC's China Economic Database for the 30 missing prefectures, four municipalities and nine county-level cities.

Measuring economic activity at small geographic scale can be challenging in any country. In addition to the usual data collection and processing errors that arise for national accounts, subnational accounts are particularly prone to non-statistical errors. These include imputations, conceptual differences, index construction, sectoral definitions, and the scope of the exclusions (such as home production, subsistence farming, illegal activity and smuggling) [33]. Often, subnational estimates of GDP require triangulating with multiple data sources, or downscaling data collected for higher-level administrative regions. This is true even in high statistical capacity countries. In the United States, the Bureau of Economic Analysis relies on the income approach to measure GDP at state and county level, computed as the sum of compensation of employees, taxes on production and imports minus subsidies, and gross operating surplus (capital income). This method has the potential to underestimate capital-intensive industries whose production relies heavily on physical or financial capital [34]. It measures GDP at people's place of work, as opposed to place of residence. Interpolation between benchmark years and downscaling from state-level data introduce other sources of measurement error. For example, years when the Economic Census is available (every 5 years) are used as benchmark years, with other years interpolated using sales data from the National Establishment Time Series (NETS) database. In addition, some state-level data on OIC (income payments other than employees and proprietors) are distributed among counties using NETS, the Quarterly Census of Employment and Wages (QCEW), Economic Census data, and industry-specific data from various sources [35].

Similar methodological challenges exist in the European Union, although analysis on revisions suggest that estimation errors are small (when countries undergo revisions they constitute less than 1% of GDP). However, there are likely to be larger errors in the historical series. Regional GDP is calculated as regional Gross Value Added plus taxes on products minus subsidies. GVA, in turn, is calculated using the production approach (value of output minus intermediate inputs) or the income approach (similar to the U.S.) depending on the member state [36]. Ongoing work in the European Union is tackling methodological questions for regional GDP such as recording foreign direct investment, non-market services, incorporating global production and integrated global accounts, the digital economy and other price and volume measures for intellectual property products [37,38].

Table 1. Sources of subnational GDP data.

Country	Geography	Units	Availability	Industrial Class.	Source
USA	Counties	3080	2001–2019	NAICS	BEA
Germany	Districts	401	1992–2018	NACE Rev. 2	ADERCO
Italy	Provinces	110	1992–2018	NACE Rev. 2	ADERCO
Spain	Provinces	58	1992–2018	NACE Rev. 2	ADERCO
Brazil	Municipalities	5569	2002–2018	Primary to tertiary	IBGE
China	Prefectures	342	1999–2018	n/a	EIU

Notes: Data on counties in the United States are from the Bureau of Economic Analysis (BEA), which publishes data on real GDP by NAICS industry and population at the county level. We define all NAICS codes starting with 11 as agriculture, 21–33 as industry and 42–92 remaining codes as services. Data on German districts (*Kreise*), Italian provinces, and Spanish provinces (*Provincias, islas, and Ceuta y Melilla*) are from the Annual Regional Database of the European Commission's Directorate General for Regional and Urban Policy (ADERCO). ADERCO publishes data on real GDP by NACE industry and population at the NUTS-3 (rev. 2016) level. We define all NACE codes starting with A as agriculture, B–E as industry and F as services. ADERCO includes historical estimates and projections for the most recent years. We only use data until 2018 which matches the availability of actual GDP figures reported by the relevant national offices. Data on Brazilian municipal GDP is from the Instituto Brasileiro de Geografia e Estatística (IBGE). GDP data exist for all 5570 municipalities, however 1 municipality had zero nightlights over the period (Marajá do Sena) and was dropped when taking logarithms. Data on Chinese prefecture-level GDP is available from the Economics Intelligence Unit (EIU) for 289 prefectures, and augmented with data for 53 more prefectures from Provincial Yearbooks or the CEIC Database.

Less is known about the quality of subnational accounts in Brazil and China. Brazilian municipal GDP is based on Gross Value Added calculated at state level with the production approach, where state GDP is distributed among municipalities using various methods depending on the good or service [39]. China officially uses both the production and income methods for national accounts; value-added of agriculture, forestry, animal husbandry and fishery is calculated by the production method, while the current value-added of other industries is calculated using the income method. Regional GDP is measured by local governments using the production approach from major surveys on large industrial firms, large service sector firms, and some construction firms. These data are supplemented by surveys of smaller firms and administrative data from government departments. In addition, they estimate expenditure by household surveys and investment project surveys. Since local Chinese governments are rewarded for meeting growth targets, the Chinese National Bureau of Statistics revises local GDP estimates in computing national GDP [40,41].

We obtain nighttime lights within each geography from 1992 until 2013 from the Defense Meteorological Satellite Program (DMSP) Operation Line Scan (OLS) sensors. Specifically, we use annual composites which report yearly average "stable lights" as a 6-bit digital number (DN) from 0 to 63, after observations affected by cloud cover, background noise and other disturbances have been removed. We follow common practice to delete gas flares from the annual composites. Gas flares are disproportionately bright in relation to the change in output they represent and create significant overglow into neighboring pixels. Like Henderson et al. [3], we also set all pixels that are not on land to zero. For each region-year pair, we calculate the sum of lights and total area of all pixels. Our outcome of interest is the logarithm of lights per area (ln DN/km^2). Table 2 provides the distribution of nightlight density values across subnational regions by country.

Physical detection limits of the DMSP sensors and the difficulty of separating background noise and transient lights from permanent light sources effectively impose a bottom-coding threshold where pixels with DNs of 1–2 and small clusters of pixels with values of less than 4 DN are removed in the stable lights composite. Solutions to this problem range from adding the minimum detection threshold to recorded lights in a region [5] to using auxiliary data to distinguish background noise from "human lights" [42]. The problem is most severe in Sub-Saharan Africa, where a lack of consistent electrification implies that mid-sized settlements can be missed. Rural electrification rates are close to 100% in Brazil and China today—the two emerging economies in our data—so we consider this source of error to be less important in our study than others. In fact, very few observations record zero light (see Table 2).

Table 2. Distribution of nightlight density values across subnational regions.

Lights/Area	USA	Germany	Italy	Spain	Brazil	China
0	0 (0%)	0 (0%)	0 (0%)	0 (0%)	367 (0.6%)	0 (0%)
0–1	4179 (10.4%)	0 (0%)	0 (0%)	0 (0%)	25,551 (39.8%)	889 (18.5%)
1–5	10,192 (25.5%)	42 (0.5%)	10 (0.41%)	224 (17.3%)	24,567 (36.8%)	1873 (39.0%)
5–10	9354 (23.4%)	346 (3.9%)	125 (5.2%)	393 (30.3%)	7809 (11.7%)	894 (18.6%)
10–20	8779 (21.9%)	2888 (32.7%)	759 (31.4%)	323 (24.9%)	4271 (6.4%)	756 (15.7%)
20–50	5676 (14.2%)	3040 (34.5%)	1316 (54.4%)	290 (22.3%)	2249 (3.4%)	335 (7.0%)
>50	1859 (4.6%)	2506 (28.41%)	210 (8.7%)	68 (5.2%)	952 (1.4%)	55 (1.2%)

Notes: Lights/area refers to the sum of nighttime lights in a subnational region in a given year in the sample divided by the region's area. Values refer to the number of region-years with the given level of nightlights per area, and the percentage in parentheses is the percentage of all observations from that country with that level of nightlight density.

A potentially serious issue in subnational analyses is that the DMSP data are heavily topcoded. Topcoding primarily occurs in city centers as the sensor gradually reach its saturation limit and affects values well below 63 DN in the yearly averages [21,23]. The intensity of topcoding is correlated with our variables of interest, such as average GDP, density, and industrial structure. To deal with this issues, one of our tests uses topcoding corrected data [23] which applies a Pareto-correction to the stable lights composites (and builds on the radiance-calibrated data available in selected years [21]).

The DMSP data exhibit overglow or blurring effects for a variety of reasons. The nominal resolution of the data provided by the National Centers for Environmental Information (NCEI) is 30 arc seconds. However, the effective instantaneous field of view (EIFOV) expands from 2.2 km to about 5.4 km at the edge of the scan and the system "smoothes" these data on-board by forming pixel blocks that are 2.7 km by 2.7 km (with different location offsets for each nightly image [22]). As a result, the same light source will show up in several 30 arc second pixels. On-board processing also magnifies the blur effect for brighter light sources [22]. Geolocation errors which displace lights by about 3 km exacerbate this problem [20]. We discuss the biases introduced by overglow *within* administrative units when presenting the results and allow for spatial correlation *across* units in a robustness check.

A final challenge in using the DMSP data is that the recorded DN cannot be mapped to a physical quantity (radiance). This occurs because the sensors dynamically adjust their low-light detection ability over the lunar cycle but the sensor's 'variable gain' settings are not stored [21]. Efforts have been made to calibrate using light emitted from islands (such as Sicily) and using active targets for some nights [43]. However, there are simply no permanently constant light sources on Earth that can unequivocally solve this problem in the historical data, while ad hoc calibration adjustments have the potential to introduce more noise. Another source of variability across years is that the orbit of the DMSP satellites slowly degrades over time, recording lights at a slightly earlier time each day. This feature has recently been used to extend the DMSP data from 2012 to 2019 using pre-dawn data from older satellites that crossed back into a dawn-dusk orbit [44]. We do not use the extended series as the orbital shift to pre-dawn hours introduces an additional source of measurement error. Following the economics literature [3], we average the data whenever two years are available and include year fixed effects in all regressions. This accounts for differences in average sensor settings in each year which affect all regions in the same manner. If all pixels were illuminated and topcoding did not exist, then a constant shift in each year would fully account for this problem. However, since differences in gain settings also imply that some pixels cross the detection threshold and others become topcoded before on-board averaging occurs, there is likely to be some residual region-year specific error which cannot be accounted for.

As motivation, Figure 1 illustrates the raw correlations between light density and economic density. We observe a strong degree of nonlinearity whenever the correlation is based on highly disaggregated data (as in the case of US counties, municipalities in Brazil, and, to a lesser extent, districts in Germany) and less nonlinearity when the data are more aggregated (provinces and prefectures in Italy, Spain and China).

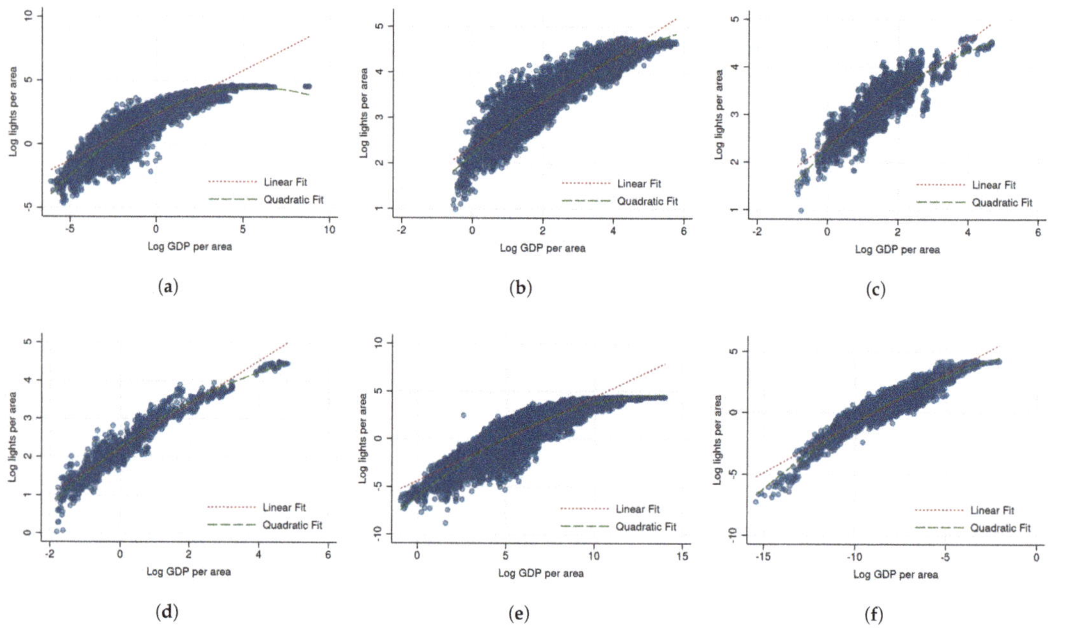

Figure 1. Raw correlation of nighttime lights with real GDP, in logarithms. (**a**) USA, (**b**) Germany, (**c**) Italy, (**d**) Spain, (**e**) Brazil, (**f**) China.

3. Results

3.1. Analysis of Estimates by Income Group

We begin by investigating potential heterogeneity in the light-output relationship using regressions estimated on samples split according to quartiles of average GDP. For each country in our data set, Figure 2 shows the estimated income elasticities of nightlights using four equal-sized groups of the data, with β^1 referring to the coefficient for regions up to the lowest GDP quartile, and β^4 for the regions above the highest quartile.

The results in Figure 2a,b show the patterns for the U.S. and Germany, the countries with the highest statistical capacity in our sample. Rather than estimates which are constant or rise with average incomes, we observe the highest elasticities among regions in lowest quartile ($\hat{\beta}^1 = 0.482$ for the United States and $\hat{\beta}^1 = 0.499$ for Germany) followed by a steady decline as we move up in the distribution of average incomes which matches the raw correlations presented in Figure 1. In fact, the estimates for the group with the highest incomes are either approaching zero in the case of the United States ($\hat{\beta}^4 = 0.158$) or cannot be distinguished from zero in the case of Germany ($\hat{\beta}^4 = 0.050$). Of course, the presence of measurement errors implies that we cannot take this evidence at face value. However, it is difficult to rationalize this pattern with measurement errors alone since it implies that the errors in GDP would have to be increasingly severe as incomes rise if the constant elasticity model were correct.

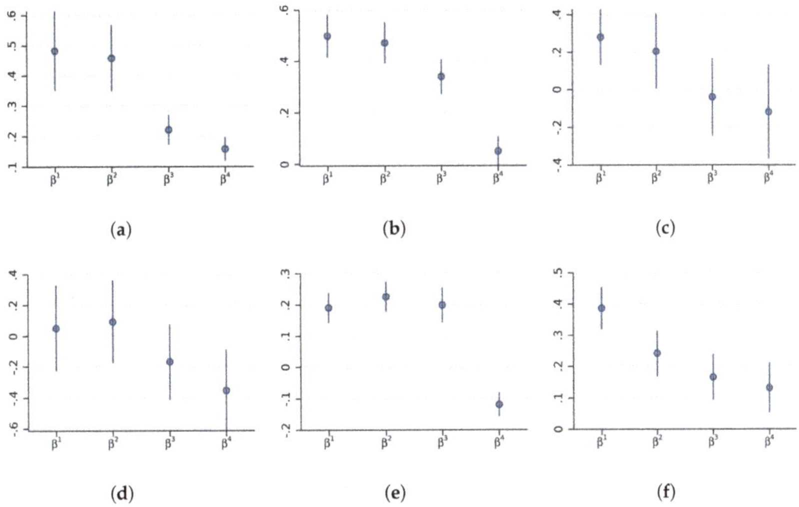

Figure 2. Income elasticity of nightlights for samples split by quartiles of average GDP. (**a**) USA, (**b**) Germany, (**c**) Italy, (**d**) Spain, (**e**) Brazil, (**f**) China.

Figure 2c,d report the results for Italy and Spain which, like Germany, use EU reporting standards but had somewhat lower statistical capacity and a larger informal sector in the 1990s. For Italy, the estimated elasticities are lower at every split of the data (consistent with greater attenuation throughout) but they also show a decreasing pattern as incomes rise. For Spain, we find elasticities that are indistinguishable from zero for the first two GDP quartiles and even negative in the highest quartile. While some of the lack of statistical significance is likely due to having few cross-sectional units for Italy and Spain, the pattern is consistent with the findings for the United States and Germany.

Figure 2e,f present the estimates for the two emerging economies in our data set where measurement error in GDP likely plays a larger role. The estimates for Brazil suggest that the relationship is approximately constant (with estimates around 0.2) until we reach the highest quartile of aggregate income (where we estimate a negative elasticity of −0.119). Somewhat remarkably, the estimates for Chinese prefectures follow the same decreasing pattern of the US counties or German districts, and even have comparable magnitudes ($\hat{\beta}^1 = 0.386$ and $\hat{\beta}^4 = 0.132$).

Figure 3 plots the implied relative attenuation effects and their 95% confidence intervals, or more specifically, the ratio of the coefficient (β^k) estimated for each GDP quartile to the coefficient estimated in the data up to the first quartile (β^1). The first ratio θ^1 is equal to one by construction, and we are interested in substantively large and statistically significant deviations from one in the other three quarters of the data.

Figure 3a,b show the results for the United States and Germany. As a result of the similar coefficient estimates up to the median of average GDP, we find no evidence suggesting that the signal-to-noise ratios differ in these two quarters of the data (see $\hat{\theta}^2$s). However, we estimate large differences the more we move up in the distribution of average GDP. For the United States, the coefficient in the third (fourth) GDP quarter would need to be more than two (three) times as attenuated as the coefficient estimated up to the first GDP quartile. For Germany, the ratio of coefficients in the third to the first quartile implies that the former is about 1.5 times more attenuated than the latter, while our estimate of $\hat{\theta}^4$ is about 0.1, suggesting that the coefficient in the highest quartile is 10 times more attenuated than that of the lowest quartile if the structural elasticity were truly constant. Given the high quality of the GDP data in these two countries and the fact that we have sufficient units in each group to estimate the uncertainties of these ratios relatively precisely, we consider these implied differences in measurement errors across income groups too large to be plausible. We are not aware of studies suggesting that such differences in measurement errors across subnational units are likely, especially since regional accounts in the United States and Germany are developed by a single federal authority (the Bureau of Economic Analysis for United States and the Federal Statistical Office of Germany). Moreover, the suggested pattern of increasing measurement error with higher incomes appears unlikely. Instead, these results strongly suggest that the structural elasticity declines as GDP rises.

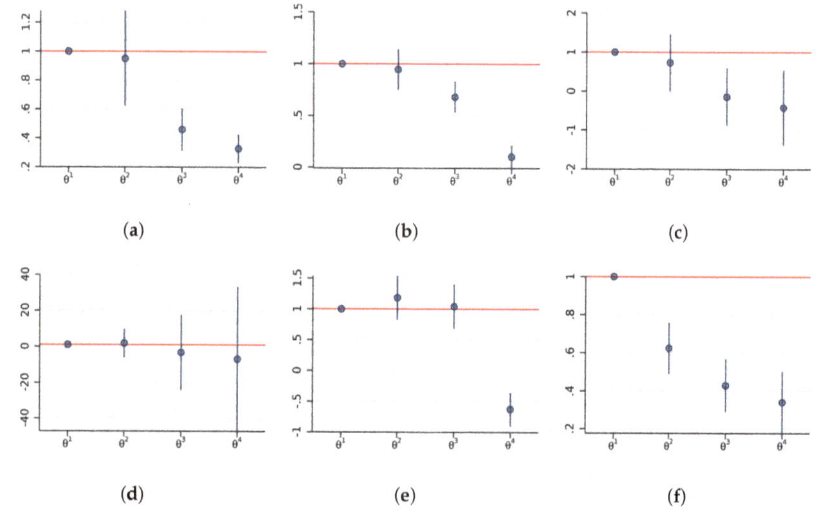

Figure 3. Estimates of relative measurement errors for samples split by quartiles of average GDP. (**a**) USA, (**b**) Germany, (**c**) Italy, (**d**) Spain, (**e**) Brazil, (**f**) China.

Figure 3c–f repeat this exercise with data from Italy, Spain, Brazil and China. The estimates for Italy and Spain also suggest that measurement errors rise with GDP but the confidence interval of these ratios are comparatively wide. For Italy the upper bound of the 95% confidence intervals for $\hat{\theta}^4$ is 0.546 and the confidence intervals for Spain always include one. We find near constancy in Brazil up to the third quartile but then the estimate of $\hat{\theta}^4$ falls below zero which cannot be meaningfully interpreted in terms of relative attenuation factors that are theoretically bounded by zero from below. The estimates for China mirror the results for the United States and Germany. They suggest that if the constant elasticity model were correct, measurement errors would have to be strictly increasing in average GDP. Moreover, the coefficient in the highest quarter would have to be close to three times more attenuated than that in the first.

An important robustness check, given that we observe this decreasing pattern of elasticities, is whether these differences are simply driven by topcoding in the lights data. If topcoding is severe, then changes in GDP do not translate into changes in observed lights beyond some threshold of GDP and this effect might gradually become more pronounced as more pixels in a region reach the topcoding threshold [21]. Figure S1 in the Supplementary Materials repeats this analysis using the topcoding corrected data [23]. While this somewhat moderates the steep decline of the estimated elasticities in the last quarter of the data, we still observe the same pattern of declining coefficients across all four quarters and estimate relative measurement errors that too large to be plausible. For example, $\hat{\theta}^4$ is around 0.485

in the United States and 0.500 in Germany, suggesting that the coefficients would still need to be twice as attenuated in the fourth quartile of GDP than in the first for the constant elasticity model to be correct.

Another concern is that overglow and geolocation errors lead to light being recorded in pixels adjacent to a light source, which can cause significant measurement errors when regions are small. This is more likely in urban areas since they generate more overglow and cities are often their own (geographically small) administrative region. Since we are studying changes in light intensities within regions and control for year fixed effects, higher baseline levels of overglow in urban regions are not a concern. Bias would arise if the change in overglow over time rose differently in regions with higher GDP. However, this bias would work in the opposite direction of the nonlinearity we observe (by driving up the change in observed light in periurban regions for the same change in GDP). Spillovers in light to adjacent regions, however, do create spatial correlation in the error terms even after netting location and time fixed effects. Figure S2 in the Supplementary Materials re-estimates the relative attenuation factors allowing for spatial correlation in the error terms up to 500 km (and arbitrary correlation within units over time). Only some of the confidence intervals increase marginally and none of our substantive findings are affected.

3.2. Analysis of Interactions with Population Density

We now move to investigate the nonlinearity in the relationship using the regression framework in Equation (2). Panel A in Table 3 shows the income elasticity of nightlights by country, using nightlight density per square kilometer and GDP density per square kilometer. Regressions in Panel B show the variation in the elasticity across levels of population density in the initial year. The logarithm of population density has been normalized such that the average for each country is set to zero, meaning that the coefficients on GDP represent the elasticities at the mean population density for the country.

The results in Panel A show a statistically significant association between GDP and nightlights in four of the six countries, with the highest elasticity in the U.S., followed by Germany and China, and Brazil having the lowest significant coefficient at 0.1. The U.S. elasticity of 0.4 indicates that a 10% increase in a county's GDP is associated with a 4% increase in nightlights. Italy and Spain have insignificant coefficients with the largest standard errors (notably, these countries have the smallest number of subnational regions, a point we will return to in Section 3.4). Of course, these estimates are biased downward by measurement errors in GDP and we make no attempt to assess the absolute size of these errors. The results in Panel B show that the income elasticity at the mean population density is similar (though slightly smaller) than the uninteracted estimates in Panel A. In the case of US counties and German districts, the income elasticity of nightlights is around 0.3 at the mean population density. In all countries, the interaction with population density is negative and significant, indicating that the effect of increasing GDP on nightlights becomes smaller at higher population densities. The magnitudes of the coefficient on the interaction term suggests that the effect of GDP changes on nightlights approaches zero at population densities around 3–4 times higher than the mean (for the U.S. and China), and around double the population density mean in Germany and Brazil. While the interaction terms are significant in the case of Italy and Spain, the noisy estimates of the baseline GDP coefficient makes interpretation difficult over the upper half of the population density domain. For the two emerging economies in our sample, Brazil and China, we find lower elasticities (0.06 and 0.2, respectively) at the mean population density but they also exhibit variation in the elasticity across levels of population density that is statistically significant at all conventional levels. In sum, the evidence in Panel B suggests that there is some cross-country variation in the elasticity of nighttime lights with respect to GDP at mean population density (as would be expected when measurement error differs across countries) but also strong evidence that this elasticity varies with population density within all six countries.

Table 3. Income elasticity of nighttime lights by population density.

	(1) USA	(2) Germany	(3) Italy	(4) Spain	(5) Brazil	(6) China
			Panel A: Real GDP density without interaction			
GDP	0.405 ***	0.348 ***	0.096	−0.092	0.100 ***	0.292 ***
	(0.043)	(0.034)	(0.081)	(0.124)	(0.015)	(0.042)
Regions	3080	392	110	58	5569	342
Observations	40,039	8624	2420	1276	66,398	4802
			Panel B: Real GDP density interacted with Pop Dens			
GDP	0.278 ***	0.291 ***	0.072	−0.037	0.061 ***	0.195 ***
	(0.026)	(0.025)	(0.058)	(0.126)	(0.013)	(0.037)
GDP × PopDens	−0.094 ***	−0.206 ***	−0.275 ***	−0.165 ***	−0.108 ***	−0.058 ***
	(0.019)	(0.015)	(0.032)	(0.031)	(0.007)	(0.009)
Regions	3080	392	110	58	5569	342
Observations	40,039	8624	2420	1276	66,398	4802

Notes: Nightlights, GDP, and population density are all in logarithms. GDP refers to GDP density (GDP divided by region area) in constant prices. All regressions use the sum of lights divided by region area as the dependent variable. Panel A shows the constant elasticity model, while Panel B interacts GDP with the population density in the first year that region is included in the regression. All regressions include region and year fixed effects. Standard errors in parenthesis are clustered at the region level, and significance levels are denoted at conventional levels *** $p < 0.01$.

The implied difference in the elasticity across the observed range of population density is economically significant. In the case of the United States, the elasticity at the 10th percentile of population density is 0.489 (95% CI is 0.387–0.587) while the elasticity at the 90th percentile is 0.09 (0.006–0.175). The elasticity in Germany at the 10th percentile of population density is 0.531 (0.465–0.598) while at the 90th percentile it is −0.053 (−0.115–0.008). These estimates are very close to the estimates by the income group reported earlier, suggesting that average incomes and initial population density capture similar variation across groups of regions.

Figure S3 in the Supplementary Materials shows the income elasticity by population density. The elasticity declines in all countries as population density rises, reaching zero at the highest population density levels (it rarely becomes negative and negative estimates are supported by very few observations). Table S1 shows that the patterns are the same when using the topcoding corrected lights, indicating that topcoding in the DMSP data is not driving the variation in the relationship we are documenting here.

3.3. Analysis by Economic Sector

Having documented variation in the income elasticity of nighttime lights across initial levels of population density, we proceed to explore whether the elasticity varies systematically by economic structure. Measurement errors still bias these elasticities downwards, so that we continue to give more weight to the results from the United States and Germany where statistical capacity and reporting quality is highest. We also note that this analysis cannot be run for China given that we do not have data on sectoral composition for all prefectures.

We first estimate the overall elasticities by sector without explicitly allowing for variation across regions. We note that the number of regions in the United States decreases from those in Table 3 and varies across sectors because some counties do not report the share of GDP across economic sectors (for privacy reasons if the sector is too small). The results are shown in Table 4. We find that the income elasticity of nightlights is smallest for agricultural activity (Panel A), higher for industrial and construction activity (Panel B), and highest for the service sector (Panel C) in the United States, Germany, and Brazil. In the case of the United States, for example, the elasticity for agricultural GDP is close to zero (0.01), 0.15 for industrial GDP, and 0.46 for the service sector. Italy follows this pattern for the first two sectors but not services (where the estimate has a wide 95% confidence interval which includes 0 but also elasticities up to around 0.18). As above, Italy and Spain result in the noisiest estimates, likely for reasons we will explore in the next section. The difference in magnitudes across economic sectors is also telling: if the constant elasticity model were correct, the 45-to-1 difference in magnitude between the service and the agricultural sector in the United States would have to be due to differences in measurement error between sectors. In unreported results, the estimates change somewhat when we include all sectors at the same time but the differences in magnitudes across sectors remains similar. While variation in measurement error across sectors is plausible even in high statistical capacity countries, such high ratios in the United States or Germany suggest that the structural elasticity varies by sector.

Table 4. Income elasticity of nighttime lights across economic sectors.

	(1) USA	(2) Germany	(3) Italy	(4) Spain	(5) Brazil
			Panel A: Agricultural GDP density		
GDP	0.008 ***	−0.039 ***	0.006	0.069 **	0.012 **
	(0.002)	(0.011)	(0.020)	(0.030)	(0.005)
Regions	2982	392	110	58	5567
Observations	31,428	8624	2420	1276	66,372
			Panel B: Industrial GDP density		
GDP	0.146 ***	0.157 ***	0.073 *	0.014	0.058 ***
	(0.017)	(0.016)	(0.037)	(0.077)	(0.006)
Regions	2939	392	110	58	5569
Observations	30,681	8624	2420	1276	66,363
			Panel C: Service sector GDP density		
GDP	0.458 ***	0.315 ***	0.034	−0.085	0.140 ***
	(0.065)	(0.035)	(0.078)	(0.121)	(0.015)
Regions	2872	392	110	58	5569
Observations	27,608	8624	2420	1276	66,396

Notes: Nightlights and GDP are in logarithms. All regressions use the sum of lights divided by region area as the dependent variable. Panel A uses GDP in the agricultural sector (multiplying regional GDP by the share of agriculture in GDP) in constant prices, divided by region area to produce a GDP density. Panel B uses real GDP density in the industrial and construction sectors, and Panel C uses real GDP density in the service sector. All regressions include region and year fixed effects. For all regressions, standard errors in parenthesis are clustered at the region level. Significance levels denoted at conventional levels *** $p < 0.01$, ** $p < 0.05$, * $p < 0.1$.

Table 5 shows the results from adding an interaction with initial population density. The regressions are based on same specification in Equation (2) above, now separating out the analysis for agricultural GDP (Panel A), industrial GDP including construction (Panel B), and service sector GDP (Panel C). The results show that the variation in the elasticity across population density is evident in all countries and nearly all sectors. The United States and Germany exhibit curvature in all three sectors, while four of five countries show curvature in industrial GDP. The curvature tends to be strongest in the services sector, where the size of estimated interaction terms is at least half of the elasticity at mean population density but often considerably larger and statistically significant for all countries. The results for the service sector most closely approximate the results in Table 3, which is not surprising given that services are the dominant sector in all of these economies.

3.4. Spatial Aggregation in the United States & Brazil

Our results so far illustrate that the structural relationship between nighttime lights and GDP varies with level of GDP, population density, and industrial composition. Since nightlights are used to proxy for economic activity at multiple subnational scales, we proceed to explore how the structural relationship between nightlights and GDP varies according to the size and number of subnational partitions in a country. Specifically, we test whether partitioning a country into subnational regions of different shapes and sizes affects how nightlights respond to increases in economic activity. We use disaggregated data from two countries, the United States and Brazil, that are at opposite ends of the income spectrum covered by our sample but are physically large and exhibit a varied pattern of regional economic specialization. Figure 4 shows an example of how we randomly partition the United States and Brazil into 50 administrative units.

Table 5. Income elasticity of nighttime lights by population density and by economic sector.

	(1) USA	(2) Germany	(3) Italy	(4) Spain	(5) Brazil
	Panel A: Agricultural GDP density				
GDP	0.006 ***	−0.011	0.001	0.028	0.011 **
	(0.002)	(0.012)	(0.022)	(0.036)	(0.005)
GDP × PopDens	−0.005 ***	−0.035 ***	0.054 **	0.033 **	−0.007 *
	(0.001)	(0.009)	(0.023)	(0.013)	(0.004)
Regions	2982	392	110	58	5567
Observations	31,428	8624	2420	1276	66,372
	Panel B: Industrial GDP density				
Real GDP	0.113 ***	0.159 ***	0.088 **	0.064	0.045 ***
	(0.012)	(0.016)	(0.042)	(0.072)	(0.005)
GDP × PopDens	−0.045 ***	−0.046 ***	0.043	−0.084 ***	−0.033 ***
	(0.012)	(0.013)	(0.042)	(0.019)	(0.004)
Regions	2939	392	110	58	5569
Observations	30,681	8624	2420	1276	66,363
	Panel C: Service sector GDP density				
Real GDP	0.232 ***	0.237 ***	0.036	−0.038	0.105 ***
	(0.032)	(0.031)	(0.047)	(0.122)	(0.013)
GDP × PopDens	−0.115 ***	−0.197 ***	−0.231 ***	−0.167 ***	−0.118 ***
	(0.029)	(0.013)	(0.022)	(0.026)	(0.007)
Regions	2872	392	110	58	5569
Observations	27,608	8624	2420	1276	66,396

Notes: Nightlights, GDP and population density are all in logarithms. All regressions use the sum of lights divided by region area as the dependent variable. Panel A uses GDP in the agricultural sector (multiplying regional GDP by the share of agricultural share in GDP), divided by region area to produce a GDP density. Panel B uses real GDP density in the industrial and construction sectors, and Panel C uses real GDP density in the service sector. PopDens refers to the population density in the first year that region is included in the regression. All regressions include region and year fixed effects. For all regressions, standard errors in parenthesis are clustered at the region level. Significance levels denoted at conventional levels *** $p < 0.01$, ** $p < 0.05$, * $p < 0.1$.

Figure 5 shows the distribution of estimates of Equation (2) for 16,200 simulations of partitions of the United States and 28,000 simulations of partitions of Brazil over which we aggregate nighttime lights, GDP, and population. For a given number of total units (ranging from 50–3000 for the United States and 50–5400 for Brazil), we simulate 1000 partitions in which the spatial aggregation of original units is based on a unique random starting seed in each iteration. We report results for the elasticity at the mean of population density ($\hat{\beta}$) and the interaction with population density ($\hat{\gamma}$) for the United States in Figure 5a,b and Brazil in Figure 5c,d. We observe that the overall average elasticity (at mean population density) is around 0.3 in the case of the US and 0.053 in the case of Brazil. Both of these values are close to the regression results presented in columns (1) and (5) of Table 3. For the US, almost every single estimated value using simulated partitions is larger than zero (99.97% of simulations) and does not contain zero in its 95% confidence interval (98.23%), while we estimate elasticities below zero only in 4.1% of the sampled partitions in the case of Brazil.

We find strong evidence of nonlinearity in nearly all simulated partitions. Figure 5b,d illustrate that the average of all simulated interaction coefficients is −0.104 for the US and −0.11 for Brazil (again close to Table 3). Less than 1% of the US estimates are positive and only 8.7% of the estimates include zero in their 95% confidence interval. Similarly, only 9 out of 28,000 simulations for Brazil yield a positive interaction coefficient and only about 1% cannot be distinguished from zero.

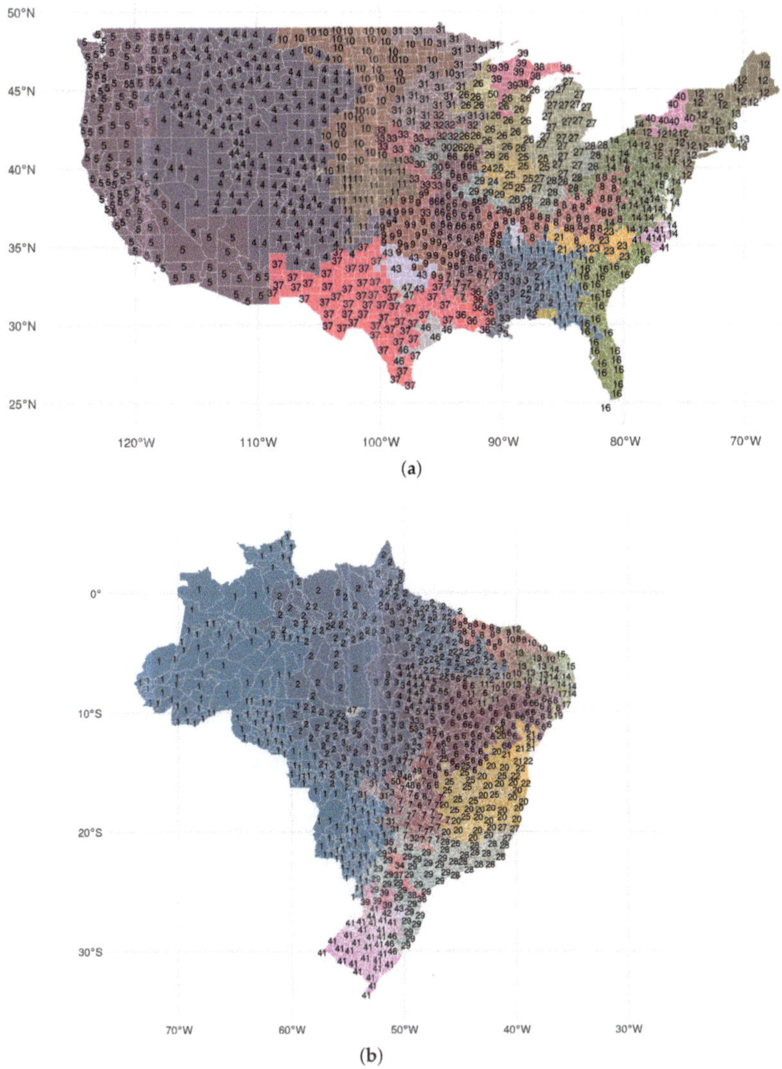

Figure 4. Examples of spatially contiguous random partitions. (**a**) USA: Partition 1 for 50 administrative units, (**b**) Brazil: Partition 1 for 50 administrative units.

Figure 6 presents results conditional on the scale of spatial aggregation, that is, the number of administrative units. The random aggregation is simulated 1000 times for each total number of administrative units. We present the regression results as box-and-whisker plots (outliers outside the lower and upper adjacent values are indicated as red dots). Several features stand out. The elasticity at the mean of population density, shown in Figure 6a,c, is remarkably stable in both countries (a result which carries over to specifications without an interaction term, see Figure S4 in the Supplementary Materials). The conditional means fluctuate only moderately around the overall means documented above. While the mean is stable, the variance of the estimates increases markedly as the number of partitions become smaller and the average size of each unit increases. For relatively coarse levels of spatial aggregation, it is not difficult to find partitions at which we observe no relationship between nighttime lights and GDP over time. For example, at 50 artificial US "states", large parts of the distribution of estimates cross zero and around 31% of the underlying coefficients contain zero in their 95% confidence interval, while almost every single draw for partitions of size 200 and more yields a coefficient for which we can reject the null hypothesis of zero. We observe a similar relationship in the data for Brazil, where around 89% of all estimated coefficients with partitions using 50 units have zero in their 95% confidence intervals but this figure drops to less than 10% at 2800 units and less than 1% at 3400 units.

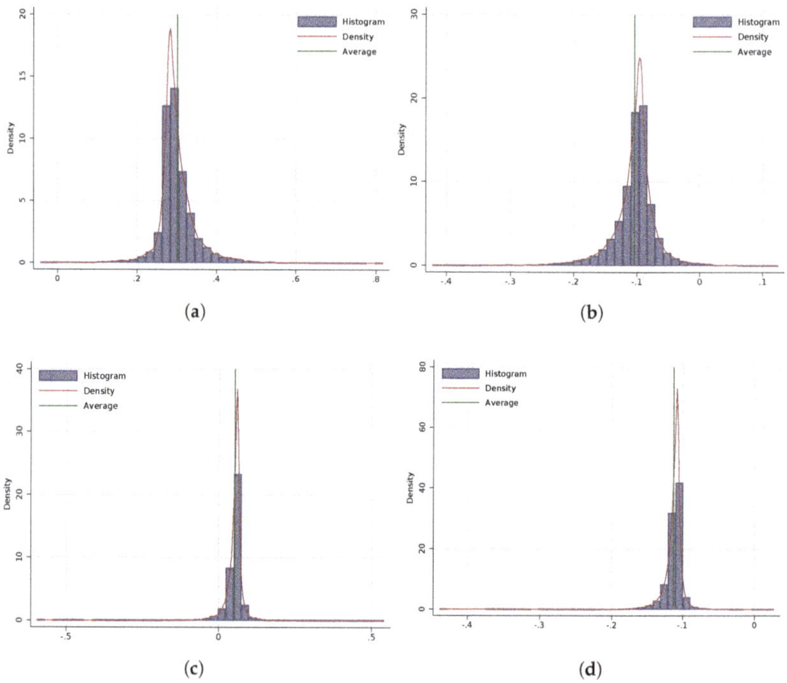

Figure 5. Unconditional distribution of estimates across partition simulations. (**a**) USA: Elasticity at average ($\hat{\beta}$), (**b**) USA: Interaction ($\hat{\gamma}$), (**c**) Brazil: Elasticity at average ($\hat{\beta}$), (**d**) Brazil: Interaction ($\hat{\gamma}$).

The interaction effects in Figure 6b,d follow the same pattern of stable mean and increasing variance as the number of administrative units falls. A total of 59.7% of the partitions of the United States into 50 "states" yield estimates of the interaction term that cannot be distinguished from zero. This is particularly interesting in light of an estimate of -0.23 (with a standard error of 0.05) which we obtain when running our baseline regression on data aggregated to the *actual* US states. This value sits at the 10th percentile of the distribution of simulated results for partitions into 50 units and is more than 2.5 times larger than the average elasticity across all simulations. At 200 units, 37.4% contain zero in their 95% confidence interval, and at 800 units, this falls to 5.5%.

Taken together, these simulations demonstrate a robust association of lights with GDP and provide strong evidence of nonlinearity across different geographic scales. As aggregation reduces measurement errors, we take this as another indication that measurement errors are not the driving force behind this nonlinearity. Moreover, we show that it is easy to obtain insignificant results at high levels of aggregations, where the estimated elasticities depend on how larger regions align with the spatial structure of production and density.

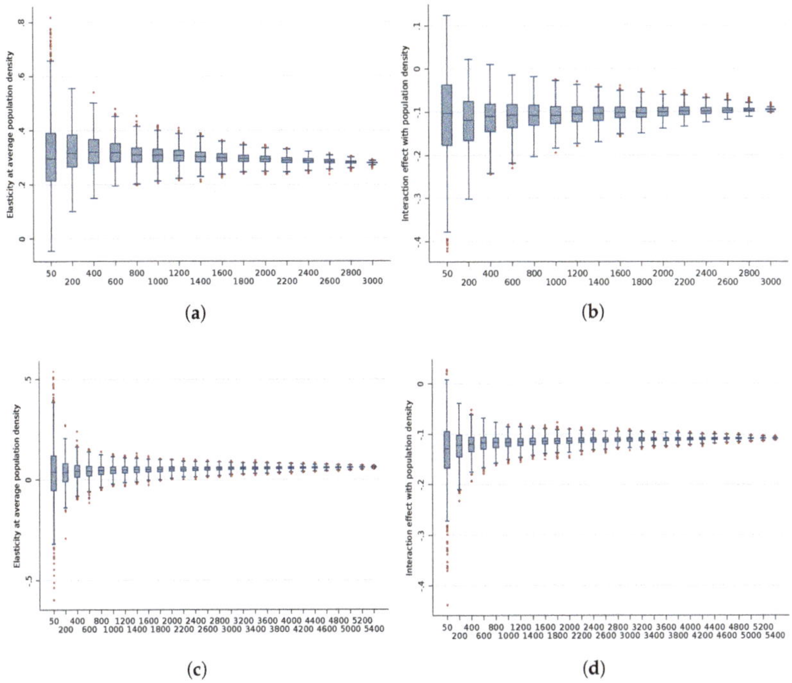

Figure 6. Distribution of estimates across partition simulations, by number of units. (**a**) USA: Elasticity at average ($\hat{\beta}$), (**b**) USA: Interaction ($\hat{\gamma}$), (**c**) Brazil: Elasticity at average ($\hat{\beta}$), (**d**) Brazil: Interaction ($\hat{\gamma}$).

4. Discussion

This article investigates whether the ability of the nighttime lights to pick up changes in GDP varies across subnational regions and their characteristics. We show theoretically that any variation in the structural light-output relationship spills over into estimating policy-relevant parameters in applied work using nightlights as an outcome. Measurement errors in both nightlights and economic activity complicate inference about this relationship. We develop a framework for assessing the implied, relative magnitude of measurement errors across subnational regions, and apply it to study heterogeneity in the light-output relationship in several countries with varying degrees of statistical capacity.

Our findings document significant variation in the relationship between economic activity and nightlights at the subnational level which cannot be explained by variation in measurement errors alone. Variation in the elasticity persists whether we estimate the elasticities by income group, economic sector, or specify interactions with population density. Moreover, the elasticities with respect to industry and service GDP declines as population density rises. The elasticity in the agricultural sector is much smaller (in the United States it is 20 times smaller than the industry elasticity and over 50 times smaller than the service sector elasticity) and less consistent across countries. Since services dominate the economies we study, pooling across sectors results in the nonlinearity by population density exhibited in the service sector. The evidence favoring these nonlinearities is most robust in countries with the highest statistical capacity. However, the relationship follows a similar, albeit noisier, pattern in the other countries in our sample.

The second contribution of our study is on the stability of the nonlinear nightlight-GDP relationship over different levels of spatial aggregation. We find that the nonlinear relationship is remarkably stable over many alternative administrative divisions of varying shapes and sizes. However, our estimates of the GDP elasticities are more stable under random partitions where the total number of regions is large (keeping their size small).

At the same time, these smaller regions exhibit stronger nonlinearity since smaller units are more homogeneous in terms of population density, economic activity, and economic structure. At larger geographies, estimates are less stable, and there is a higher likelihood of drawing statistically insignificant estimates. This may help explain why other studies have failed to find significant relationships between nightlights and economic activity at some levels of aggregation [18]. It is also worth noting that the less precise estimates for Italy and Spain in our analysis are consistent with these two countries having the fewest number of regions.

We note that our findings provide a framework for reconciling other empirical research. Papers documenting the relationship between nightlights and economic activity at subnational scale have not found a consistent statistical relationship [18], and at smaller scales there is evidence for nonlinearity as well [2,7]. Our work helps scholars adjudicate between measurement error and structural nonlinearity in explaining observed nonlinearity in the NTL-GDP elasticity. We offer observable covariates such as population density, GDP density, and economic structure as predictors for the local NTL-GDP elasticity. We also show that larger subnational scales exhibit larger variation in the estimated elasticity.

Researchers using nightlights as a proxy for economic activity at small geographies, for example, to study the effects of conflict or measure economic inequality, need to be conscious of the variation we find in this paper. The income elasticity of nighttime lights may be considerably smaller in agricultural regions, regions with higher GDP, and regions with high population density. Changes in economic activity in such areas may result in small changes in nightlights that are, perhaps, not distinguishable from zero. As a result, a researcher may erroneously conclude that a policy or investment did not affect economic growth or inequality because the change in nightlights is insignificant (a null finding, of course, which includes a wide range of potential effects sizes in addition to zero). Moreover, even when the policy has an effect, additional analyses of treatment effect heterogeneity across regions may be driven solely by the heterogeneity of the light-output elasticity we document in this paper. The nonlinearity also implies that large changes in nightlights might occur in some contexts despite little change in economic activity.

In addition, our results suggest caution in taking estimates from other contexts to infer how changes in nightlights in a particular location translate into changes in GDP. The presence of a nonlinear relationship between nightlights and economic activity across subnational regions of different population densities, GDP levels, or economic structures means that researchers should consider whether taking an elasticity from the literature and applying it to a specific empirical context is appropriate. While such an elasticity might reflect the average relationship between nightlights and changes in economic activity over many subnational units, research focused on one or a small number of regions should rely on nightlights-GDP elasticities estimated from regions with similar characteristics and have some sense of the scale of measurement errors in their context. Moreover, research on agricultural, high GDP, or high population density settings may want to examine alternative proxies of economic activity and not make inferences only from nightlights.

While our work documents significant variation in the nightlights-income relationship, it has some fundamental limitations. First, we can only explore the relationship in countries with a sufficiently long panel of subnational GDP data. Our work considers a set of high-income and middle-income countries. It would have been ideal also to have data from low-income countries with robust statistical capacity that report subnational GDP because the research we are trying to inform is most often conducted in low-income settings where nightlights are one of the few available proxies for subnational GDP.

A second challenge is that the actual degree of measurement error in GDP at the subnational level is unknown. We use the elasticities across regions in countries with the highest statistical capacity—Germany and the United States—to conclude that the range is too wide to be explained by differences in measurement errors in GDP across regions. However, we can only infer this from the high statistical capacity of these countries and have no way to verify the true differences in measurement error in subnational GDP in any country. Moreover, we know very little about the scale of errors introduced by standard approaches to calculating (and usually scaling up) estimates of local GDP.

Finally, our study uses data on nighttime lights derived from a system with many known limitations. The main advantage of the dated DMSP-OLS system over the newer VIIRS data is that it allows us to study changes within regions over one or two decades, depending on the availability of GDP data. While we use a topcoding corrected version of this data to shut down one of the most likely sources of nonlinearity (which correlates with population density and GDP), we cannot entirely rule out that unfortunate features of the data generating process contribute to our findings.

Supplementary Materials: The following are available online at https://www.mdpi.com/article/10.3390/rs14051190/s1, Table S1: Income elasticity of topcoding corrected nighttime lights by population density. Figure S1: Estimates of relative measurement errors using topcoding corrected nighttime lights. Figure S2: Estimates of relative measurement errors using spatially correlated standard errors. Figure S3: Income elasticity of nighttime lights by log population density. Figure S4: Unconditional and conditional distribution of estimated elasticities across simulated partitions.

Author Contributions: R.B. and G.C.M. contributed equally to this work. All authors have read and agreed to the published version of the manuscript.

Funding: Richard Bluhm acknowledges financial support from the Alexander von Humboldt Foundation.

Data Availability Statement: Data and code are available at www.github.com/rbluhm/ntl_gdp (accessed on 22 February 2022).

Acknowledgments: We are grateful to two anonymous referees for providing thoughtful comments on the manuscript. Moreover, we have greatly benefited from discussions with Denis de Crombrugghe on the measurement error framework and have received helpful comments from Prashant Bharadwaj and Teevrat Garg.

Conflicts of Interest: The authors declare no conflict of interest.

Appendix A. Analytic Background

This section derives in detail some of the equations used in the main text and provides the relevant background for our estimation strategy. All true (unobserved) quantities are denoted by '*' and all observed variables are demeaned across units and time.

Appendix A.1. Estimating the Relative Sizes of the Measurement Errors Using Single Regressions

We first formalize the set up described in the main text. Both observed lights and GDP are observed with error:

$$y_{it}^{G} = y_{it}^{G*} + e_{it}^{G} \tag{A1}$$

$$y_{it}^{L} = y_{it}^{L*} + e_{it}^{L} \tag{A2}$$

In both equations, we assume that the the measurement errors are mean independent conditional on the values of the latent variable(s) and that there is no systematic bias in observed GDP or lights, such that $\mathbb{E}[e_{it}^{G}|y_{it}^{G*}] = \mathbb{E}[e_{it}^{L}|y_{it}^{L*}] = 0$,

Our structural model of interest is the constant elasticity model:

$$y_{it}^{L*} = \beta y_{it}^{G*} + \epsilon_{it} \tag{A3}$$

which is the relationship that is typically assumed in the related literature on measurement error in the light-output relationship and applied papers using a constant elasticity to translate their effects from changes in nighttime lights to changes in GDP. By the same logic of the assumptions made above, we require $\mathbb{E}[e_{it}^{G}\epsilon_{it}|y_{it}^{G*}] = \mathbb{E}[e_{it}^{L}\epsilon_{it}|y_{it}^{G*}] = 0$.

In addition, a key assumption in our analysis is that the error term of GDP is not correlated with the error term in lights conditional on unobserved GDP, i.e., $\mathbb{E}[e_{it}^{G}e_{it}^{L}|y_{it}^{G*}] = 0$. This assumption has been made in virtually all of the related literature on measurement errors in nighttime lights [2,3,26] but is not innocuous. While lights and GDP are quantified in very different ways (satellites vs. subnational accounting), their errors could still be correlated if, for example, topcoding in urban areas is (inversely) correlated with measurement errors in GDP. We use topcoding corrected data as a robustness check to rule out one potential source of such a correlation.

Substituting observed for unobserved quantities yields:

$$y_{it}^{L} = \beta y_{it}^{G} - \beta e_{it}^{G} + e_{it}^{L} + \epsilon_{it} \tag{A4}$$

which we estimate as

$$y_{it}^{L} = \beta y_{it}^{G} + v_{it} \tag{A5}$$

where the composite error $v_{it} = \epsilon_{it} - \beta e_{it}^{G} + e_{it}^{L}$ does not satisfy the usual assumptions for consistency. Instead, it is well known that

$$\hat{\beta} \xrightarrow{p} \beta \frac{\sigma_{y^{G*}}^{2}}{(\sigma_{y^{G*}}^{2} + \sigma_{e^{G}}^{2})} = \beta\lambda \tag{A6}$$

where $\sigma_{y^{G*}}^{2}$ is the variance true output and $\sigma_{e^{G}}^{2}$ the variance of the error in GDP. λ is known as the signal-to-noise ratio or the attenuation factor and, since $0 < \lambda < 1$, the bias is towards zero.

If we additionally assume that $\sigma_{e^{G}}^{2}$ behaves like a decreasing step-function in y^{G*}, then we should also observe decreasing attenuation factors if we group the data by observed GDP. This assumption mirrors the good data country and bad data country samples used in [3] but conjectures that the variance of these errors varies across subnational units (and indicates statistical capacity). If the constant elasticity assumption is correct, then the true β is the same in different sub-samples of the data. Therefore, any ratio of coefficients where each coefficient is estimated separately on ordered samples of y^{G*}, say high GDP and low GDP, identifies the relative size of the measurement errors:

$$\frac{\hat{\beta}^{h}}{\hat{\beta}^{l}} \xrightarrow{p} \frac{(\sigma_{y^{G*}}^{2} + \sigma_{e_{l}^{G}}^{2})}{(\sigma_{y^{G*}}^{2} + \sigma_{e_{h}^{G}}^{2})} = \frac{\lambda^{h}}{\lambda^{l}} = \theta^{h}. \tag{A7}$$

In the main text we present estimates of these ratios and their standard errors for samples split along quartiles of average GDP, denoted as $\theta^{k} = \lambda^{k}/\lambda^{1}$ for $k = 1, 2, \ldots, 4$. In fact, the precise shape of the conditional heteroskedasticity in $\sigma_{e^{G}}^{2}(y^{G*})$ is not important. Any difference in the error variance across different subsamples of GDP will translate into differences in the estimated θs. If we observe large and statistically significant deviations in these relative signal-to-noise ratios *within* countries with uniformly high statistical capacity, then we can interpret this as evidence against the constant elasticity model and in favor of variation in the structural elasticity.

Appendix A.2. Estimating the Coefficients in the Interacted Model

Suppose that instead of Equation (A3), the structural model of interest is

$$y_{it}^{L*} = \beta y_{it}^{G*} + \gamma y_{it}^{G*} z_{it} + \epsilon_{it} \tag{A8}$$

where z_{it} is observed without error. In our application, z_{it} is population density in the initial year, which is considerably easier to observe than GDP. Of course, with $\gamma = 0$ we are back to the constant elasticity model assumed earlier.

Substituting observed for unobserved quantities results in

$$y_{it}^{L} = \beta y_{it}^{G} - \beta e_{it}^{G} + \gamma y_{it}^{G} z_{it} - \gamma e_{it}^{G} z_{it} + e_{it}^{L} + \epsilon_{it} \tag{A9}$$

which we estimate as

$$y_{it}^{L} = \beta y_{it}^{G} + \gamma y_{it}^{G} z_{it} + v_{it} \tag{A10}$$

where the new composite error is $v_{it} = \epsilon_{it} - \beta e_{it}^{G} - \gamma e_{it}^{G} z_{it} + e_{it}^{L}$.

In general, estimates of coefficients in models with more than one independent variable where one or more of these variables are measured with error are biased in an unknown direction. To see this, consider the formulae for the regression coefficients estimated on the basis of Equation (A10):

$$\hat{\beta} = \frac{var(y_{it}^{G} z_{it}) cov(y_{it}^{L}, y_{it}^{G}) - cov(y_{it}^{G} z_{it}, y_{it}^{G}) cov(y_{it}^{L}, y_{it}^{G} z_{it})}{var(y_{it}^{G}) var(y_{it}^{G} z_{it}) - (cov(y_{it}^{G} z_{it}, y_{it}^{G}))^2} \tag{A11}$$

$$\hat{\gamma} = \frac{var(y_{it}^{G}) cov(y_{it}^{L}, y_{it}^{G} z_{it}) - cov(y_{it}^{G}, y_{it}^{G} z_{it}) cov(y_{it}^{L}, y_{it}^{G})}{var(y_{it}^{G}) var(y_{it}^{G} z_{it}) - (cov(y_{it}^{G}, y_{it}^{G} z_{it}))^2} \tag{A12}$$

Typically, both $\hat{\beta}$ and $\hat{\gamma}$ depend on the true β, γ and several (co)variances, so that the direction of the bias is indeterminate.

In our case, however, it could be plausible that $cov(y_{it}^{G}, y_{it}^{G} z_{it}) = 0$, if we are willing to make the stronger assumptions that e_{it}^{G} is independent of y_{it}^{G*}, z and the structural error, and $cov((y_{it}^{G*})^2 z_{it}) = 0$ (which restricts the relationship between population density and unobserved GDP to be linear).

It is straightforward to see that under these conditions

$$\hat{\beta} \xrightarrow{p} \beta \frac{\sigma_{y^{G*}}^2}{(\sigma_{y^{G*}}^2 + \sigma_{e^G}^2)} \tag{A13}$$

which is the same as Equation (A6) and does not depend on the value of γ.

Similarly, we have

$$\hat{\gamma} \xrightarrow{p} \gamma \frac{\sigma_{y^{G*}z}^2}{(\sigma_{y^{G*}z}^2 + \sigma_{e^G}^2 \sigma_z^2)} \tag{A14}$$

which is attenuated by the signal-to-noise ratio in the product of observed GDP with population density. In the constant elasticity case, $\gamma = 0$, so $\hat{\gamma} \xrightarrow{p} 0$. Hence, any evidence rejecting the null of $\gamma = 0$ suggests that the constant elasticity model is incorrect (or that one or more of the assumptions made here is violated). Again, it is important to note that these results rely on substantially stronger assumptions than the single regression estimates presented above. We only view the resulting estimates as additional (and weaker) evidence on the whether the constant elasticity model is empirically supported.

References

1. Elvidge, C.D.; Baugh, K.E.; Kihn, E.A.; Kroehl, H.W.; Davis, E.R.; Davis, C.W. Relation between satellite observed visible-near infrared emissions, population, economic activity and electric power consumption. *Int. J. Remote Sens.* **1997**, *18*, 1373–1379. [CrossRef]
2. Chen, X.; Nordhaus, W.D. Using luminosity data as a proxy for economic statistics. *Proc. Natl. Acad. Sci. USA* **2011**, *108*, 8589–8594. [CrossRef] [PubMed]
3. Henderson, J.V.; Storeygard, A.; Weil, D.N. Measuring economic growth from outer space. *Am. Econ. Rev.* **2012**, *102*, 994–1028. [CrossRef] [PubMed]
4. Hodler, R.; Raschky, P.A. Regional Favoritism. *Q. J. Econ.* **2014**, *129*, 995–1033. [CrossRef]
5. Storeygard, A. Farther on down the Road: Transport Costs, Trade and Urban Growth in Sub-Saharan Africa. *Rev. Econ. Stud.* **2016**, *83*, 1263–1295. [CrossRef]
6. Jean, N.; Burke, M.; Xie, M.; Davis, W.M.; Lobell, D.B.; Ermon, S. Combining satellite imagery and machine learning to predict poverty. *Science* **2016**, *353*, 790–794. [CrossRef]

7. Asher, S.; Lunt, T.; Matsuura, R.; Novosad, P. Development Research at High Geographic Resolution: An Analysis of Night-Lights, Firms, and Poverty in India Using the SHRUG Open Data Platform. *World Bank Econ. Rev.* **2021**, *35*, 845–871. [CrossRef]
8. Kocornik-Mina, A.; McDermott, T.K.J.; Michaels, G.; Rauch, F. Flooded Cities. *Am. Econ. J. Appl. Econ.* **2020**, *12*, 35–66. [CrossRef]
9. Lee, Y.S. International isolation and regional inequality: Evidence from sanctions on North Korea. *J. Urban Econ.* **2018**, *103*, 34–51. [CrossRef]
10. Cook, C.J.; Shah, M. Aggregate Effects from Public Works: Evidence from India. *Rev. Econ. Stat.* **2020**, 1–38. [CrossRef]
11. Shenoy, A. Regional development through place-based policies: Evidence from a spatial discontinuity. *J. Dev. Econ.* **2018**, *130*, 173–189. [CrossRef]
12. Weidmann, N.B.; Schutte, S. Using night light emissions for the prediction of local wealth. *J. Peace Res.* **2017**, *54*, 125–140. [CrossRef]
13. Bruederle, A.; Hodler, R. Nighttime lights as a proxy for human development at the local level. *PLoS ONE* **2018**, *13*, 1–22. [CrossRef] [PubMed]
14. Henderson, J.V.; Squires, T.; Storeygard, A.; Weil, D. The global distribution of economic activity: Nature, history, and the role of trade. *Q. J. Econ.* **2018**, *133*, 357–406. [CrossRef]
15. Falchetta, G.; Pachauri, S.; Parkinson, S.; Byers, E. A high-resolution gridded dataset to assess electrification in sub-Saharan Africa. *Sci. Data* **2019**, *6*, 110. [CrossRef]
16. Alder, S. Chinese Roads in India: The Effect of Transport Infrastructure on Economic Development. 2016. Available online: https://ssrn.com/abstract=2856050 (accessed on 1 February 2022).
17. Hu, Y.; Yao, J. Illuminating economic growth. *J. Econom.* **2021**. [CrossRef]
18. Bickenbach, F.; Bode, E.; Nunnenkamp, P.; Söder, M. Night lights and regional GDP. *Rev. World Econ.* **2016**, *152*, 425–447. [CrossRef]
19. Gibson, J.; Olivia, S.; Boe-Gibson, G.; Li, C. Which night lights data should we use in economics, and where? *J. Dev. Econ.* **2021**, *149*, 102602. [CrossRef]
20. Tuttle, B.T.; Anderson, S.J.; Sutton, P.C.; Elvidge, C.D.; Baugh, K. It Used To Be Dark Here. *Photogramm. Eng. Remote Sens.* **2013**, *79*, 287–297. [CrossRef]
21. Hsu, F.C.; Baugh, K.E.; Ghosh, T.; Zhizhin, M.; Elvidge, C.D. DMSP-OLS Radiance Calibrated Nighttime Lights Time Series with Intercalibration. *Remote Sens.* **2015**, *7*, 1855–1876. [CrossRef]
22. Abrahams, A.; Oram, C.; Lozano-Gracia, N. Deblurring DMSP nighttime lights: A new method using Gaussian filters and frequencies of illumination. *Remote Sens. Environ.* **2018**, *210*, 242–258. [CrossRef]
23. Bluhm, R.; Krause, M. *Top Lights: Bright Cities and Their Contribution to Economic Development*; SoDa Laboratories Working Paper Series 2020-08; SoDa Laboratories, Monash University: Melbourne, VIC, Australia, 2020.
24. Devarajan, S. Africa's Statistical Tragedy. *Rev. Income Wealth* **2013**, *59*, S9–S15. [CrossRef]
25. Martinez, L.R. *How Much Should We Trust the Dictator's GDP Growth Estimates?* Becker Friedman Institute for Economics Working Paper 2021-78; University of Chicago: Chicago, IL, USA, 2021.
26. Pinkovskiy, M.; Sala-i Martin, X. Lights, Camera ... Income! Illuminating the National Accounts-Household Surveys Debate. *Q. J. Econ.* **2016**, *131*, 579–631. [CrossRef]
27. Heston, A.; Summers, R.; Aten, B. *Appendix for a Space-Time System of National Account aka Penn World Table 5.6*; Technical Report; Center for International Comparisons of Production, Income and Prices at the University of Pennsylvania: Philadelphia, PA, USA, 1994.
28. Heston, A.; Summers, R.; Aten, B. *Data Appendix for a Space-Time System of National Accounts: Penn World Table 6.1 (PWT 6.1)*; Technical Report; Center for International Comparisons of Production, Income and Prices at the University of Pennsylvania: Philadelphia, PA, USA, 2002.
29. Gibson, J.; Boe-Gibson, G. Nighttime Lights and County-Level Economic Activity in the United States: 2001 to 2019. *Remote Sens.* **2021**, *13*, 2741. [CrossRef]
30. Addison, D.M.; Stewart, B.P. *Nighttime Lights Revisited: The Use of Nighttime Lights Data as a Proxy for Economic Variables*; Policy Research Working Paper Series 7496; The World Bank: Washington, DC, USA, 2015.
31. Briant, A.; Combes, P.P.; Lafourcade, M. Dots to boxes: Do the size and shape of spatial units jeopardize economic geography estimations? *J. Urban Econ.* **2010**, *67*, 287–302. [CrossRef]
32. Openshaw, S.; Taylor, P. A million or so correlation coefficients: Three experiments on the modifiable area unit problem. In *Statistical Applications in the Spatial Sciences*; Wrigley, H., Ed.; Pion: London, UK, 1979; pp. 127–144.
33. Fixler, D. Measurement error in the national accounts. In *Measurement Error: Consequences, Applications and Solutions*; Emerald Group Publishing Limited: Bingley, UK, 2009.
34. Guci, L.; Mead, C.I.; Panek, S.D. *A Research Agenda for Measuring GDP at the County Level*; Bureau of Economic Analysis Working Paper; Bureau of Economic Analysis: Suitland, MD, USA, 2016.
35. Aysheshim, K.; Hinson, J.R.; Panek, S.D. A Primer on Local Area Gross Domestic Product Methodology. *Surv. Curr. Bus.* **2020**, *100*, 1–13.
36. European Commission. *Eurostat Manuals and Guidelines: Manual on Regional Accounts Methods*; European Commission: Brussels, Belgium, 2013.
37. European Commission. *Quality Report on National and Regional Accounts: 2016 Data*; European Commission: Brussels, Belgium, 2018.

38. European Commission. *Quality Report on National and Regional Accounts: 2019 Data*; European Commission: Brussels, Belgium, 2020.
39. Instituto Brasileiro de Geografia e Estatística. *Produto Interno Bruto dos Municípios, Ano de Referência 2010*; Instituto Brasileiro de Geografia e Estatística: Rio de Janeiro, Brazil, 2016.
40. National Bureau of Statistics of China. Announcement of the National Bureau of Statistics on the Final Verification of GDP in 2019. 2019. Available online: http://www.stats.gov.cn/english/PressRelease/202012/t20201231_1811928.html (accessed on 15 December 2020).
41. Chen, W.; Chen, X.; Hsieh, C.T.; Song, Z. A Forensic Examination of China's National Accounts. In *Brookings Papers on Economic Activity*; National Bureau of Economic Research: Cambridge, MA, USA, 2019.
42. Määttä, I.; Lessmann, C. Human Lights. *Remote Sens.* **2019**, *11*, 2194. [CrossRef]
43. Tuttle, B.T.; Anderson, S.; Elvidge, C.; Ghosh, T.; Baugh, K.; Sutton, P. Aladdin's Magic Lamp: Active Target Calibration of the DMSP OLS. *Remote Sens.* **2014**, *6*, 12708–12722. [CrossRef]
44. Ghosh, T.; Baugh, K.E.; Elvidge, C.D.; Zhizhin, M.; Poyda, A.; Hsu, F.C. Extending the DMSP Nighttime Lights Time Series beyond 2013. *Remote Sens.* **2021**, *13*, 5004. [CrossRef]

remote sensing

MDPI

Article

Urbanization Level in Chinese Counties: Imbalance Pattern and Driving Force

Baifa Zhang [1,2], Jing Zhang [1] and Changhong Miao [1,2,*]

1 Key Research Institute of Yellow River Civilization and Sustainable Development & Collaborative Innovation Center on Yellow River Civilization Jointly Built by Henan Province and Ministry of Education, Henan University, Kaifeng 475001, China; zbf@henu.edu.cn (B.Z.); 104753181181@vip.henu.edu.cn (J.Z.)
2 College of Geography and Environmental Science, Henan University, Kaifeng 475004, China
* Correspondence: chhmiao@henu.edu.cn

Abstract: Urbanization level is a key indicator for socioeconomic development and policy making, but the measurement data and methods need to be discussed further due to the limitation of a single index and the availability and accuracy of statistical data. China is urbanizing rapidly, but the urbanization level at the county scale remains a mystery due to its complexity and lack of unified and effective measurement indicators. In this paper, we proposed a new urbanization index to measure the Chinese urbanization level at the county scale by integrating population, land, and economic factors; by fusing remote sensing data and traditional demographic data, we investigated the multi-dimensional unbalanced development patterns and the driving mechanism from 1995 to 2015. Results indicate that: The average comprehensive urbanization level at the Chinese county scale has increased from 31.06% in 1995 to 45.23% in 2015, and the urbanization level in the permanent population may overestimate China's urbanization process. There were significant but different spatial and temporal dynamic patterns in population, land, and economic levels as well as at a comprehensive urbanization level. The comprehensive urbanization level shows the pattern of being high in the south-east and low in the north-west, divided by "Hu line". The urbanization of registered populations presents high in the northern border and the eastern coastal areas, which is further strengthened over time. Economic urbanization based on lighting data presents high in the east and low in the west. Land urbanization based on remote sensing data shows high in the south and low in the north. The registered population urbanization level is lower than economic and land urbanization. County urbanization was driven by large population size, reasonable industrial structure, and strong government capacity; 38% and 59% of urbanization levels can be regarded as the key nodes of the urbanization process. When the urbanization rate is lower than 38%, the secondary industry plays a strong role in powering urbanization; when the urbanization rate is higher than 38% but less than 59%, the promotion effect of the tertiary industry is more obvious, and the secondary industry is gradually weakened. When the urbanization rate exceeds 59%, the tertiary industry becomes the major driver.

Keywords: county urbanization; population urbanization; land urbanization; economic urbanization; comprehensive urbanization; China

Citation: Zhang, B.; Zhang, J.; Miao, C. Urbanization Level in Chinese Counties: Imbalance Pattern and Driving Force. *Remote Sens.* **2022**, *14*, 2268. https://doi.org/10.3390/rs14092268

Academic Editors: Ran Goldblatt, Steven Louis Rubinyi and Hogeun Park

Received: 28 March 2022
Accepted: 5 May 2022
Published: 8 May 2022

1. Introduction

Urbanization has developed rapidly in the 20th and 21st centuries and has been regarded as an important development strategy [1]. With increasing global urbanization, the Earth is gradually becoming an urban planet [2,3]. The proportion of the global urban population has increased from 33% in 1950 to 55% in 2018, and urban land expanded at twice the rate of population growth [4]. Rapid urbanization in China requires more attention from the world. Stiglitz, a Nobel laureate in economics, once predicted that China's urbanization was one of the two major engines driving world economic development in the

21st century [5,6]. However, due to the complexity and comprehensiveness of urbanization, the development and change of urbanization have always been a hot issue for governments, scholars, and the public [7,8]. So, how to measure the level of urbanization development accurately has strong theoretical and practical significance [9].

The measurement of urbanization level is particularly important in the stage of rapid urbanization development. Northam believes that urbanization development presents an "s"-shaped curve and can be divided into three stages and that the rapid urbanization stage is when the urbanization level is between 25% and 70% [10]. At this stage, one of the research objectives in China and abroad is how to measure the level of urbanization scientifically. Right now, China is at this stage [7–9]. As the fastest growing economy in the world, China's urban population increased from 200 million in 1980 to 800 million in 2018 [11]. In 2019, the urbanization rate in China exceeded 60%, and the urbanization rate of household registration reached 44.38% [12]. However, scholars have certain disputes over the current level of urbanization [13] and believe that the current level of urbanization is overestimated [14,15]. Therefore, how to evaluate the level of urbanization in China scientifically is an important issue [16–18].

Through a literature review, the selection of indicators is an important issue in the study of urbanization. The existing studies are mainly based on a single indicator or composite indicator system, which cannot reflect the urbanization process comprehensively and truly. In terms of a single indicator, the research on measuring the level of urbanization based on population data is the most used and the most representative. As early as the 1930s, the "Hu Huanyong Line" showed the characteristics of the spatial pattern of China's population was extremely dense in the east and sparse in the west. Although these findings are from a population density map, they also reflects the differentiation of urbanization levels to a certain extent [19]. Since then, the research on population urbanization has gradually increased [20–22]. Moreover, there is also much research on India, Central Asia, and other countries and regions using population data [23–25]. With the intensification of population mobility, the measure data also changes [26]. Qi believed that the urbanization rate of registered populations could more scientifically reflect the quality of China's population urbanization, so he used the data of the sixth population census to measure the urbanization rate of the permanent resident population and registered population at the county and city scales in China [27].

It is also the content of single index research to analyze the urbanization process from the dimension of land use. With the rapid development of urbanization, it is not only manifested in the increase of urban population but also manifested in the disorderly expansion of out of control urban construction land in space. Important resources such as arable land and water are overconsumed, the environment is seriously polluted, and urban infrastructure is greatly wasted. Therefore, in 2007, Mr. Lu proposed to use land urbanization for the first time in China and pointed out that the speed of land urbanization was too fast and much faster than population urbanization [28]. The research on land urbanization has become a hot spot. Some scholars [29–31] proposed that the land-use area of urban built-up areas in a country or region reflects the local urbanization level. Moreover, other scholars believed that the proportion of industrial and mining land in urban areas could better reflect the process of land urbanization and propose that the ratio of urban and rural construction land to urban construction land can be used to express the land urbanization rate [32–34]. Furthermore, Asabere pointed out that the spatial change process caused by urbanization can be better revealed by analyzing the change in regional land use and cover quantity, quality, and structure [35]. Vogler conducted a case study of the United States and showed that per capita land change could be a new spatial index for measuring urbanization [36]. In addition, the use of per capita GDP, the proportion of secondary and tertiary output values and composite economic indicators to measure economic urbanization, is also one of the focuses of urbanization research [37–41].

Although the single index measure has strong brevity, it cannot fully reveal the connotation of urbanization, and there are certain errors in judging the development stage of urbanization [42–44]. To solve these problems, scholars have constructed a multi-standard comprehensive evaluation index system to measure the level of urbanization (i.e., population and land, population and economy, land and economy, population-land and economy) [5,6,45–52]. Furthermore, although the multi-dimensional index system is more scientific than a single index in measuring urbanization, the data source of the multi-dimensional index system is still mainly the traditional social and economic statistical data, while the traditional data has many limitations (data quality is relatively low; temporal continuity is weak; spatial scale is limited, etc.) [1,53]. With the digital wave sweeping the world, using remote sensing images to do urbanization research has gradually become a new research direction [54,55], and the fusion of social economy, satellite remote sensing, social media, and other multi-source data is the mainstream trend of future research [56,57]. Among them, lighting data is the most widely used and provides a good opportunity to monitor annual urbanization activities. The light data provided by Operational Line-scan System (OLS) on the Defense Meteorological Satellite Program (DMSP) can detect city lights with a low-light detecting capability [58–60]. Due to the advantages of rich historical archived data and wide spatial coverage [61,62], lighting data has been widely used in socio-economic estimation [63,64] and urban agglomeration development [65]. For example, Sutton et al. [66] estimated the global human population using the statistical relationship between nighttime lighted areas and the urban population.

In order to solve the limitation of a single index and the data traditionally attributed to the composite index system based on the connotation of urbanization, we combine multi-dimensional (population, land, and economy) and multi-source data (population statistics, land use, and night light), and propose a new urbanization measurement indicator—comprehensive urbanization. Meanwhile, China is urbanizing rapidly, but the urbanization level at the county scale remains a mystery due to the complexity and lack of unified and effective measurement indicators. Therefore, we take China's counties as the research object and use the newly constructed comprehensive urbanization index to answer three questions: What is the level of county urbanization in China from 1995–2015? What are the characteristics of geographical imbalance? What are the driving forces and mechanisms? By answering these questions, we try to provide new ideas and methods for urbanization measurement, and we try our best to show the real and comprehensive process of county urbanization in China so as to provide guidance for future development of urbanization.

2. Research Methods and Data Sources

2.1. Method of Urbanization Measurement

Population urbanization: Human behavior is generally regarded as one of the direct driving factors affecting and changing urbanization. Non-agricultural population refers to the population engaged in the secondary and tertiary industries and the part of the population supported by them. Due to the small population flow in the early years, the high urban-rural dual barrier and the large difference in treatment between the non-agricultural population and agricultural population, the ratio of the non-agricultural population to the total population can better reflect the population urbanization rate in China. After 2000, the ratio of the permanent urban population to total permanent population was used to represent the urbanization rate, which was better reduced to the error effect caused by population flow. However, the permanent population statistics include the floating population that has not really achieved the process of settling down [21,67]. Therefore, after considering the availability of data, this paper uses the urbanization rate of the registered residence population to show the urbanization development pattern of the county population in China. The urban rate of registered population is the ratio of the non-agricultural population to the total registered population of a county, explained in Formula (1).

$$\text{Urban}_{\text{pop}} = \frac{\text{pop}_{\text{non\_agri}}}{\text{pop}_{\text{all}}} * 100\% \tag{1}$$

In Formula (1), $Urban_{pop}$ represents the urbanization rate of the registered population; pop_{non_agri} represents the non-agricultural population of a county; and pop_{all} represents the total registered population of the county.

Land urbanization: Land is the carrier of urbanization and the basis of human activities. Land urbanization is the process of transforming agricultural land into non-agricultural land. In China, due to the existence of ecological protection red line, permanent basic farmland protection red line, and urban development boundary, all the areas of a region cannot be completely transformed into urban land. At the same time, it is reasonable to use the land urbanization index and land non-agricultural index to construct the county land urbanization rate [34]. Therefore, this paper uses the ratio of urban land, industrial and mining land, and transportation land to the total size of urban and rural construction land to reflect the land urbanization Formula (2).

$$Urban_{land} = \frac{ul + il + tl}{ul + il + tl + rl} * 100\% \tag{2}$$

In Formula (2), $Urban_{land}$ represents land urbanization rate; ul represents urban land area; il represents industrial and mining land area; tl represents transportation land area; and rl represents the rural residential land area.

Economic urbanization: Industry plays a crucial role as the internal driving force of urbanization [68]. Due to the close relationship between lighting data and economic development [69,70], light data provided a valuable data source for elucidating the dynamics of China's urbanization [71–73]. Referring to land urbanization and population urbanization models, "economic activity area" and "economic activity intensity carried by land" are incorporated into the economic urbanization model. Based on the existing studies, this paper constructs a model reflecting economic urbanization from two aspects of night light: (a) the intensity attribute of regional average light; (b) the area attribute of regional lighting [74–77]. Finally, the model of economic urbanization is constructed by combining the two indexes in a linear way, shown in Formula (3).

$$Urban_{eco} = \left(\frac{TDN}{N * 63} * W_1 + \frac{Area_N}{Area} * W_2\right) * 100\% \tag{3}$$

In Formula (3), $Urban_{eco}$ represents the economic urbanization rate; TDN represents the total value of bright elements in a certain county; N represents the number of bright elements; and 63 represents the maximum value of a single bright element. The first formula of the model represents the intensity attribute of regional average light, that is, the ratio of the actual total value of the light element in a county to the theoretical maximum value of the light element in that county. $Area_N$ refers to the total area with a pixel value greater than 0 in a region, and Area refers to the area of a county. The second formula in the model represents the area attribute of regional lighting, the ratio of the total area of all light elements in the county to the total area of the county. By verifying the weight combination, it is finally determined that W_1 is 0.5 and W_2 is also 0.5.

Comprehensive urbanization: The urban-rural system contains three important elements: population, land, and industry [41]. Urbanization is usually used to refer to three related but different processes, including the transformation of an agricultural population into a non-agricultural population, an agricultural region into a non-agricultural region, and agricultural activity into non-agricultural activity [42,78]. Therefore, this paper studies urbanization from the perspective of population, land, and industry [47]. By comprehensively considering population urbanization, land urbanization, and economic urbanization, the comprehensive urbanization rate is constructed by assigning the three weights, respectively, expressed in Formula (4).

$$T = \alpha * Urban_{pop} + \beta * Urban_{land} + X * Urabn_{eco} \tag{4}$$

In Formula (4), T represents the comprehensive urbanization rate. A, β, and X are the undetermined coefficients of population urbanization rate, land urbanization rate, and economic urbanization rate, respectively. Since population, land, and economic urbanization are equally important, the three are assigned the same weights in this paper, which are all 1/3.

2.2. Fixed Effects Panel Regression Model

This paper selects four indicators from three aspects of population size, industrial structure, and government capacity, namely population density, the proportion of primary industry in GDP, secondary and tertiary industrial structure, and the level of fiscal transfer payments (Table 1). Specifically, first, as the core element of urbanization, population size provides basic conditions for the development of urbanization. Secondly, economic development is an important driving force of urbanization, and industrial structure is an important representation of economic development. As a largely agricultural country, the development of primary industry may restrict the development of urbanization in China. At the same time, the tertiary industry has a stronger employment absorption capacity than the secondary industry, and its development may promote the rapid increase of the urbanization rate. Therefore, the proportion of primary industry in GDP and the structure of secondary and tertiary industries are selected as the indicators of industrial structure. Thirdly, China has a unique administrative system, and the government has played a strong role in the development of urbanization. The level of fiscal transfer payments can better characterize the government's ability. The panel regression model is constructed based on the selection of variables, and the expression is as follows:

$$T_{it} = \alpha_{i,t} + \beta_1 \ln pd_{i,t} + \beta_2 ppi_{i,t} + \beta_3 tos_{i,t} + \beta_4 gov_{i,t} + \varepsilon_{it} \quad (5)$$

Table 1. Index selection of influencing factors of county comprehensive urbanization.

Primary Index	Secondary Index	Indicator Description
Population size	Pd	Population size/county area
Industrial structure	Ppi	Primary industry value/GDP
	Tos	Tertiary industry value/Second industry value
Government capacity	Gov	(Expenditure-Revenue)/GDP

In Formula (5), T stands for comprehensive urbanization; pd is the population density; ppi is the proportion of primary industry in GDP; Tos represents secondary and tertiary industrial structure; and gov represents the level of fiscal transfer payments (Table 1). $\alpha_{i,t}$ is the individual fixed effect; β_1~β_4 is the regression coefficient of each variable; $\varepsilon_{i,t}$ is the random disturbance term; and i and t represent region and time, respectively.

In order to ensure the accuracy of the regression results, the model is processed and tested as follows: (a) In order to ensure the data stability and weaken the influence of collinearity and heteroscedasticity of the sequence on the estimation results, some indexes of the model are treated with logarithm; (b) through F-test and Hausman test, it is found that the chi-square *p* value in the model is less than 0.05. Therefore, this paper selects the fixed effect as the estimation method of the model.

2.3. Threshold Regression Model

We take population urbanization, land urbanization, and economic urbanization as threshold variables to test the threshold effect of Tos on comprehensive urbanization. Then determine how secondary industry and tertiary industry contribute to urbanization at different stages of urbanization.

$$Y_{it} = \beta_1 x_{I,t} I(I \leq \gamma_1) + \beta_2 I_{i,t} I(\gamma_1 < I \leq \gamma_2) + I x_{i,t} I\left(I < g_{i,t} \leq \gamma_n\right) + \beta_4 coltrol_{i,t} + \varepsilon_{it} \quad (6)$$

In Formula (6), i stands for the county, and t stands for the year; Y_{it} is the dependent variable (comprehensive urbanization), x_{it} is the explanatory variable (tos); β_1, β_2, and β_3 are the corresponding coefficient variables; g_{it} is the threshold variable; population urbanization rate, land urbanization rate, and economic urbanization rate are selected as the threshold variables respectively; γ_1, γ_2, and γ_n are the specific threshold values; and I is the index function. When the conditions in parentheses are met, I is 1, otherwise, it is 0; control is the set of control variables (pd, ppi, and gov); and ε_{it} is the random disturbance term.

2.4. Data Sources

Due to the long research period, in order to facilitate the analysis and research, the administrative division of 2015 is taken as the standard, and the data of each year are unified to the administrative unit of that year. The research data mainly includes five aspects: (a) China's provincial, municipal, and county-level administrative units are China's 1:250,000 basic geographic data provided by the Resources and Environment science data of the Chinese Academy of Sciences (http://www.resdc.cn/, accessed on 16 July 2020). Excluding Hong Kong, Macao, and Taiwan, there are 31 provincial-level administrative units and 2851 county-level units in the study area; (b) the registered population data of each county in the calculation of population urbanization come from the National Population Statistics of Counties and Cities of the People's Republic of China in 1995, 2000, 2005, and 2010. Since the data source only counted until 2012 and the division of "agricultural and non-agricultural" household registration was canceled in 2015, this paper used the "agricultural and non-agricultural" population data from 2014 to replace 2015. The data are from the official websites of statistical bureaus of all provinces, cities, and counties; (c) the land use data for 1995, 2000, 2005, 2010, and 2015 are obtained from the Data Center of Resources and Environmental Sciences, Chinese Academy of Sciences, with a resolution of 30 m (http://www.resdc.cn/); (d) the social statistics for regression are obtained from the statistical yearbook. (e) the stable night light data for economic urbanization calculation is provided by the NGDC website (https://ngdc.noaa.gov/eog/dmsp, last accessed on 16 July 2020), including 34 stable night light images without radiation calibration (1992–2013) from six DMSP satellites: F10, F12, F14, F15, F16, and F18. Background noises of the stable night data were identified and replaced with values of zero, and the final DN values for lit pixels ranged from 1 to 63. Since the data source is only updated to 2013, this paper uses the 2013 night light image to replace the 2015 data.

Due to the insufficient accuracy of the original noctilucent image (discontinuous problem and multi-sensor problem), we used the existing methods to correct it [56,57]. Firstly, the obtained global light images were extracted according to the Chinese regional boundaries, and then we obtained the Chinese light image data from 1992 to 2013. Secondly, the projection of light data was defined as the Lambert Azimuthal Equal projection, and data were resampled to a pixel size of 1 km. Thirdly, we chose "the method of invariant target region". The core idea of this method is to use a small area with a wide range of light values (0–63) but little change over the years to correct all images. Moreover, this method mainly consists of three steps: intercalibration, intra-annual composition, and inter-annual series correction. The specific operation is as follows: (a) Selection of the invariant target region. Of all the areas examined, Hegang City of Heilongjiang Province was selected as the "constant target area" due to its characteristics mentioned above. (b) Intercalibration. In order to solve the continuity problem, we chose the F16-2006 light image of Hegang City as the reference image and corrected all the images by constructing a quadratic equation of one variable. However, we found that the change in Hegang City from 1992 to 2013 was slightly larger, and the fitting degree of lighting value in 1992 and 2013 was only 0.78, less than 0.80. Therefore, after completing the whole process and comparing it with Lu, we thought that the coefficients provided by Lu might be more scientific, so we finally chose it (Table 2). (c) Intra-annual composition. Light data for some years will be provided by two sensors (F14-2000 and F15-2000; F15-2005 and F16-2005). In order to make full use of the images of the same year obtained by multiple sensors and ensure the stability and continuity of night light image data, we averaged the two-image data of the same year. (d) Series correction. In order to solve the problem of individual pixel value mutation, we assumed that the DN value of a pixel in the DMSP/OLS lighting image should not be small in the following year DN value of the pixel in the previous year. According to the operation, we obtained the final Chinese lighting data set from 1992 to 2013.

Table 2. Parameters of a quadratic equation with one variable provided by Lu [79].

Sensor	Year	a	b	c	R^2
F12	1995	0.034	0.513	0.485	0.851
F14	2000	0.024	0.606	0.346	0.873
F15	2000	0.028	0.578	0.485	0.878
	2005	0.026	0.872	0.123	0.855
F16	2005	0.022	0.919	0.096	0.887
F18	2010	0.021	0.510	0.901	0.851
	2013	0.016	0.271	0.680	0.850

3. Temporal and Spatial Pattern Evolution of County Urbanization in China

3.1. Characteristics of Spatio-Temporal Differentiation of Population-Land-Economy Urbanization

In order to describe the urbanization development pattern of China's counties in detail, the data of registered populations, nighttime lights, and land use were used to show the pattern of the registered population urbanization rate, economic urbanization rate, and land urbanization rate of 2851 counties in China during 1995–2015 (Figure 1).

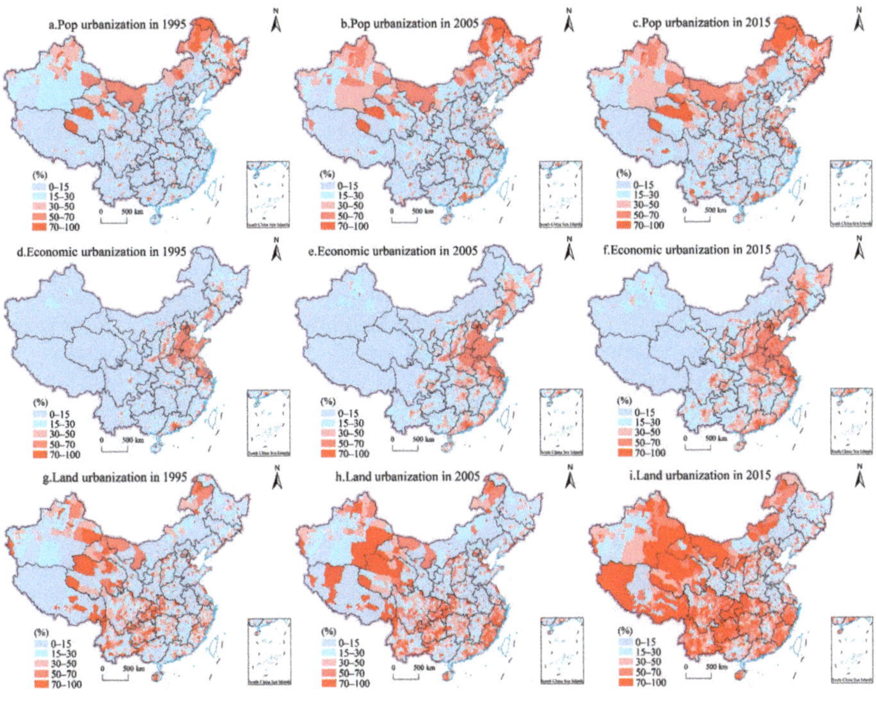

Figure 1. Spatial distribution of population-economy-land urbanization rates in China from 1995 to 2015.

County urbanization was represented by registered population data (Figure 1a–c). The average urbanization rate of the registered population of 2851 counties was 27.49% in 1995, 34.31% in 2005, and 38.62% in 2015, with an average annual growth rate of 2.19%. From the perspective of spatial pattern, the urbanization rate in 1995 was generally low. Counties with high urbanization rates were mainly concentrated in Northern Xinjiang, border areas of Inner Mongolia, and Northeast China. The prototype of a high urbanization belt on the northern border had basically appeared. In addition, the eastern coastal areas of Beijing-Tianjin-Hebei, the Yangtze River Delta, and the Pearl River Delta also had small-scale high urbanization agglomeration areas with Beijing, Shanghai, and Guangzhou as the core. It could be seen that the dot high urbanization areas with provincial capitals and municipal districts as the core were also more prominent. In 2015, the Beijing-Tianjin-Hebei region, the Yangtze River Delta, and the Pearl River Delta along the eastern coast spread rapidly, forming a large area with a high urbanization rate. At the same time, high urbanization rate areas also appeared in the Shandong Peninsula, Jiangsu, Fujian, and other places. The high urbanization belt along the eastern coast had been basically formed, and a "herringbone" pattern had been formed with the high urbanization belt along the northern border, which is consistent with Liu's research results [21].

Economic urbanization was represented by light data (Figure 1d–f). The average urbanization rate of 2851 counties was 31.22% in 1995, 39.68% in 2005, 46.68% in 2015, and the average annual growth rate was 3.10%. From the perspective of spatial pattern, in 1995, the distribution of counties with high urbanization rate showed a pattern of "three points and one side", that was, ultra-high urbanization rate concentration points with Beijing, Shanghai, and Guangzhou as the core, and high urbanization rate areas along the eastern coast mainly with Shandong, Hebei, and Jiangsu. Moreover, the point regions with provincial capitals and municipal districts as the core were also areas with high economic urbanization rates. There was a significant difference in the rate of economic urbanization between the east and the west. In 2015, the difference in economic urbanization rate between east and west was more significant. In the east, three high-rate areas of Beijing-Tianjin-Hebei, Yangtze River Delta, and Pearl River Delta with Beijing, Shanghai, and Guangdong as the core had been formed. Furthermore, North China Plain was also a high-value concentration area of China's economic urbanization.

Land urbanization was represented by land-use data (Figure 1g–i). The average land urbanization rate of 2851 counties was 34.46% in 1995, 39.16% in 2005, and 50.37% in 2015, with an average annual growth rate of 2.63%. From the perspective of spatial pattern, in 1995, counties with a high land urbanization rate mainly were concentrated in the eastern coastal Yangtze River Delta, Fujian, Guangdong, and other places. Chongqing and Guizhou in southwest China formed concentrated and continuous high-value areas, and small high-value areas appeared in western Xinjiang, Qinghai, and Gansu. The difference between the north and the south in land urbanization was obvious. In 1995, the average land urbanization rate of counties was 35.51% in the east of the "Hu Line", while in the west, it was 28.49%, with a difference of 7.02%. However, the average land urbanization rate in the southern and northern counties was 38.68% and 29.90%, respectively, with a difference of 8.78%. In 2015, the gap between the north and the south in land urbanization rate is more significant, which is consistent with Gao's research results [34]. The difference in the average land urbanization rate between east and west counties of "Hu line" was 0.11%, and the difference between North and South was 18.58%. The southwest high-value area with Chongqing as the core was connected with the southeast coastal high-value area. Only some counties in western Jiangxi and central Guangxi had low land urbanization rates. The land urbanization rate of the county in the south had reached a high level, while the northeast plain and north China Plain in the north formed a wide range of low land urbanization rate regions.

In terms of spatial pattern, the urbanization of the registered population presents a "herring-shaped" pattern composed of high-value areas in the northern border area and high-value areas in the eastern coastal area, which further strengthens over time. The pattern of economic urbanization is "high in the east and low in the west". Land urbanization presents a pattern of "high in the south and low in the north", which is more obvious than the differentiation between the east and the west. In terms of development level, the urbanization rate from highest to lowest is land, economy, and population. In terms of development speed, the urbanization rate from high to low is economy, land, and population. In general, the development level of population urbanization is the lowest, and the development rate is the slowest. However, urbanization is a development process with people as the core, so in the future, we must pay attention to the urbanization of "people".

3.2. Characteristics of Spatio-Temporal Differentiation of Comprehensive Urbanization

Based on the population, land, and economic development, a comprehensive urbanization rate index is constructed to show the spatio-temporal pattern of urbanization development at the county level in China (Figure 2). The spatial and temporal dynamics of the comprehensive urbanization level at the county level in China are significantly different, and the overall distribution still follows the "Hu Line". However, with the development, the comprehensive urbanization rate in western inland areas gradually increased. Three high-value areas of Beijing-Tianjin-Hebei, Yangtze River Delta, and Pearl River Delta, with Beijing, Shanghai, and Guangdong as the core have been formed along the eastern coast. BesidesIn addition, in 1995, the average comprehensive urbanization rate in 2851 counties was 31.06%. In 2015, it was 45.23%, with an average annual growth rate of 2.67%. It showed that the high-value areas with provincial capitals and municipal districts as the core form a multi-point distribution throughout the country.

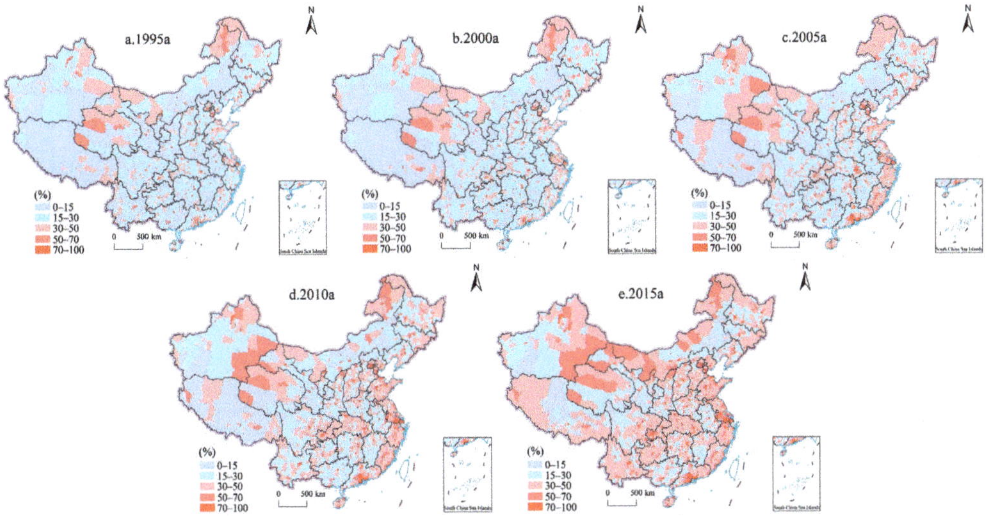

Figure 2. Spatial distribution of comprehensive urbanization rate of China's counties from 1995 to 2015.

In 1995, the overall comprehensive urbanization rate of counties in China was low. The average urbanization rate of 2851 counties was 31.06%, with obvious regional differences. A total of 41% of the 2851 counties in China had a comprehensive urbanization rate of 15–30%, followed by the total number of counties with an urbanization rate of 0–15% (685), indicating that China's county urbanization rate was extremely low. First, counties with a high urbanization rate were mainly concentrated in the eastern coastal areas, forming the high-value belt in the eastern coastal areas, among which Beijing, Shanghai, and Guangzhou were the three high-value points of urbanization rate. Furthermore, provincial capitals and municipal districts were the areas with high urbanization rates and scattered in points. There were also some counties in the western inland areas with relatively high comprehensive urbanization rates. Second, the areas with a low comprehensive urbanization rate were mainly concentrated in the western inland and other rough terrain areas, as well as non-provincial capitals and non-municipal districts.

In 2015, the average comprehensive urbanization rate of counties in China was 45.23%. Among the three core regions, the Pearl River Delta and Yangtze River Delta had achieved better development, followed by the Beijing-Tianjin-Hebei region. The urbanization rate in the inland areas of central and western China was growing rapidly, such as counties in Hubei, Chongqing, and Henan. The comprehensive urbanization rate of most counties was concentrated in the range of 30–50%. There were only 67 counties with a comprehensive urbanization rate of 0–15%, accounting for only 3% of the total number of counties in China. The number of counties with a comprehensive urbanization rate of 15–30% accounted for 22%. The number of counties with a comprehensive urbanization rate of 30–50% increased rapidly to 1291, accounting for 45% of the total number of counties in China. The number of counties with a comprehensive urbanization rate of 50–100% was also gradually increasing, accounting for 30%, indicating that China's county urbanization had reached the above medium level.

4. Results

4.1. Fixed Effects Panel Model Regression Results

After answering the first three questions mentioned above: "How high is the level of county urbanization in China? How to measure the level of urbanization? What are the characteristics of geographical imbalance?", we will further answer the question "What are the driving forces and mechanisms of county urbanization in China?". We selected four indicators to explain this problem, including population density, the proportion of primary industry in GDP, secondary and tertiary industrial structure, and the level of fiscal transfer payments. The regression results are shown in Table 3.

Table 3. Regression results of comprehensive urbanization.

	All	East	Mid	West
Pd	6.42 ***	5.97 ***	2.76 ***	9.31 ***
	(0.58)	(0.99)	(0.91)	(0.97)
Ppi	−0.31 ***	−0.41 ***	−0.28 ***	−0.28 ***
	(0.01)	(0.02)	(0.01)	(0.01)
Tos	0.08 *	1.19 ***	0.13 *	−0.01
	(0.05)	(0.28)	(0.07)	(0.07)
Gov	6.59 ***	23.84 ***	15.88 ***	5.04 ***
	(0.37)	(1.85)	(0.99)	(0.46)
Constant	4.00	6.10	18.49 ***	−5.39 **
	(2.90)	(5.84)	(4.98)	(3.99)
F-value	12.80 ***	12.77 ***	13.01 ***	12.05 ***
R^2	0.37	0.44	0.43	0.38

Standard errors in parentheses; *** $p < 0.01$, ** $p < 0.05$, * $p < 0.1$

In order to explore the development mechanism of urbanization, the full sample is incorporated into the model and regression analysis is conducted on each factor. At the same time, there is a large difference among different counties with strong heterogeneity. Therefore, the counties are divided into eastern, central, and western regions according to their geographical location, and the regression results are shown in Table 3. Overall, a large population scale, reasonable industrial structure, and strong government capacity play a good role in promoting the development of county urbanization. Specifically, (a) When all counties are included in the model, the regression coefficients of pd and gov are 6.42 ($p < 0.01$) and 6.59 ($p < 0.01$), respectively, indicating that population size and government capacity have promoted the improvement of the county urbanization rate. In terms of industrial structure, the regression coefficient of ppi is −0.31 ($p < 0.01$), indicating that the development of primary industry may inhibit the improvement of county urbanization rate. However, the regression coefficient of Tos is 0.08 ($p < 0.1$), indicating that the tertiary industry could better promote the development of urbanization than the secondary industry. (b) In the eastern region, the regression results are similar to those at the national level. The regression coefficients of pd, ppi, Tos, and gov are 5.97 ($p < 0.01$), −0.41 ($p < 0.01$), 1.19 ($p < 0.01$), and 23.84 ($p < 0.01$), respectively. Among them, Tos has a stronger significance and a larger regression coefficient, indicating that the tertiary industry plays a greater role in improving the urbanization rate in eastern China. (c) In central China, the regression coefficients of pd, proportion of primary and secondary industries, proportion of secondary and tertiary industries, and government capacity were 2.76 ($p < 0.01$), −0.28 ($p < 0.01$), 0.13 ($p < 0.1$), and 15.88 ($p < 0.01$), respectively. (d) In the western region, Tos shows particularity, and its regression coefficient is −0.01, which does not pass the significance test. It shows that the development of the tertiary industry cannot contribute to the improvement of county urbanization rate in western inland areas, while the development of the secondary industry may be more conducive to the development of urbanization.

China is a vast country with complex and diverse landforms. Different types of county urbanization may have different driving mechanisms. Therefore, according to the terrain and geomorphology, the counties are divided into plain counties, hill counties, and mountain counties, and regression is conducted respectively (Table 4). Whether in a plain county, hill county, or mountain county, the population scale and government capacity can promote the urbanization rate of the county, while the development of the primary industry will limit the urbanization development. However, Tos shows differences among the three types. In plain counties, the Tos regression coefficient is 0.07, but it does not pass the significance test, indicating that its promotion effect on the urbanization rate is not obvious. In hilly counties, Tos regression coefficient is −0.45 ($p < 0.01$), indicating that it inhibits the development of urbanization. It shows that the development of the tertiary industry cannot contribute to the improvement of the county urbanization rate in hill countries, while the development of the secondary industry may be more conducive to the development of urbanization. In mountainous counties, Tos regression coefficient is 0.12 ($p < 0.1$), indicating that the development of the tertiary industry will significantly promote the improvement of the urbanization rate.

Table 4. Regression results of different types of areas.

	Plain County	Hill County	Mountain County
Pd	6.86 ***	6.82 ***	6.71 ***
	(0.99)	(1.17)	(0.90)
Ppi	−0.34 ***	−0.27 ***	−0.30 ***
	(0.01)	(0.01)	(0.01)
Tos	0.07	−0.45 ***	0.12 *
	(0.08)	(0.15)	(0.07)
Gov	9.23 ***	18.73 ***	5.61 ***
	(1.27)	(1.42)	(0.42)
Constant	0.33	−0.80	4.28
	(5.60)	(6.09)	(3.96)
F-value	11.48 ***	13.39 ***	13.08 ***
R^2	0.37	0.37	0.40

Standard errors in parentheses; *** $p < 0.01$, ** $p < 0.05$, * $p < 0.1$.

Finally, based on t from the official documents published by each Chinese province, we divide the counties into different functional zones, including key development zones, optimized development zones, main grain-producing areas, and ecological protection zones, and then carry out regression for the four types (Table 5). In the four subregions, both population size and government capacity play a significant positive role in the improvement of urbanization, while the development of the primary industry significantly inhibits the improvement of the urbanization rate. However, Tos shows great differences in different functional zones, specifically: in the key development zone and ecological protection area; Tos regression coefficients are 0.50 and 0.03, respectively, which both fail the significance test, indicating that Tos has no significant promoting effect on urbanization. In the optimized development zone, the Tos regression coefficient is 5.66, and through the significance test of 1%, which shows that the good development of the tertiary industry will better promote the increase of urban rate. In the main grain-producing areas, Tos also shows a positive effect, and its regression coefficient is 0.14 ($p < 0.1$).

Table 5. Regression results of different functional zones.

	Key Development Zone	Optimized Development Zone	Major Grain-Producing Zone	Ecologic Protection Zone
Pd	9.09 ***	12.30 ***	5.78 ***	4.70 ***
	(1.28)	(2.45)	(0.89)	(0.97)
Ppi	−0.46 ***	−0.62 ***	−0.28 ***	−0.25 ***
	(0.02)	(0.09)	(0.01)	(0.01)
Tos	0.50	5.66 ***	0.14 *	0.03
	(0.34)	(1.67)	(0.08)	(0.06)
Gov	14.97 ***	27.40 *	13.35 ***	5.45 ***
	(1.74)	(15.50)	(0.84)	(0.43)
Constant	−9.71	−29.95 *	2.77	12.46 **
	(7.31)	(16.00)	(4.81)	(3.92)
F-value	12.16 ***	9.75 ***	10.95 ***	13.34 ***
R^2	0.46	0.47	0.44	0.33

Standard errors in parentheses; *** $p < 0.01$, ** $p < 0.05$, * $p < 0.1$.

In conclusion, a large population size and strong government capacity will promote the rapid improvement of county urbanization rate in China, but a high proportion of primary industry will inhibit the development of urbanization. In different types of counties, the performance of Tos is quite different, which shows that relying solely on the tertiary industry or the secondary industry cannot promote the development of urbanization effectively. Each county should adjust its industrial structure reasonably according to local conditions. For example, an optimized development zone should vigorously develop the tertiary industry, while the development of the secondary industry in hilly counties may be more conducive to the development of urbanization.

4.2. Threshold Model Regression Results

The empirical results of the panel model have shown that there are great differences in the impact of the Tos on urbanization in different regions, which preliminarily shows that the tertiary industry plays a stronger role in promoting the urbanization rate in developed regions, while the secondary industry plays a greater role in relatively developing regions. However, can the regression results of different regions show that with the increase in urbanization rate, the promotion effect of the tertiary industry increases while the promotion effect of the secondary industry decreases? This paper then uses the threshold model to test the threshold effect of Tos on comprehensive urbanization by taking the population urbanization rate, land urbanization rate, and economic urbanization rate as threshold variables.

In Table 6, when population urbanization is used as the threshold variable, within the 95% confidence interval, the whole sample passes the double threshold test, and the double threshold values are 38.56% and 58.73%, respectively; when land urbanization is used as the threshold variable, within the 95% confidence interval, the whole sample passes the double threshold test, and the double threshold values are 0.18% and 38.14%, respectively; when economic urbanization is used as the threshold variable, within the 95% confidence interval, the whole sample passes the double threshold test, and the double threshold values are 32.27% and 59.06%, respectively. Based on this, this paper takes population urbanization (Model 1), land urbanization (Model 2), and economic urbanization (Model 3) as threshold variables for regression, and the estimation results are shown in Table 7.

Table 6. Threshold estimates and test results.

Threshold Variable	Threshold Test	Threshold Value	*p* Value	95% Confidence Interval
Pop_urban	Single threshold test	38.56	0.00	[38.135, 38.958]
	Double threshold test	58.73	0.00	[57.657, 60.185]
	Triple threshold test	—	—	—
Land_urban	Single threshold test	0.18	0.00	[0.000, 1.851]
	Double threshold test	38.14	0.00	[37.621, 38.313]
	Triple threshold test	—	—	—
Eco_urban	Single threshold test	32.27	0.00	[29.369, 32.430]
	Double threshold test	59.06	0.00	[58.617, 60.480]
	Triple threshold test	—	—	—

—indicates that there is no triple threshold.

Table 7. Threshold panel model estimation results.

Variable	Model 1	Model 2	Model 3
Tos (γ_1)	−0.02	−0.45 ***	−0.21 ***
	(0.05)	(0.08)	(0.05)
Tos (γ_2)	2.55 ***	−1.24 ***	0.74 ***
	(0.20)	(0.10)	(0.09)
Tos (γ_3)	6.49 ***	0.84 ***	4.75 ***
	(0.34)	(0.06)	(0.30)
Pd	6.79 ***	6.00 ***	5.13 ***
	(0.57)	(0.57)	(0.58)
Ppi	−0.31 ***	−0.29 ***	−0.30 ***
	(0.01)	(0.01)	(0.01)
Gov	6.19 ***	6.48 ***	6.61 ***
	(0.36)	(0.36)	(0.36)
_cons	1.73	5.98 ***	9.60 ***
	(2.85)	(2.84)	(2.89)
F-test	11.16 ***	11.20 ***	12.16 ***
_cons	1.73	5.98 ***	9.60 ***

Standard errors in parentheses; *** $p < 0.01$, ** $p < 0.05$, * $p < 0.1$.

According to the results of the three models, there is an obvious nonlinear relationship between the Tos and the comprehensive urbanization in the process of urbanization. In the early and middle stages of urbanization, the secondary industry plays a strong role in promoting the urbanization rate. With the development, the role of the secondary industry is gradually weakened, and the tertiary industry occupies a dominant position. Specifically: (a) Taking population urbanization as the threshold variable, when it is at the first threshold (pop_urban $\leq$ 38.56%), the Tos regression coefficient is −0.02, indicating that the urbanization rate will not increase with the increase of Tos, and the indicator Tos has a negative correlation with the secondary industry and a positive correlation with the tertiary industry, so the secondary industry has an obvious effect on the urbanization rate; when it is in the second threshold range (38.56% < pop_urabn $\leq$ 58.73%), the Tos regression coefficient is 2.55 ($p < 0.01$), indicating that the comprehensive urbanization rate will increase with the increase of Tos, and the promotion effect of the tertiary industry will gradually exceed that of the secondary industry; when crossing the second threshold (58.73% $\leq$ pop_urban), Tos is still positively correlated with the comprehensive urbanization rate (the correlation coefficient is 6.49), and the tertiary industry promotes the urbanization rate more obviously. (b) Taking land urbanization as the threshold variable, when it is in the first threshold interval (land_urban $\leq$ 0.18%) and the second threshold interval (0.18% < land_urabn $\leq$ 38.14%), the Tos regression coefficients are −0.45 ($p < 0.01$) and −1.24 ($p < 0.01$) respectively, indicating that there is an obvious negative correlation between Tos and the comprehensive urbanization rate, and the development of the secondary industry will greatly promote the improvement of the comprehensive urbanization rate; when crossing the second threshold (38.14% $\leq$ land_urban), the regression coefficient of Tos is

0.84 ($p < 0.01$). The comprehensive urbanization rate will increase with the increase of Tos, and the tertiary industry will begin to play a stronger role in promoting it. (c) Taking economic urbanization as the threshold variable, when it is at the first threshold (eco_urban $\leq$ 32.27%), the Tos regression coefficient is −0.21 ($p < 0.01$), indicating that the secondary industry can better promote the improvement of the comprehensive urbanization rate; when it is located in the second threshold range (32.27% < eco_urban $\leq$ 59.06%) and crosses the second threshold (59.06% $\leq$ eco_urban), the Tos regression coefficients are 0.74 ($p < 0.01$) and 4.75 ($p < 0.01$) respectively, indicating that the role of the tertiary industry in promoting the urbanization rate has exceeded that of the secondary industry.

Generally, when the urbanization rate is lower than 38%, the role of the secondary industry in promoting the urbanization rate is very obvious, and the role of the tertiary industry is weak; when the urbanization rate exceeds 38%, the role of the tertiary industry is gradually enhanced, while the role of the secondary industry is gradually weakened; when the urbanization rate exceeds 59%, the role of the tertiary industry in promoting the urbanization rate is more obvious, and its role is far more than that of the secondary industry.

5. Discussion

This study comprehensively reveals the temporal and spatial dynamic pattern of county urbanization in China from 1995 to 2015. We answered four questions: "How to measure the level of urbanization?"; "How high is the level of county urbanization in China?"; "What are the characteristics of geographical imbalance?"; and "What is the driving mechanism of urbanization?".

5.1. The Rationality of Index Construction

The comprehensive urbanization index we constructed can more scientifically measure the real urbanization development level. On the one hand, closely adhering to the definition of urbanization, we combined multi-dimensional (population, land, and economy) and multi-source data (population statistics, land use, and night light), which solved the limitations of single indicators [22,79] and the traditional attribute of comprehensive indicators [80] in the existing literature, and finally provided a new urbanization measurement index. On the other hand, in order to test the objectivity of comprehensive urbanization constructed, the indexes of permanent resident population urbanization [81,82] commonly used in scholar and government statistics are selected and compared with comprehensive urbanization to reflect the differences between the two measurement methods.

In 2000, the average permanent resident urbanization rate of China's 2851 counties was 37.80 %. In 2010, it was 46.65%, with an average annual growth rate of 4.03%. According to the comprehensive urbanization rate measurement, the average urbanization rate of 2851 counties was only 33.54% in 2000, 41.39% in 2010, and the average annual growth rate was 2.80%. As can be seen from Figure 3, the urbanization rate of permanent residents was higher than the comprehensive urbanization rate in both 2000 and 2010. At the same time, in the permanent population statistics, the northwest border area belonged to the area with a high urbanization rate. However, in the comprehensive urbanization based on the combination of population, land and economy, the urbanization rate of the northwest border region declined to a certain extent, and the "false high urbanization error" in the northwest border region was reduced. In addition, the comprehensive urbanization measurement method highlighted the core status of the Beijing-Tianjin-Hebei region, the Yangtze River Delta, and the Pearl River Delta.

In order to make a more detailed distinction between the two types of urbanization, the urbanization rate is divided into five intervals, and the number of counties in each interval is counted (Figure 4). In 2000, the number of counties with a resident population urbanization rate of 15–30% was 884, accounting for 31% of the number of counties in China; the number of counties with a comprehensive urbanization rate of 15–30% was 1246, accounting for 44% of the number of countries in China. The number of counties in the urbanization rate of 70–100% also had a large difference between the two calibers. Among them, the number of counties in the ordinary population statistics caliber and the comprehensive caliber were respectively 573 and 306. This showed that the urbanization rate of permanent residents in 2000 was higher than the comprehensive urbanization rate. In 2010, the number of counties with a resident population urbanization and a comprehensive urbanization rate of 30–50% were respectively 1072 (accounting for 37.6%) and 1016 (accounting for 35.6%). However, the number of comprehensive urbanization counties with an urbanization rate in the range of 50–100% was 750, which was less than 936 counties with a permanent population. It further indicated that the urbanization rate of a permanent resident population overestimated the urbanization process at the county level in China.

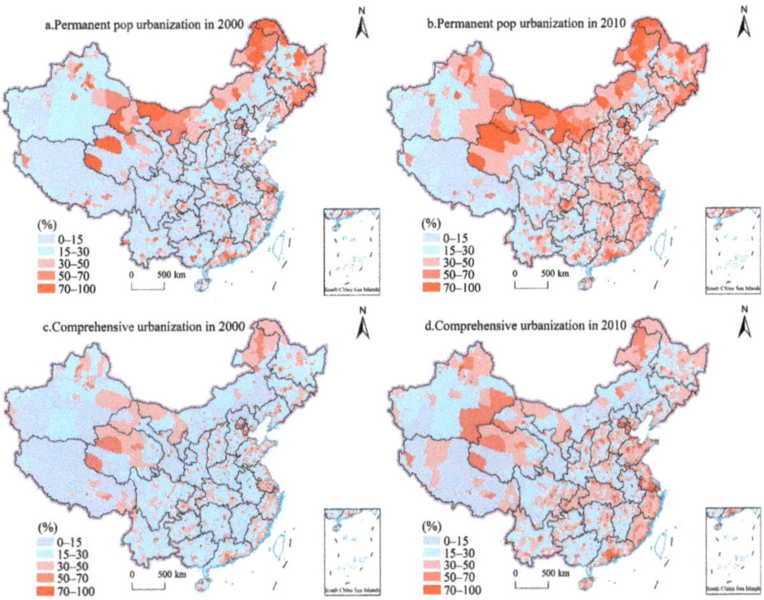

Figure 3. Comparison between permanent resident population urbanization and comprehensive urbanization.

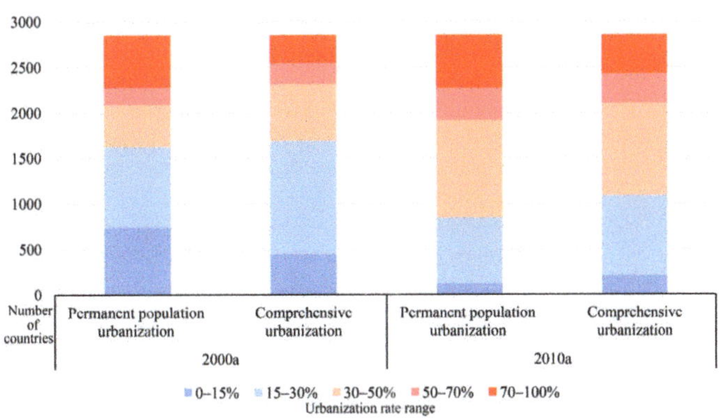

Figure 4. Number of counties with permanent population urbanization and comprehensive urbanization in different urbanization rates.

The measurement results of the two methods are similar. However, the comparison results show the urbanization level of the permanent population may overestimate China's urbanization process, which is also consistent with existing research results [14,15,27].

5.2. The High-Quality Development of Urbanization

First, narrowing regional differences is the primary task for achieving high-quality urbanization development. Our study found that there are obvious regional differences in population, economy, land, and comprehensive urbanization [83,84]. Although China has adopted regional strategies (such as the western development strategy and rising strategy in the central region), the regional gap is still large. On the basis of the difference between east and west, the widening of the difference between north and south has become a new regional development problem in China [85,86]. In the future, in-depth analysis of the formation mechanism of regional differences is an important research direction.

Second, "people" urbanization should be the core of high-quality urbanization, but it is still underestimated. By measuring the rate of population, economy, and land urbanization, respectively, it is found that the rate of population urbanization is lower than that of economic urbanization and land urbanization both in terms of development level and speed. Overemphasis on economic and land development and neglect of population urbanization will easily cause great pressure on the environment [87] and even widen the gap between the rich and the poor [17,66]. In the future, the government should continue to firmly promote the "three hundred million" policy, promote the settlement of 100 million agricultural transfer population in cities and towns, improve the living conditions of 100 million people, and guide 100 million people to urbanize nearby in the central and western regions.

Third, the optimization of industrial structure is one of the necessary ways to realize high-quality urbanization. The government promoted the development of early urbanization [88,89]. With economic development and marketization, the government's role in promoting urbanization has gradually weakened, and county economic development and industrial structure have become important factors for the improvement of urbanization. At the same time, there is an obvious nonlinear relationship between the impact of Tos on urbanization, and 38% and 59% can be used as important nodes of urbanization [90]. Each county at different stages of development should adjust its industrial structure reasonably according to local conditions. Therefore, in the early and middle stages of urbanization, we should vigorously develop the secondary industry; with the increase in urbanization rate, counties should gradually pay attention to the development of the tertiary industry [91,92].

5.3. Application of Night Light Data

Due to the close relationship between lighting data and economic development [69,70], light data provided a valuable data source for elucidating the dynamics of China's urbanization [71–73,93]. At the same time, due to the attributes of a long time scale and high spatial resolution, light data can not only estimate the urbanization development at the national, regional, and city scales but also play a role at the county and grid scales. For example, Amaral [94] used DMSP/OLS data to estimate the population of the Amazon region and realized the population estimation in the areas lacking statistical data, which showed that light data has strong application at the regional scale; Yu [95] used light data to build a 500-m population grid, proving that light data can also be applied on a microscopic scale. Through the literature review, the research on using light to represent urbanization is mainly divided into two aspects. On the one hand, based on the area attribute of light data, by distinguishing between bright cities and dark villages, the information on the urban boundary is extracted [96], and finally, the study of urbanization is realized. For example, Yang [97] proposed a method to measure land urbanization level based on DMSP/OLS nighttime light data of long time series and takes the Bohai Rim region as an example to study the spatio-temporal measurement of land urbanization level from 1992 to 2010. Liu [98] explored the urbanization process of Shanxi, Shaanxi, and Mongolia by using the area attribute of lighting. On the other hand, the intensity attribute of light data is used to reflect the urbanization process. For example, Gao [76] constructed urbanization indicators through lighting brightness attributes and analyzed the dynamics of urbanization levels in China from 1992 to 2012. However, due to the simple information contained in light data, it cannot fully reflect the information of the research object, and its advantages cannot be fully utilized, but multi-source data fusion can solve these problems [99]. Therefore, we use multi-source data fusion for research, which not only makes full use of the advantages of lighting data but also can more comprehensively and scientifically reflect the process of urbanization in China.

Although light data has advantages in reflecting urbanization development, the shortcomings of the data weaken the accuracy of the nighttime light data in quantitative analysis. The first problem is the time scale of the data. Although DMSP/OLS provided a long time data set, it was only updated to 2013; NPP/VIIRS data are dated, but the earliest data date is only 2012, which makes it difficult to carry out research on a long time scale. The second problem is the spatial scale of data. The grid size of DMSP/OLS data is about 1 km, while the grid size of NPP/OLS data is about 500 m, so the fusion of the two kinds of data is limited to some extent. At the same time, DMSP/OLS with a 1 km grid size will have large errors when studying small scales. The third problem is the brightness value of the light. Due to sensor problems, the maximum value of DMSP/OLS data is only 63, which will lead to large errors in estimating economic and urbanization development.

In the future, solving the problems of lighting data is a research focus. The first is the fusion of DMSP/OLS data and NPP/VIIRS data. In order to construct light data at a longer time scale, it is of great significance to combine the two data sources scientifically and effectively in future research on urbanization. The second is light data correction. A good solution to the data quality problem (the value range of DMSP/OLS data) can reduce the error in the application research. The third is the fusion of multi-source data. The nighttime light data information is relatively simple. Compared with single lighting data, the fused data can give full play to the respective advantages of multi-source data (such as statistical data and lighting data, vector data and lighting data) to expand the application scenarios of data.

5.4. Limitations and Future Work

This study still had some limitations. Firstly, due to the limitation of data availability, the time period for this study was from 1995 to 2015, and the development of county urbanization in China after 2015 was not measured and analyzed. Secondly, in the process of using social statistics, night light, and land use remote sensing data to construct comprehensive urbanization indicators at the county level in China, the weights of all three are set at 1/3, which may cause errors in some regions. Thirdly, because of the difficulty of obtaining population data, this paper uses registered residence populations to construct a comprehensive urbanization rate. However, the number of floating populations in China has increased over the past years, and the use of registered residence populations may cause some errors. Further, in addition to the transformation of population, land, and economy, the transformation of rural culture and urban culture is also an important symbol of the process of urbanization. Limited by the data, we do not measure the transformation of cultural elements. Finally, since DMSP lighting data is only updated to 2013, we used 2013 data instead of 2015, which would also cause some errors. In future research, the extension of the time scale, the acquisition of population data and the scientific determination of weight are the key points. For the measurement of urbanization, incorporating more elements such as culture and society into the model will more comprehensively and truly reflect the process of urbanization. For lighting data, it is also one of the key points to integrate DMSP data and NPP data more scientifically to form a set of longer time data sets and apply it to the research of social and economic activities.

6. Conclusions

In order to measure and characterize the county urbanization in China from 1995 to 2015 in detail, this paper constructs a comprehensive urbanization index by fusing the demographic statistics, nighttime lights data, and land use remote sensing data. Moreover, we use the regression model to analyze the factors affecting the county urbanization level. Finally, the following conclusions could be drawn:

(a) In 1995, the average urbanization rate of 2851 counties of the registered population urbanization rate, economic urbanization rate, and land urbanization rate was 27.49%, 31.22%, and 34.46% respectively, and in 2015, they reached 38.62%, 46.68%, and 50.37%, respectively. The urbanization of registered populations presented a "herring-shaped" pattern consisting of the high-value areas on the northern border and the high-value areas in the eastern coastal areas, and it would further strengthen as time went on; economic urbanization presented a pattern of "high in the east and low in the west"; and land urbanization presented a pattern of "high in the south and low in the north", which was more obvious than the differentiation between the east and the west. In terms of development level, population urbanization rate was lower than economic urbanization rate, while economic urbanization rate was lower than land urbanization rate. In terms of development speed, the economic urbanization rate was faster than the land urbanization rate, and the land urbanization rate was faster than the population urbanization rate.

(b) By integrating the three elements of population, economy, and land, the comprehensive urbanization rate was constructed and calculated. The average comprehensive urbanization rate of 2851 counties in China was 31.06% in 1995 and 45.23% in 2015. The spatial and temporal dynamics of comprehensive urbanization level at the county level in China were significantly different, and the overall distribution still followed the "Hu Line". The comprehensive urbanization rate east of the line was higher, while the comprehensive urbanization rate west of the line was lower. However, with the passage of time, the comprehensive urbanization rate in western inland areas gradually increased. Counties with high urbanization rates were mainly concentrated in the eastern coastal areas, forming a high-value belt along the eastern coastal areas. Beijing, Shanghai, and Guangzhou were the top three.

(c) The urbanization rate of permanent resident populations overestimates the urbanization process at the county level in China. In 2000, the average permanent urbanization rate of 2851 counties was 37.80%, and the comprehensive urbanization rate was 33.54%. In 2010, the two types of urbanization rates were 46.65% and 41.39%, respectively.

(d) Large population size, reasonable industrial structure, and strong government capacity can play a positive role in promoting urbanization. In the industrial structure, the primary industry will limit the development of urbanization, but the promotion effect of secondary and tertiary industries on urbanization is different in different counties

(e) There is an obvious nonlinear relationship between urbanization rate and Tos. When the urbanization rate is lower than 38%, the role of the secondary industry in promoting the urbanization rate is obvious; when the urbanization rate exceeds 38%, the role of the tertiary industry is gradually enhanced, and the secondary industry is gradually weakened; when the urbanization rate exceeds 59%, the role of the tertiary industry is more obvious and far more than that of the secondary industry.

Through the research of this paper, we found it was feasible to use the multi-source data fusion of demographic statistics, night lights, and land use remote sensing to construct a comprehensive urbanization index, as well as meaningful to comprehensively and truly reflect China's urbanization at the county scale and guide the future development of urbanization.

Author Contributions: Conceptualization, B.Z. and C.M.; methodology, B.Z. and C.M.; software, B.Z. and J.Z.; validation, B.Z. and C.M.; formal analysis, B.Z. and J.Z.; resources, B.Z. and C.M.; data curation, B.Z. and C.M.; writing—original draft preparation, B.Z. and C.M.; writing—review and editing, B.Z., C.M. and J.Z.; visualization, B.Z., C.M. and J.Z.; supervision, C.M.; project administration, C.M.; funding acquisition, C.M. All authors have read and agreed to the published version of the manuscript.

Funding: This research was funded by the National Natural Science Foundation of China (Fund-project host: Miao Changhong), grant number 42171186; The Major Project of China National Social Science Fund in Art (Fundation sub-project host: Miao Changhong), grant number 21ZD03.

Data Availability Statement: (a) China's provincial, municipal, and county-level administrative units are China's 1:250,000 basic geographic data provided by the Resources and Environment science data of the Chinese Academy of Sciences at http://www.resdc.cn/; (b) The registered population data of each county in the calculation of population urbanization come from the National Population Statistics of Counties and Cities of the People's Republic of China in 1995, 2000, 2005 and 2010 at https://data.cnki.net/yearbook; (c) The land use data of 1995, 2000, 2005, 2010 and 2015 are obtained from the Data Center of Resources and Environmental Sciences, Chinese Academy of Sciences, with a resolution of 30 m at http://www.resdc.cn/; (d) The stable night light data for economic urbanization calculation is provided by NGDC website at https://ngdc.noaa.gov/eog/dmsp; (e) The social statistics for regression are obtained from statistical yearbook at https://data.cnki.net/yearbook. All data accessed on 16 July 2020.

Conflicts of Interest: The authors declare no conflict of interest.

References

1. Shen, L.; Zhou, J. Examining the effectiveness of indicators for guiding sustainable urbanization in China. *Habitat Int.* **2014**, *44*, 111–120. [CrossRef]
2. Katz, C. Splanetary Urbanization. *Int. J. Urban Reg. Res.* **2021**, *45*, 597–611. [CrossRef]
3. Ruddick, S.; Peake, L.; Tanyildiz, G.S.; Patrick, D. Planetary urbanization: An urban theory for our time? *Environ. Plan. D Soc. Space* **2018**, *36*, 387–404. [CrossRef]
4. Yu, S.; Zhang, Z.; Liu, F. Monitoring Population Evolution in China Using Time-Series DMSP/OLS Nightlight Imagery. *Remote Sens.* **2018**, *10*, 194. [CrossRef]
5. Zhang, L.; Wang, Y.H.; Guo, Y.N. A Research on the Coordination of Land Urbanization and Economic Urbanization in China. *East China Econ. Manag.* **2016**, *30*, 111–117. [CrossRef]
6. Zhang, R.T.; Jiao, H.F. A review on new urbanization research in China. *World Reg. Stud.* **2016**, *25*, 59–66. [CrossRef]
7. Pu, B.; Qiu, Y. A Bibliometric Analysis on Urbanization Research From 1984 to 2013. *Open House Int.* **2015**, *40*, 37–43. [CrossRef]
8. Wang, H.J.; He, Q.Q.; Liu, X.J.; Zhuang, Y.H.; Hong, S. Global urbanization research from 1991 to 2009: A systematic research review. *Landsc. Urban Plan.* **2012**, *104*, 299–309. [CrossRef]
9. Liu, X.L.; Wang, Y.; Li, Y.; Wu, J.S. Quantifying the Spatio-Temporal Process of Township Urbanization: A Large-Scale Data-Driven Approach. *ISPRS Int. J. Geo-Inf.* **2019**, *8*, 389. [CrossRef]
10. Northam, R.M. *Urban Geography*; John Wiley & Sons: New York, NY, USA, 1967; pp. 12–32.
11. Hamnett, C. Is Chinese urbanisation unique? *Urban Stud.* **2020**, *57*, 690–700. [CrossRef]
12. Yang, Y.; Tang, Y.L.; Jia, Y.Y.; Zhan, Q.Q. Spatial-temporal coupling coordination and driving factors of population-land-economy urbanization in the Yangtze River Basin. *World Reg. Stud.* **2020**, *30*, 978–990. [CrossRef]
13. Gao, X.; Cao, M.Q.; Zhang, Y.R.; Liu, Y. Towards sustainability: An assessment of an urbanisation bubble in China using a hierarchical—Stochastic multicriteria acceptability analysis—Choquet integral method. *J. Clean. Prod.* **2021**, *279*, 123650. [CrossRef]
14. Duan, C.; Zou, X. China's urban population exceeding half: Challenges and responses. *Popul. Res.* **2012**, *36*, 45–49.
15. Hu, B.L.; Chen, C.L. New Urbanisation under Globalisation and the Social Implications in China. *Asia Pac. Policy Stud.* **2015**, *2*, 34–43. [CrossRef]
16. Gu, C.L.; Wu, L.Y. Studies of China's urbanization (II). *J. Urban Reg. Plan.* **2008**, *1*, 100–163.
17. Lu, D.D.; Song, L.F.; Ren, P. China's urbanization development model: How to develop scientifically. *J. Suzhou Univ. (Philos. Soc. Sci.)* **2007**, *2*, 1–7. [CrossRef]
18. Lu, D.D.; Yao, S.M. A scientific thought about urbanization progress in China. *Hum. Geogr.* **2007**, *4*, 1–5. [CrossRef]
19. Hu, H.Y. Population distribution in China with statistical table and density map. *Acta Geogr. Sin.* **1935**, *2*, 33–74.
20. Gao, S.Z.; Zhu, N. Regional differences and regional development models of urbanization in China. *Popul. Sci. China* **1993**, *1*, 7–16.
21. Liu, Y.S.; Yang, R. The spatial characteristics and formation mechanism of the county urbanization in China. *Acta Geogr. Sin.* **2012**, *67*, 1011–1020. [CrossRef]
22. Zhu, C.G.; Sun, S.S.; Li, Z.J. The influencing factors and spatial distribution of population urbanization in China. *Geogr. Res.* **2008**, *27*, 13–22. [CrossRef]
23. Ma, H.; Sun, Z. Comprehensive urbanization level and its dynamic factors for five Central Asian countries. *Acta Geogr. Sin.* **2020**, *76*, 367–382. [CrossRef]

24. Megeri, M.N.; Kadi, A.S.; Kengnal, P. Measures of Urbanization in India: A Statistical Analysis. *Int. J. Agric. Stat. Sci.* **2012**, *8*, 585–596.
25. Onda, K.; Sinha, P.; Gaughan, A.E.; Stevens, F.R. Missing millions: Undercounting urbanization in India. *Popul. Environ.* **2019**, *41*, 126–150. [CrossRef]
26. Yuan, Y. A Discussion on Urbanization Level of Registered Population. *Sci. Econ. Soc.* **2015**, *33*, 115–120.
27. Qi, W.; Liu, S.H.; Jin, H.R. Calculation method and pattern of urbanization rate of population in China. *Geogr. Res.* **2017**, *36*, 616–632. [CrossRef]
28. Long, H.L.; Qu, Y. Land use transitions and land management: A mutual feedback perspective. *Land Use Policy* **2018**, *74*, 111–120. [CrossRef]
29. Jiang, A.L. Discuss on different measurement methods of urbanization level. *J. Cent. Univ. Financ. Econ.* **2002**, *8*, 76–80.
30. Lv, P.; Zhou, T.; Zhang, Z.Z.; Tian, Z. Construction and application of land urbanization and corresponding measurement index system. *China Land Sci.* **2008**, *22*, 24–28.
31. Zhang, K.H.; Qian, Q.L.; Yang, Q.S. An analysis of multilevel variables influencing China's land urbanization process. *Acta Geogr. Sin.* **2020**, *75*, 179–193. [CrossRef]
32. Lin, J. *Urban-Rural Construction Land Growth in China*; The Commercial Press: Beijing, China, 2009; pp. 4–32.
33. Ren, Y. Balanced urbanization and connotative path to future urbanization. *J. Tongji Univ. (Soc. Sci. Sect.)* **2018**, *29*, 58–68.
34. Gao, J.L.; Bao, J.W.; Liu, Y.S.; Chen, J.L. Regional disparity and the influencing factors of land urbanization in China at the county level. *Acta Geogr. Sin.* **2018**, *73*, 2329–2344. [CrossRef]
35. Asabere, S.B.; Acheampong, R.A.; Ashiagbor, G.; Beckers, S.C.; Keck, M.; Erasmi, S.; Schanze, J.; Sauer, D. Urbanization, land use transformation and spatio-environmental impacts: Analyses of trends and implications in major metropolitan regions of Ghana. *Land Use Policy* **2020**, *96*, 104707. [CrossRef]
36. Vogler, J.B.; Vukomanovic, J. Trends in United States Human Footprint Revealed by New Spatial Metrics of Urbanization and Per Capita Land Change. *Sustainability* **2021**, *13*, 12852. [CrossRef]
37. Malaque, I.I.R.; Yokohari, M. Urbanization process and the changing agricultural landscape pattern in the urban fringe of Metro Manila, Philippines. *Environ. Urban.* **2007**, *19*, 191–206. [CrossRef]
38. Sun, P.J.; Ding, S.B. Spatial differentiation on northeast China urbanization based on the perspective of population -economy-space. *Econ. Geogr.* **2011**, *31*, 1094–1100. [CrossRef]
39. Cheng, L.; Zhou, Z. Research on the Coordination and Interaction between population urbanization and economic urbanization. *Theory Mon.* **2014**, *1*, 119–122. [CrossRef]
40. Wang, Y.L.; Peng, B.F.; Xiong, J.X.; Wang, Q. The economy urbanization and population urbanization of Dongting lake area in China since 2001. *Sci. Geogr. Sin.* **2014**, *34*, 67–75. [CrossRef]
41. Liu, F.W.; Xu, H.Z.; Wang, S. Analysis on spatial-temporal coupling coordinate degree among population, land and economy urbanization: Based on China provincial panel data. *Urban Dev. Stud.* **2014**, *21*, 7–11.
42. Fox, S.; Bloch, R.; Monroy, J. Understanding the dynamics of Nigeria's urban transition: A refutation of the 'stalled urbanisation' hypothesis. *Urban Stud.* **2018**, *55*, 947–964. [CrossRef]
43. Chen, M.; Lu, D.; Zhang, H. Comprehensive evaluation and the driving factors of China's urbanization. *Acta Geogr. Sin.* **2009**, *64*, 387–398. [CrossRef]
44. Xiong, X.H.; Xu, Z.Y. Research on level and mechanical under the guidance of new urbanization. *J. Quant. Tech. Econ.* **2018**, *35*, 44–63. [CrossRef]
45. Guo, F.Y.; Li, C.G.; Chen, C. Spatial-Temporal coupling characteristics of population urbanization and land urbanization in northeast China. *Econ. Geogr.* **2015**, *35*, 49–56. [CrossRef]
46. Wu, Y.F.; Liu, Y.S.; Li, Y.R. Spatio-temporal coupling of demographic-landscape urbanization and its driving forces in China. *Acta Geogr. Sin.* **2018**, *72*, 1865–1879. [CrossRef]
47. Yang, Y.Y.; Liu, Y.S.; Li, Y.R. Measure of urban-rural transformation in Beijing-Tianjin-Hebei region in the new millennium: Population-land-industry perspective. *Land Use Policy* **2018**, *79*, 595–608. [CrossRef]
48. He, S.W.; Shao, X. Spatial clustering coupling coordination of population-lan-economic urbanization in Beijing-Tianjin-Hebei region. *Econ. Geogr.* **2018**, *38*, 95–102. [CrossRef]
49. Shi, Y.J.; Zhu, Q.; Xu, L.H. Independent or Influential? Spatial-Temporal Features of Coordination Level between Urbanization Quality and Urbanization Scale in China and Its Driving Mechanism. *Int. J. Environ. Res. Public Health* **2020**, *17*, 1587. [CrossRef]
50. Wang, G.G. Urbanization: Core of China Economic Development Mode Transition. *Econ. Res. J.* **2010**, *45*, 70–81.
51. Zhong, L.N.; Li, X.N.; Law, R.; Sun, S. Developing Sustainable Urbanization Index: Case of China. *Sustainability* **2020**, *11*, 4585. [CrossRef]
52. Zeng, C.; Deng, X.; Dong, J.; Hu, P. Urbanization and Sustainability: Comparison of the Processes in "BIC" Countries. *Sustainability* **2016**, *8*, 400. [CrossRef]
53. Li, L.; Zhao, K.; Wang, X.; Zhao, S. Spatio-Temporal Evolution and Driving Mechanism of Urbanization in Small Cities: Case Study from Guangxi. *Land* **2022**, *11*, 415. [CrossRef]
54. Fan, P.L.; Ouyang, Z.T.; Nguyen, D.D.; Nguyen, T.T. Urbanization, economic development, environmental and social changes in transitional economies: Vietnam after Doimoi. *Landsc. Urban Plan.* **2019**, *187*, 145–155. [CrossRef]
55. Fan, F.; He, H.; Hu, T.; Yan, S. An assement method of urban degree in Syria War. *Sci. Surv. Mapp.* **2018**, *43*, 129–134.

56. Andersen, H.T.; Møller-Jensen, L.; Engelstoft, S. The End of Urbanization? Towards a New Urban Concept or Rethinking Urbanization. *Eur. Plan. Stud.* **2011**, *19*, 595–611. [CrossRef]
57. Gao, J.; Zhang, Y.-C.; Zhou, T. Computational socioeconomics. *Phys. Rep.* **2019**, *817*, 1–104. [CrossRef]
58. Elvidge, C.D.; Baugh, K.E.; Kihn, E.A.; Kroehl, H.W.; Davis, E.R. Mapping city lights with nighttime data from the DMSP operational linescan system. *Eng. Remote Sens.* **1997**, *63*, 727–734. [CrossRef]
59. Elvidge, C.D.; Ziskin, D.; Baugh, K.E.; Tuttle, B.T.; Ghosh, T.; Pack, D.W.; Erwin, E.H.; Zhizhin, M. A Fifteen Year Record of Global Natural Gas Flaring Derived from Satellite Data. *Energies* **2009**, *2*, 595–622. [CrossRef]
60. Croft, T.A. Nighttime Images of the Earth from Space. *Sci. Am.* **1978**, *239*, 86–98. [CrossRef]
61. Li, D.; Li, X. An Overview on Data Mining of Nighttime Light Remote Sensing. *Acta Geod. Cartogr. Sin.* **2015**, *44*, 591–601. [CrossRef]
62. Yu, B.L.; Wang, C.X.; Gong, W.K.; Chen, Z.Q. Nighttime light remote sensing and urban studies: Data, methods, applications, and prospects. *Natl. Remote Sens. Bull.* **2021**, *25*, 342–364. [CrossRef]
63. Elvidge, C.D.; Sutton, P.C.; Ghosh, T.; Tuttle, B.T. A global poverty map derived from satellite data. *Comput. Geosci.* **2009**, *35*, 1652–1660. [CrossRef]
64. Wang, M.M.; Wang, J.L. Spatialization of township-level population based on nighttime light and land use data in Shandong province. *J. Geo-Inf. Sci.* **2019**, *21*, 699–709. [CrossRef]
65. Lin, Z.L.; Xu, H.Q.; Huang, S.L. Monitoring of the urban expansion dynamics in China's east coast using DMSP/OLS nighttime light imagery. *J. Geo-Inf. Sci.* **2019**, *21*, 1074–1085. [CrossRef]
66. Sutton, P.; Roberts, D.; Elvidge, C.D.; Baugh, K. Census from Heaven: An estimate of the global human population using night-time satellite imagery. *Int. J. Remote Sens.* **2001**, *22*, 3061–3076. [CrossRef]
67. Liu, S.W.; Zhang, P.Y.; Jiang, X.L. Measuring sustainable urbanization in China: A case study of the coastal Liaoning area. *Sustain. Sci.* **2013**, *8*, 585–594. [CrossRef]
68. Gibson, C. Rural Transformation and Cultural Industries: Popular Music on the New South Wales Far North Coast. *Aust. Geogr. Stud.* **2002**, *40*, 337–356. [CrossRef]
69. Keola, S.; Andersson, M.; Hall, O. Monitoring Economic Development from Space: Using Nighttime Light and Land Cover Data to Measure Economic Growth. *World Dev.* **2015**, *66*, 322–334. [CrossRef]
70. Ghosh, T.; Anderson, S.; Powell, R.L.; Sutton, P.C.; Elvidge, C.D. Estimation of Mexico's Informal Economy and Remittances Using Nighttime Imagery. *Remote Sens.* **2009**, *1*, 418–444. [CrossRef]
71. Zhuo, L.; Shi, P.J.; Chen, J.; Ichinose, T. Application of compounded night light index derived from DMSP/OLS data to urbanization analysis in China in the 1990s. *Acta Geogr. Sin.* **2003**, *58*, 893–902.
72. Yue, W.; Gao, J. Estimation of Gross Domestic Product Using Multi-Sensor Remote Sensing Data: A Case Study in Zhejiang Province, East China. *Remote Sens.* **2014**, *6*, 7260–7275. [CrossRef]
73. Shao, Z.; Liu, C. The Integrated Use of DMSP-OLS Nighttime Light and MODIS Data for Monitoring Large-Scale Impervious Surface Dynamics: A Case Study in the Yangtze River Delta. *Remote Sens.* **2014**, *6*, 9359–9378. [CrossRef]
74. Chen, J.; Zhuo, L.; Shi, P. The study on urbanization process in China based on DMSP/OLS data: Development of a light index for urbanization level estimation. *J. Remote Sens.* **2003**, *7*, 168–175. [CrossRef]
75. Zhou, Y.; Li, L.; Liao, F. Analysis the spatial-temporal pattern of the urbanization of Chengdu metropolitan region using nighttime light image. *Urban Dev. Stud.* **2015**, *22*, 28–32.
76. Gao, B.; Huang, Q.X.; He, C.Y. Dynamics of Urbanization Levels in China from 1992 to 2012: Perspective from DMSP/OLS Nighttime Light Data. *Remote Sens.* **2015**, *7*, 1721–1735. [CrossRef]
77. Li, Y.; Ye, H.P.; Gao, X.; Sun, D. Spatiotemporal Patterns of Urbanization in the Three Most Developed Urban Agglomerations in China Based on Continuous Nighttime Light Data (2000–2018). *Remote Sens.* **2021**, *13*, 2245. [CrossRef]
78. Fox, S.; Goodfellow, T. *Cities and Development*; Routledge: London, UK, 2016; pp. 5–21.
79. Lu, X.; Li, J.; Duan, P.; Li, C. Correction of nighttime light images of DMSP /OLS in China. *Bull. Surv. Mapp.* **2019**, *7*, 127–131. [CrossRef]
80. Zhu, J.G.; Xu, J.W.; Li, X.J.; Lou, F. Impact of land urbanization and population urbanization on economic growth in China. *Sci. Geogr. Sin.* **2020**, *40*, 1654–1662. [CrossRef]
81. Ye, C.; Yang, D.Y.; Zhao, J.N. An empirical research of the registered population transformation in China's megacities. *Acta Geogr. Sin.* **2022**, *77*, 369–380. [CrossRef]
82. Wang, J.; Li, Y. Spatial pattern and influencing factors of urbanization development in China at county level: A quantitative analysis based on 2000 and 2010 census data. *Acta Geogr. Sin.* **2016**, *71*, 621–636. (In Chinese) [CrossRef]
83. Li, J.M.; Lu, D.D.; Xu, C.D.; Li, Y. Spatial heterogeneity and its changes of population on the two sides of Hu Line. *Acta Geogr. Sin.* **2017**, *72*, 148–160. [CrossRef]
84. Chen, M.; Gong, Y.; Li, Y.; Lu, D.; Zhang, H. Population distribution and urbanization on both sides of the Hu Huanyong Line: Answering the Premier's question. *J. Geogr. Sci.* **2016**, *26*, 1593–1610. [CrossRef]
85. Dong, X.B.; Chi, R.N. Characteristics of the Temporal and Spatial Pattern of the Economic Disparity and Convergence Between Different Regions in China. *Econ. Geogr.* **2020**, *40*, 11–21. [CrossRef]
86. Fan, J.; Zhao, J.; Guo, R. The New Trend and Coping Strategies of Regional Development Gap in China. *Econ. Geogr.* **2022**, *42*, 1–11. [CrossRef]

87. Fan, J.; Zhao, D.T. Measurement of coordination between land urbanization and population urbanization and its influencing factors. *Economist* **2012**, *161*, 61–67. [CrossRef]
88. Ong, L.H. State-Led Urbanization in China: Skyscrapers, Land Revenue and "Concentrated Villages". *China Q.* **2014**, *217*, 162–179. [CrossRef]
89. Qiu, Y. The Role of Local Government in China's Urbanization: The Relationship between Local Land Finance And Government-Led Urbanization. Available online: https://ir.lib.uwo.ca/lgp-mrps/148 (accessed on 1 July 2020).
90. Chen, X.; Xi, Q.; Li, G. Urbanization level and spatial distribution of manufacturing industry: An empirical research based on provincial panel data. *Sci. Geogr. Sin.* **2015**, *35*, 259–267. [CrossRef]
91. Chen, M.Y. Moving Forward into Quality 6 oriented Development: A Preliminary Study on China's Urbanization from the Seventh National Population Census. *Popul. Dev.* **2022**, *28*, 93–103. Available online: https://kns.cnki.net/kcms/detail/detail.aspx?dbcode=CJFD&dbname=CJFDAUTO&filename=SCRK202202008&uniplatform=NZKPT&v=5OKka5ltV4Wa8MV0wg1fzAkKKx7mJyTSW30EdPmEvnEd5cwL-Djh7JS5KiJpHvxs (accessed on 1 July 2020).
92. Fang, C.L.; Liu, X.L.; Lin, X.Q. Stages correction and regularity analysis of urbanization course of China. *Arid. Land Geogr.* **2008**, *31*, 512–523. [CrossRef]
93. Li, X.; Ge, L.; Chen, X. Detecting Zimbabwe's Decadal Economic Decline Using Nighttime Light Imagery. *Remote Sens.* **2013**, *5*, 4551–4570. [CrossRef]
94. Amaral, S.; Monteiro, A.M.V.; Camara, G.; Quintanilha, J.A. DMSP/OLS night-time light imagery for urban population estimates in the Brazilian Amazon. *Int. J. Remote Sens.* **2006**, *27*, 855–870. [CrossRef]
95. Yu, B.L.; Lian, T.; Huang, Y.X.; Yao, S. Integration of nighttime light remote sensing images and taxi GPS tracking data for population surface enhancement. *Int. J. Geogr. Inf. Sci.* **2018**, *33*, 687–706. [CrossRef]
96. He, C.Y.; Shi, P.J.; Li, J.G. Restoring urbanization process in China in the 1990s by using non-radiance-calibrated DMSP/OLS nighttime light imagery and statistical data. *Chin. Sci. Bull.* **2006**, *51*, 1614–1620. [CrossRef]
97. Yang, Y.; Huang, X.Q.; Zhang, L.L. The Spatial-Temporal Measurement on the Land Urbanization Level Using DMSP/OLS Nighttime Light Data—A Case Study of Bohai Rim. *Econ. Geogr.* **2015**, *35*, 141–148+168. [CrossRef]
98. Liu, Y.X.; Wu, W.H.; Wen, X.J.; Zhang, D.H. Urban process and its eco-environmental impact in Shanxi-Shaanxi-Inner Mongolia energy area. *Geogr. Res.* **2013**, *32*, 2009–2020. [CrossRef]
99. Chen, Y.B.; Zheng, W.H.; Wen, X.J.; Zhang, D.H. Review and prospect of application of nighttime light remote sensing data. *Prog. Geogr.* **2019**, *38*, 205–223. [CrossRef]

remote sensing

MDPI

Technical Note

Exploring VIIRS Night Light Long-Term Time Series with CNN/SI for Urban Change Detection and Aerosol Monitoring

Changyong Cao [1,*], Bin Zhang [2], Frank Xia [3] and Yan Bai [2]

1 Center for Satellite Applications and Research, NOAA/NCWCP, College Park, MD 20740, USA
2 CISESS, University of Maryland, 5825 University Research Ct., College Park, MD 20740, USA; bin.zhang@noaa.gov (B.Z.); yan.bai@noaa.gov (Y.B.)
3 ESRI DC Regional Office, 8615 Westwood Center Dr., Vienna, VA 92373, USA; fxia@esri.com
* Correspondence: changyong.cao@noaa.gov

Abstract: There is a great need to study the decadal long-term time series of urban night-light changes since the launch of Suomi NPP, NOAA-20, to future JPSS-2, 3, and 4 in the next decades. The recently recalibrated and reprocessed Suomi NPP VIIRS/DNB dataset overcomes a number of limitations in the operational data stream for time series studies. However, new methodologies are desirable to explore the large volume of historical data to reveal long-term socio-economic and environmental changes. In this study, we introduce a novel algorithm using convolutional neural network similarity index (CNN/SI) to rapidly and automatically identify cloud-free observations for selected cities. The derived decadal clear sky mean radiance time series allows us to study the urban night light changes over a long period of time. Our results show that the radiometric changes for some metropolitan areas changed on the order of 29% in the past decade, while others had no appreciable change. The strong seasonal variation in the mean radiance appears to be highly correlated with seasonal aerosol optical thickness. This study will facilitate the use of recalibrated/reprocessed data, and improve our understanding of urban night light changes due to geophysical, climatological, and socio-economic factors.

Keywords: Suomi NPP VIIRS; recalibrated/reprocessed historical radiance data; CNN/SI; urban night light long-term time series; urban growth; aerosols

Citation: Cao, C.; Zhang, B.; Xia, F.; Bai, Y. Exploring VIIRS Night Light Long-Term Time Series with CNN/SI for Urban Change Detection and Aerosol Monitoring. *Remote Sens.* **2022**, *14*, 3126. https://doi.org/10.3390/rs14133126

Academic Editors: Ran Goldblatt, Steven Louis Rubinyi and Hogeun Park

Received: 23 May 2022
Accepted: 25 June 2022
Published: 29 June 2022

1. Introduction

The launch of the Suomi NPP satellite with the Visible Infrared Imaging Radiometer Suite (VIIRS) has inspired many novel studies in both geophysical and social economic research. In particular, the VIIRS Day/Night Band (DNB) enabled unprecedented capabilities in low light imaging at night, with the ability to detect faint lights emitted from the atmosphere or the so-called air glows at a radiance level on the order of 1.0×10^{-10} W/(cm^2-sr). Recent studies also suggest that it might be able to detect bio-luminescence if the instrument is well calibrated [1]. This is in addition to the well-publicized fishing boat detection [2], light outages [3], and night fires which are at higher radiance levels. It can also be used to detect the internal waves and possibly other oceanic surface features under sun and moon glint [4]. For social economic applications, numerous published results have correlated economic activities with night lights. Ten years of valuable global observations of night light from Suomi NPP VIIRS/DNB have been accumulated at the NOAA CLASS archive. Recent efforts at NOAA Center for Satellite Applications and Research (STAR) have recalibrated and re-processed the Suomi NPP sensor data records (or level 1b) which is made available currently through the NOAA cooperative institute at University of Maryland, and will be made available on the NOAA CLASS later [5]. In addition, NOAA-20 VIIRS was launched in November 2017 and has been operating successfully for 4+ years, and it will be followed by JPSS-2, 3, and 4 in the next decades, all equipped with VIIRS. Therefore, the night light observation data will grow significantly and greater opportunities will exist to use the night light data for geophysical, environmental, climatological, and social economic studies.

Despite the tremendous progress, the use of VIIRS DNB data is still somewhat limited for scientific research and great potential exists. There are several known challenges. First, most social economic studies use the NOAA NGDC-produced monthly composite clear sky average radiances. While this dataset works well for correlation studies with social economic variables where there is no stringent need for stability in the time series, it has drawbacks when used for time series analysis, because the NGDC data was created based on operationally produced VIIRS SDRs at NOAA operations which had a number of inconsistencies, and that makes it difficult for time series studies where stable calibration is a prerequisite. In addition, extracting time series data for a small urban area from a large global dataset can be challenging. Second, for studies using the VIIRS SDR data, obtaining clear sky images is a major challenge. This is because the traditional method relies on using the cloud masks generated from other channels of VIIRS, which is a downstream product produced separately in the data production. Applying the VIIRS cloud mask to DNB requires some skills in spatial matching especially at high scan angles since the spatial resolution difference between DNB and other bands grows with scan angle. Unfortunately, a large percentage of the satellite observations always have cloud coverage. As a result, time series analysis using the VIIRS DNB data becomes difficult. Another challenge for data access is that NOAA NGDC has ceased the production of the monthly mean VIIRS night light data as of 15 October 2019, and the follow-on effort has been transferred to the academic sector at the Colorado School of Mines (https://ngdc.noaa.gov/eog/index.html, accessed on 21 June 2022).

In this study, we present an Artificial Intelligence CNN/SI method for separating clear sky from cloudy DNB images automatically so that a time series can be generated for analyzing the long-term urban night light changes, and we demonstrate its use for selected metropolitan areas. The recalibrated/reprocessed Suomi NPP VIIRS/DNB data used in this study has much improved data quality, stability, and accuracy, and allows us to detect long term urban night light changes. It is known that urban night lights are affected by a number of other factors, such as moon light which varies greatly for a given urban area, clouds and aerosols which may attenuate and/or diffuse urban lights. In the following sections, we present the methodology and data, followed by applications for selected cities. The time series for these cities are analyzed with concluding remarks and recommended future studies.

2. Methodology and Data

The Suomi NPP and NOAA-20 satellites have an equator crossing time of ~1:30 a.m. descending local time, or near mid-night. There are advantages and disadvantages for this observation time. At midnight, the night lights are more stable compared to early evening when there would still be varying light changes due to daily human activities. Therefore, the observations from the VIIRS/DNB would be mainly for late night lights which are mostly lit overnight, including street lights. On the flip side, the late-night light does not capture the range of human activities and most likely would underestimate the light intensity over the areas of study because of the observation time at mid-night.

This study addresses two major issues in time series analysis using VIIRS DNB datasets. The first is long-term calibration stability, achieved by using recalibrated VIIRS/DNB SDR, and the second is the development of a method to perform rapid identification of clear sky images using the AI CNN/SI approach for night time images. The traditional cloud mask-based approach is rather complex since it requires a separate data set (cloud mask product), and the pixel by pixel matching is not straightforward since VIIRS DNB has a near constant spatial resolution across scan while the cloud mask is based on radiometric band observations (also known as VIIRS M-bands) which have a growing pixel size towards the high scan angles from nadir, from 750 m to 1.6 km to the edge of the scan [6]. In addition, the cloud mask is a separate product with a latency in production which means one would have to find the matching data files from the archive. There are also issues related to different versions of the cloud masks with different algorithms that make the task more complex.

To address these challenges, we present a novel methodology using artificial intelligence. In our study, we experimented with two separate AI based approaches. The first approach is an atypical method, in which image similarity comparisons are performed between a known clear sky image of a given city and all candidate images, which produces a similarity index (SI) for each pair of images based on CNN/SI. The clear sky images are found based on the similarity index in the subsequent classification, much like face matching based on photographs from a database of photographs, or genetic identification based on DNAs.

The second approach uses the typical procedure of training the CNN model to learn cloudy vs. clear night light images of selected cities. The model is then validated and used to predict whether a given night light image is clear or cloudy within a certain accuracy. Based on experiments comparing the two approaches, we found that the second approach has greater potential because once it is well trained, it is able to find clear sky images of any city; however, the current results show that the prediction accuracy is not ideal (<64%) for cloudy images. In contrast, the CNN/SI image similarity approach is more accurate although it has to be done for each city separately and with georeferencing. As a result, in this study, the CNN/SI approach is used.

The AI CNN/SI model used in this study is the popular Tensorflow/Keras with Python. The MobileNet V2 CNN [7] is used for image identification and similarity analysis between VIIRS DNB images. Images with high SI similarity are clustered and grouped together. A simpler explanation of this process is that a clear sky night light image of a city has a distinct spatial pattern, while cloudy images have a different, blurred pattern; therefore, the similarity index (SI) between the clear sky and cloudy images are lower than those between clear sky images. In addition, since cloud patterns are usually random, cloudy night light images typically do not have high similarities to other night light observations, either clear sky or cloudy. However, exceptions do exist, in which two cloudy night light images could be very similar. Therefore, in this study we introduce the known "seed" clear sky night light image as a reference in comparisons based on similarity index (SI). Similarly, lunar-contaminated night light images would have a different pattern than clear sky images acquired during the new moon, although we found that clear sky images during the new moon may have higher similarities to those during full or partial moon. Therefore, another criterion is used to screen out lunar-illuminated images, which is based on the minimum radiance within the study area (typically over large water bodies). If the minimum radiance is lower than a threshold value, then the image is considered unaffected by the lunar light. We have experimented with a threshold value between 0.5 and 1.0 nW/(cm^2-sr), and choose to use 0.6 nW/(cm^2-sr), which is the same value used by Elvidge [8] as the DNB noise floor in their study.

Figure 1 provides an overview of the method and procedure used in this study. First, a time series of near nadir VIIRS/DNB night light images over a particular city is collected from the reprocessed database. In our case, the reprocessed VIIRS SDR (or level 1b) is used, as discussed earlier. Second, the CNN/SI model is applied to the areas of interest for the city selected to generate a unique identification called "feature vector" for each image, with multiple dimensions. Third, based on the unique identification of the images, a similarity index (SI) is computed between every pair of images in this multi-dimensional space. Fourth, our algorithm then searches for images with high similarity to the known clear sky "seed" image based on similarity scores recursively. Finally, the mean radiances for the selected clear sky images are computed for the area of interest, which becomes the time series for analysis of urban night light changes.

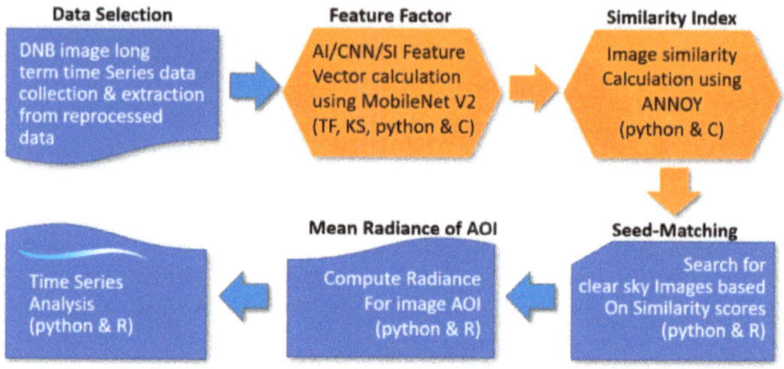

Figure 1. Algorithm Flowchart for Urban Night Light Long-Term Time Series Analysis (blue = developed by the authors; orange = adopted and integrated).

While the procedure and methodology are relatively straightforward, as presented in Figure 1, the key component of CNN/SI requires further explanation. In this study, we leverage several state-of-the-art key components that have become available in the field of artificial intelligence in the last few years, which includes Tensorflow, Keras, MobileNet V2, image feature vector, and ANNOY as discussed below:

Tensorflow is an end-to-end open-source platform for machine learning developed by Google. It has tools, libraries, and community resources that let developers easily build and deploy machine learning applications. Keras is a deep learning API written in Python, running on top of the machine learning platform Tensorflow. TensorFlow Hub provides many reusable machine learning models online. It makes transfer learning very easy, as it provides pre-trained models for different problems and tasks, such as image classification and segmentation. In this study, it is used as the basis and interface to the MobileNet V2 as discussed later.

A key concept in the AI/CNN/SI approach used in this study is the image feature vector which is a unique list of numbers or identifiers that represents a particular image, used for image similarity calculations or image classification tasks. An analogy of image feature vector or signature to an image would be the DNA to a human. It more closely resembles DNA than a signature, because each feature vector has categorical or inheritance information embedded. Low-level image features are minor details of the image, such as lines, edges, corners, or dots, while high-level features are built on top of low-level features to detect objects and larger shapes in the image. Both types of features can be detected using CNN. Some convolutional layers can capture low-level features, while the later layers learn to recognize common shapes and objects.

In this study, the CNN MobileNet V2 which is available on the Tensorflow Hub is used to extract features of VIIRS/DNB night light images. In searching for the CNN model among a number of candidates, we found that this model works well for VIIRS/DNB images with a general-purpose laptop, instead of relying on high power computers as many AI applications require. The spatial resolution of a VIIRS DNB image is 750×750 m^2, compared to the area of a typical large city <100×100 km^2. Therefore, a large city can be covered by a VIIRS DNB image with approximately 100×100 pixels. This works well with MobileNet V2 which allows array size up to 224×244. In this study, MobileNet V2 is essential in generating the image feature vector for each of the VIIRS/DNB images.

MobileNet V2 is widely used for computer vision applications in many real-world applications, including object detection, fine-grained classifications, face attributes, and localization. It is a deep neural network architecture suitable for mobile and resource constrained applications. Therefore, we envision that the methodology presented here can be implemented in cell phones or PDA devices in the future to tell whether a city is clear or cloudy in near real time. MobileNet V2 is 16 blocks (96 layers) deep. It features depth-wise separable convolution with kernel size of 3×3, with 3.4 million parameters (Figure 2).

Benchmark testing shows that its accuracy is comparable to other state-of-the-art CNN models such as VGG 16.

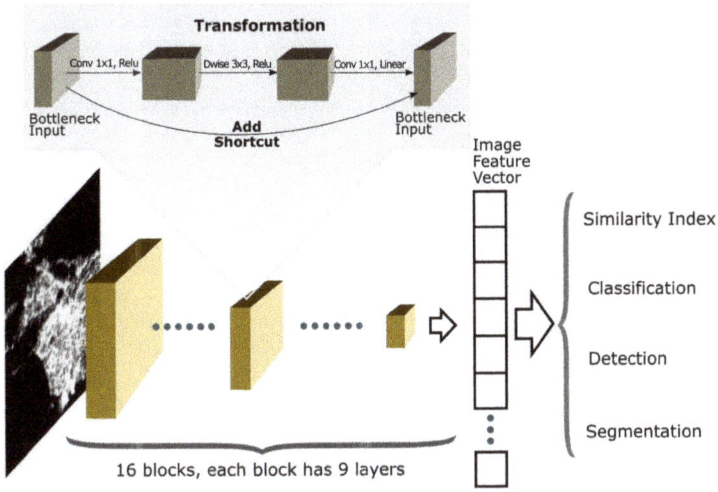

Figure 2. MobileNet V2 Architecture and Workflow.

Another key component in this AI approach is the ANNOY (Approximate Nearest Neighbor Oh Yeah), which is an open-source library for approximating nearest neighbor based on the image feature vector and similarity indices generated in the previous step. ANNOY can quickly compute the similarity index in the multidimensional space between a given pair of images based on their image feature vectors.

The following are detailed procedures for our AI/CNN/SI-based algorithm:

1. The VIIRS/DNB night light images for a city of interest are obtained from the reprocessed VIIRS/DNB dataset [5]. Only the observations near nadir for the city of interest are included in this study to avoid scan angle-dependent observation issues. Typically, there are two nadir passes near the city, one on the left side of the city center, and one on the right side. Given the orbital repeating cycle of 16 days, there are two qualified observations for each cycle within the selected area, which amounts to ~45 images per year. For the time period from 2012 to 2020, we typically obtained about ~365–380 images, which are re-projected or georeferenced onto a common grid with a constant 750×750 m^2 resolution for the given city with nearest neighbor resampling of the radiance pixels (without interpolation to preserve the radiance values). The reprojection makes it easier to perform quantitative radiometric comparisons between images and further facilitates the AI/CNN/SI similarity-based image feature vector comparisons.
2. The image feature vectors are generated for each image using Tensor-flow/Keras/MobileNet V2 so that each image has a unique feature vector ID based on the characteristics or "Image DNA". Then a cluster analysis is performed based on the similarity index (SI) of the feature vector.
3. The image pairs are grouped together based on similarity index (SI). A similarity of 1.0 means the two images are identical. All other image pairs have similarities less than 1.0, typically between 0.6 and 0.98. We found that images with the highest similarities are typically clear sky images near new moon days. This is because clear sky night light images for a given city have a fairly fixed pattern, while cloudy and lunar illuminated ones typically vary (Figure 3). Therefore, using the similarity index can effectively filter out the images contaminated by cloud and lunar illumination. However, we also found that there are cases where two cloudy images may have high

similarities in cases where the cloud is spatially uniform over the city, or the cloud optical depth is large enough to obscure the urban night lights significantly. Therefore, we introduced a step to identify a known clear sky image as the "seed" image for the clustering so that images with high similarities to the "seed" image are identified as clear sky images. The algorithm searches for images with similarity to those with clear sky recursively with a given threshold value of similarity.

4. To validate the results from the AI based algorithm, sample data are taken to compare with the traditional cloud-mask based method, which relies on a separate product generated using cloud-mask algorithms. The errors of commission vs. omission are estimated.
5. The next step is to compute the total radiance for the city of interest for the selected images in the time series. Then the total radiance change over time can be plotted as a time series, and statistical analysis using the Python and R statistical package is performed.
6. As an option, a movie loop of the selected images can be made to confirm the clear sky images selected in the study visually.

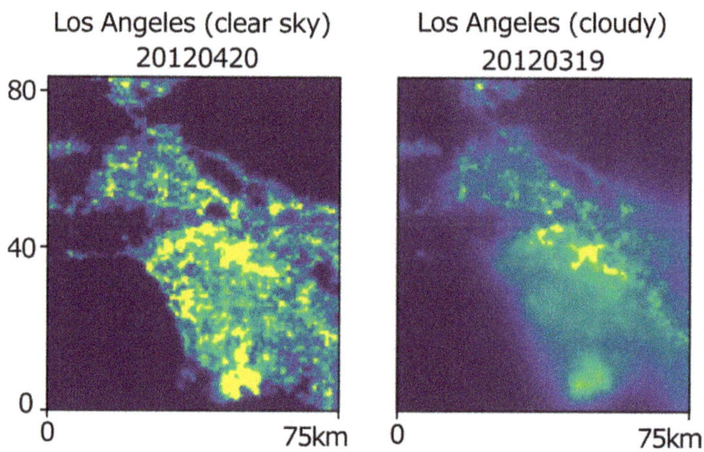

Figure 3. Clear vs. cloudy VIIRS/DNB night light image samples over Los Angeles show dissimilar patterns (unit: km).

The metropolitan areas used for this study to demonstrate our methodology included Los Angeles, Beijing, and Delhi (Figure 4). These cities are chosen based on a combination of cloud frequency, urban growth, and climate types. We believe that the methodology is applicable to all major metropolitan areas with a size greater than 50×50 km^2 in area, given the spatial resolution of VIIRS DNB of 750×750 m^2. The metropolitan areas of interest (AOI) in this study are somewhat arbitrary but selected based on several considerations. (1) The exact same area (based on latitude/longitude coordinates) is used for long term time series analysis by projecting the VIIRS DNB pixels onto the same fixed grid with cell size of 750×750 m^2 as discussed previously (Figure 4). (2) The AOI covers the downtown areas as well as adjacent suburbs where bright lights are observed from the VIIRS/DNB. (3) The AOI typically do not coincide with the administrative or political boundaries (which might change over time), compared to the fixed grid, which is based on latitude/longitude; also, the aerosols and night light distributions are not likely to follow administrative/political boundaries. (4) AOI should include dark areas such as water bodies or dark forest areas in order to exclude lunar-illuminated pixels using the minimum radiance threshold algorithm.

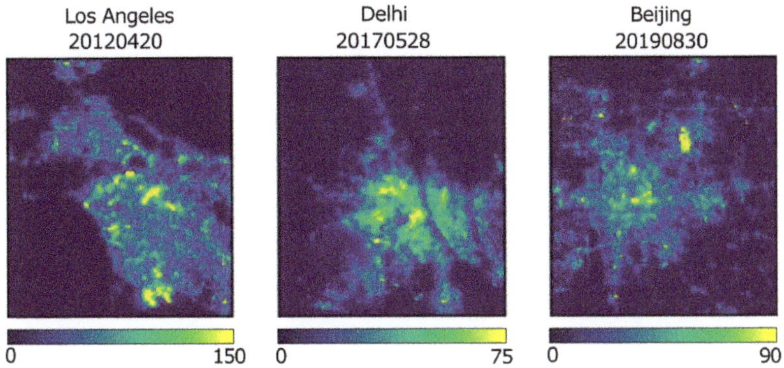

Figure 4. Sample VIIRS/DNB nightlight images (color indicates light intensity, unit: nW/cm^{-2}-sr).

3. Results and Discussion

In this section, we first evaluate the accuracy of the AI/CNN/SI method for identifying clear sky VIIRS DNB night time images in terms of error of commission vs. omission. Then the clear sky mean radiance time series from VIIRS DNB over selected cities are analyzed in relation to social-economic and environmental factors. Finally, the suitability of VIIRS DNB for urban night time aerosol monitoring, based on a combination of its spectral response and the active light spectral characteristics, is analyzed.

3.1. Assessing the Accuracy of the Algorithm Separating Clear from Cloudy Images

To assess the accuracy in separating clear sky from cloudy images, we performed two tests on the Los Angeles dataset. The first is visual inspection of the movie loops and images for the processed images. By comparing the results with the original images, we found that most images are identified correctly, although some images identified as clear may have scattered clouds in the peripheral areas of the city (or typically suburbs).

The second is a quantitative test where we took random samples of the images from the time series and obtained their corresponding VIIRS cloud masks. The cloud masks coded each pixel of the image into four categories: 0 = confidently clear; 1 = probably clear; 2 = probably cloudy; 3 = confidently cloudy. It is known that a given image may contain a mixture of these four categories, and it is difficult to obtain totally clear images over a study region because there are always some clouds in any part of the world at a given time. As a result, we developed a mean cloudiness index (MCI) which takes the mean value of the cloud mask code of the pixels for the area of study. For example, a MCI of <1.05 is considered clear, while MCI > 1.05 is considered cloudy in this study. We evaluate the results based on the error of commission vs. error of omission. Here, errors of commission are those cloudy images (with MCI > 1.05) that are mistakenly identified as clear sky images. In contract, error of omission are those clear sky images (with MCI < 1.05) that are mistakenly identified as cloudy images. Since our objective is to use clear sky VIIRS DNB images, the error of commission is more significant than the error of omission. In other words, skipping some clear sky images in the time series is not as critical as including some cloudy images.

Our results show that among the 65 random samples from the 380 images (17%), the error of commission is ~3.1%, in which two cases were identified as clear but they were probably cloudy according to the cloud mask test (with MCI scores of 1.4 and 1.7). Put another way, 96.9% of the selected VIIRS DNB images are truly clear sky. We found that cloud optical thickness may play a role here. Night lights can penetrate optically thin clouds, so the image may appear clear but the cloud mask may report it as cloudy. In comparison, 14 were labeled as cloudy but were actually clear sky (error of omission 21.5%). Overall, 32 (or 75.4%) are fully identified correctly as clear. However, this is assuming that the cloud mask represents the truth, which may not necessarily be the case, especially for urban areas, due to limitations of the cloud mask algorithm [9].

Further inspection of the error of omission cases suggest that these were affected by two factors: one is the reference seed image used for the similarity analysis. In this test, images on days 21 February 2012, (#7) and 15 April 2020 (#375) were used as seed images, and were both identified as clear in both cloud mask and visual inspection tests. It is likely that if more seed images are used in the analysis, the accuracy rate could further increase, although increasing the number of seed images would reduce the efficiency of the AI method. The other factor is the cutoff threshold for the similarity scores, for which there is a trade-off between number of samples selected vs. the degree of similarity. In our study, the SI index ranged from 0.5 to 0.99 (on a scale of 0–1.0) for all samples, and we narrowed it down to the 0.90 to 0.95 range. We found that for SI >0.95 (only limit to highly similar data), there will be too few samples remaining, while SI <0.90 (which includes more less similar data), the time series becomes noisy. The SI acts as an AI tuning parameter in this study and as a result, 0.90 was used in the final analysis. Nevertheless, the result ensured that the vast majority of the images (96.9%) were correctly identified as clear sky and for the purpose of clear sky VIIRS DNB images time series in this study, we believe that the AI/CNN/SI method is successful.

In the early phases of testing the algorithm presented in Section 2, we found that there were a few pairs of images with similarity of 1.0 which was unexpected. After visual inspection of those images, we were positively surprised that they were either duplicated images or blank images due to human errors in the data collection process. This is encouraging because this method could be used for data quality assurance of satellite images. It is also worth noting that the quantitative accuracy evaluation using cloud masks took significant effort and some of the cloud masks had to be regenerated because they were not all available in the archive. This is also the same reason that a more rapid method of clear sky image identification algorithm is much more desirable, as developed in this study.

3.2. Long-Term Time Series Analysis and Potential for Aerosol Monitoring

Using the AI-based clear sky image identification method presented in the previous section, we found that the data processing is relatively fast as it takes approximately five minutes to process each city with about 350 to 400 images for the time series from 2012 to 2021 (about 2 GB in storage space for each city). The clear sky images are identified, and a time series of the mean radiances for those are plotted and compared for all images in the time series for the city. Figure 5 compares the mean radiances between the identified clear sky vs. all images for Los Angeles. It shows that the time series, without applying the AI algorithm, would produce results with data points so scattered that an erroneous downward trend in the time series would result. However, after applying the AI algorithm, the identified clear sky image radiances form a very distinct pattern in the time series, and a slight positive trend emerges.

Further examination of the clear sky time series in Figure 5 suggests two interesting patterns: the first pattern is that there appears to be an annual oscillation (hormonic fit curve in Figure 5) with a high mean radiance (22 nW/(m^2-sr)) in the winter months, and a low (17 nW/(m^2-sr)) in the summer months or a variability of approximately 26% which has far exceeded the calibration uncertainties documented in previous studies [5,10,11]. Since all observations are near nadir and scan angles are nearly identical with 16 days orbital repeating cycles, the scan angle effect [12,13] can also be ruled out. This pattern is especially clear from 2017 to 2020 but also identifiable for other years. An analysis of the climatology in Los Angeles and other possible causes suggest that this is likely due to seasonal changes in the aerosol optical depth (AOD). In the winter months the AOD is lower than in the summer months. We have also investigated other possible factors, such as residual moon lights, seasonal changes in night-time human activities, daylight saving time changes, and orbital variations, but none of them can explain this seasonal pattern.

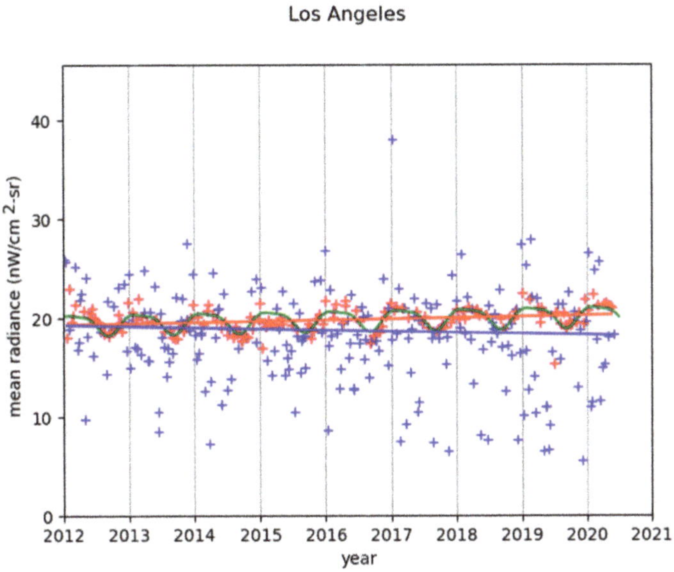

Figure 5. Suomi NPP VIIRS/DNB Observed Urban Mean Radiance Time Series (9 years from 2012 to 2020) before (blue cross with linear fit) and after (red cross with linear fit) applying AI based algorithm for Los Angeles (lunar-contaminated images are also screened out). Green curve shows hormonic fit to the time series. Each data point in the plot is the mean radiance value within the AOI calculated from the VIIRS/DNB image for that day.

Previous aerosol studies have shown similar seasonal patterns in AOD for selected areas and cities. In study [14] (Figure 3), when comparing MODIS and VIIRS daytime retrieved aerosols over land, it was shown that a strong seasonal change in the AOD at 550 nm exists and is in very good agreement between MODIS Terra, Aqua, and Suomi NPP VIIRS retrievals. The peak values in AOD appear to be in the summer months around June, while the trough occurs in later fall to early winter months, although the exact time varied from year to year between 2012 to 2019. Similarly, in a study comparing aerosol retrievals from CALIPSO and MODIS, study [15] (Figure 5) showed strong seasonal variations in AOD over several regions including S.E. Asia, Sahara, N.W. China, and S. America. Although the seasonal cycles are not synchronized among all the areas, there appears to be a pattern with a low AOD in the winter months and high AOD in the summer months in most of the regions presented in their study. Of course, the study period [15] (2006–2011) does not overlap with our current study; nonetheless, this seasonal pattern appears to be persistent, according to their study. Another study [16] (Figure 3) also shows a seasonal pattern in AOD over north India, although it is not always lowest in the winter months. Finally, AERONET [17] measurements in Beijing show that the peak AOD appears in summer month (Figure 6), and the troughs occur in winter months, although the AOD values can have a large dynamic range for any given month. It should be noted that AOD and PM2.5 are not the same. AOD measures the total column aerosol between the ground light source and the satellite, while PM2.5 measures the aerosol near the surface. These two quantities have a complex relationship depending on many factors for a particular location. It is also noted that many of the AERONET stations have data gaps for months or even years, such as Los Angeles, and India, probably due to instrument issues.

Studies using VIIRS DNB observations to retrieve night AOD under lunar illumination have shown good agreement with those from daytime retrievals using other VIIRS bands [18,19]. If the method presented in this study can monitor aerosols at night, it would be a significant complement to global aerosol monitoring over urban areas because the

VIIRS/DNB data has been available since 2012 on Suomi NPP, and has been succeeded by NOAA-20 (J1), as well as the followon satellites on J2, J3, and J4 well into 2040 time frame.

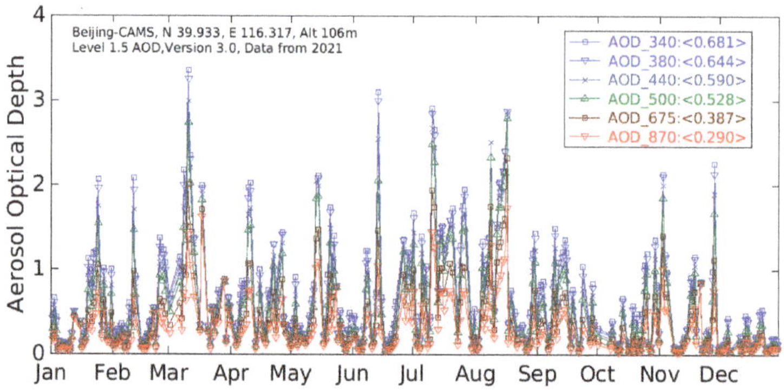

Figure 6. In situ measurements of Aerosol Optical Depth (AOD, daily average) at the NASA AERONET Beijing station shows high AOD in summer months and lower AOD in winter months [20].

The second pattern in Figure 5 is that for Los Angeles, there was a minor increase of 0.13%/year over time in the mean radiance in this time series during the period, based on the trend from linear regression. The increasing trend is clearer from 2016 and onward, which represents urban growth (population increased from 3.9 M to 4 M or 2.5% during the period), while before 2016 this was not the case, in fact, a downward trend may be observed from 2012 to 2016 (with a slope of −0.06/year in linear trend). However, one caveat in interpreting this light intensity change is that there are two opposing effects affecting light change: urbanization which would make it brighter [21]; while aerosol increase would make it dimmer. In other words, in the VIIRS/DNB observations of nightlight intensity, the following equation might apply:

$$L = f\,(NL, -AOD) \tag{1}$$

where L is the mean radiance observed by VIIRS/DNB over the city; NL is the light intensity of the urban night lights; AOD is the aerosol optical depth. Note here that NL and AOD have opposite effects. A high AOD would attenuate more NL which would make L smaller. In other words, the radiance increases if more light is radiated. However, if there is significant increase in AOD, the VIIRS/DNB observed radiance would decrease. Similarly, if NL is constant over time, a decrease in AOD would lead to an increase in L. Unfortunately, these two factors cannot be easily separated, except for cities that have known unchanging nightlight intensity in the time series.

There is also an additional factor that might impact the time series of night radiance: the type of night lamp or light bulb used. There has been a change from the traditional HPS (High Pressure Sodium Lamp) to LED (Light Emitting Diode) during this time period [22] in US cities and abroad. As a result, the downward trend in VIIRS/DNB observed radiance could also be affected by changing light sources which would affect the nightlight intensity.

The same approach presented earlier for the Los Angeles time series is extended to the analysis for Beijing and Delhi. The nightlight time series for these three cities is presented in Figure 7. Note that the absolute mean value differences between different cities are affected by the area of interest selected for each city, so their absolute value differences are not critical; therefore, this study focuses on the time series analysis of selected cities to investigate the light intensity changes over time.

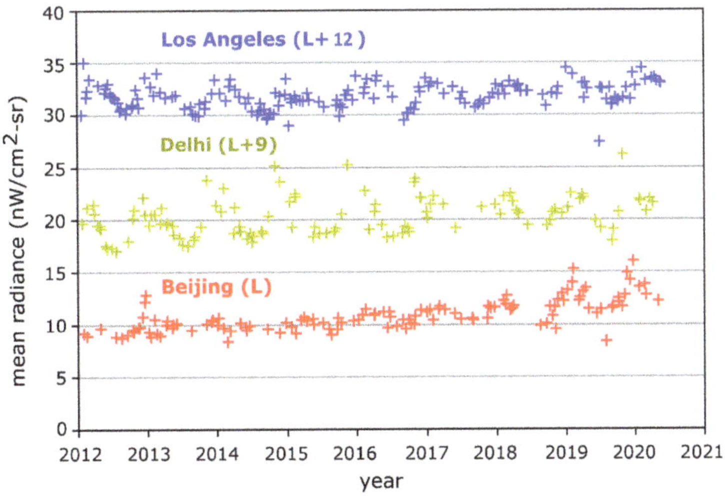

Figure 7. Urban night light mean radiance (L) time series (2012–2020) for Los Angeles [centered at: 34.05°N, 118.24°W] (blue: L + 12), Delhi [28.7°N, 77.1°E] (yellow, L + 9), and Beijing [39.9°N, 116.41°E] (red, L). Each data point in the plot is the mean radiance value within the AOI calculated from the VIIRS/DNB image for that day.

We found that all three cities have an annual variation in the light intensity. For Delhi, the amplitude of the oscillation appears to be larger than for the other two cities, and the peak appears more often in spring than in winter. Compared to the time series of Los Angeles, the nightlight radiance time series of Beijing has similar seasonal oscillations, with a high in the winter when the AOD is lower and a low in the summer months when the AOD is higher, similar to the AERONET data shown in Figure 6. However, the oscillation for the Beijing time series appears to become stronger (~50%) in recent years after 2019, having been weaker pre-2016 (~25%). This could be due to the fact that in the early years, pollution was much more severe and aerosol optical thickness was high all year round (e.g., maximum AOD before 2015 was >4, while it was <3 in 2019), according to AERONET measurements (https://aeronet.gsfc.nasa.gov/cgi-bin/data_display_aod_v3?site=Beijing-CAMS&nachal=2&level=3&place_code=10, accessed on 21 June 2022), a UN Environment report (https://wedocs.unep.org/bitstream/handle/20.500.11822/27645/airPolCh_EN.pdf?sequence=1&isAllowed=y, accessed on 21 June 2022), as well as Beijing AOD studies [23]. As the air became cleaner in recent years, the seasonal variation became stronger. In other words, the aerosol optical depth and PM2.5 have changed due to the cleaner air initiatives in recent years.

The other obvious trend is that the Beijing's mean radiance has increased signifi-cantly from 2012 to 2020, from ~9.97 to ~12.89 nW/(cm^2-sr), an increase of 29.3% for the time period (or ~3.66%/year). This is likely due to urban growth and urbanization. Population statistics show that the Beijing population has grown from 17.3 M to 20.5 M (or 18%) from 2012 to 2020 (https://worldpopulationreview.com/world-cities/beijing-population, accessed on 21 June 2022).

A summary of the time series data used in this study is listed in Table 1. In this study we only analyzed the near nadir observations, and the total number of images would be significantly increased if off nadir data are also included. However, as discussed earlier, the off-nadir effect would also become a variable which would require future studies. The clear sky images without moon were selected in the final time series plot in Figure 7. In this study, the same CNN/SI threshold is used for all three cities, and future studies should explore a city-specific threshold which might further improve the results.

Table 1. SNPP/VIIRS/DNB images used in this study.

	Total # of Images in the Time Series	Clear Sky without Moon
Los Angeles	380	161
Delhi	380	151
Beijing	364	123

3.3. Physical Basis for Using VIIRS/DNB as Active Light Source for Aerosol Monitoring

The idea of using the VIIRS DNB night light as an active light source for monitoring aerosols is not new. In [24], a preliminary study of VIIRS/DNB calibration identified the need for a ground-based calibration light source to improve the calibration accuracy, and recognized the potential for aerosol monitoring. In [11], a ground-based LED point light source with a well-constructed integrating sphere as a calibration source for VIIRS DNB was developed to establish NIST traceability in a NOAA SBIR project, and a second objective of that study was to explore its use for aerosol monitoring, for which comparisons with AERONET measurements at a site near South Dakota were performed, together with radiative transfer simulations. Results showed that the ground-based light source radiance agreed with that of NOAA-20 measurements better than 11% of the faint light, and approximately half of the difference was likely due to the atmosphere, including aerosols (the other half was due to measurement differences between ground and satellite). Therefore, aerosol is a significant contributor to the radiometric measurements by the VIIRS/DNB of the ground-based light source. Previous studies have also explored the measurements of aerosols with satellite observations of night lights [21]. The theoretical basis has also been studied with models from light pollution perspectives [25–27]

It is known that the VIIRS DNB spectral response covers a wide range from 0.4 to 0.9 um, in which aerosol, water vapor, and oxygen absorptions can all have an impact on the measured radiances. However, as Figure 8 shows, the urban night light spectra are dominated by HPS (High Pressure Sodium Lamp) and LED (Light Emitting Diode), which have a much narrower spectral range concentrated between 0.4 and 0.65 um [22,24], and therefore are largely unaffected by water vapor and oxygen absorptions. Therefore, the absorption and scattering from aerosol become a major factor affecting the light intensity of night light measurements from VIIRS/DNB. Effectively, the center spectra in the VIIRS/DNB measurement of nightlights are in the 0.45 to 0.6 um spectral region as shown in Figure 8, which is appropriate for aerosol measurements.

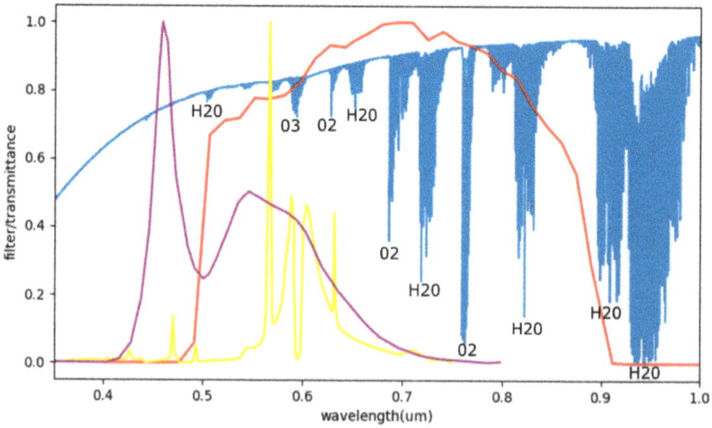

Figure 8. VIIRS/DNB spectral response (red), HPS (yellow) and LED (purple) spectra, and Atmospheric Spectral Transmittance (blue).

Using urban night light to monitor aerosols has distinct advantages and is a unique type of active remote sensing. At night, the anthropogenic aerosols are more stable, and using night light to detect aerosols over land is largely free from background noise. The approach complements the traditional approach of aerosol retrievals with solar bands during daytime under solar illumination over land, where a number of challenges still exist, such as background noise, landcover reflectance, volumetric large particle-size scattering, as well as the reliance on climatology information. It is noted that there is a growing scientific community using VIIRS/DNB for nightlight research. In addition to the social-economic and geophysical studies discussed earlier, there is also the astronomy community for the study of light pollution at night (https://lighttrends.lightpollutionmap.info/, accessed on 21 June 2022), where aerosol is also a major factor. Finally, the CNN/SI approach presented in this study focuses on city specific analysis, which distinguishes it from the traditional cloud mask-based approach, as discussed earlier. It complements the latter for night-time aerosol studies with rapid identification of clear sky images for specific cities, and its applicability to large scale studies require further exploration.

4. Summary and Future Studies

In this paper we presented an artificial intelligence (CNN/SI)-based algorithm for finding clear sky night light VIIRS/DNB images for urban areas. Since the urban night light pattern is relatively stable at local midnight, and many observations are contaminated by clouds, a high similarity index compared to a seed clear sky night image allows us to identify and search for clear sky images for a given city. The algorithm is based on similarity calculations using CNN with MobileNet V2 module in Tensorflow which works efficiently on personal laptop computers without the need for intensive computation. We demonstrated that using this algorithm, a time series of night light images can be developed for a given city. Examples are provided for selected cities including Los Angeles, Delhi, and Beijing, which show distinct seasonal variations, as well as long-term increases in the night light mean radiances. We found that the long-term changes in the night light intensity and seasonal variations are likely caused by a combination of urban development and changes in aerosol optical depth.

Future studies will expand the analysis to additional world-wide cities using the reprocessed Suomi NPP VIIRS night light images to study the time series for those cities, as well as the possible relationship with aerosols and urban growth trends. Further tuning of the CNN/SI threshold for specific cities may also improve the results. Using VIIRS observed night lights as an active light source to monitor aerosol optical depth is very promising. We also plan to expand the AI-based clear sky image identification algorithm to other satellite data for more applications.

Author Contributions: C.C. is responsible for the conceptualization, data analysis, and writing the manuscript; B.Z. is responsible for the VIIRS data reprocessing, data extraction, preparation, and analysis; F.X. contributed to the interpretation, experiments and analysis methods using convolutional neural network; Y.B. contributed to the graphics work, data collection, verification, and manuscript editing. All authors have read and agreed to the published version of the manuscript.

Funding: This research was funded by the NOAA JPSS JSTAR program [NOAA grant NA19NES4320002 (Cooperative Institute for Satellite Earth System Studies—CISESS) at the University of Maryland/ESSIC].

Data Availability Statement: The original Suomi NPP VIIRS/DNB data are available from http://www.class.noaa.gov (all accessed on 21 June 2022). The reprocessed Suomi NPP VIIRS/DNB data can be obtained from https://ncc.nesdis.noaa.gov/VIIRS/index.php, and will be made available on http://www.class.noaa.gov by 2023. All data are in the public domain and freely available.

Acknowledgments: We thank the anonymous reviewers (external), as well as Douglas Howard (internal) for a critical review of this manuscript with constructive suggestions and comments. We also thank Brent Holben and his staff for establishing and maintaining the Beijing-CAMS AERONET site used in this investigation. Thanks are extended to NOAA Lapenta Summer Intern Madison Lytle of California Polytechnic State University for cross verification of satellite overpasses, and summer intern Clarence Lam from University of Maryland/CISESS for CNN/SI code verification. The views and opinions in this paper are solely those of the authors and do not constitute a statement of policy, decision, or position on behalf of NOAA or the U.S. Government.

Conflicts of Interest: The authors declare no conflict of interest.

References

1. Miller, S.D.; Haddock, S.H.D.; Straka, W.C.; Seaman, C.J.; Combs, C.L.; Wang, M.; Shi, W.; Nam, S.-H. Honing in on bioluminescent milky seas from space. *Sci. Rep.* **2021**, *2021*, 15443. [CrossRef] [PubMed]
2. Elvidge, C.D.; Zhizhin, M.; Baugh, K.; Hsu, F. Automatic Boat Identification System for VIIRS Low Light Imaging Data. *Remote Sens.* **2015**, *7*, 3020–3036. [CrossRef]
3. Cao, C.; Shao, X.; Uprety, S. Detecting Light Outages after Severe Storms Using the S-NPP/VIIRS Day/NightBand Radiances. *IEEE Geosci. Remote Sens. Lett.* **2013**, *10*, 1582–1586. [CrossRef]
4. Liu, B.; D'Sa, E.J. Oceanic Internal Waves in the Sulu–Celebes Sea under Sunglint and Moonglint. *IEEE Trans. Geosci. Remote Sens.* **2019**, *57*, 6119–6129. [CrossRef]
5. Cao, C.; Zhang, B.; Shao, X.; Wang, W.; Uprety, S.; Choi, T.; Blonski, S.; Gu, Y.; Bai, Y.; Lin, L.; et al. Mission-Long Recalibrated Science Quality Suomi NPP VIIRS Radiometric Dataset Using Advanced Algorithms for Time Series Studies. *Remote Sens.* **2021**, *13*, 1075. [CrossRef]
6. Cao, C.; De Luccia, F.J.; Xiong, X.; Wolfe, R.; Weng, F. Early On-Orbit Performance of the Visible Infrared Imaging Radiometer Suite Onboard the Suomi National Polar-Orbiting Partnership (S-NPP) Satellite. *IEEE Trans. Geosci. Remote Sens.* **2014**, *52*, 1142–1156. [CrossRef]
7. Sandler, S.; Howard, A.; Zhu, M.; Zhmoginov, A.; Chen, L. MobileNetV2: Inverted Residuals and Linear Bottlenecks. In Proceedings of the 2018 IEEE/CVF Conference on Computer Vision and Pattern Recognition, Salt Lake City, UT, USA, 18–23 June 2018; pp. 4510–4520. [CrossRef]
8. Elvidge, C.D.; Zhizhin, M.; Ghosh, T.; Hsu, F.-C.; Taneja, J. Annual Time Series of Global VIIRS Nighttime Lights Derived from Monthly Averages: 2012 to 2019. *Remote Sens.* **2021**, *13*, 922. [CrossRef]
9. Li, Z.; University of Wisconsin, Madison, WI, USA. Personal Communication, 2022.
10. Liao, L.; Weiss, S.; Mills, S.; Hauss, B. Suomi NPP VIIRS day-night band on-orbit performance. *J. Geophys. Res.* **2013**, *118*, 12705–12718. [CrossRef]
11. Ryan, R.E.; Pagnutti, M.; Burch, K.; Leigh, L.; Ruggles, T.; Cao, C.; Aaron, D.; Blonski, S.; Helder, D. The Terra Vega Active Light Source: A First Step in a New Approach to Perform Nighttime Absolute Radiometric Calibrations and Early Results Calibrating the VIIRS DNB. *Remote Sens.* **2019**, *11*, 710. [CrossRef]
12. Bai, Y.; Cao, C.; Shao, X. Assessment of scan-angle dependent radiometric bias of Suomi-NPP VIIRS day/night band from night light point source observations. *Proc. SPIE Earth Obs. Syst.* **2015**, *9607*, 960727. [CrossRef]
13. Li, X.; Ma, R.; Zhang, Q.; Li, D.; Liu, S.; He, T.; Zhao, L. Anisotropic characteristic of artificial light at night—Systematic investigation with VIIRS DNB multi-temporal observations. *Remote Sens. Environ.* **2019**, *233*, 111357. [CrossRef]
14. Sawyer, V.; Levy, R.C.; Mattoo, S.; Cureton, G.; Shi, Y.; Remer, L.A. Continuing the MODIS Dark Target Aerosol Time Series with VIIRS. *Remote Sens.* **2020**, *12*, 308. [CrossRef]
15. Ma, M.; Bartlett, K.; Harmon, K.; Yu, F. Comparison of AOD between CALIPSO and MODIS: Significant differences over major dust and biomass burning regions. *Atmos. Meas. Tech.* **2013**, *6*, 2391–2401. [CrossRef]
16. Abish, B.; Mohanakumar, K. A Stochastic Model for Predicting Aerosol Optical Depth over the North Indian Region. *Int. J. Remote Sens.* **2013**, *34*, 1449–1458. [CrossRef]
17. Holben, B.N.; Eck, T.F.; Slutsker, I.; Tanre, D.; Buis, J.P.; Setzer, A.; Vermote, E.; Reagan, J.A.; Kaufman, Y.J.; Nakajima, T.; et al. AERONET—A federated instrument network and data archive for aerosol characterization. *Remote Sens. Environ.* **1998**, *66*, 1–16. [CrossRef]
18. Zhou, M.; Wang, J.; Chen, X.; Xu, X.; Colarco, P.R.; Miller, S.D.; Reid, J.S.; Kondragunta, S.; Giles, D.M.; Holben, B. Nighttime smoke aerosol optical depth over U.S. rural areas: First retrieval from VIIRS moonlight observations. *Remote Sens. Environ.* **2021**, *267*, 112717. [CrossRef]
19. Johnson, R.S.; Zhang, J.; Hyer, E.J.; Miller, S.D.; Reid, J.S. Preliminary investigations toward night time aerosol optical depth retrievals from the VIIRS Day/Night Band. *Atmos. Meas. Tech.* **2013**, *6*, 1245–1255. [CrossRef]
20. Holben, Goloub and Chen. AERONET. 2021. Available online: https://aeronet.gsfc.nasa.gov/cgi-bin/data_display_aod_v3?site=Beijing-CAMS&nachal=2&level=3&place_code=10 (accessed on 21 June 2022).
21. Cavazzani, S.; Ortolani, S.; Bertolo, A.; Binotto, R.; Fiorentin, P.; Carraro, G.; Zitelli, V. Satellite measurements of artificial light at night: Aerosol effects. *Mon. Not. R. Astron. Soc.* **2020**, *499*, 5075–5089. [CrossRef]

22. Cao, C.; Bai, Y. Quantitative analysis of VIIRS DNB nightlight point source for light power estimation and stability monitoring. *Remote Sens.* **2014**, *6*, 11915–11935. [CrossRef]
23. Kondragunta, S.; NOAA/Center for Satellite Applications and Research, College Park, MD, USA. Personal Communication, 2022.
24. Cao, C.; Zong, Y.; Bai, Y.; Shao, X. Preliminary study for improving the VIIRS DNB low light calibration accuracy with ground based active light source. In *Earth Observing Systems XX*; Butler, J.J., Xiong, X., Eds.; International Society for Optics and Photonics: San Diego, CA, USA, 2015; pp. 96070D1–96070D8.
25. Kocifaj, M. Light pollution model for cloudy and cloudless night skies with ground-based light sources. *Appl. Opt.* **2009**, *48*, 4650. [CrossRef]
26. Kocifaj, M.; Komar, L. A role of aerosol particles in forming urban skyglow and skyglow from distant cities. *Mon. Not. R. Astron. Soc.* **2016**, *458*, 438–448. [CrossRef]
27. Kocifaj, M. Multiple scattering contribution to the diffuse light of a night sky: A model which embraces all orders of scattering. *J. Quant. Spectrosc. Radiat. Transf.* **2018**, *206*, 260–272. [CrossRef]

remote sensing

MDPI

Article

Monitoring the Distribution and Variations of City Size Based on Night-Time Light Remote Sensing: A Case Study in the Yangtze River Delta of China

Yuan Ding [1,†], Jia Hu [1,†], Yingbao Yang [1,*], Wenyu Ma [1], Songxiu Jiang [1], Xin Pan [1], Yong Zhang [1,2], Jingjing Zhu [1] and Kai Cao [3,4]

1 School of Earth Sciences and Engineering, Hohai University, Nanjing 211100, China; dingyuanhhu@hhu.edu.cn (Y.D.); jiahu@hhu.edu.cn (J.H.); mawy@hhu.edu.cn (W.M.); 181309040002@hhu.edu.cn (S.J.); px1013@hhu.edu.cn (X.P.); zyrs@hhu.edu.cn (Y.Z.); 181309030007@hhu.edu.cn (J.Z.)

2 College of Geography and Planning, Chizhou University, Chizhou 247000, China

3 School of Geographic Sciences, East China Normal University, Shanghai 200241, China; kcao@geo.ecnu.edu.cn

4 Key Laboratory of Geographic Information Science (Ministry of Education), East China Normal University, Shanghai 200241, China

* Correspondence: yyb@hhu.edu.cn

† These authors contributed equally to this work.

Citation: Ding, Y.; Hu, J.; Yang, Y.; Ma, W.; Jiang, S.; Pan, X.; Zhang, Y.; Zhu, J.; Cao, K. Monitoring the Distribution and Variations of City Size Based on Night-Time Light Remote Sensing: A Case Study in the Yangtze River Delta of China. *Remote Sens.* **2022**, *14*, 3403. https://doi.org/10.3390/rs14143403

Academic Editors: Ran Goldblatt, Steven Louis Rubinyi and Hogeun Park

Received: 16 June 2022
Accepted: 14 July 2022
Published: 15 July 2022

Abstract: Effectively monitoring the size of a city in real time enables the scientific planning of urban development. Models that utilize the distribution and variations in city size generally use population data as inputs, which cannot be obtained in a timely and rapid manner. However, night-time light (NTL) remote sensing may be an alternative method. A case study was carried out on the Yangtze River Delta (YRD) in China, and the rank–size rule, the law of primate cities, and the Gini coefficient were employed to monitor the variation in city size in the study area. The urban areas extracted based on NTL remote sensing were utilized instead of the traditionally used population data to evaluate the variations in city size from 2012 to 2017. Considering the empiricism and subjectivity of the thresholding method, urban areas were extracted from NTL data combined with the normalized differential vegetation index and land-surface temperature data based on the artificial neural network algorithm. Based on the results, the YRD did not fit the distribution of the primate cities from 2012 to 2017. However, this region satisfied the rank–size rule well, which indicated that the development of medium–small cities was more prominent than that of larger cities, and the dispersed force was larger than the concentrated force. Notably, the city size reached a relatively balanced level in the study area. Further, sensitivity analysis revealed that the relatively low extraction accuracy of urban areas of few small cities had little effect on the results of city size variations. Moreover, the validation of city size computed from statistical population data and its comparison with results calculated based on the statistical data of urban areas aligned with the results of this study, which indicates the rationality and applicability of monitoring the variations in city size using the urban areas extracted from NTL remote sensing instead of population data.

Keywords: night-time light data; extraction of urban areas; rank–size rule; law of primate city

1. Introduction

Dramatic urbanization has occurred globally, and is mainly due to rapid economic development and population growth [1]. According to statistical data, more than 54% of land is populated by humans worldwide. In fact, the total population in urban areas is predicted to exceed 2 billion by 2050 [2]. Rapid urbanization can lead to several environmental issues, including air pollution, urban heat islands, climate change, shortage of resources, and pressure for sustainability worldwide [3–6]. Rapid urbanization may also result in unhealthy development in urban systems, especially developing countries. For instance, the development of big cities was better than that of small and medium cities. Accordingly, the development of city size in an urban system was recognized to be unbalanced. A comprehensive understanding of variations in city size is thus useful for urban planning and decision making, and monitoring the distribution and variations in city size, which is also helpful for understanding the development model and rational planning of urban systems, is of great significance.

Several popular measurement methods for analyzing the distribution and variations in city size include the rank–size rule [7], the law of the primate city [8], and the Gini coefficient [9]. Briefly, the urban population or urban area is input into the above models [10], and the distribution and variation trends of city size are analyzed based on the calculated indices. Although the most direct and effective monitoring method is population data, there are some problems with this method [11–13]. First, the statistical standards for population data vary among countries. For instance, population statistics in China are mainly based on administrative units, as urban and rural areas are divided according to the administrative boundaries of urban districts and towns. Socioeconomic units in other countries are generally based on functional areas, such as metropolitan areas in the United States. Moreover, considerable time and labor are required to obtain population data. Night-time light (NTL) remote sensing, which has emerged in recent years, has enabled new directions for research on city monitoring, as it can provide real-time information about the earth with the same observation standard. Several studies employed remote-sensing data to monitor the urbanization of cities and assess socioeconomic activities [14–16]. Nitsch [17] showed that urban area and population have similar distribution trends when measuring urban size distribution. Therefore, in this study, urban areas extracted from NTL remote sensing were used to replace the traditionally used population data in these city size analysis models to monitor the distribution and variations in city size.

Many satellite images of different resolutions can be used to extract urban areas for cities that are suitable for different scenarios. High- and medium-high-resolution images, such as Landsat series [18], are suitable for exploring urbanization at the city scale, but are not suitable for large scales, such as urban and global levels. Mapping urban areas on a large scale requires several high-resolution images with cloud-free images, which also requires significant time and labor for processing. Coarse-resolution satellite images, such as MODIS and NTL data, are suitable for large-scale urban area extraction [1,19,20]. However, NTL data can be utilized to monitor human activities, as it captures city light at night, which separates urban areas from the surrounding suburbs [21,22]. According to prior studies, the NTL value has a highly close connection with population, gross domestic product, and built-up areas [23,24]. Thus, this value is more suitable for selecting NTL data for the analysis of variations in city size compared to other optimal satellite images.

Currently, commonly used NTL data are obtained from two NTL satellites, the Defense Meteorological Satellite Program-Operational Linescan System (DMSP-OLS) and Suomi National Polar-Orbiting Partnership Visible Infrared Imaging Radiometer Suite (NPP-VIIRS). NPP-VIIRS has better quality than DMSP-OLS, as it decreases the problem of the saturation effect that exists in DMSP-OLS [25–27]. Thus, NPP-VIIRS is better choice for extracting urban built-up areas. Recently, methods for extracting urban areas based on NPP-VIIRS data mainly include two types: thresholding and classification. The thresholding method has been used by many researchers, as it is simple and easy to conduct [28–32]. Nevertheless, this method is too subjective and empirical to determine the thresholds for extracting urban areas, owing to considerable uncertainty among cities at different development levels [33]. To resolve such limitations of the thresholding method, some researchers have utilized machine learning methods for the extraction of urban areas. For instance, Jing et al. [34] applied four machine learning algorithms to extract urban areas of eastern China. Based on their results, the machine learning methods achieved good accuracy and lower sensitivity than the thresholding method. Xu et al. [35] extracted urban areas from NTL data by utilizing an artificial neural network algorithm, which led to a more accurate quality result than the thresholding method. Hence, extracting urban areas based on machine learning is better than extraction achieved with the traditional thresholding approach. The selection of training samples is another technical problem in urban area extraction using a classification method based on machine learning. Extracting training samples from NTL data alone may lead to the overestimation or underestimation of urban areas. To improve the extraction of urban areas, some researchers have added other remote-sensing data for sample selection, including normalized differential vegetation index (NDVI) and land-surface temperature (LST) data [34,35]. Notably, a better extraction result was achieved when the NTL data were combined with multi-source remote-sensing data to select training samples to extract urban areas based on the machine learning method.

The objectives of this study were to extract urban built-up areas in the study region using a combination of NTL data and NDVI and LST data based on the machine learning algorithm, and to monitor the distribution and variations in city size in the study area by utilizing the models of rank–size rule, the law of primate city, and the Gini coefficient, using extracted urban areas instead of the traditional population data.

2. Methodology

2.1. Study Area

The Yangtze River Delta (YRD) is located in eastern China and is adjacent to the East China Sea (Figure 1). This region has a subtropical monsoon climate, with an annual mean temperature of 16.9 °C. The region also has three provinces, namely Jiangsu, Zhejiang, and Anhui, and one international city, Shanghai city, with a total of 26 cities. The YRD covers an area of 217,700 km^2 and has a population of 150 million. Owing to its unique geographical location and natural resources, the YRD urban agglomeration has become the most developed and largest economic zone in China, and it is one of six metropolitan agglomerations in the world. Further, this area has experienced tremendous urban expansion over the past few decades.

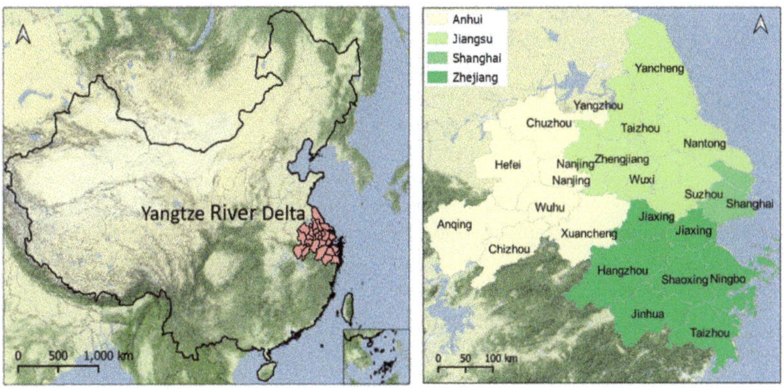

Figure 1. Location of the YRD in China and demarcation of the study area.

2.2. Data Collection and Preprocessing

Multiple datasets were used in this study (Table 1), including NTL data and four MODIS datasets. Version 1 NPP-VIIRS NTL data, on a monthly basis from 2012 to 2017, were obtained from the National Oceanic and Atmospheric Administration (NOAA) (https://ngdc.noaa.gov/eog/viirs/download_dnb_composites.html (accessed on 11 November 2020)). Additionally, the four MODIS data from the same time series, which were 16-day Normalized Differential Vegetation Index (NDVI) product (MOD13A1), 8-day MODIS land-surface temperature (LST) product (MOD11A2), annual global land water mask product (MOD 44w), and annual MODIS Land Cover Type product (MCD12Q1), were obtained from National Aeronautics and Space Administration (NASA) (https://ladsweb.modaps.eosdis.nasa.gov/search/ (accessed on 17 November 2020)). The NPP-VIIRS, MOD 13A1

NDVI, and MOD 11A2 LST data were used to select samples for urban area extraction for the study area. MCD12Q1 land-cover data were used to determine the optimal combination of coefficients for urban area extraction. The MOD 44w data were employed to remove the light intensity of water bodies from the NPP-VIIRS images. In this study, all remote-sensing data were reprojected to the Albers Conical Equal Area projection and resampled to a resolution of 500 m for consistency with the NTL and land-cover data. The annual NPP-VIIRS and LST data were reproduced by the mean value composition in this study, while the annual NDVI data were reproduced based on the maximum value composition to decrease the influence of cloud contamination [36]. The population and actual urban area data from 2012 to 2017 were obtained from the statistical yearbooks of provinces and cities. The population data were utilized to calculate the indices for city size analysis, and the statistical urban areas were used for comparison with urban areas extracted using our NTL remote-sensing-based method. In this study, the permanent urban population at the end of the year was regarded as the urban population of each city in the study area.

Table 1. Data used in this study.

Data	Description	Resolution	Time
NPP VIIRS	Night-time light	500 m/Month	
MOD 13A1	Normalized Differential Vegetation Index	500 m/16 Days	2012–2017
MOD 11A2	Land-surface temperature	1 km/8 Days	
MOD 44w	Global land water mask	250 m/Year	2012–2017
MCD 12Q1	Land-cover type	500 m/Year	2012–2017
Population		Year	2012–2017
Statistical urban areas		0.01 km^2	2012–2017

2.3. *Monitoring of the Distribution and Variations in City Size*

In this study, the rank–size rule, the law of the primate city, and the Gini coefficient were employed to analyze the distribution and variations in city size for the study area. The rank–size rule is usually represented by Zipf's index, and the law of the primate city is calculated using the urban primacy index. Zipf's index, the urban primacy index, and the Gini coefficient were calculated from the acreage of the extracted urban areas. The three indices were also computed from the population data to validate the results obtained from the extracted urban areas. If the variation trend of the three indices calculated from both data sources is the same, urban areas extracted from NTL remote sensing can replace the population to analyze the variation in the differences in city size.

2.3.1. Rank–Size Rule

The rank–size rule was utilized to reveal the size distribution in an urban system based on the relationship between the size of cities and the rank of city sizes, which can also reflect the development level of a country or an urban system. In 1949, Zipf provided a theoretical foundation [7], which can be expressed as follows:

$$\lg P_i = \lg P_1 - q \lg R_i \tag{1}$$

where lg is the basic logarithm with base 10, R_i is the rank of city i according to city size among all cities, P_1 is the city size of the largest city, P_i is the city size of city i, and q is Zipf's index.

The ideal Zipf index should be 1, indicating balanced urban development within the urban system. However, in general, Zipf's index is not equal to one. An index greater than 1 indicates that the development of big cities is prominent, while that of small and medium-sized cities is underdeveloped. On the contrary, an index lower than 1 indicates that the development of big cities is not prominent, whereas that of small- and medium-sized cities is better. An increasing trend in Zipf's index indicates that the concentrated force in the distribution of city size is larger than the dispersed force, which means that the development of city size in large cities is faster than that in small cities. Conversely, a decrease in Zipf's index indicates that the dispersed force in the urban system is larger than the concentrated force, and the development of city size in small cities is faster than in big cities.

2.3.2. Law of Primate City

The urban primacy index was utilized to describe the extent of concentration of the urban population in the primate city in an urban system, which can largely reflect the development features of the urban system. This index was proposed by Mark [8], which indicates that the size of the largest city is markedly greater than that of other cities in an urban system. The index is equal to the ratio of the population of the largest city to that of the second largest city in the study area, which can be described by the following formula:

$$S_2 = P_1/P_2 \tag{2}$$

where S_2 is a 2-city index and P_1 and P_2 are the populations of the largest and second-largest cities, respectively.

According to some researchers, a 2-city index is too one-sided. Thus, the 4-city index and 11-city index were proposed, and can be calculated as follows:

$$S_4 = P_1/(P_2 + P_3 + P_4) \tag{3}$$

$$S_{11} = 2P1/\sum_{i=2}^{11} Pi \tag{4}$$

where S_4 and S_{11} are the 4-city index and 11-city indices, respectively, P_1 is the population of the largest city in the urban system, and P_2 to P_{11} is the population of the second to the eleventh city based on city size.

The threshold of the 2-city index should be 2, while the thresholds of the 4-city and 11-city indices should be 1. The greater the number of cities involved in the calculation of the urban primacy index, the more reliable the results. Accordingly, the 11-city index was selected to compute the urban primacy index for the study area. In this study, the urban primacy index was calculated from the age of the extracted urban areas and the population in the study area.

2.3.3. Gini Coefficient

The Gini coefficient was originally used to measure the income gap between residents in a country or region, as proposed by Gini in 1912 [9]. Marshall applied the Gini model to cities to analyze city development based on differences in size by analyzing the degree of population aggregation on the whole urban agglomeration, and proposed the concept of the city Gini coefficient for the first time, which can be calculated as follows:

$$G = T/2S(n-1) \tag{5}$$

where n is the total number of cities in the urban system, S is the total population of these cities in the urban system, and T is the sum of the absolute differences in population size among these cities in the urban system.

As an indicator of the relative difference, the Gini coefficient can reflect the balance degree of the population distribution in various cities, ranging from 0 to 1. The closer the Gini coefficient to the value of 0, the more dispersed the city size, while the closer the coefficient to the value of 1, the more concentrated the city size. The value range of the Gini coefficient can be divided into five groups: values below 0.2, 0.2, 0.3, 0.3, 0.4, 0.4, 0.5, and greater than 0.5. These groups represent the absolute balance of city size, relative balance of city size, relative reasonable city size, significant difference in city size, and abnormal disparity, respectively.

2.4. Urban Area Extraction

In this study, NPP-VIIRS NTL, NDVI, and LST data were utilized to extract the urban areas of the study area based on an artificial neural network (ANN) algorithm. The main steps for extracting urban areas are described below. First, the pixels occupied by water bodies were removed using global land water mask data, and potential urban areas were regarded as pixels with NTL values greater than 1, according to Yang et al. [37].

The thresholds of the three datasets utilized to select samples were subsequently determined based on the standard deviation from the mean value. Notably, urban areas were extracted city by city in the study area, as the development and economic levels markedly vary in the study area, and the accuracy of the extracted results may be low if urban areas of different cities are extracted as a whole. The equations are as follows:

$$\begin{cases} T_j^{VIIRS} = Mean_j^{VIIRS} + aStd_j^{VIIRS} \\ T_j^{NDVI} = Mean_j^{NDVI} + bStd_j^{NDVI} \\ T_j^{LST} = Mean_j^{LST} + cStd_j^{NDVI} \end{cases} \quad (6)$$

where T_j^{VIIRS}, T_j^{NDVI}, and T_j^{LST} are the thresholds of the NPP-VIIRS, NDVI, and LST data for city j, respectively; $Mean_j^{VIIRS}$, $Mean_j^{NDVI}$, and $Mean_j^{LST}$ are the mean values of the three datasets for city j; Std_j^{VIIRS}, Std_j^{NDVI}, and Std_j^{NDVI} are the standard deviations of these data for city j; and a, b, and c are coefficients, which are normally set to -1, -0.5, 0, 0.5, and 1, respectively. Different values of T_j^{VIIRS}, T_j^{NDVI}, and T_j^{LST} can be obtained according to the values of a, b, and c.

Based on the features of high NTL values, high temperatures, and low vegetation coverage in urban areas, the training samples of urban and non-urban areas were selected based on the following formulas:

$$Sample(i,j) = \begin{cases} VIIRS(i,j) > T_j^{VIIRS} \& NDVI(i,j) < T_j^{NDVI} \& LST(i,j) > T_j^{LST}, Urban \\ VIIRS(i,j) < T_j^{VIIRS} \& NDVI(i,j) > T_j^{NDVI} \& LST(i,j) < T_j^{LST}, Non-urban \end{cases} \quad (7)$$

where $Sample(i,j)$ represents the selected samples of pixel i of city j, and $VIIRS(i,j)$, $NDVI(i,j)$, and $LST(i,j)$ are the pixel values of the NPP-VIIRS, NDVI, and LST data for pixel i of city j, respectively.

Therefore, there are 125 combinations of coefficients a, b, and c. The optimal set of combinations was selected based on the following standard:

$$\min|S_k - S_0| \& \max\{g\_mean_k\} \quad (8)$$

where S_k and S_0 are the number of total pixels of the extracted urban areas based on the NTL data and the number of total pixels of the urban areas in the land-cover data, respectively, and g_mean_k is the geometric mean precision, which is introduced below.

Based on the optimal set of coefficients confirmed above, the optimal thresholds for the three datasets used to select samples for urban area extraction were determined in the third step. Thereafter, the samples were selected and randomly divided into approximately 10% training samples and 90% test samples. An ANN algorithm was employed to extract the urban areas of the study area based on the training samples.

ANN is a type of machine learning model that has been widely utilized in many studies owing to its self-adapting, self-organizing, and self-learning advantages. A feedforward network structure composed of an input layer, a hidden layer, and an output layer was utilized in this study [35]. The input layer consists of three nodes: NTL, LST, and NDVI. Based on a sensitivity analysis, the number of hidden nodes was set to five. The output nodes have two possible results: urban and non-urban areas. The pixels that met the expectations of urban and non-urban areas were outputted as (1, 0) and (0, 0), respectively.

After training, the test samples were utilized to validate the accuracy of urban area extraction for the study area. In this study, in addition to the Kappa coefficient used, the geometric mean accuracy was also used [38]. The geometric mean accuracy was calculated as follows:

$$g\_mean = \sqrt{\frac{d}{m+d} \times \frac{d}{n+d}} \quad (9)$$

where g_mean is the geometric mean accuracy, d is the number of matched urban area pixels between NTL data and land-cover data, m is the number of urban area pixels for which NTL data do not match the land-cover data, and n is the number of urban area pixels for which the land-cover data do not match the NTL data.

3. Results and Discussion

3.1. Urban Expansion of the YRD

3.1.1. Accuracy Assessment of Urban Area Extraction

In this study, the kappa coefficient and geometric mean accuracy were used to validate the accuracy of the urban area extraction of the YRD. Table 2 shows the average geometric mean accuracy and kappa coefficient for the YRD. As depicted in the table, the urban area extraction of YRD based on the method proposed in this study achieved high accuracy, with an average geometric mean accuracy and kappa coefficient ranging from 79.21% to 84.88% and 0.761 to 0.828, respectively. The geometric mean accuracy of all years was between 82% and 85%, except for 2015, which had a value of 79.21%. In 2012 and 2015, the Kappa coefficients were 0.793 and 0.761; however, in other years, this value was approximately 0.82. Figures 2 and 3 show the distribution of geometric mean accuracy and Kappa coefficient of urban area extraction for all cities in the YRD from 2012 to 2017. This result indicates that most cities in the YRD have a high geometric mean accuracy and Kappa coefficient for six years, with values above 90% and 0.80, respectively. However, the accuracy of several medium- and small-sized cities, such as Anqing, Chizhou, Xuancheng, and Chuzhou of the Anhui province, which had relatively lower economic development, was relatively low, with a geometric mean accuracy and Kappa coefficient of less than 60% and 0.60, respectively.

3.1.2. Spatial–Temporal Variations of Urban Areas

Figure 4 displays the distribution of the extracted urban areas of the study area from 2012 to 2017 and highlights an obvious spatial discrepancy in urban areas. This finding indicates that most urban areas within the study area were distributed in international metropolises (Shanghai, China), metropolitan circles (Suzhou, Changzhou, and Wuxi, China), and provincial capitals (Nanjing, Ningbo, Hangzhou, and Hefei, China) during the six years. On the contrary, only a small proportion of urban areas were scattered in medium- and small-sized cities. As shown in Figure 5, the acreage of all cities in the study area increased gradually from 2012 to 2017. The urban areas of all cities in the study area extended along the central urban areas to the surrounding suburbs (Figure 4). However, the increase in the urban areas of big cities was small, such as Shanghai, Suzhou, Wuxi, and Nanjing, the urban areas of which increased from 2012 to 2014. In contrast, the extension force of the urban areas of medium- and small-sized cities, such as Xuancheng, Shaoxing, and Nantong, increased in recent years, especially from 2015 to 2017.

Table 2. Mean accuracy of urban area extraction in the YRD.

Accuracy	2012	2013	2014	2015	2016	2017
g-mean (%)	82.10	84.88	83.06	79.21	83.94	84.10
Kappa	0.793	0.828	0.805	0.761	0.816	0.817

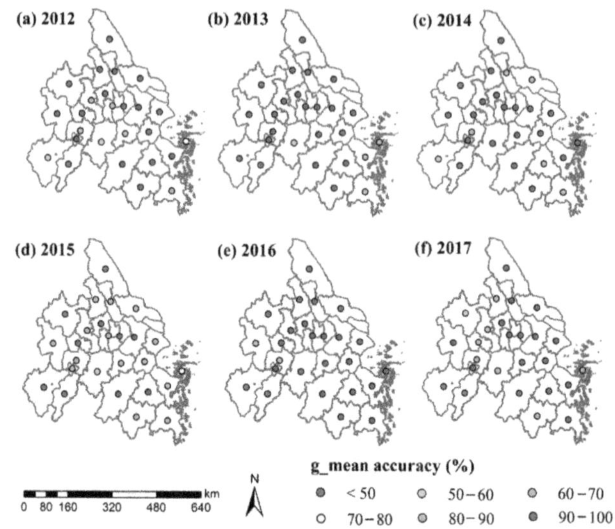

Figure 2. Geometric mean accuracy for urban area extraction in the YRD.

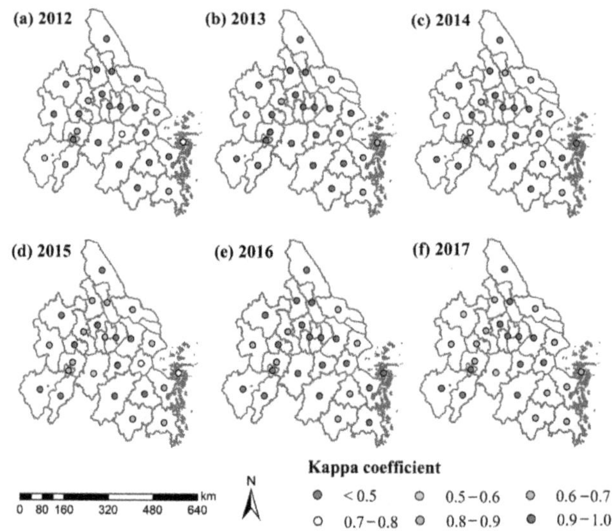

Figure 3. Kappa coefficient for urban area extraction in the YRD.

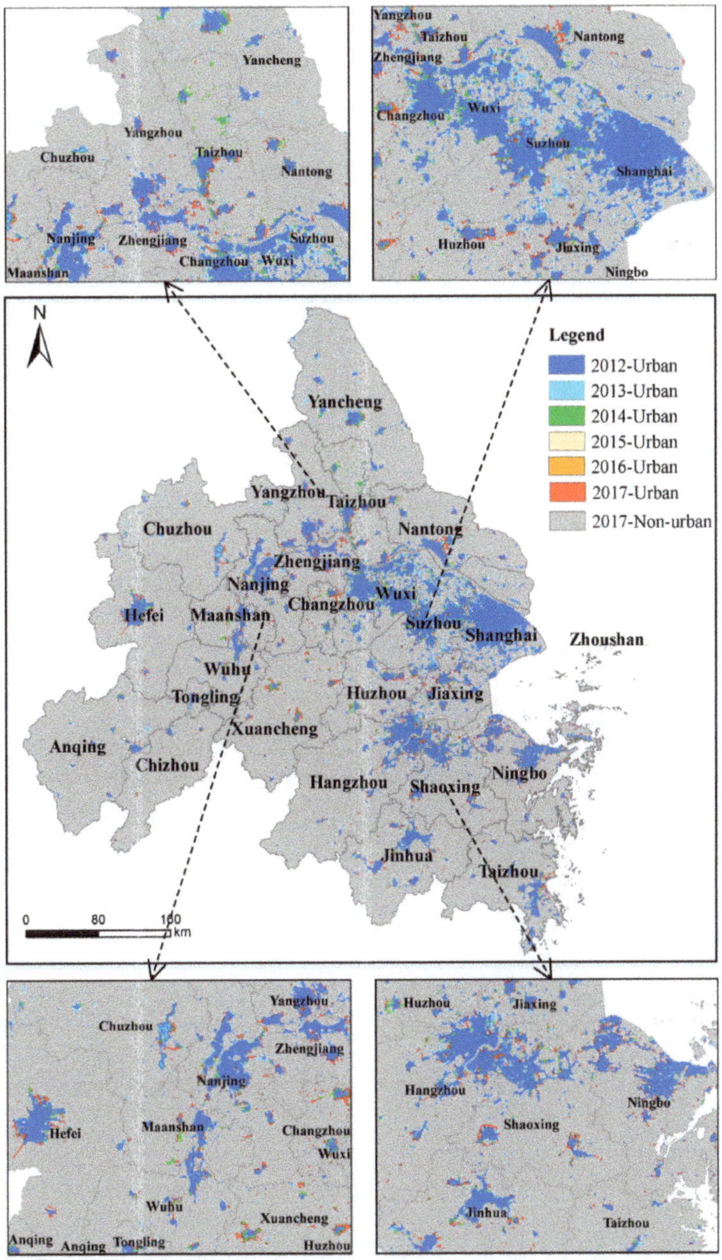

Figure 4. Urban expansion of the study area from 2012 to 2017.

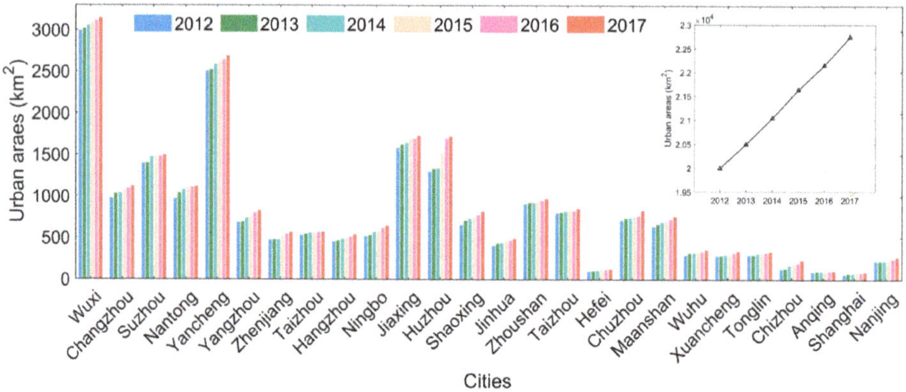

Figure 5. The increase in urban areas in all cities in the study area from 2012 to 2017.

3.2. Variations in City Size in the YRD

3.2.1. Variations in Rank–Size

Table 3 shows the fitting results of the double logarithm of the rank–size model in the YRD from 2012 to 2017. These results indicate that both the urban areas extracted from remote-sensing data and the population data can fit the model well, with R^2 values of approximately 0.79 and greater than 0.83 from 2012 to 2017, respectively. Although the value of R^2 fitted from urban areas was lower than that fitted from population size, the value met the requirements of the rank–size model. Therefore, a subsequent analysis of the variations in rank size of the YRD from 2012 to 2017 was performed.

Table 3. Fitting results of rank–size in the YRD from 2012 to 2017 (y represents Zipf's index; x in the left and right columns represent urban area and urban population, respectively).

Year	Urban Areas		Population	
	Fitting Equations	R^2	Fitting Equations	R^2
2012	y = −1.048x + 3.782	0.796	y = −0.884x + 3.347	0.840
2013	y = −1.034x + 3.784	0.801	y = −0.880x + 3.352	0.838
2014	y = −1.019x + 3.784	0.795	y = −0.875x + 3.357	0.835
2015	y = −1.008x + 3.788	0.798	y = −0.856x + 3.351	0.860
2016	y = −1.010x + 3.793	0.797	y = −0.852x + 3.357	0.860
2017	y = −0.982x + 3.791	0.789	y = −0.845x + 3.361	0.858

Zipf's index was calculated from urban areas and the population size of the YRD from 2012 to 2017, respectively; the values are shown in Figure 6. The Zipf's index calculated from the extracted urban areas from 2012 to 2017 was below the value of one, which indicates that the development of big cities in the YRD was not prominent, while that of small- and medium-sized cities was better. In addition, a decreasing Zipf's index trend calculated from the extracted urban areas was found, with a value of 0.704 in 2012, which decreased to 0.651 in 2017. Such findings indicate that the dispersed force in the YRD was larger than the concentrated force. Thus, the development of the city size of small cities is faster than that of big cities in the YRD. However, from 2010 to 2013, the urban area in the Yangtze River Delta expanded rapidly, while the resident population increased slowly, especially in small and medium-sized cities. Thus, the Zipf's index calculated from urban area is bigger than that calculated by population.

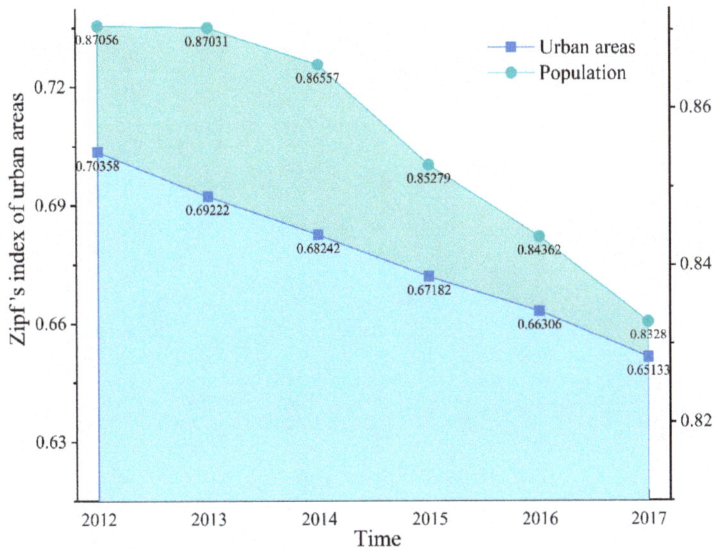

Figure 6. Variations in Zipf's index for the extracted urban areas and the statistical population data of the YRD from 2012 to 2017.

The Zipf's index calculated from the population data of the YRD displayed the same distribution as that calculated from the extracted urban areas, which indicates the feasibility of analyzing the variation in rank–size by utilizing NTL remote sensing.

3.2.2. Variations in Primate City

In this study, the 11-city index of urban primacy was utilized to analyze the variations in urban primacy in the study area. Figure 7 shows the distribution of the YRD urban primacy index from 2012 to 2017. As depicted in the figure, the urban primacy index calculated from the extracted urban areas of the YRD was lower than one from 2012 to 2017, which indicates that the YRD agglomeration did not belong to the distribution of the primate city during this period. Additionally, the urban primacy index of the study area decreased gradually, declining from 0.506 in 2012 to 0.471 in 2017. Such a decrease indicates that the size distribution of the primate city in the YRD economic region had a relatively decreasing trend compared with other cities, and the status and function of the primate city in the study area gradually reduced during this period. Such findings indirectly illustrate that the size of medium- and small-sized cities in the study area increased gradually during this period.

The urban primacy index computed from population data also displayed a similar distribution and variation trend to that calculated from the extracted urban areas, which was also below the value of one and decreased gradually from 2012 to 2017. This finding indicates that the application of NTL remote sensing for the analysis of variations in urban primacy is a feasible method.

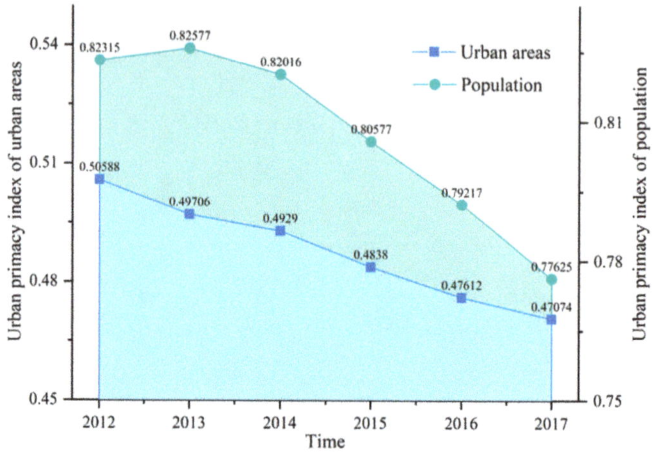

Figure 7. Variations in urban primacy index for extracted urban areas and the statistic population data of the YRD from 2012 to 2017.

3.2.3. Variations in the Gini Coefficient

In this study, the Gini coefficient for cities was used to analyze the degree of balance in city development in the study area. Figure 8 shows the variations in the Gini coefficient computed from the extracted urban areas using NTL remote-sensing and population data from 2012 to 2017. The value of the Gini coefficient calculated from the extracted urban areas ranged from 0.2 to 0.3, which indicates that the development of the city size in the study area occurred at a relatively balanced level during this period. Furthermore, the Gini coefficient displayed a gradually decreasing tendency from 2012 to 2017, with the value declining from 0.226 in 2012 to 0.215 in 2017. Such a decrease indicates that the city size of the YRD displayed a scattering trend from 2012 to 2017. Thus, the dispersed force in the study area was greater than the concentrated force, implying that the development of medium- and small-sized cities was faster than that of big cities, and the development of big cities reached saturation in the study area during this period.

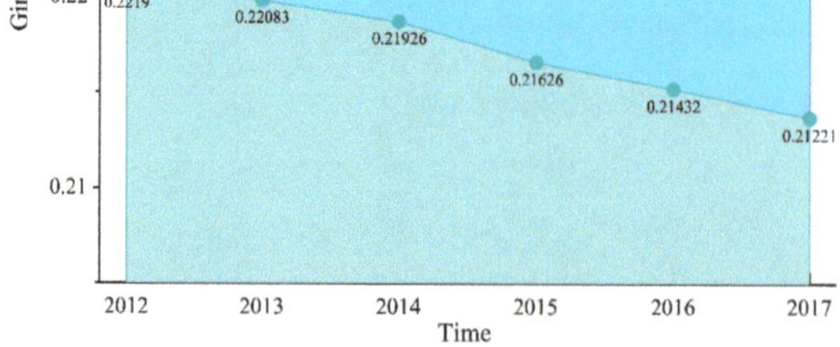

Figure 8. Variations in the Gini coefficient for extracted urban areas and the statistic population data of YRD from 2012 to 2017.

The Gini coefficient calculated from the population data was used to validate the value computed from the extracted urban areas of the study area. The value of the Gini coefficient calculated from population data also ranged from 0.2 to 0.3, and displayed a decreasing tendency from 2012 to 2017 (Figure 8), aligning with the variations in the Gini coefficient computed from the extracted urban areas. This result also highlights the feasibility of using extracted urban areas instead of population data to study the distribution and variation in city size.

3.3. Sensitive Analysis

In the accuracy assessment of urban area extraction presented in Section 3.1.1, five small to medium-sized cities with lower levels of economic development, namely Chizhou, Xuancheng, Anqing, Chuzhou, and Yancheng, were found to have relatively low accuracy, with Kappa coefficients and geometric mean accuracy below 60% and 0.6, respectively. This finding was mainly due to cities with lower levels of economic development having relatively lower signals of NTL and LST images, which may affect the urban extraction results. However, the purpose of this study was to prove that urban areas extracted using remote-sensing technology can be utilized to analyze variations in city size instead of statistical population data. Whether the extraction accuracy can affect the results of the city size in the study area must be determined. Therefore, we adjusted the urban areas of the five cities by increasing or decreasing 5% of the extracted urban areas to explore the effect of extraction accuracy on city size monitoring. When the urban areas of one city were adjusted, those of the other four cities remained unchanged. Zipf's index, the urban primacy index, and the Gini coefficient were computed from the originally extracted urban areas; the urban areas increased by 5%, and urban areas decreased by 5%. The top 11 cities ranked by the acreage of urban areas remained unchanged regardless of whether they decreased or increased by 5%, which caused very slight changes in the Gini coefficient (between 0.0001–0.0003). Thus, only Zipf's index and the Gini coefficient were calculated (Figure 9a–e) and (Figure 10a–e).

As depicted in Figure 9a–e, the Zipf's index calculated from the urban areas increasing by 5% or decreasing by 5% was consistent with that computed from the original urban areas, with values below 1.0 from 2012 to 2017, presenting a gradually decreasing tendency. This finding indicates that the development of small and medium-sized cities was more prominent than that of big cities in the YRD, and the dispersed force was larger than the concentrated force in the YRD. Further, as depicted in Figure 10a–e, the Gini coefficient calculated from both urban areas increasing by 5% or decreasing by 5% was consistent with that computed from the original urban areas, with all values ranging from 0.2 to 0.3, highlighting a gradually decreasing tendency from 2012 to 2017. This finding indicates that the development of YRD has reached a relatively balanced level. Further, the city size of YRD tended to be scattered from 2012 to 2017. Regardless of a decrease or increase by 5% in the extracted urban areas, the difference between the two indices calculated from the original urban areas and the adjusted urban areas was very small, with values below 0.001 (the three lines of the two indices calculated from the original and adjusted urban areas were very close). Consequently, the relatively low accuracy of urban area extraction of some cities has little effect on the monitoring of variations in city size in this study.

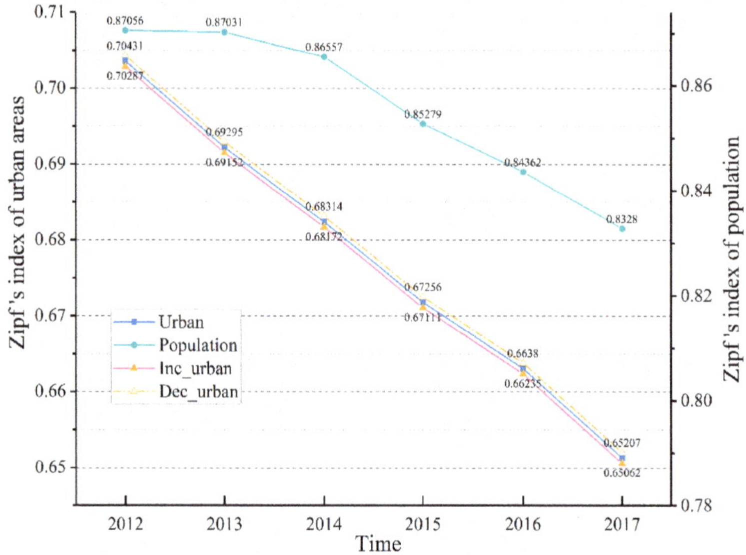

(a) Yancheng

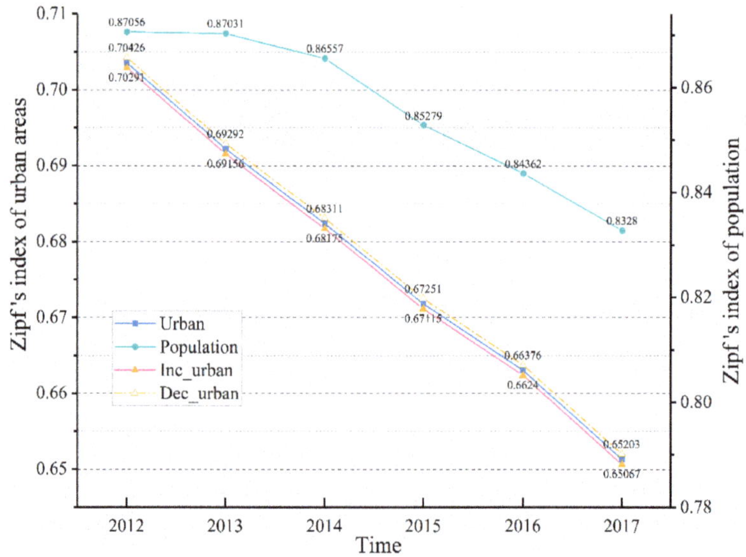

(b) Chuzhou

Figure 9. *Cont.*

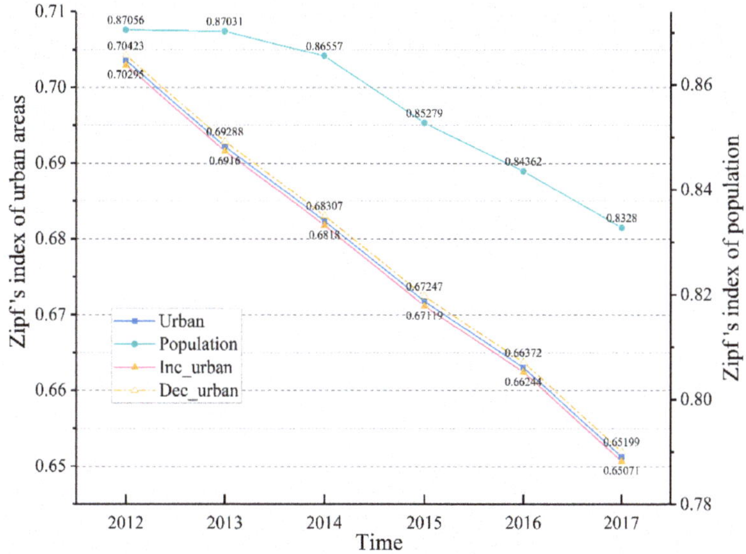

(c) Xuancheng

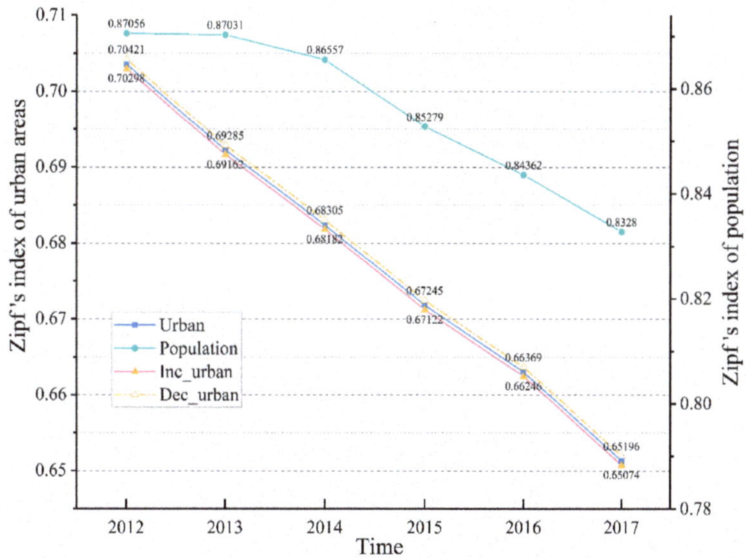

(d) Chizhou

Figure 9. *Cont.*

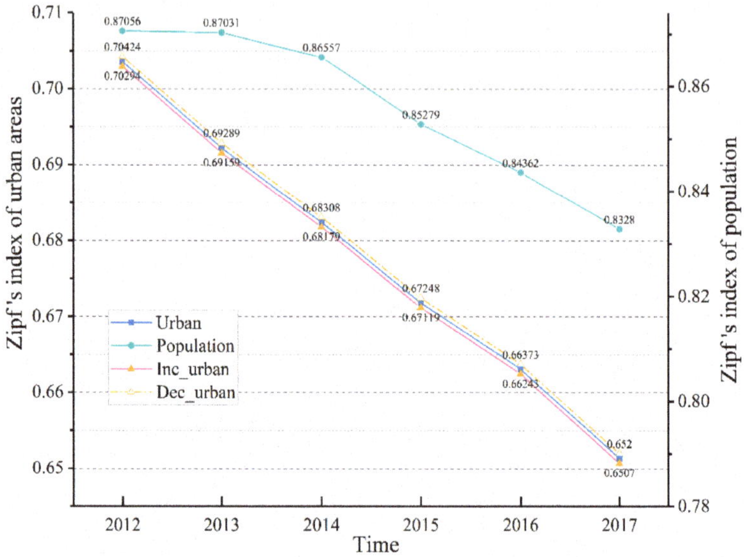

(e) Anqing

Figure 9. Variations in the Zipf's index of the YRD from 2012 to 2017; the "Inc_urban" and "Dec_urban" represent the results calculated from urban areas of five cities that increased by 5% and decreased by 5%, respectively.

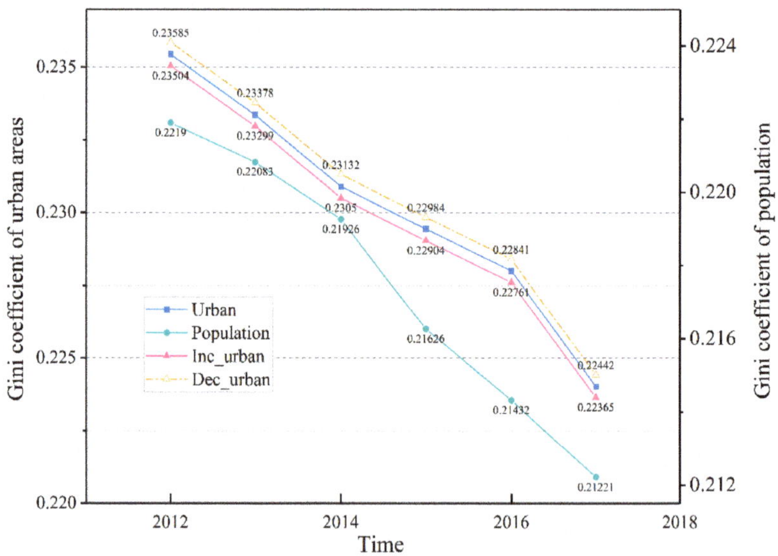

(a) Yancheng

Figure 10. *Cont.*

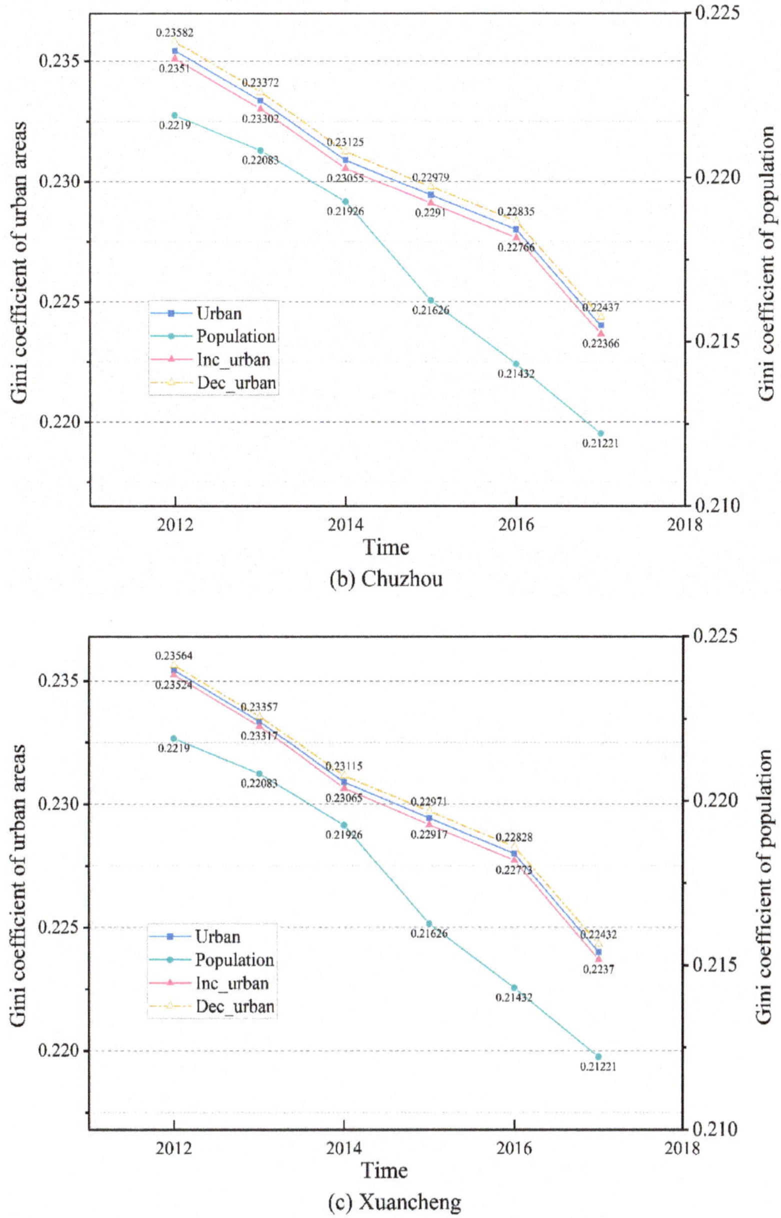

Figure 10. *Cont.*

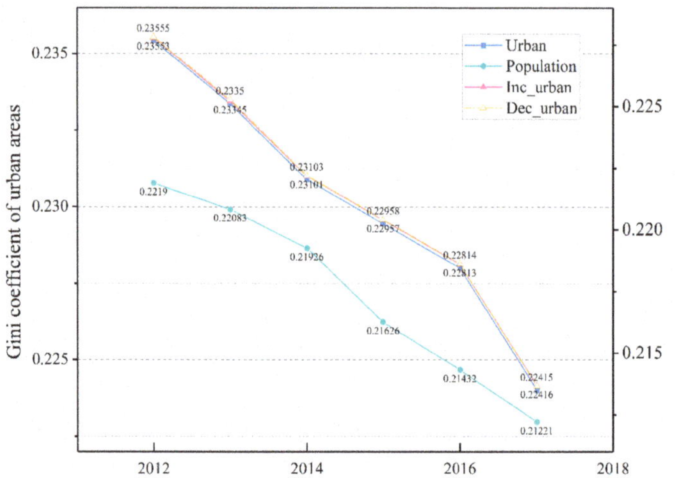

(d) Chizhou

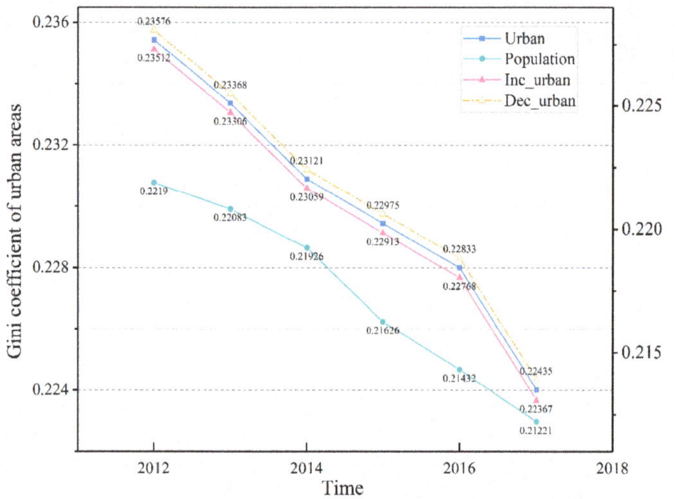

(e) Anqing

Figure 10. Variations in the Gini coefficient of the YRD from 2012 to 2017; The "Inc_urban" and "Dec_urban" represent the results calculated from urban areas of five cities that increased by 5% and decreased by 5%, respectively.

3.4. Comparison with Other Results

In this study, Zipf's law, the urban primacy index, and the Gini coefficient were computed from the statistical urban areas, and the distribution and variations in city size were compared with those calculated from the urban areas extracted from NTL remote sensing (Figures 11–13). As shown in Figure 11, Zipf's index calculated from statistical urban areas was consistent with that computed from the urban areas extracted from remote-sensing technology, with values ranging from 0.6 to 0.72 and a gradually decreasing tendency. Moreover, in Figure 12, the urban primacy index calculated from the statistic urban areas showed a decreasing trend, with values ranging from 0.41 to 0.51, thereby aligning with the values calculated from the extracted urban areas. The Gini coefficient calculated from statistical urban areas and extracted urban areas also showed some variation tendency in Figure 13, with values ranging from 0.2 to 0.3. Therefore, all three indices calculated from statistics and extracted urban areas revealed that the YRD did not fit the distribution of the primate city, the dispersed force was stronger than the concentrated force in the study area, and the development of the YRD reached a relatively balanced level. Consequently, the city size analyzed from the extracted urban areas had the same variation trend as that analyzed from the statistical urban areas, which proved the feasibility of utilizing NTL remote sensing to analyze the distribution and variation of city size instead of statistics.

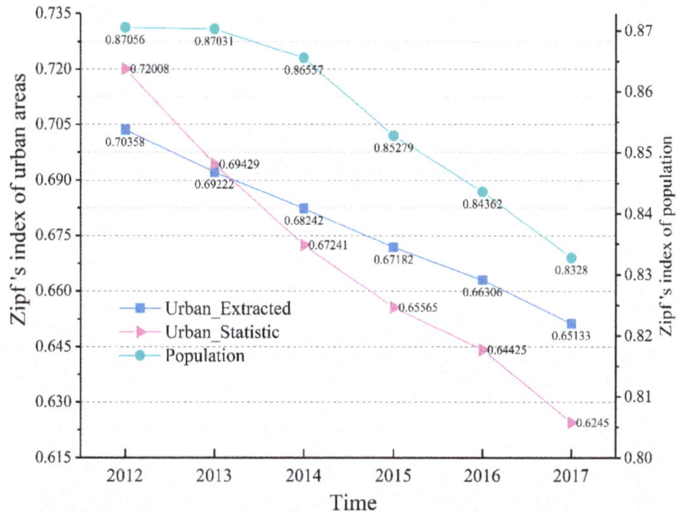

Figure 11. Variations in the Zipf's index for extracted urban areas, statistic urban areas, and population data of the YRD from 2012 to 2017.

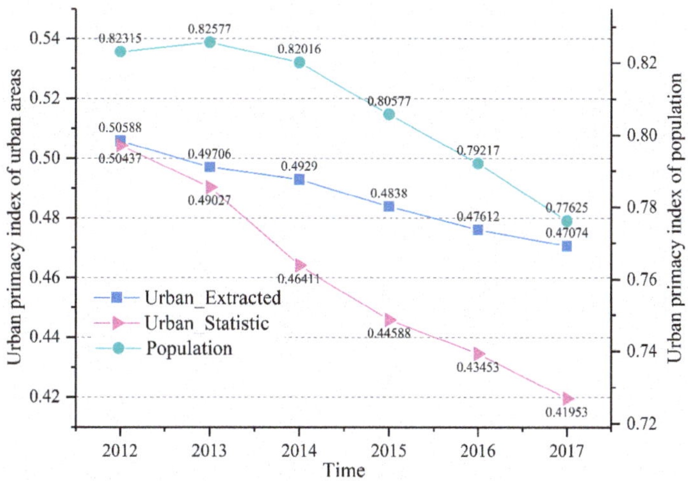

Figure 12. Variations in the urban primacy index for extracted urban areas, statistic urban areas, and population data of the YRD from 2012 to 2017.

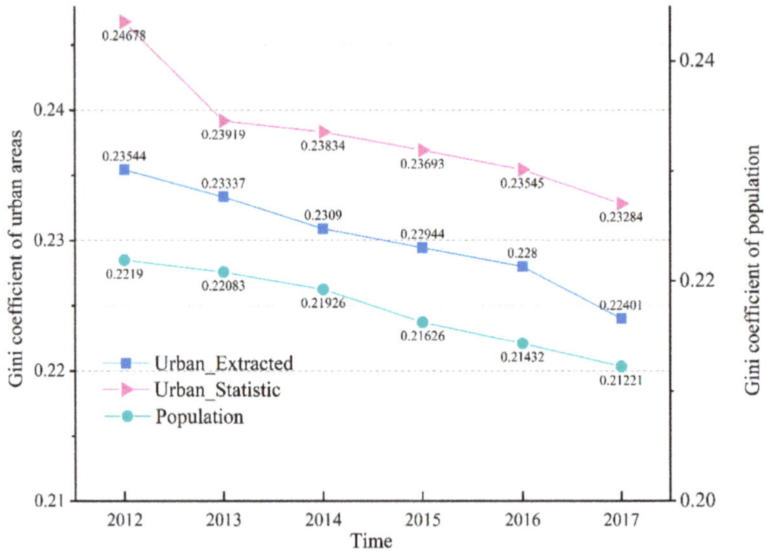

Figure 13. Variations in the Gini coefficient for extracted urban areas, statistic urban areas, and population data of the YRD from 2012 to 2017.

4. Conclusions

The YRD of China was selected as the study area for this assessment and multi-source remote-sensing data based on an ANN algorithm were employed to extract urban areas for the analysis of variations in city size from 2012 to 2017 in the study area, instead of the traditional approach of using population data. The main conclusions are as follows.

By combining NTL data with land-surface temperature and normalized differential vegetation index data, the method employed in this study could be used to efficiently extract the urban areas of the study area based on the ANN algorithm. The geometric mean accuracy and Kappa coefficient ranged from 79.21% to 84.88% and 0.761 to 0.828, respectively. In addition, the urban areas of the study area expanded gradually from 2012 to 2017, with the entire region and all cities in the region undergoing expansion. Moreover, the rank–size rule, urban primacy index, and Gini model were utilized to assess the development of city size in the study area from 2012 to 2017. Based on the results, the study area did not fit the distribution of the primate city, and the status and function of the primate city in the YRD gradually decreased during this period. The development of big cities was not prominent compared with that of small and medium-sized cities, and the dispersed force was greater than the concentrated force in the study area. The development of city size in the YRD urban agglomeration reached a relatively balanced level during this period. Additionally, the sensitivity analysis revealed that the relatively low extraction accuracy of urban areas for a few small cities had little effect on the results of city size in this study. The three indices calculated from the statistical population data and the statistical urban areas also displayed the same distribution and variation tendency as those computed from the extracted urban areas via NTL remote sensing. These findings indicate the applicability and feasibility of analyzing the variation in city size using NTL remote sensing.

Author Contributions: Conceptualization, Y.D. and J.H.; methodology, Y.D. and J.H.; investigation, Y.D., J.H. and W.M.; validation, Y.Y., S.J., Y.Z. and J.Z.; writing—original draft preparation, Y.D., J.H. and W.M.; writing—review and editing, Y.Y., S.J., X.P., Y.Z., K.C. and J.Z.; funding acquisition, Y.D. and Y.Y. All authors contributed to the analysis of the results and reviewed the manuscript. All authors have read and agreed to the published version of the manuscript.

Funding: This work was supported in part by the Fundamental Research Funds for the Central Universities (Grant numbers B200202017, B210201013, and B220203008), the National Natural Science Foundation of China (Grant number 42071346), the Natural Science Foundation of Jiangsu Province under (Grant number BK20190495), and the Postgraduate Research and Practice Innovation Program of the Jiangsu Province (No. KYCX21_0529).

Conflicts of Interest: The authors declare no conflict of interest.

References

1. Schneider, A.; Friedl, M.A.; Potere, D. Mapping global urban areas using MODIS 500-m data: New methods and datasets based on 'urban ecoregions'. *Remote Sens. Environ.* **2010**, *114*, 1733–1746. [CrossRef]
2. United Nations. World Urbanization Prospects: The 2014 Revisio. Available online: https://population.un.org/wup/publications/files/wup2014-report.pdf (accessed on 15 May 2020).
3. Muhammad, S.; Lean, H.H. Does Financial Development Increase Energy Consumption? The Role of In-dustrialization and Urbanization in Tunisia. *Energy Policy* **2012**, *40*, 473–479.
4. Walsh, C.J. Urban Impacts on the Ecology of Receiving Waters: A Framework for Assessment, Conservation and Restoration. *Hydrobiologia* **2000**, *2*, 107–114. [CrossRef]
5. Kalnay, E.; Cai, M. Impact of urbanization and land-use change on climate. *Nature* **2003**, *423*, 528–531. [CrossRef] [PubMed]
6. Weng, Q.; Firozjaei, M.K.; Sedighi, A.; Kiavarz, M.; Alavipanah, S.K. Statistical analysis of surface urban heat island intensity variations: A case study of Babol city, Iran. *GIScience Remote Sens.* **2018**, *56*, 576–604. [CrossRef]
7. Zipf, G.K. *Human Behavior and the Principle of Least Effort: An Introduction to Human Ecology*; Addison-Wesley: New York, NY, USA, 1949.
8. Mark, J. The Law of the Primate City. *Geogr. Rev.* **1939**, *2*, 226–232.
9. Gini, C. Measurement of Inequality of Incomes. *Econ. J.* **1921**, *31*, 124. [CrossRef]
10. Huang, Q.; He, C.; Gao, B.; Yang, Y.; Liu, Z.; Zhao, Y.; Dou, Y. Detecting the 20 year city-size dynamics in China with a rank clock approach and DMSP/OLS nighttime data. *Landsc. Urban Plan.* **2015**, *137*, 138–148. [CrossRef]
11. Anderson, G.; Ying, G. The Size Distribution of Chinese Cities. *Reg. Sci. Urban Econ.* **2005**, *6*, 756–776. [CrossRef]
12. Fan, C.C. The Vertical and Horizontal Expansions of China's City System. *Urban Geogr.* **1999**, *6*, 493–515. [CrossRef]
13. Xu, Z.; Zhu, N. City Size Distribution in China: Are Large Cities Dominant? *Urban Stud.* **2009**, *10*, 2159–2185. [CrossRef]
14. Levin, N.; Zhang, Q. A global analysis of factors controlling VIIRS nighttime light levels from densely populated areas. *Remote Sens. Environ.* **2017**, *190*, 366–382. [CrossRef]
15. Yao, Y.; Chen, D.; Chen, L.; Wang, H.; Guan, Q. A time series of urban extent in China using DSMP/OLS nighttime light data. *PLoS ONE* **2018**, *13*, e0198189. [CrossRef]
16. Zou, Y.; Peng, H.; Liu, G.; Yang, K.; Xie, Y.; Weng, Q. Monitoring Urban Clusters Expansion in the Middle Reaches of the Yangtze River, China, Using Time-Series Nighttime Light Images. *Remote Sens.* **2017**, *10*, 1007. [CrossRef]
17. Nitsch, V. Zipf Zipped. *J. Urban Econ.* **2005**, *1*, 86–100. [CrossRef]
18. Fu, Y.; Li, J.; Weng, Q.; Zheng, Q.; Li, L.; Dai, S.; Guo, B. Characterizing the spatial pattern of annual urban growth by using time series Landsat imagery. *Sci. Total Environ.* **2019**, *666*, 274–284. [CrossRef]
19. Huang, X.; Schneider, A.; Friedl, M.A. Mapping sub-pixel urban expansion in China using MODIS and DMSP/OLS nighttime lights. *Remote Sens. Environ.* **2016**, *175*, 92–108. [CrossRef]
20. Xie, Y.; Weng, Q.; Fu, P. Temporal Variations of Artificial Nighttime Lights and Their Implications for Ur-banization in the Conterminous United States, 2013–2017. *Remote Sens. Environ.* **2019**, *225*, 160–174. [CrossRef]
21. Yu, B.; Shu, S.; Liu, H.; Song, W.; Wu, J.; Wang, L.; Chen, Z. Object-based spatial cluster analysis of urban landscape pattern using nighttime light satellite images: A case study of China. *Int. J. Geogr. Inf. Sci.* **2014**, *28*, 2328–2355. [CrossRef]
22. Su, Y.; Chen, X.; Wang, C.; Zhang, H.; Liao, J.; Ye, Y.; Wang, C. A new method for extracting built-up urban areas using DMSP-OLS nighttime stable lights: A case study in the Pearl River Delta, southern China. *GIScience Remote Sens.* **2015**, *52*, 218–238. [CrossRef]
23. Small, C.; Elvidge, C.D. Night on Earth: Mapping Decadal Changes of Anthropogenic Night Light in Asia. *Int. J. Appl. Earth Obs. Geoinf.* **2013**, *22*, 40–52. [CrossRef]
24. Small, C.; Elvidge, C.D.; Balk, D.; Montgomery, M. Spatial scaling of stable night lights. *Remote Sens. Environ.* **2011**, *115*, 269–280. [CrossRef]
25. Elvidge, C.D.; Baugh, K.E.; Zhizhin, M.; Hsu, F.-C. Why VIIRS data are superior to DMSP for mapping nighttime lights. *Proc. Asia-Pac. Adv. Netw.* **2013**, *35*, 62. [CrossRef]
26. Miller, S.D.; Straka, W.; Mills, S.P.; Elvidge, C.D.; Lee, T.F.; Solbrig, J.; Walther, A.; Heidinger, A.K.; Weiss, S.C. Illuminating the Capabilities of the Suomi National Polar-Orbiting Partnership (NPP) Visible Infrared Imaging Radiometer Suite (VIIRS) Day/Night Band. *Remote Sens.* **2013**, *5*, 6717–6766. [CrossRef]
27. Cao, C.; Bai, Y. Quantitative Analysis of VIIRS DNB Nightlight Point Source for Light Power Estimation and Stability Monitoring. *Remote Sens.* **2014**, *6*, 11915–11935. [CrossRef]
28. Zhou, Y.; Smith, S.J.; Elvidge, C.D.; Zhao, K.; Thomson, A.; Imhoff, M. A cluster-based method to map urban area from DMSP/OLS nightlights. *Remote Sens. Environ.* **2014**, *147*, 173–185. [CrossRef]

29. Xie, Y.; Weng, Q. Updating Urban Extents with Nighttime Light Imagery by Using an Object-Based Thresh-olding Method. *Remote Sens. Environ.* **2016**, *187*, 1–13. [CrossRef]
30. Henderson, M.; Yeh, E.; Gong, P.; Elvidge, C.; Baugh, K. Validation of urban boundaries derived from global night-time satellite imagery. *Int. J. Remote Sens.* **2003**, *24*, 595–609. [CrossRef]
31. Liu, Z.; He, C.; Zhang, Q.; Huang, Q.; Yang, Y. Extracting the dynamics of urban expansion in China using DMSP-OLS nighttime light data from 1992 to 2008. *Landsc. Urban Plan.* **2012**, *106*, 62–72. [CrossRef]
32. Ma, T.; Zhou, Y.; Zhou, C.; Haynie, S.; Pei, T.; Xu, T. Night-Time Light Derived Estimation of Spa-tio-Temporal Characteristics of Urbanization Dynamics Using Dmsp/Ols Satellite Data. *Remote Sens. Environ.* **2015**, *158*, 453–464. [CrossRef]
33. Cao, X.; Chen, J.; Imura, H.; Higashi, O. A SVM-based method to extract urban areas from DMSP-OLS and SPOT VGT data. *Remote Sens. Environ.* **2009**, *113*, 2205–2209. [CrossRef]
34. Jing, W.; Yang, Y.; Yue, X.; Zhao, X. Mapping Urban Areas with Integration of DMSP/OLS Nighttime Light and MODIS Data Using Machine Learning Techniques. *Remote. Sens.* **2015**, *7*, 12419–12439. [CrossRef]
35. Xu, T.; Coco, G.; Gao, J. Extraction of urban built-up areas from nighttime lights using artificial neural network. *Geocarto Int.* **2019**, *35*, 1049–1066. [CrossRef]
36. Lu, D.; Tian, H.; Zhou, G.; Ge, H. Regional mapping of human settlements in southeastern China with multisensor remotely sensed data. *Remote Sens. Environ.* **2008**, *112*, 3668–3679. [CrossRef]
37. Yang, Y.; He, C.; Zhang, Q.; Han, L.; Du, S. Timely and accurate national-scale mapping of urban land in China using Defense Meteorological Satellite Program's Operational Linescan System nighttime stable light data. *J. Appl. Remote Sens.* **2013**, *7*, 073535. [CrossRef]
38. Kubat, M.; Holte, R.C.; Matwin, S. Machine Learning for the Detection of Oil Spills in Satellite Radar Images. *Mach. Learn.* **1998**, *30*, 195–215. [CrossRef]

remote sensing

Article

Night-Time Light Remote Sensing Mapping: Construction and Analysis of Ethnic Minority Development Index

Fei Zhao [1,2], Lu Song [1], Zhiyan Peng [1], Jianqin Yang [3], Guize Luan [1], Chen Chu [1], Jieyu Ding [4], Siwen Feng [5], Yuhang Jing [1] and Zhiqiang Xie [1,2,*]

1 School of Earth Sciences, Yunnan University, Kunming 650500, China; cartographer@ynu.edu.cn (F.Z.); songlu@mail.ynu.edu.cn (L.S.); pengzy834@mail.ynu.edu.cn (Z.P.); luanguize@mail.ynu.edu.cn (G.L.); chuchen2017@mail.ynu.edu.cn (C.C.); yuhang@mail.ynu.edu.cn (Y.J.)
2 Engineering Research Center of Domestic High-Resolution Satellite Remote Sensing Geology for Universities of Yunnan Province, Kunming 650500, China
3 Kunming Municipal People's Congress Standing Committee, Kunming 650500, China; yjqshe@163.com
4 School of Resources and Environmental Science, Wuhan University, Wuhan 430079, China; jadeding@whu.edu.cn
5 Institute of International Rivers and Eco-Security, Yunnan University, Kunming 650500, China; fengsiwen1004@mail.ynu.edu.cn
* Correspondence: xzq_2019@ynu.edu.cn

Abstract: Using toponym data, population data, and night-time light data, we visualized the development index of the Yi, Wa, Zhuang, Naxi, Hani, and Dai ethnic groups on ArcGIS as well as the distribution of 25 ethnic minorities in the study area. First, we extracted the toponym data of 25 ethnic minorities in the study area, combined with night-time light data and the population proportion data of each ethnic group, then we obtained the development index of each ethnic group in the study area. We compared the development indexes of the Yi, Wa, Zhuang, Naxi, Hani, and Dai ethnic groups with higher development indexes. The results show that the Yi nationality's development index was the highest, reaching 28.86 (with two decimal places), and the Dai nationality's development index was the lowest (15.22). The areas with the highest minority development index were concentrated in the core area of the minority development, and the size varied with the minority's distance. According to the distribution of ethnic minorities, we found that the Yi ethnic group was distributed in almost the entire study area, while other ethnic minorities had obvious geographical distribution characteristics, and there were multiple ethnic minorities living together. This research is of great significance to the cultural protection of ethnic minorities, the development of ethnic minorities, and the remote sensing mapping of lights at night.

Keywords: night-time light remote sensing; ethnic minorities; core ethnic minority areas; development index; toponym data

Citation: Zhao, F.; Song, L.; Peng, Z.; Yang, J.; Luan, G.; Chu, C.; Ding, J.; Feng, S.; Jing, Y.; Xie, Z. Night-Time Light Remote Sensing Mapping: Construction and Analysis of Ethnic Minority Development Index. *Remote Sens.* **2021**, *13*, 2129. https://doi.org/10.3390/rs13112129

Academic Editors: Ran Goldblatt, Steven Louis Rubinyi and Hogeun Park

Received: 22 April 2021
Accepted: 27 May 2021
Published: 28 May 2021

Publisher's Note: MDPI stays neutral with regard to jurisdictional claims in published maps and institutional affiliations.

1. Introduction

Ethnic minorities refer to ethnic groups other than the main ethnic group in a multi-ethnic country. The proportion of their population is smaller than that of the main ethnic group. There are currently more than 2000 ethnic groups in the world, and the total number of Asian ethnic groups is more than 1000, accounting for about half of the total number of ethnic groups in the world. Among them, the total number of ethnic groups in China, India, the Philippines, and Indonesia exceeds 50. There are about 170 ethnic groups in Europe, and there are about 20 basically single-ethnic countries. There are 55 ethnic minorities in China except for the main ethnic group. The distribution of ethnic minorities in China is relatively wide, mainly showing the distribution of "large mixed residences and small settlements". The indicators to measure the development level of a region include education level [1], regional GDP [2–4], population [5,6], poverty index [7], etc. Among them, the most direct and quantifiable one is economic development. The most direct connection between a nation and a country is the consistency of economic interests [8]. The distribution of ethnic minorities is different, their ecological environment, cultural diversity (such as living habits, languages, religious beliefs, etc.), the technology used in production, the allocation of resources is different, so their economic development is also different [9]. The economic development of ethnic minorities is part of the country's economic development and contributes to the economic development of the entire country. If there is a problem with the economic development of ethnic minorities, it will directly affect the country's economic development to a certain extent. Due to differences in living environment and life concepts, there are different economic development models in economic development, leading to better ethnic development in some places and poorer ethnic development in other places. However, the economic development of China's ethnic minority areas is generally unbalanced. China is a multi-ethnic country, and the common development and mutual assistance of all ethnic groups can make our country stronger and more prosperous. However, due to the different levels of economic development of different ethnic groups, studying the development of ethnic minorities plays an important role in formulating and adjusting corresponding policies. It is very important to understand and discover the development status of each ethnic group. This study helps to understand the development status of ethnic minorities through a simple and quick method.

At present, it mainly studies the economic development index of ethnic minorities from gross domestic product (GDP). A study of the economic development status of the five western ethnic autonomous regions in Inner Mongolia, Guangxi, Tibet, Ningxia, and Xinjiang found that the GDP of the five ethnic minorities regions lagged behind the national level, and there were also significant differences in the economic development level of ethnic minorities in the prefectures regions. The urban–rural per capita income ratio exceeded 2.5:1, and the highest urban–rural per capita income ratio reached 5.6:1, which far exceeded the international standard (according to the general international situation, the per capita GDP is between US$800 and US$1000, and the urban–rural per capita income ratio is 1.7:1 or so) [10]. Li [11] found that the income gap between urban and rural areas in ethnic minorities regions is large, as was the gap between GDP and the national level. The absolute difference in the per capita GDP of the ethnic minorities in Northwest China is gradually expanding, and the absolute difference in the economic development level of the ethnic minorities is expanding [12,13]. Zheng [14] pointed out in his research that both in terms of innovation and economic development, ethnic minority areas lagged behind the national level, and there were large differences in economic development among ethnic minority areas. Luo and Zhuang [15] conducted research on the economic development of the two provinces of Guangxi and Yunnan in the past 15 years, and found that the higher the proportion of the minority population in the total population, the lower the economic development level of the county-level region. Although there are many studies on the development of ethnic minorities, there are very few studies on the development index of ethnic minorities, and the research on the GDP of ethnic minorities only stays at the level of statistical yearbook research and qualitative analysis. The use of more scientific methods to study the development index of ethnic minority regions is of reference significance for understanding the development of ethnic minority regions, the development differences of various ethnic minorities, and the state's formulation of corresponding policies.

Night-time light data refer to the capture of town lights, fishery lights, etc. at night without clouds [16]. The currently widely used night-time light data include: (1) The Defense Meteorological Satellite Program's Operational Linescan System (DMSP-OLS) satellite, which provides data from 1992 to 2013; (2) The Suomi National Polar-Orbiting Partnership's Visible Infrared Imaging Radiometer Suite (NPP-VIIRS), which provides data from 2012 to the present; and (3) China's first professional night-time light remote sensing satellite "Luojia-1", jointly developed and produced by the Wuhan University team and related institutions, which provides data from 2018 to the present.

The level of human activities and economic development can be better reflected by night-time light remote sensing data, so it is widely used in social and economic fields [17–19] such as economic activity monitoring [20] and economic development research [21]. Doll et al. [22] used night-time light data to assess socio-economic development and found that it was highly correlated with GDP on a national scale (R^2 = 0.85, when R^2 is greater than 0.8, it can be considered that the two variables are highly correlated), and simulated the spatial distribution of GDP. Elvidge et al. [23] used DMSP-OLS data to analyze the relationship between night lighting area and GDP in 200 countries and found that there was a good linear relationship between night-time light area and GDP. Henderson et al. [24] used a DMSP stabilized light source and radiometric correction images, which correctly reflected the differences in the social and economic development levels of San Francisco, Beijing, and Lhasa. Henderson et al. [25] found that the brightness of night lights in a country had an obvious linear relationship (R^2 = 0.8) with the country's GDP development level. Michalopoulos et al. [26] used a similar method (similar to Henderson et al.) to study the correlation between night-time light data and GDP in Africa, and got good results. Wu et al. [27] used DMSP-OLS data to estimate GDP and the results were satisfactory. Jiang et al. [28] used DMSP-OLS data and NPP-VIIRS data to perform regression simulations on multiple socio-economic parameters, and found that using NPP-VIIRS night-time light data to regress with the whole city's GDP, R^2 reached 0.9102. This proves that night-time light data have a good linear correlation with GDP and power consumption, and found that NPP-VIIRS had higher accuracy and more advantages. Zhu et al. [29] found that compared with traditional socio-economic indicators (GDP, oil and gas production, etc.), night light data are more sensitive and more intuitively reflects social and economic development.

Some scholars have also used night-time light data to study the poverty index of a region. This method can also reflect the development status of the region to a certain extent. Li et al. [30] used the method of machine learning, combined with the robust features of the night light image spatial characteristics to identify China's high-poor counties. The overall accuracy of the results was greater than 82%, and the user accuracy was greater than 63%. Andreano et al. [31] used DMSP-OLS data to perform spatial classification and continuous time estimation of poverty gap, number of people, and Gini index in 20 Latin American and Caribbean countries. It was found that combining night-time light data helped to better understand poverty and its temporal and spatial dynamics. Pokhriyal et al. [32] used environmental data and call data records to accurately predict the global multidimensional poverty index. This method has high accuracy in predicting health, education, and living standards (Pearson's correlation coefficient is 0.84–0.86). Li et al. [33] used the principal component analysis method to establish a comprehensive multi-dimensional poverty index, and showed the temporal and spatial heterogeneity of multi-dimensional poverty in 2311 counties in China. It was found that the mountainous areas of Southwest, North China, Northwest China, and the plateau areas of Southeast China had higher levels of economic development.

A large number of studies have proven that the night-time light data reflect the development level of a region, so it is feasible to use it to construct a development index. Compared with traditional statistical yearbook research and qualitative analysis, this paper used night-time light data to construct the development index of ethnic minority areas, which is more accurate and saves resources.

2. Materials and Methods

2.1. Materials

2.1.1. Study Area

Yunnan Province is located on the border of southwestern China. Its geographic location is between 21°8′–29°15′ N and 97°31′–106°11′ E. Yunnan Province is the province with the largest number of ethnic minorities in China. According to the statistics of the sixth national census in 2010, there are 25 ethnic minorities in Yunnan Province, among which the population of Yi, Bai, and Dai are larger. Among the 25 ethnic minorities in Yunnan Province, 15 ethnic minorities are unique to Yunnan such as the Bai, Hani, Lisu, Dulong, etc.

The development of ethnic minorities in Yunnan Province has made great contributions to the socio-economic development of the entire Yunnan Province. Yunnan Province is a mountainous plateau. Compared with provinces in plain areas, its topographic features are unfavorable for its development. However, at the same time, Yunnan Province is located on the border of southwest China and is a key area for the development of the "Belt and Road" initiative. There are 16 prefecture-level administrative regions in Yunnan Province including eight prefecture-level cities, eight autonomous prefectures, 17 county-level cities, and 129 county-level districts. Among the 16 prefecture-level administrative regions, there are eight ethnic minority core areas. The administrative division and specific geographical location of Yunnan Province are shown in Figure 1.

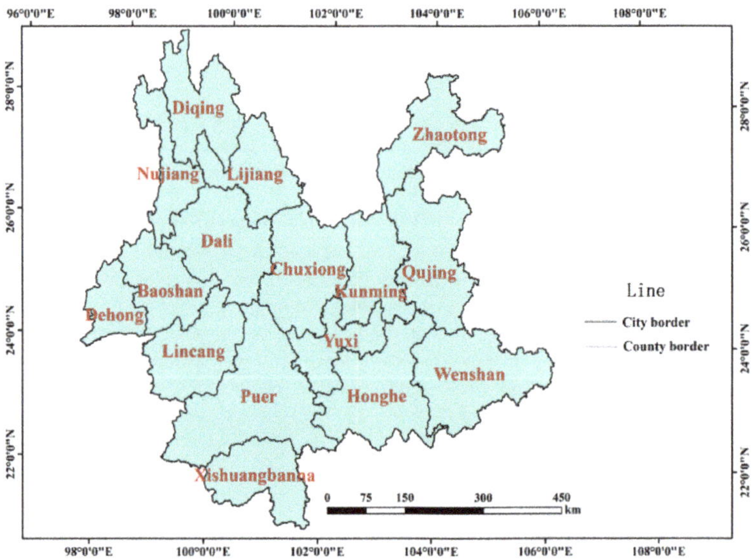

Figure 1. Geographical location and administrative boundaries of Yunnan Province.

2.1.2. Data Sources

The data used in this article are as follows (Table 1): (1) NPP-VIIRS composite data; (2) toponym data; (3) Yunnan Province census statistics; (4) Yunnan Province county level Administrative division boundaries; and (5) Yunnan Statistical Yearbook Data.

Table 1. Details of the data sources in this study.

Data	Data Information	Year	Source
NPP-VIIRS	NPP-VIIRS cloudless DNB compound monthly average data	2018	Earth Observation Group (EOG) (https://eogdata.mines.edu/download_dnb_composites.html, accessed on 28 May 2020)
Toponym	Results of the Second National Toponymic Census of China	2019	China National Geographical Names Database (http://dmfw.mca.gov.cn/, accessed on 20 May 2020)
Statistics of Yunnan Province Census	Data from the Sixth Census of Yunnan Province	2010	China Social Big Data Research Platform (http://data.cnki.net/, accessed on 14 June 2020)
Boundaries of county-level administrative divisions in Yunnan Province	County-level vector data in Yunnan Province	2017	National Basic Geographic Information Center (http://www.ngcc.cn/ngcc/, accessed on 13 May 2019)
Yunnan Statistical Yearbook Data	Socio-economic indicators of Yunnan Province	2013–2018	People's Government of Yunnan Province (www.yn.gov.cn, accessed on 2 May 2021)

The NPP-VIIRS night-time light data adopt the monthly average data of the global cloudless Day–Night Band (DNB) composite data in 2018, and the spatial resolution of NPP-VIIRS data is 500 m. Studies have shown that the DNB of the NPP satellite system is widely used to estimate social and economic parameters, and the in-orbit radiation correction can improve data quality [34,35]. Finally, monthly average data were used to synthesize annual average data for research. The data were downloaded from the Earth Observation Group (EOG) (https://eogdata.mines.edu/download_dnb_composites.html, accessed on 28 May 2020).

The toponym data used in the study come from the results of the second national toponym data census, which mainly includes the meaning of toponyms, that is, the ethnic types of toponyms, the feature type of toponyms, the historical sources of toponyms, the spatial location, and other information, which can be downloaded from the China National Geographical Names Information Database (http://dmfw.mca.gov.cn/, accessed on 20 May 2020).

The census statistics of Yunnan Province use the data of the sixth national census, and the data can be downloaded from the sixth census data of Yunnan Province on the China Social Big Data Research Platform (http://data.cnki.net/, accessed on 14 June 2020). In the data, detailed statistics are made on the population of all ethnic groups in the county-level regions of Yunnan Province.

The county-level administrative divisions of Yunnan Province are derived from the 1:4 million vector data provided by the National Basic Geographic Information Center. In order to make the research more convenient, all the data in this paper were converted into the Lambert projection (Asia_Lambert_Conformal_Conic) based on WGS_1984. In order to make the research more accurate, combined with the geographic location of the study area, the central meridian was set to 102°, the first standard latitude was 22°, and the second standard latitude was 28.3°.

The statistical yearbook data contain a large amount of socio-economic data such as regional GDP per capita, regional total GDP, and regional employees. The development data and production methods of a region can be obtained from the statistical yearbook. The statistical yearbook data of Yunnan Province from 2013 to 2018 was used to verify the feasibility of the method in this paper.

2.2. Methods

Using the 2018 NPP-VIIRS night-time light data to construct the Yunnan Minority Development Index requires the following three steps. First, preprocess the downloaded NPP-VIIRS cloudless DNB composite monthly average data to obtain stable night light data. Second, extract the toponym data that contain minority information in the toponym data to obtain the Yunnan Province minority toponym dataset, and conduct a kernel density analysis on each type of ethnic minority toponym data in Yunnan Province. Calculate the minority development index using the results of kernel density analysis combined with the results of the minority population proportion grid results and the NPP-VIIRS night-time light data. Finally, in order to more clearly reflect the distribution of ethnic minorities, combine the toponym data and the results of the minority development index to obtain the research area distribution of 25 ethnic minorities. The specific process is shown in Figure 2.

2.2.1. NPP-VIIRS Data Preprocessing

In order to avoid the influence of grid deformation, sensors, and other factors on the research results, first, geometric correction was performed on the 2018 NPP-VIIRS monthly cloudless DNB composite data using the geometric correction tool in ENVI. Since the geographic coordinate system of the acquired NPP-VIIRS data is WGS_1984, set the projection parameter to the WGS_1984 geographic coordinate system, set the output pixel size to 1000 m, and select the cubic convolution method as the resampling method. The NPP-VIIRS night-time light data obtained include fires, aurora, and other noises. Therefore, it needs to be radiated to eliminate the influence of background noise. The process of radiant correction can be referred to in [36]. Load the data to be corrected in ENVI and use the RPC orthorectification workflow tool for correction. First, select the average radiance value of the cloud in the low reflectivity area of the sea surface as the calibration value for removing scattered light, and then subtract the calibration value from the entire image to remove the cloud scattering. Second, using the method of adjacent aberrations, a threshold was set to obtain a stable surface area, and the obtained stable surface area was used as a mask, and the radiation value of the mask area was statistically analyzed. Finally, three times the average radiation value of the statistical analysis was taken as the confidence interval to remove the surface scattered light. After radiant correction, effective night-time light data can be obtained. Then, use the data after geometric correction and radiometric correction to synthesize the 2018 annual average data. The calculation formula is:

$$DN_j = \frac{\sum_{i=1}^{12} DN_i}{12}, \quad (1)$$

where DN_i represents the light brightness value in month i, and DN_j represents the average light brightness value in year j.

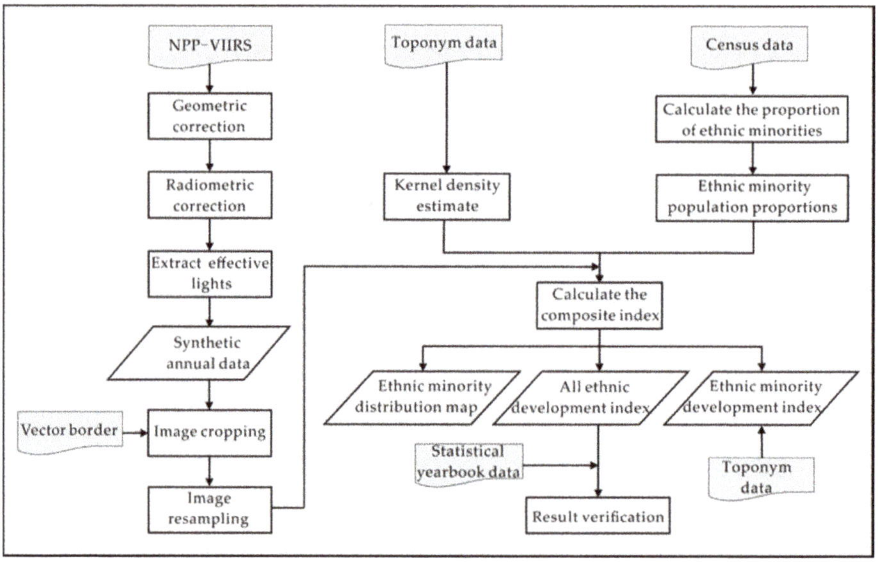

Figure 2. Flowchart of the methodology.

After synthesizing the 2018 NPP-VIIRS annual data, we used the administrative divisions of Yunnan Province as a mask to trim the night-time light data to obtain the study area. In order to make subsequent research more convenient, the coordinates were unified into the Lambert projection based on WGS_1984. Finally, using the cubic convolution interpolation method to resample the NPP-VIIRS data to a grid size from the original pixel size of 500 m × 500 m to 1000 m × 1000 m, and obtained stable night light data in 2018. The results are shown in Figure 3.

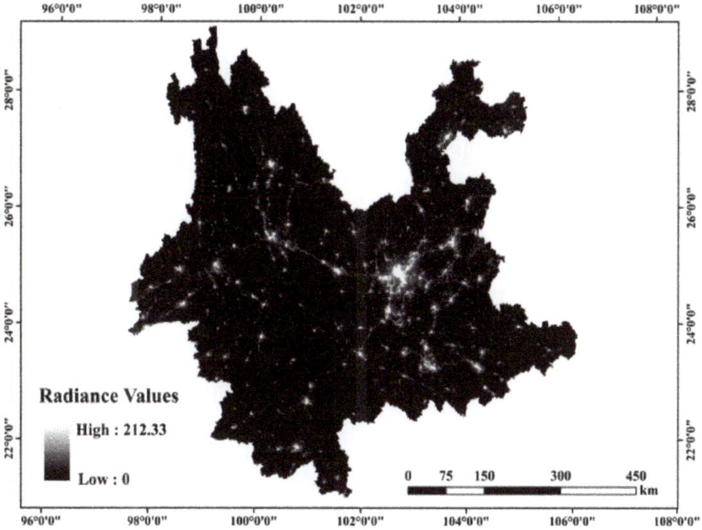

Figure 3. Data processing results of NPP-VIIRS in 2018.

2.2.2. Construction of the Development Index of Various Ethnic Minorities

Gelling stated that toponyms are "road signs to understand the past" [37] as toponyms are used to indicate the names of specific geographic areas and contain rich information such as the ethnic type of the local residents and the interpretation of the geographical environment by local people at the time of naming [38,39]. Studying toponyms is the basis for understanding the national culture and local characteristics of a region [40]. From toponym data, the ethnic types, language and culture, and religious beliefs of a region [41], spatial location, and the environmental evolution process related to history [42,43], environment, and landforms [44] can be extracted. This is of great significance for understanding ethnic minority settlements and the distribution of ethnic minorities.

The national census is a census about the population of the whole country. The contents of the census mainly include gender, age, ethnicity, etc. The subjects of the census are mainly natural persons living in the People's Republic of China (except Hong Kong, Macau, and Taiwan). From the census data, information about ethnic minorities can be extracted such as the place of residence of the ethnic minority population, and information about the proportion of the ethnic minority population can also be further extracted.

The distribution of ethnic minorities in China mainly shows the distribution of "large mixed residences and small settlements". Therefore, the toponym of ethnic minorities will be unevenly distributed, and the toponym data obtained are discrete measured values. Kernel density estimation (KDE) is used to calculate the unit density of the measured value of point and line elements within a specified area. It can intuitively reflect the distribution of discrete measured values in a continuous area. Kernel density estimation can obtain the weighted average density of all data points in the study area [45]. The weight assigned is related to the distance of the center point of the data point. The farther away from the center point, the smaller the weight is assigned, and vice versa [46]. The formula for calculating the kernel density P_i at any point i in space is:

$$P_i = \frac{1}{n\pi R^2} \times \sum_{j=1}^{n} K_j \left(1 - \frac{D_{ij}^2}{R^2}\right)^2, \tag{2}$$

where R is the search radius (bandwidth) of the selected area ($D_{ij} < R$); K_j is the weight of the research data point j; D_{ij} is the distance between the space point i and the research data point j; and n is the number of research data points j within the search radius R. The search radius R has a direct impact on the results of kernel density analysis [47].

In this study, 25 ethnic minority geographic names were used for kernel density analysis. Because the area of an ethnic minority gathering area in the study area is about one square kilometer. According to this feature, through comparative analysis, the search radius of kernel density estimation is constantly changed, and finally, it was found that when the search radius was 1000 m, the effect was better, and can distinguish ethnic minority gathering areas. Considering that there are places with ethnic minority toponyms, but no ethnic minorities living in them, this paper used census data to calculate the proportion of 25 ethnic minorities in the study area, and obtained a grid map of the proportion of 25 ethnic minorities for future use.

The development of a region or a nation is often affected by many factors such as population, economy, environment, geographical location, etc. In addition, there are differences in the development of different regions of the same ethnic group and between different ethnic groups in the same region. Therefore, it is necessary to construct a development index that can reflect this difference in order to quantitatively analyze the development of ethnic minorities. This article used population, toponym data, and NPP-VIIRS data combined with the literature [48,49] as well as the formula form of the spatialization of population data to propose a method to calculate the development index of various ethnic minorities. The calculation formula is shown in Equation (3):

$$CPS_i = \sqrt{PR_i \times KDE_i \times NPP_i} \tag{3}$$

where CPS_i is the development index of minority i; PR_i is the population proportion of minority i; KDE_i is the kernel density analysis result of minority i; and NPP_i is the night light radiance value of minority i.

2.2.3. Distribution of Ethnic Minorities

In order to clearly understand the distribution of each ethnic group, we used the obtained ethnic development index combined with ethnic toponym data. We used the 2018 NPP-VIIRS data as a base map, and used the point method to show the distribution of 25 ethnic minorities. Due to the large number of ethnic minorities, it was difficult to distinguish between ethnic groups using only different colors. This paper applied the literature [50] on the classification of language affiliation, and used the language branches of different ethnic minority languages to classify 25 ethnic minorities into 13 categories. Since the 13 categories were difficult to distinguish on the map, the 13 categories were merged into six categories based on the language branch classification. The specific classification is shown in Table 2.

Table 2. Language branch classification.

Branch	Ethnic Minority
Yi Branch	Yi, Lisu, Naxi, Bai, Lahu, Hani, Jinuo
Zhuang and Dai Branch	Zhuang, Buyi, Dai
Tibetan Branch	Tibetan
Jingpo Branch	Jingpo, Dulong
Chinese Branch	Hui, Manchu
Other Languages	Achang (Burmese branch), Shui (Dong Shui branch), Pumi; Nu; Mongolian, Deang (Undecided language), Miao (Miao branch), Yao (Yao branch), Wa; Bulang (Benglong language branch)

Since the development index of the Yi nationality was the highest, but less than 30, the 0–30 was divided into five categories by the equal interval: higher development index, high development index, medium development index, low development index, and lower development index. According to the development index range of each type of development level, 25 ethnic minority development indexes were classified.

3. Results and Accuracy Verification

3.1. Ethnic Minority Development Index

In order to better reflect the development index of ethnic minorities, this article selecteed the Yi, Wa, Zhuang, Naxi, Hani, and Dai, six ethnic minorities with higher development indexes, for cartographic analysis. By comparing the development index calculated by Equation (3), and reference [49], the natural fracture method can most appropriately group similar values and maximize the difference between each class, so we compared the three methods of using the natural breaks method, average classification method, and manual breaks method, and found that the method using natural breaks method worked the best. This article divided the development index into five categories. The first category indicates areas with extremely poor development of the ethnic minorities, which are directly regarded as areas without the distribution of ethnic minorities. The second category indicates areas with poor development of the ethnic minority. The third category indicates areas with a moderate development. The fourth category indicates areas where the ethnic minority has developed well. The fifth category indicates areas with excellent development of the ethnic minority. The results are shown in Figures 4–9.

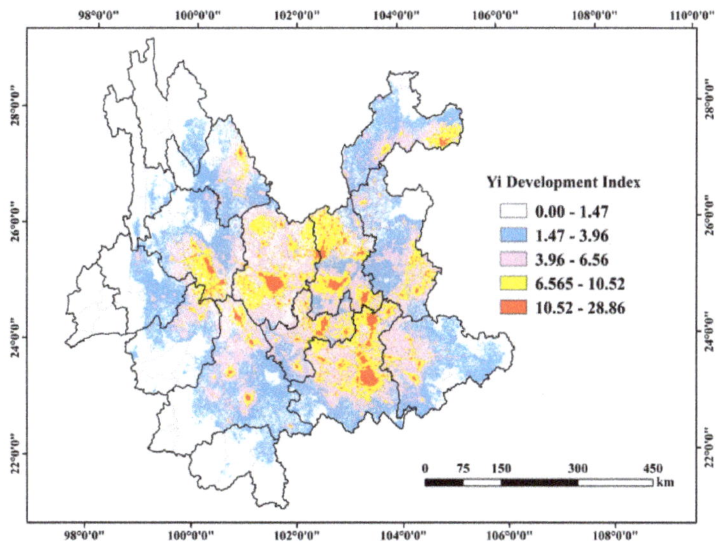

Figure 4. Yi nationality development index classification results.

It can be seen from Figures 4–9 that among the six ethnic minorities with high development indexes in Yunnan Province, the development indexes from high to low were: Yi, Wa, Zhuang, Naxi, Hani, and Dai. Moreover, the development index of Hani and Dai, Zhuang and Naxi were not much different. In other words, among the six ethnic minorities, the Yi ethnic group had the best development (the Yi ethnic group had the highest development index, and there were many areas with high development indexes), and the Dai ethnic group had the worst development compared to the other five ethnic minorities.

It can be seen from Figure 4 that the Yi nationality was distributed almost throughout Yunnan Province. The areas with higher Yi development index were: (1) the northeast area of Nanjian Yi County in Dali Bai Prefecture and the east area of Weishan Yi Hui County; (2) the eastern part of Chuxiong City, Chuxiong Yi Prefecture; (3) the junction of Wuhua District, Xishan District, Guandu District, and Panlong District of Kunming City, the western part of Shilin Yi County, and the southern part of Luquan Yi and Miao County; (4) the southeast area of Eshan Yi County, Yuxi City, and the east area of Yuxi City; and (5) the northern area of Mile County, the western area of Kaiyuan City, the western area of Mengzi County, and the eastern area of Gejiu City in Honghe Hani and Yi Prefecture.

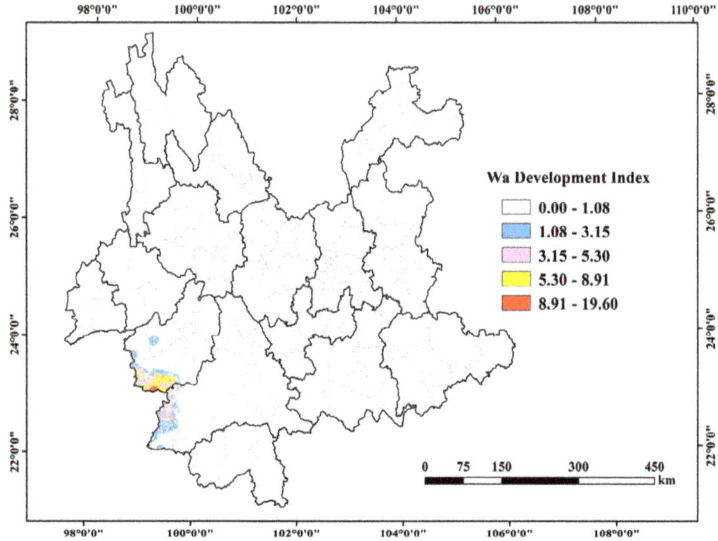

Figure 5. Wa nationality development index classification results.

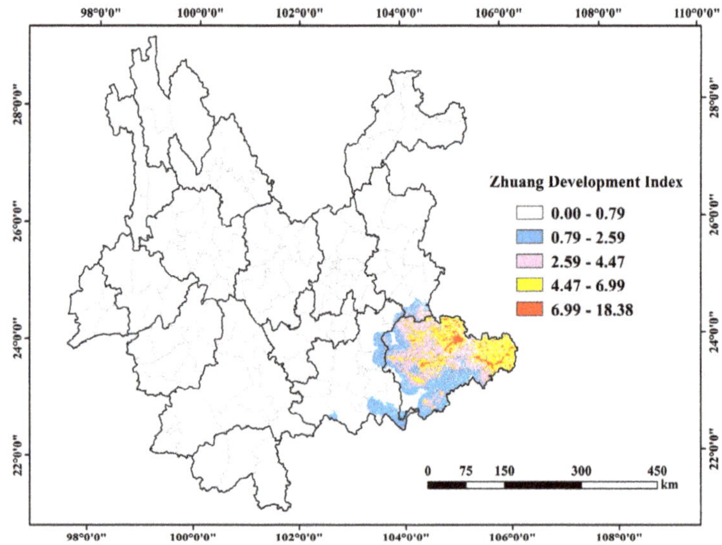

Figure 6. Zhuang nationality development index classification results.

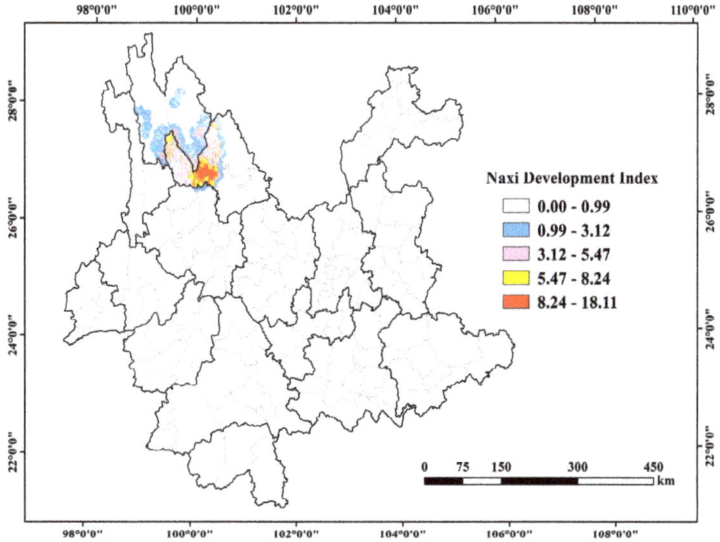

Figure 7. Naxi nationality development index classification results.

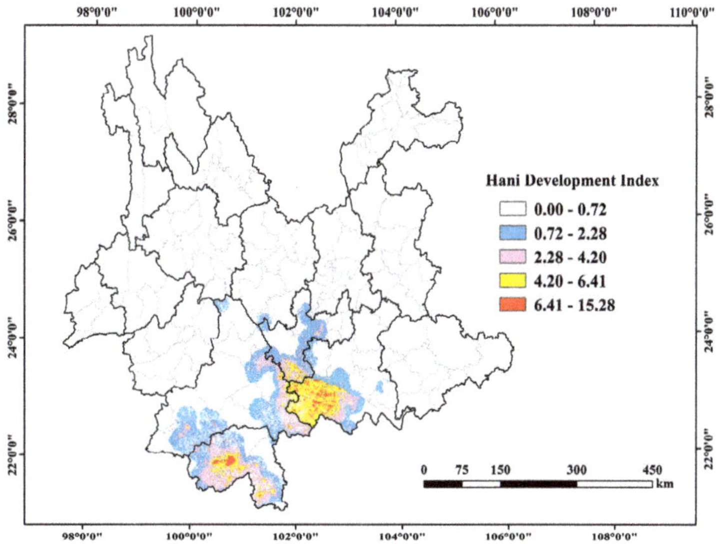

Figure 8. Hani nationality development index classification results.

It can be seen from Figure 5 that the distribution of the Wa nationality had regional characteristics, mainly in Cangyuan Wa County in Lincang City and Ximeng Wa County in Pu'er City. Between them, the Wa development index was the highest in the southern area of Cangyuan Wa County.

It can be seen from Figure 6 that the Zhuang nationality was mainly distributed in Wenshan Zhuang and Miao Prefecture. The areas with higher Zhuang development index were: (1) Qiubei County and the central area of Yanshan County; (2) the northern part of Funing County; and (3) the northwestern part of Guangnan County.

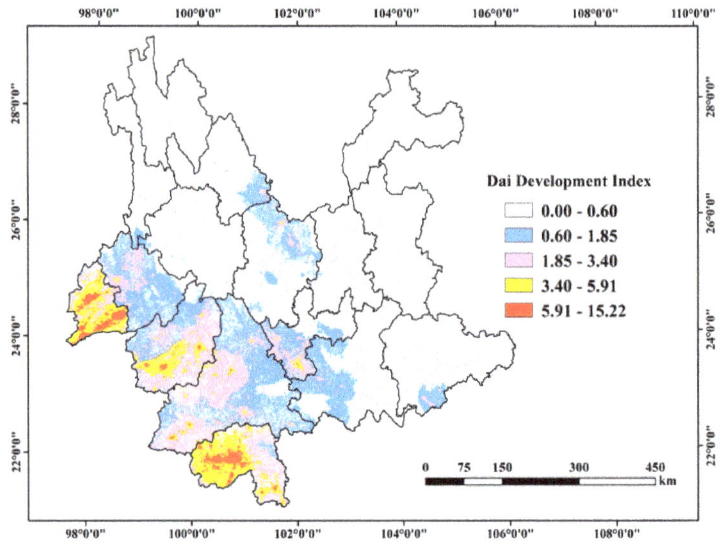

Figure 9. Dai nationality development index classification results.

It can be seen from Figure 7 that the Naxi nationality was mainly distributed in the western region of Lijiang City. The development index of the Naxi nationality was higher in the southern area of Lijiang urban and the southern area of Yulong Naxi County.

It can be seen from Figure 8 that the Hani nationality was mainly distributed in Xishuangbanna Dai Prefecture, southwest of Honghe Hani and Yi Prefecture, and southeast of Pu'er City. The areas with higher Hani development index were: (1) the western area of Jinghong City and the eastern area of Menghai County; (2) the central area of Hani and Yi County in Jiangyu; and (3) Honghe County, Yuanyang County, and Luchun County. It has the characteristics of not being concentrated and more scattered.

It can be seen from Figure 9 that the Dai nationality was mainly distributed in Dehong Dai Jingpo Prefecture, Lincang City, Xishuangbanna Dai Prefecture, Pu'er City, Baoshan City, and the western area of Yuxi City. The areas with higher Dai development index were: (1) Yingjiang County and Ruili City's southern area, and Mang City's central area; (2) the central area of Menghai County and Jinghong City; (3) Lincang city center and the central area of Gengma Dai and Wa County; (4) the central area of Yuanjiang County; and (5) the central areas of Menglian County, Lancang County, and Jinggu County.

3.2. Ethnic Minority Distribution Results

We used the method of in Section 2.2.3 to obtain the distribution results of 25 ethnic minorities in Yunnan Province (Figure 10).

It can be seen from Figure 10 that the coverage of the Yi ethnic group was the widest, involved the most counties, and was concentrated in Chuxiong Prefecture, the southeastern area of Qujing City, and the northern area of Kunming. The Jingpo branch is mainly distributed in Dehong Prefecture and Gongshan County. Zhuang Dai language branch was mainly distributed in Wenshan Prefecture, Dehong Prefecture, Xishuangbanna Prefecture, and Lincang Prefecture. The Chinese branch was mainly distributed in Zhaotong City and Baoshan City. Other language branches were mainly distributed in the east of Zhaotong and the north of Zhaotong, the east of Wenshan Prefecture, Nujiang Prefecture, Lijiang City, Dehong Prefecture, Xishuangbanna Prefecture, and the south of Lincang Prefecture.

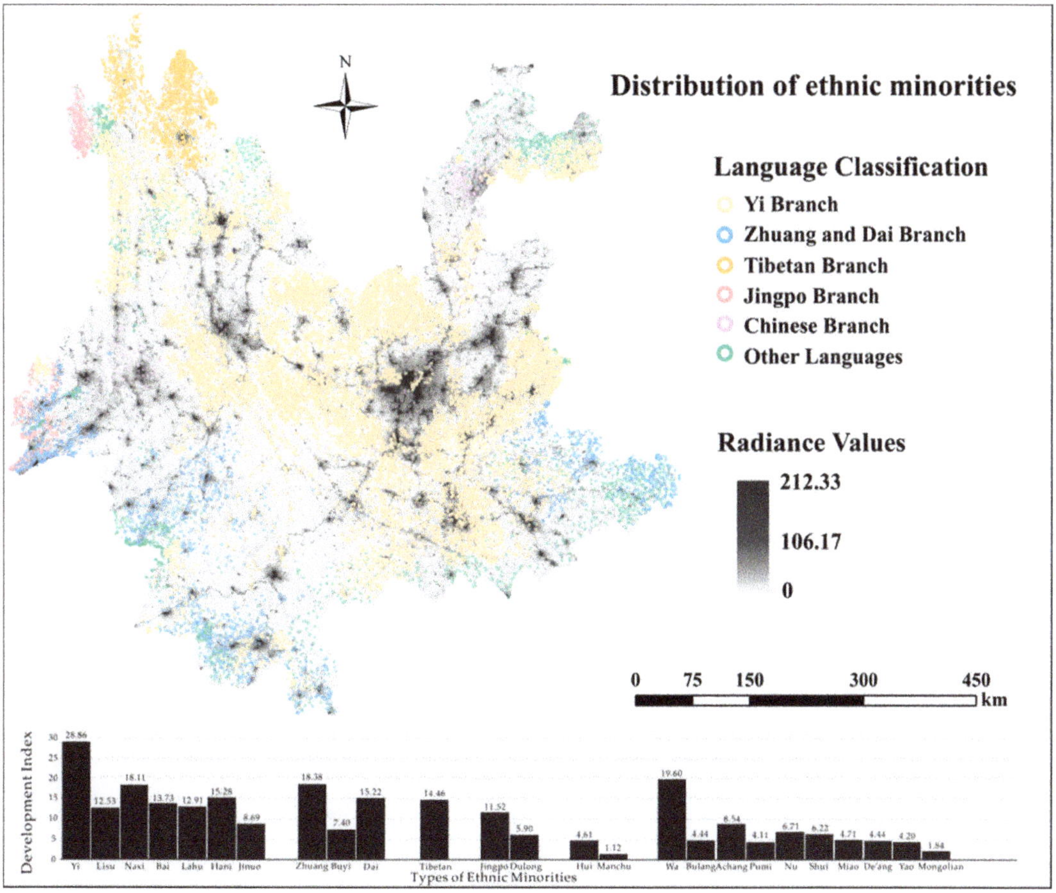

Figure 10. Distribution of ethnic minorities in Yunnan Province.

From the perspective of development index, the higher developed ethnic minority was the Yi. The high-developed index ethnic minorities included the Naxi, Zhuang, and Wa. The medium-developed ethnic minorities included the Hani, Bai, Lahu, Lisu, Dai, and Tibetan. The low-developed ethnic minorities included the Jingpo, Jinuo, Buyi, Achang, Nu, and Shui. The lower-developed ethnic minorities included Dulong, Bulang, Pumi, Miao, Hui, Manchu, De'ang, Yao, and Mongolian.

From the perspective of the language branch, the overall development index of ethnic minorities in the Yi, Zhuang, Dai, and Tibetan branch was relatively high, which may be related to the inheritance and development of these language branches.

Several ethnic minorities lived together in most areas. Among them, the mixed living of ethnic minorities in Dehong Prefecture was more obvious. Areas where the phenomenon of multi-ethnic mixed living was more obvious were: (1) Zhaotong City has mixed living of Yi branch, Chinese Branch, and other languages; (2) Funing County and Guangnan County in Wenshan Prefecture had mixed living of the Zhuang and Dai branch, and other languages; (3) Fumin County in Kunming City had mixed living of the Yi branch, Chinese Branch, and other languages; (4) Mengla County in Xishuangbanna Prefecture had mixed living of the Zhuang and Dai branches and other languages. (5) Jinghong City in Xishuangbanna Prefecture has mixed living of Zhuang and Dai branches and other languages, Yi branch, and Jingpo branch; (6) Longchuan County in Dehong Prefecture had mixed living of Zhuang and Dai branches, other languages, and Jingpo branch; (7) Yingjiang County in Dehong Prefecture had mixed living of Zhuang and Dai branches, other languages, Yi branch, and Jingpo branch; and (8) Lushui County in Nujiang Prefecture had mixed living of other languages and the Yi branch.

There are currently 25 ethnic minorities in Yunnan Province, among which the Yi nationality is the most widely distributed and relatively scattered. Among the six selected ethnic minorities with the highest development index, from the perspective of each development index, the Yi nationality's development index was the highest, reaching 28.86 (to two decimal places). The Wa nationality had the second development index, reaching 19.60, but was far from the Yi nationality, which had the highest ranked development index. The Zhuang nationality had the third development index, reaching 18.38. The Naxi nationality had the fourth development index, reaching 18.11. The Hani nationality had the fifth development index, reaching 15.28. The Dai nationality's development index was the lowest at 15.22.

From the perspective of the relationship between the development index of each ethnic group and the geographic location of the ethnic group: the six areas with higher development indexes of ethnic minorities were located in the corresponding ethnic minority states, counties, or the city center of each city. Ethnic minorities had the highest development index in their corresponding minority prefecture or county, and the further the distance from the minority prefecture or county, the smaller the development index. The minority development index decreased as the distance between the minority nationality and its core development zone increased.

From the perspective of the relationship between each ethnic development index and the corresponding ethnic minority prefecture and county, areas with a higher ethnic development index were concentrated in the ethnic minority prefecture or county, but the development of the ethnic minority in the prefecture was better than that in the county.

3.3. Accuracy Verification

In order to verify the correctness of the development index calculated by the method used in this article, the method used in this article was compared with the method of the traditional research statistical yearbook. Considering that the development of a region is affected by many factors such as rural population, urban population, employment rate, average resident salary, etc., it is difficult to verify the correctness of the results of this article by selecting only one indicator. Comprehensively referenced in [51–53], this article selected eight indicators for the study area from 2013 to 2018. These were the total output value of agriculture, forestry, animal husbandry, and fishery in each county, and the per capita disposable income of rural residents in each county. County GDP per capita, GDP index of each county, rural employees in each county, rural population in each county, average salary of employees in each county, and number of employees in each county. Since the magnitudes of the eight indicators were different, the indicators were normalized first. The normalized formula is shown in Equation (4):

$$X = \frac{x - \min}{\max - \min}, \quad (4)$$

where X is the standardized result of the index; x is the original value of the index; max is the maximum value of the sample data; and min is the minimum value of the sample data.

After obtaining the normalized results of the indicators, a comprehensive development index was established according to the method of establishing a comprehensive poverty index in the literature [53,54]. First, the entropy method was used to determine the weight of the eight indicators. In the entropy method, the larger the amount of information, the smaller the uncertainty of the information, and the smaller the entropy value, so the greater the weight. Using the entropy method to calculate the weight of each indicator, we can obtain the comprehensive development index. The calculation formula is shown in Equations (5)–(8):

$$Z = \sum_{i=1}^{n} w_j * X_i, \tag{5}$$

$$f_{ij} = \frac{y_{ij}}{\sum\limits_{i=1}^{m} y_{ij}}, \tag{6}$$

$$H_j = -(1/\text{lnm}) \sum_{i=1}^{m} f_{ij} \ln f_{ij}, \tag{7}$$

$$w_j = \frac{(1 - H_j)}{\sum\limits_{j=1}^{n} (1 - H_j)}, \tag{8}$$

where f_{ij} is the index value weight of the i evaluation object under the j index; m is the 129 counties included in the study area; n is the eight indicators to construct the comprehensive development index; Z is the comprehensive development index; X_i is the standardized result of i evaluation object; H_j is the entropy value of the j index; and w_j is the weight of the j index.

The weights of the eight indicators in 2013–2018 calculated by the formula are shown in Table 3 (with four decimal places).

Table 3. The calculation results of the weight of each indicator from 2013 to 2018.

Indicators \ Year	2013	2014	2015	2016	2017	2018
GDP per capita	0.1446	0.1401	0.1350	0.1304	0.1321	0.1330
GDP Index	0.0419	0.0152	0.0252	0.0183	0.0144	0.0277
Number of employees	0.3597	0.3630	0.3752	0.3732	0.3755	0.3716
Average salary of employees	0.0057	0.0059	0.0052	0.0086	0.0090	0.0093
Per capita disposable income of rural residents	0.0171	0.0239	0.0232	0.0230	0.0225	0.0222
Total output value of agriculture, forestry, animal husbandry and fishery	0.1569	0.1503	0.1456	0.1453	0.1444	0.1438
Rural population	0.1636	0.1646	0.1649	0.1667	0.1641	0.1665
Rural workers	0.1520	0.1507	0.1507	0.1526	0.1523	0.1533

According to the calculation results of the weight of each index, the comprehensive development index of each county in the study area was obtained, as shown in Table 4.

Table 4. The results of the comprehensive development index of each county from 2013 to 2018.

County \ Year	2013	2014	2015	2016	2017	2018
Wuhua	0.5565	0.4184	0.4553	0.5595	0.5919	0.5626
Panlong	0.4935	0.5719	0.4175	0.5943	0.4999	0.5973
Guandu	0.4058	0.5501	0.5886	0.5818	0.5571	0.5577
Xishan	0.4362	0.5920	0.3979	0.3641	0.5711	0.3423
Dongchuan	0.2069	0.1855	0.1970	0.1763	0.1263	0.1595
Chenggong	0.1706	0.2412	0.3289	0.5510	0.5790	0.5906
Jinning	0.2200	0.1880	0.1278	0.1258	0.1740	0.1603
Fumin	0.0309	0.1454	0.1275	0.0793	0.0285	0.1881
Yiliang	0.0802	0.1650	0.1813	0.2292	0.2224	0.2432
Shilin	0.1635	0.3860	0.0723	0.1983	0.1256	0.1756
Songming	0.2307	0.1250	0.1700	0.1531	0.2550	0.2424
Luquan	0.1190	0.0354	0.0404	0.2198	0.0635	0.0625
Xundian	0.0482	0.0228	0.1243	0.1005	0.0759	0.0975
Anning	0.1812	0.0925	0.1538	0.2232	0.2606	0.2492

Table 4. *Cont.*

County \ Year	2013	2014	2015	2016	2017	2018
Qilin	0.2941	0.2193	0.2126	0.3072	0.2386	0.2395
Malong	0.1123	0.1928	0.1693	0.0839	0.1293	0.2136
Luliang	0.1289	0.1170	0.0310	0.0507	0.1856	0.0680
Shizong	0.0929	0.0283	0.1104	0.0220	0.0602	0.1154
Luoping	0.0533	0.0200	0.1049	0.1091	0.1217	0.1467
Fuyuan	0.1455	0.1421	0.1275	0.1070	0.1327	0.1951
Huize	0.1337	0.1262	0.0275	0.0728	0.0610	0.0458
Zhanyi	0.1868	0.1564	0.1065	0.0506	0.0582	0.1565
Xuanwei	0.1228	0.1014	0.1931	0.0926	0.1288	0.1064
Hongta	0.3898	0.2574	0.2017	0.2177	0.3029	0.2844
Jiangchuan	0.2124	0.2147	0.1954	0.1948	0.2097	0.2263
Chengjiang	0.2890	0.2553	0.2037	0.1633	0.2272	0.2198
Tonghai	0.1425	0.0818	0.1842	0.2408	0.2071	0.2459
Huaning	0.0806	0.0789	0.1427	0.0529	0.1255	0.1587
Yimen	0.0270	0.1556	0.1092	0.0388	0.0239	0.1186
Eshan	0.1073	0.1150	0.0641	0.1003	0.0377	0.1089
Xinping	0.0507	0.1436	0.0201	0.0203	0.1204	0.1154
Yuanjiang	0.0563	0.0205	0.0623	0.0751	0.0713	0.1310
Longyang	0.0422	0.0615	0.1098	0.0220	0.0541	0.0646
Shidian	0.0196	0.0898	0.0342	0.0203	0.0542	0.1068
Tengchong	0.0290	0.0575	0.0766	0.0439	0.0550	0.1588
Longling	0.1197	0.0751	0.0888	0.0423	0.0244	0.0209
Changning	0.0450	0.0403	0.0283	0.0782	0.0783	0.1131
Zhaoyang	0.1454	0.0551	0.1440	0.1380	0.1399	0.1832
Ludian	0.0938	0.1802	0.1376	0.1312	0.1831	0.2062
Qiaojia	0.1017	0.0723	0.0385	0.0712	0.0211	0.1004
Yanjin	0.0391	0.0204	0.0759	0.0422	0.1075	0.1131
Daguan	0.1040	0.0895	0.0613	0.1594	0.1380	0.1172
Yongshan	0.1071	0.1135	0.0575	0.1412	0.0964	0.0843
Suijiang	0.1458	0.0706	0.0896	0.0864	0.0877	0.0756
Zhenxiong	0.0600	0.0408	0.0878	0.0424	0.0297	0.0558
Yiliang	0.1195	0.1198	0.1100	0.0821	0.0885	0.1049
Weixin	0.0859	0.0209	0.0809	0.1667	0.1223	0.0757
Shuifu	0.0223	0.0302	0.0634	0.0380	0.0706	0.1554
Gucheng	0.0924	0.0613	0.0887	0.0946	0.1542	0.1332
Yulong	0.0197	0.0201	0.0578	0.0211	0.0302	0.0342
Yongsheng	0.0447	0.0215	0.0626	0.0393	0.0251	0.0568
Huaping	0.0271	0.0917	0.0322	0.0201	0.0556	0.0243
Ninglang	0.0247	0.0216	0.0632	0.0393	0.1433	0.0790
Simao	0.0651	0.0913	0.0317	0.0801	0.0302	0.0564
Ning'er	0.0214	0.0308	0.0895	0.0216	0.0480	0.1134
Mojiang	0.0253	0.0327	0.0297	0.0344	0.0605	0.0199
Jingdong	0.0474	0.1037	0.0204	0.1062	0.1195	0.0205
Jinggu	0.1055	0.1084	0.0374	0.0449	0.0464	0.0528
Zhenyuan	0.0360	0.0397	0.0409	0.0687	0.0601	0.0623
Jiangcheng	0.0264	0.0302	0.0202	0.0435	0.1008	0.0237
Menglian	0.0294	0.0202	0.0457	0.0207	0.0453	0.0267
Lancang	0.0671	0.0458	0.0204	0.0510	0.0640	0.1807
Ximeng	0.0813	0.0196	0.0208	0.0767	0.0207	0.2507
Linxiang	0.0295	0.0204	0.1297	0.0450	0.0535	0.0206
Fengqing	0.0199	0.1145	0.0513	0.0888	0.0207	0.1274
Yunxian	0.0334	0.0656	0.0211	0.0209	0.0321	0.0962
Yongde	0.0273	0.0264	0.0206	0.0403	0.1061	0.1245
Zhenkang	0.1196	0.0242	0.1059	0.0640	0.0198	0.0520
Shuangjiang	0.0833	0.0486	0.0206	0.0440	0.0528	0.0883
Gengma	0.0389	0.0459	0.0480	0.0325	0.0205	0.0563
Cangyuan	0.0317	0.0201	0.0312	0.0276	0.0352	0.0569
Chuxiong	0.0251	0.0231	0.1089	0.0338	0.2816	0.1152

Table 4. *Cont.*

County \ Year	2013	2014	2015	2016	2017	2018
Shuangbo	0.0299	0.0536	0.0210	0.1057	0.0997	0.0580
Mouding	0.1196	0.0204	0.1000	0.1567	0.2166	0.1957
Nanhua	0.0581	0.0204	0.1086	0.0968	0.1148	0.1205
Yao'an	0.0496	0.0685	0.0392	0.0620	0.0336	0.0608
Dayao	0.0471	0.0208	0.0578	0.0443	0.0625	0.0219
Yongren	0.0384	0.0339	0.0499	0.0897	0.0200	0.0800
Yuanmou	0.0200	0.0208	0.0824	0.0914	0.0456	0.0410
Wuding	0.0621	0.0281	0.0825	0.0366	0.0553	0.2346
Lufeng	0.0287	0.0322	0.0937	0.0561	0.1170	0.1709
Mengzi	0.0761	0.0203	0.0214	0.0760	0.1079	0.0890
Gejiu	0.0284	0.1278	0.1015	0.0870	0.1264	0.1201
Kaiyuan	0.0455	0.1137	0.1624	0.1265	0.1662	0.1792
Mile	0.0906	0.1955	0.1257	0.1705	0.1895	0.1721
Pingbian	0.0353	0.0507	0.0206	0.0807	0.1048	0.0269
Jianshui	0.0201	0.0705	0.0245	0.0418	0.1063	0.0564
Shiping	0.0199	0.0380	0.1395	0.0611	0.1143	0.0946
Luxi	0.1447	0.0652	0.0994	0.0891	0.0717	0.0992
Yuanyang	0.0199	0.0638	0.0708	0.0309	0.0922	0.0706
Honghe	0.0230	0.1195	0.0217	0.0256	0.0238	0.1657
Jinping	0.0509	0.0202	0.0208	0.0207	0.0317	0.0206
Luchun	0.0477	0.0621	0.0290	0.0359	0.0618	0.0369
Hekou	0.0678	0.0488	0.0540	0.0390	0.0466	0.0206
Wenshan	0.1196	0.0734	0.0784	0.0527	0.1758	0.1454
Yanshan	0.0590	0.0655	0.0204	0.0804	0.0204	0.1157
Xichou	0.1082	0.0495	0.0737	0.0613	0.0949	0.1313
Malipo	0.1195	0.1039	0.0898	0.0211	0.1296	0.2052
Maguan	0.0538	0.0537	0.0206	0.0208	0.0878	0.1211
Qiubei	0.0212	0.0522	0.0347	0.0264	0.0746	0.0210
Guangnan	0.1195	0.0275	0.0352	0.0208	0.0579	0.0204
Funing	0.0200	0.0202	0.0390	0.0370	0.0244	0.0225
Jinghong	0.0339	0.0985	0.1048	0.1219	0.1242	0.0996
Menghai	0.0286	0.0216	0.0309	0.0324	0.0738	0.0533
Mengla	0.0546	0.1012	0.1310	0.1316	0.0516	0.1163
Dali	0.2015	0.1204	0.2399	0.2192	0.1301	0.1922
Yangbi	0.0202	0.0489	0.0407	0.0577	0.1129	0.0841
Xiangyun	0.0971	0.0307	0.0752	0.0860	0.1438	0.1869
Binchuan	0.0358	0.1710	0.0278	0.0977	0.1210	0.0878
Midu	0.0577	0.1017	0.0822	0.1511	0.0946	0.1239
Nanjian	0.0218	0.0967	0.0246	0.0510	0.1526	0.0868
Weishan	0.0534	0.1220	0.0705	0.1026	0.0599	0.0919
Yongping	0.0740	0.0848	0.0351	0.0952	0.0422	0.0622
Yunlong	0.0422	0.1352	0.0734	0.1527	0.1387	0.2176
Eryuan	0.0214	0.0419	0.0310	0.0206	0.0202	0.1260
Jianchuan	0.0379	0.0366	0.0211	0.0832	0.1196	0.0851
Heqing	0.0715	0.0213	0.0658	0.0636	0.0688	0.0851
Mangshi	0.0200	0.1012	0.1463	0.1428	0.1542	0.0607
Ruili	0.0626	0.1126	0.0557	0.0518	0.1487	0.2090
Lianghe	0.0343	0.1255	0.0221	0.0202	0.1423	0.1570
Yingjiang	0.1195	0.0694	0.1095	0.0374	0.2701	0.0613
Longchuan	0.0256	0.0949	0.0204	0.0198	0.0495	0.1184
Lushui	0.0209	0.0219	0.0874	0.1034	0.0254	0.0943
Fugong	0.0289	0.0288	0.0217	0.0427	0.0465	0.0644
Gongshan	0.0204	0.0498	0.0332	0.1214	0.0707	0.0924
Lanping	0.0272	0.0383	0.0282	0.0199	0.0686	0.0912
Shangri-La	0.3128	0.4385	0.2572	0.4639	0.4023	0.4179
Deqin	0.0506	0.0309	0.0282	0.0950	0.0798	0.0859
Weixi	0.0772	0.0261	0.0730	0.1198	0.0490	0.0460

Using the method developed in this article to calculate the comprehensive development index of all ethnic minorities and Han nationality in the study area from 2013 to 2018, we performed district statistics on the development index of each county on ArcMap, and took the average value of the development index of each county as the statistical value. The development index of each county from 2013 to 2018 is shown in Table 5.

Table 5. The development index result calculated by the method in this paper.

County \ Year	2013	2014	2015	2016	2017	2018
Wuhua	2333.1825	2339.1373	2376.7834	2394.4854	2415.8620	2448.1288
Panlong	2470.0052	2554.7521	2333.1047	2404.9305	2504.0174	2511.2501
Guandu	2170.9487	2259.7741	2353.8594	2381.0156	2449.4958	2561.8397
Xishan	1621.7904	1784.9426	1598.3018	1565.5310	1597.4138	1602.9692
Dongchuan	960.6572	994.8467	742.9193	836.3549	854.5164	934.8275
Chenggong	943.5753	1076.5617	1205.3264	2298.8631	2363.8025	2420.8939
Jinning	595.8789	627.4076	688.4572	711.8920	830.7505	850.9899
Fumin	612.8931	695.4732	746.5198	784.6463	789.6833	897.2893
Yiliang	625.2049	761.7744	830.1322	886.5681	962.0898	996.5625
Shilin	766.8347	789.6061	794.6279	825.1524	891.7881	985.6760
Songming	899.6777	927.3485	946.6784	973.7795	1107.7790	1056.7363
Luquan	320.2183	364.9809	400.2450	438.9507	448.5508	576.7864
Xundian	517.4374	522.8907	539.6836	574.8628	601.3995	746.0349
Anning	945.3103	955.8508	957.5187	1008.2649	1050.5527	1084.9657
Qilin	1069.2507	1084.1809	1090.9449	1116.4519	1172.8232	1240.0189
Malong	649.2351	693.1274	712.2659	729.2838	757.7143	860.7674
Luliang	543.4304	554.4877	576.5027	589.5539	595.1228	638.4553
Shizong	483.0773	485.1854	500.8513	519.5551	526.3722	657.7111
Luoping	491.1515	417.1506	518.5569	531.6147	554.2872	672.6678
Fuyuan	633.7029	641.5798	652.7819	663.3065	674.2953	782.7485
Huize	617.1130	647.3857	481.3886	507.0445	558.0216	605.9378
Zhanyi	600.3138	635.8879	660.7369	673.4620	696.3811	843.4202
Xuanwei	623.7029	656.5798	662.7819	673.3065	684.2953	796.7485
Hongta	1276.6486	1310.7777	1255.3083	1207.8354	1358.3035	1377.4974
Jiangchuan	897.7480	912.4791	947.5681	952.9043	1104.6690	1133.9384
Chengjiang	1004.5388	1031.805437	1047.223082	1074.157544	1149.89266	1157.818511
Tonghai	883.0552	919.5813	957.5814	990.6898	1062.5763	1117.2477
Huaning	585.7714	595.2825	640.7928	658.2505	715.3348	811.3422
Yimen	494.4001	511.3989	532.8776	551.4176	555.8165	693.6572
Eshan	489.2267	549.9414	580.4157	593.6294	615.8568	664.5724
Xinping	260.6868	421.3788	415.0625	415.7112	453.7176	537.3233
Yuanjiang	486.8459	539.9640	559.4845	596.8433	624.0441	689.0722
Longyang	433.3167	507.6858	521.9453	535.2050	590.8690	664.5847
Shidian	297.5054	455.6399	417.2362	473.0716	502.3391	582.7547
Tengchong	381.8499	569.4589	552.5687	604.0656	669.7669	753.0422
Longling	283.7437	472.8197	460.4084	499.1058	498.2598	612.5852
Changning	299.6858	441.3997	467.5553	447.8070	469.2807	565.8577
Zhaoyang	803.7647	813.1024	880.7033	856.9120	885.1878	954.2087
Ludian	640.0504	697.3706	741.4808	756.5885	772.3398	868.5031
Qiaojia	461.1212	479.4775	498.7038	540.6888	592.3084	670.8998
Yanjin	427.3850	495.8935	540.4706	557.9826	565.5458	616.2725
Daguan	529.7191	622.5744	655.1364	689.8198	755.0185	767.3721
Yongshan	696.0273	684.1769	741.3797	744.1170	810.3069	884.5728
Suijiang	626.3479	600.9896	614.0467	641.6444	706.0032	718.7469
Zhenxiong	448.4068	454.1750	578.5055	497.7722	473.2344	564.0579
Yiliang	455.5056	426.6301	535.4907	578.2334	588.6284	642.6751
Weixin	649.2141	576.1327	623.3148	665.7798	808.1880	817.0238
Shuifu	477.7044	536.3990	554.4428	565.6362	624.9817	672.7901
Gucheng	715.5696	656.3005	791.5617	844.6705	659.9559	730.8966
Yulong	249.8526	278.7471	289.6164	322.6217	348.7171	434.7898
Yongsheng	306.8789	315.7225	339.0830	358.6972	367.6797	455.4456
Huaping	398.2136	435.1668	448.9292	465.1374	473.1283	529.3849

Table 5. *Cont.*

County \ Year	2013	2014	2015	2016	2017	2018
Ninglang	483.3725	499.1192	503.1318	532.7423	568.5116	556.7700
Simao	262.0737	378.4076	411.3063	423.9178	428.7674	515.5211
Ning'er	205.7700	345.3506	377.4181	363.9567	380.2126	492.3221
Mojiang	294.4344	320.0161	345.7805	360.6149	368.5020	492.5927
Jingdong	285.6620	323.7440	442.4544	445.8952	451.9286	557.7185
Jinggu	204.1723	204.5097	286.7719	300.7873	279.4673	347.8470
Zhenyuan	316.8741	345.1779	366.7534	369.0427	369.2104	463.5557
Jiangcheng	259.2867	353.5767	385.5516	390.8240	468.7765	490.3128
Menglian	214.4475	261.5678	354.3153	353.2323	388.4604	440.2223
Lancang	111.0612	131.2043	226.3668	288.6047	273.7476	333.1441
Ximeng	112.9200	172.0400	326.1849	337.2044	349.1749	407.7943
Linxiang	306.9635	462.6050	515.0650	507.7827	502.8663	588.0734
Fengqing	380.4210	587.8980	589.8904	608.2613	611.0447	714.5013
Yunxian	315.0365	499.6152	461.8195	532.8510	521.2354	619.7606
Yongde	226.1955	423.5674	434.0769	478.3543	459.3394	528.0003
Zhenkang	297.1114	355.5066	364.0461	370.0390	376.3777	433.9040
Shuangjiang	227.5567	287.2707	326.3009	387.4047	401.0221	468.4001
Gengma	217.7960	347.9124	396.2941	372.3042	356.5077	406.0964
Cangyuan	160.4840	309.5874	338.4399	339.9750	354.0869	416.7235
Chuxiong	433.2826	645.2557	653.4692	657.8736	681.1811	837.8980
Shuangbo	252.8526	303.8850	379.6348	390.3115	407.8300	520.8493
Mouding	310.4608	560.9061	615.7034	638.1917	696.1406	832.3970
Nanhua	480.4072	505.4462	609.2866	625.6880	638.3771	782.9123
Yao'an	209.0912	390.8838	467.7279	486.7033	523.6015	651.2314
Dayao	252.5988	349.5919	369.2466	413.5369	424.8047	538.7005
Yongren	345.9574	362.8734	373.2791	374.4745	446.5159	482.9805
Yuanmou	339.6947	476.3847	520.1060	529.2911	571.1880	670.1142
Wuding	247.2879	375.7019	384.2886	436.3690	459.5321	573.1000
Lufeng	478.5784	580.9119	586.9430	591.4193	655.7322	748.1127
Mengzi	457.9591	464.1473	470.5014	506.2215	511.0077	613.6647
Gejiu	572.3553	600.1590	591.4074	642.3799	690.1944	748.6155
Kaiyuan	668.0048	673.1953	604.3685	624.1396	766.4539	826.0011
Mile	671.0291	722.1276	726.0442	747.2656	792.8220	868.9230
Pingbian	305.4681	375.1979	417.6111	443.0900	444.3564	572.0882
Jianshui	399.9278	458.7829	479.0721	504.5566	511.8208	618.5614
Shiping	321.6694	424.4328	438.0587	446.9695	470.7730	563.0486
Luxi	592.6135	609.4124	631.0997	645.3875	709.2802	801.2388
Yuanyang	333.5136	363.3037	451.0659	478.7271	523.0667	617.0672
Honghe	258.5071	372.6793	447.6026	464.3542	490.1732	609.1840
Jinping	204.2935	221.5390	339.2894	356.5964	360.8186	465.1307
Luchun	91.9753	243.7047	264.9559	309.4107	322.8647	434.3615
Hekou	272.5834	350.7499	384.7263	391.4023	396.9577	475.4815
Wenshan	474.0788	490.4182	499.7623	533.9054	594.8506	672.3611
Yanshan	421.5922	437.0946	445.0354	464.4252	494.4376	596.4948
Xichou	508.0878	515.4572	575.0413	617.1873	628.6801	727.9660
Malipo	266.5631	402.8942	460.5780	461.5752	485.8210	588.7415
Maguan	293.9284	356.8797	399.5391	404.3396	448.7799	523.6682
Qiubei	331.6157	347.3116	384.0514	401.5665	404.9738	504.4938
Guangnan	266.2608	293.2710	311.7878	352.7943	359.1112	454.3619
Funing	240.5025	324.5929	340.5920	348.3029	363.3480	485.1750
Jinghong	399.9671	410.0248	452.5865	539.5053	556.8493	620.9858
Menghai	229.7981	276.6386	298.5195	306.0052	308.7728	339.3909
Mengla	562.3500	597.2580	617.6430	697.8365	700.8417	782.5778
Dali	894.2742	929.0856	952.3336	972.2778	1017.5021	1054.9061
Yangbi	409.4086	504.7738	535.1718	541.4113	592.4125	669.4222
Xiangyun	521.8522	635.7725	637.0776	639.2351	659.8403	773.7537
Binchuan	477.8487	515.6735	532.8143	547.7162	554.5976	662.9385
Midu	515.8058	614.6749	661.6881	669.8814	676.4856	796.9106

Table 5. *Cont.*

County \ Year	2013	2014	2015	2016	2017	2018
Nanjian	410.8299	532.0153	545.4751	554.5118	581.0947	724.9630
Weishan	495.9411	521.2475	521.6193	521.8876	559.1663	680.8504
Yongping	440.5821	478.1848	494.3199	528.8982	518.7397	631.0074
Yunlong	598.6572	619.4169	655.4769	698.7190	716.7159	765.8410
Eryuan	365.9509	434.7853	446.9512	463.0531	471.9941	590.1621
Jianchuan	312.4686	377.2469	403.1285	427.2047	436.7856	538.2106
Heqing	398.3198	406.8919	437.4106	440.9969	508.3256	571.2813
Mangshi	568.9163	643.5718	655.0890	658.2708	709.8286	727.3277
Ruili	585.2299	621.4004	640.0733	722.0235	729.4622	804.0055
Lianghe	343.3566	495.8395	512.3878	525.5199	542.1922	656.7071
Yingjiang	361.4221	362.8221	365.1339	366.4192	386.2395	472.1552
Longchuan	396.3737	424.4385	447.7375	450.1162	503.5420	574.8163
Lushui	338.0632	425.1091	511.3545	528.6956	621.6195	678.0960
Fugong	281.9702	301.0313	317.7124	378.3484	428.1135	499.3321
Gongshan	449.5138	542.8045	564.2108	653.6538	657.6125	728.4643
Lanping	379.1665	394.9759	399.0575	403.1246	427.0084	492.5050
Shangri-La	1421.7904	1484.9426	1498.3018	1568.5310	1588.4138	1616.9692
Deqin	491.5996	556.1296	606.6776	630.9152	639.8685	696.2319
Weixi	378.2066	389.1270	389.1924	403.1409	415.1711	494.6600

Then, we performed linear regression analysis on the comprehensive development index calculated by the traditional method and the development index calculated by the method in this paper to obtain the regression analysis result, as shown in Equation (9) and Figure 11.

$$y = 3294.3x + 275.43 \quad (9)$$

where x is the development index calculated by the traditional method; y is the development index calculated by the method in this paper; and R^2 is the correlation coefficient of the regression.

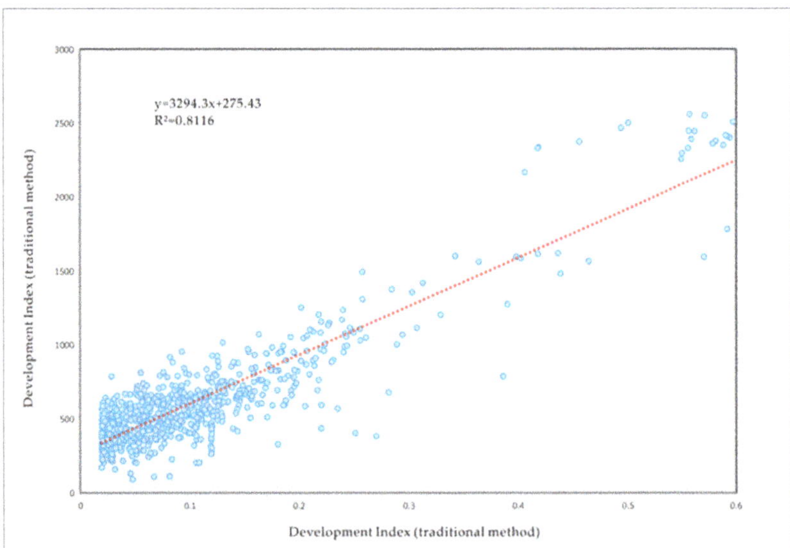

Figure 11. Linear regression results.

It can be seen from Figure 11 that the regression coefficient R^2 of the development index calculated using the method of this article and the development index calculated using the traditional method was 0.8116. When R^2 is greater than 0.8, it can be considered that the two variables are highly correlated. Therefore, the correctness of the method in this paper was proven.

4. Discussion

4.1. Significance to the Development of Ethnic Minorities

There are obvious differences in the development of different ethnic minorities and the development of the same ethnic minorities in different regions. This paper used the relationship between night-time light remote sensing data, economy, and population to establish the development index of ethnic minorities. The results can be analyzed by (1) the size of development differences among different ethnic minorities; (2) differences of the same minority in different minority prefectures and counties; and (3) the relationship between the development index of various ethnic minorities and geographical location. The factors in the ethnic development index model constructed in this paper can be changed, and more factors can be added according to different research purposes. This lays the foundation for the future development direction of ethnic minorities and the formulation of development policies.

Compared with the traditional research on statistical yearbooks, the method in this paper was faster, saved time, and could obtain the long-term national development status in time. In this way, we can quickly understand the development of each nation in time and space. For a multi-ethnic country, timely access to the development status of each ethnic group is conducive to adjusting policies on ethnic population, economic, and other fields to achieve coordinated and balanced development of all ethnic groups to the greatest extent, thereby reducing ethnic conflicts. The method studied in this article can not only target different ethnic groups, but can also be extended to different races and special groups (for example, using the method of this article to study the development of Blacks and Whites, and make a spatial distribution map), or different species. This is of great significance for the sustainable development and coordinated development of the world.

We used the method described in this article to calculate the development index of all ethnic groups in Yunnan Province, and used the natural discontinuity method to divide the development index into five categories. The first category was excellent-developed areas, the second category was well-developed areas, the third category was medium-developed areas, the fourth category was poor-developed areas, and the fifth category was very poor-developed areas. The classification results are shown in Figure 12.

4.2. The Relationship between National Development and Government

It can be roughly seen from the figure that the areas with higher national development index were mainly concentrated in the center of the county. We then counted the average distance from each type of grid to the nearest government by county. The average distance from each type of grid to the nearest government is shown in Table 6.

Table 6. The relationship between the level of national development and its average distance to the nearest government.

National Development Level	Average Distance to the Nearest Government (Unit: m)
Excellent-Developed	3352.28
Well-Developed	4695.77
Medium-Developed	6043.98
Poor-Developed	8728.84
Very Poor-Developed	12,411.90

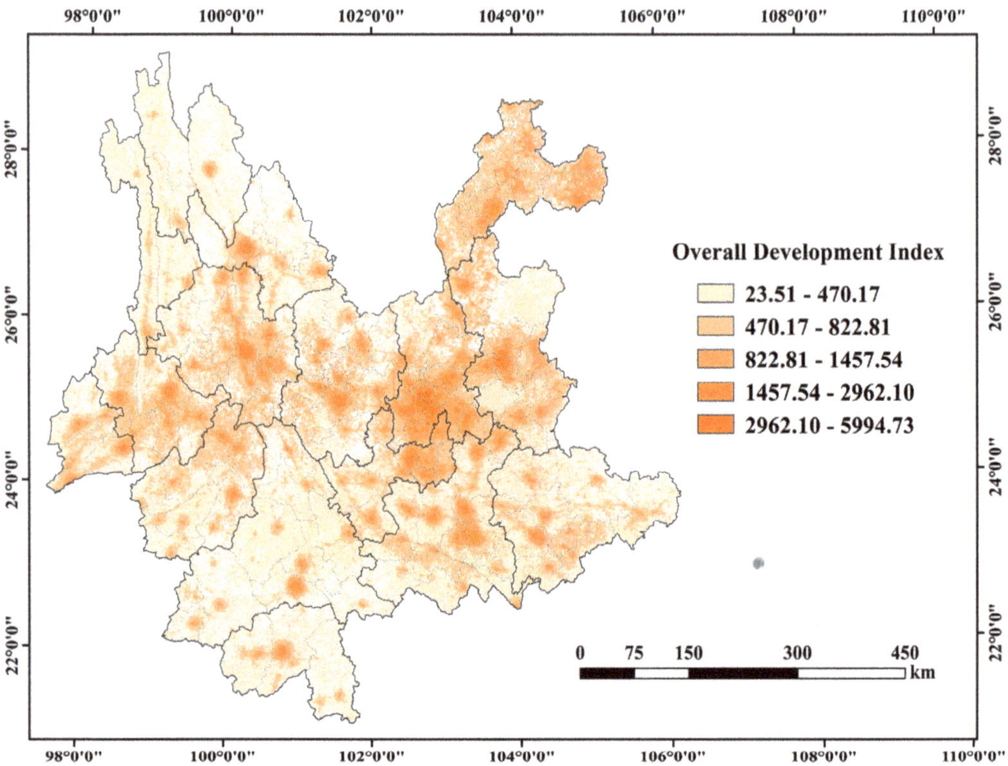

Figure 12. Yunnan Province nationality overall development index.

It can be seen from Table 6 that the area with excellent ethnic development is the closest to the local government. The farther the ethnic development zone is from the local government, the smaller the development index. The area with excellent ethnic development is about 3 km away from the local government, because in China, the development circle of a region is basically centered on the government and spreads around that. With the government as the center and a radius of 3 km, the higher the level of national development. With the continuous increase in the radius, the lower the level of development. Therefore, the government's assistance has played a very important role in the development of the nation.

First, we carried out regional statistics on the development index of each county, selected the average development index of each county as the benchmark, and classified the overall development index according to the county level. A grid map of the development of each county was obtained. Then, we extracted the best-developed grid center in each county, and calculated the distance between the grid center and the nearest local government. Finally, the development of each county and the distance between the best-developed areas of each county and the local government are shown on a map in Figure 13.

Generally speaking, the better-developed areas were closer to the local government. However, there were two situations on the map. First, the development of the region is better, but far from the government. The reason for this phenomenon is that the development strength of these regions is relatively strong, and the role of the government is not the main one relative to the development of the region. Second, the development of the region is poor, but is closer to the government. This phenomenon occurs because the government has not maximized its leading role in the development process of the region. In future development, we should pay attention to government assistance.

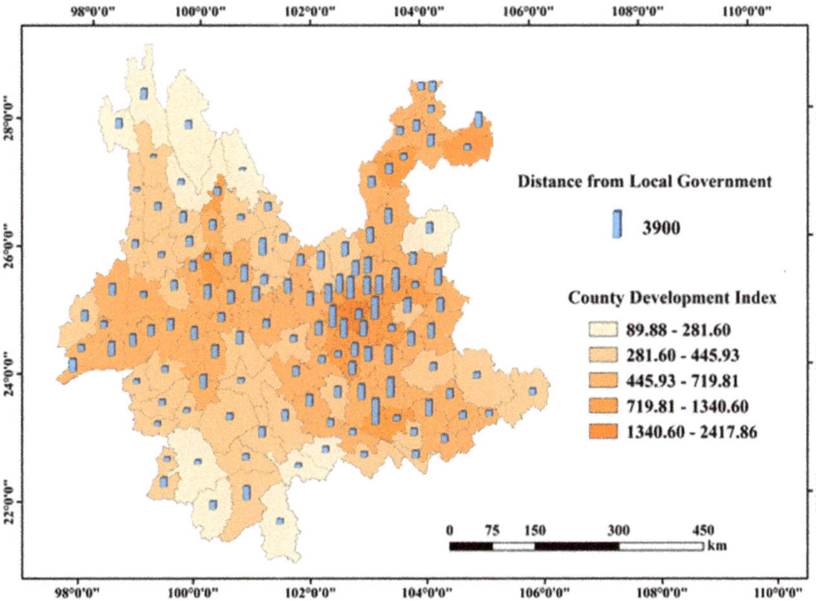

Figure 13. The development status of each county and the distance from the area with the highest development index of each county to the local government.

In the future development of nationalities, we must pay attention to giving play to the leading role of the government, mobilize the strength of all nationalities, and unite and assist each other in order to achieve better development.

4.3. Influence on Night-Time Light Remote Sensing Mapping

Night-time light remote sensing images have been widely used in economic monitoring, population mobility, environmental protection, and other fields, but there are relatively few studies [52] on night-time light remote sensing and the development of ethnic minorities. There is basically no literature on the study of ethnic minorities combined with night-time light data. This article fills this research gap to a certain extent. This paper combines toponym data, population data, and night-time light remote sensing data, considering the development of ethnic minorities from multiple perspectives. This mapping method provides a reference for subsequent similar studies. Special thematic mapping for ethnic minorities is also not common. The establishment of the ethnic minority development index plays a supporting role in dynamically monitoring the development of ethnic minorities and narrowing the development differences between ethnic minorities in various regions.

However, there are many development indexes that affect a region such as topography, population, and production patterns. Using the method in this article cannot reflect the importance of multiple variables, but can only be reflected by the brightness of night light illumination of night-time light data. The method in this article is more efficient for calculating the overall development index of a nation, but is not suitable to reflect the importance of each variable.

4.4. Significance of Cultural Protection of Ethnic Minorities

Due to industrialization and continuous economic development, people's production and lifestyles have undergone great changes, which has also caused many ethnic minority cultures to face crises. Therefore, we need to find the point of convergence between ethnic minority culture and economic development [55]. This article can understand the development of ethnic minorities by establishing the minority development index, which is conducive to summarizing the development laws of ethnic minorities, and has a positive effect on the protection and inheritance of ethnic minority cultures. It also responds to the call of General Secretary Xi Jinping to pay attention to the protection and inheritance of ethnic minority cultural heritage.

5. Conclusions

This article used ethnic toponym data, population data, and NPP-VIIRS night-time light data to obtain the development index of each ethnic group, and analyzed the six ethnic minorities with high development index as examples. The results showed that among the six ethnic minorities, the Yi nationality had the highest development index (28.86), and the Dai had the lowest development index (15.22). After in-depth analysis, we found the relationship between the minority development index and the minority prefecture, county, and geographic location, that is, the minority development index decreased as the distance between the minority nationality and its core development prefecture and county increased. According to the obtained development indexes of ethnic minorities, combined with the toponym data of ethnic minorities, the 25 ethnic minorities were divided into 13 categories according to the language branch classification method. Each ethnic minority was classified according to the level of the development index, and a map of the distribution of ethnic minorities in Yunnan Province was obtained. The Yi were distributed in almost the entire study area, and the distribution of other ethnic minorities had obvious regional characteristics. The overall development index of ethnic minorities in the Yi, Zhuang. and Dai, and Tibetan branch was higher, and the overall development index of ethnic minorities in other language branches was lower. In most areas, multiple ethnic minorities lived together. Among them, this phenomenon was most obvious in Dehong Prefecture, which may be related to the geographical location and cultural precipitation of Dehong Prefecture. In Yunnan Province, the two ethnic minorities, Yi and Dai, live together more often with other ethnic minorities.

All in all, this paper constructed a method to calculate the development index of ethnic minorities based on NPP-VIIRS night-time light data. This method is faster and more intuitive than other qualitative analysis methods that have focused on research and statistical yearbooks. On one hand, this method makes up for the lack of corresponding economic data in rural areas and ethnic minority areas to a certain extent. On the other hand, this article provides a new idea to study the mapping of ethnic minorities and night-time light remote sensing data. This is of great significance to the development of ethnic minorities and the protection of ethnic minority culture.

Author Contributions: Conceptualization, F.Z.; Methodology, F.Z.; Validation, F.Z., L.S. and Z.P.; Formal analysis, L.S., Z.P., J.Y. and G.L.; Resources, Z.X.; Data curation, F.Z., L.S., J.Y. and Z.P.; Writing—original draft preparation, F.Z. and L.S.; Writing—review and editing, F.Z., Z.X., L.S., Z.P., J.Y., J.D., G.L., C.C., S.F. and Y.J.; Visualization, L.S.; Supervision, F.Z. All authors have read and agreed to the published version of the manuscript.

Funding: This research was funded by the National Natural Science Foundation of China (Grant No. 41961064); the Yunnan Department of Science and Technology application of basic research project (Grant No. 202001BB050030); Plateau Mountain Ecology and Earth's Environment Discipline Construction Project [Grant No.C1762101030017]; Joint Foundation Project between Yunnan Science and Technology Department and Yunnan University [Grants C176240210019]; Yunnan University Graduate Research and Innovation Fund Project (Grant No. 2020188).

Acknowledgments: The authors express their sincere gratitude to the Earth Observation Team, BigeMap Downloader, the National Geomatics Center of China, and the China Social Big Data Research Platform for providing the NPP-VIIRS images, POI data, toponym data, research area vector data, and census data.

Conflicts of Interest: The authors declare no conflict of interest.

References

1. Ray, A.K. Measurement of social development: An international comparison. *Soc. Indic. Res.* **2008**, *86*. [CrossRef]
2. Fan, Q.; Lv, J.D.; Li, M. Prediction of GDP in Northeast China Based on Nighttime Light Data. *Remote Sens. Environ.* **2019**, *34*, 3–10.
3. Shi, K.F.; Yu, B.L.; Huang, Y.X.; Hu, Y.J.; Yin, B.; Chen, Z.Q.; Chen, L.J.; Wu, J.P. Evaluating the Ability of NPP-VIIRS Nighttime Light Data to Estimate the Gross Domestic Product and the Electric Power Consumption of China at Multiple Scales: A Comparison with DMSP-OLS Data. *Remote Sens.* **2014**, *6*, 1705–1724. [CrossRef]
4. Wang, J.N.; Lu, Y.T.; Zhou, J.S.; Li, Y.; Cao, D. Analysis of China resource-environment Gini coefficient based on GDP. *China Environ. Sci.* **2006**, *26*, 111–115.
5. Liang, Y.J.; Xu, Z.M. Modeling the Spatial Distribution of Population Based on Night Light Radiation and LUCC: A Case Study in Ganzhou District, Zhangye Municipality. *J. Glaciol. Geocryol.* **2012**, *34*, 999–1006.
6. Doll, C.N.; Pachauri, S. Estimating rural populations without access to electricity in developing countries through night-time light satellite imagery. *Energy Policy* **2010**, *38*, 5661–5670. [CrossRef]
7. Wang, Z.; Chen, J.H.; Cui, M.J. Prediction of GDP in Northeast China Based on Nighttime Light Data. *Mine Surv.* **2018**, *46*, 27–33.
8. Li, X.H.; Chen, J.Z. Study on the Economic Development Model of Ethnic Minorities from the Perspective of International Comparison Reg. *Reg. Econ. Rev.* **2020**, *3*, 18–24. [CrossRef]
9. Lu, M. A Study on the Status Quo and Countermeasures of Economic Development in China's Ethnic Minority Regions. *Mod. Econ. Inf.* **2020**, *3*, 186–187.
10. Jing, C.M. Status Quo of Income Gap in Western Ethnic Area and Strategy Analysis. *J. Northwest Univ. Natl.* **2005**, *6*, 80–86.
11. Li, J.J. A Case Study of Economic Disparity in Ethnic Autonomous Regions and Its Countermeasures. *J. Cent. Univ. Natl.* **2008**, *1*, 14–24. [CrossRef]
12. Gao, Z.G.; Liu, W. Regional Economic Differences and Coordinated Development in Northwest Minority Areas: Taking Xinjiang as an Example. *Nankai J.* **2016**, *3*, 147–160.
13. Dai, M.; Wang, P.; Li, Y.X. Analysis of Economic Disparities and Influencing Factors in Northwest Minority Regions. *Financ. Account. Mon.* **2017**, *12*, 108–113. [CrossRef]
14. Zheng, C.D. A Study on the "Five Development Concepts" and Comprehensive Development of the Social Economy in Ethnic Areas. *J. Ethnol.* **2017**, *8*, 1–21, 94–97. [CrossRef]
15. Luo, J.; Zhuang, Z.X. The variation trend of economic development in the ethnic minority regions of southwest borderland since the launch of western development in china—Based on Empirical Studies in Guangxi and Yunnan from 1995 to 2014. *Guangxi Ethn. Stud.* **2017**, *5*, 156–164.
16. Croft, T.A. Nighttime images of the earth from space. *Sci. Am.* **1978**, *239*, 86–98. [CrossRef]
17. Li, D.R.; Li, X. An Overview on Data Mining of Nighttime Light Remote Sensing. *Acta Surv. Mapp.* **2015**, *44*, 591–601. [CrossRef]
18. Elvidge, C.D.; Baugh, K.E.; Kihn, E.A.; Kroehl, H.W.; Davis, E.R.; Davis, C.W. Relation between satellite observed visible-near infrared emissions, population, economic activity and electric power consumption. *Int. J. Remote Sens.* **1997**, *18*, 1373–1379. [CrossRef]
19. Elvidge, C.D.; Baugh, K.E.; Dietz, J.B.; Bland, T.; Sutton, P.C.; Kroehl, H.W. Radiance calibration of DMSP-OLS low-light imaging data of human settlements. *Remote Sens. Environ.* **1999**, *68*, 77–88. [CrossRef]
20. Ghosh, T.; Anderson, S.; Powell, R.L.; Sutton, P.C.; Elvidge, C.D. Estimation of Mexico's informal economy and remittances using nighttime imagery. *Remote Sens.* **2009**, *1*, 418–444. [CrossRef]
21. Li, X.; Xu, H.M.; Chen, X.L.; Li, C. Potential of NPP-VIIRS Nighttime Light Imagery for Modeling the Regional Economy of China. *Remote Sens.* **2013**, *5*, 3057–3081. [CrossRef]
22. Doll, C.H.; Muller, J.P.; Elvidge, C.D. Night-time imagery as a tool for global mapping of socioeconomic parameters and greenhouse gas emissions. *Ambio J. Hum. Environ.* **2000**, *29*, 157–162. [CrossRef]
23. Elvidge, C.D.; Imhoff, M.L.; Baugh, K.E.; Hobson, V.R.; Nelson, I.; Safran, J.; Dietz, J.B.; Tuttle, B.T. Night-time Lights of the World: 1994–1995. *ISPRS J. Photogramm. Remote Sens.* **2001**, *56*, 81–99. [CrossRef]
24. Henderson, M.; Yeh, E.T.; Gong, P.; Elvidge, C.; Baugh, K. Validation of urban boundaries derived from global night-time satellite imagery. *Int. J. Remote Sens.* **2003**, *24*, 595–609. [CrossRef]
25. Henderson, J.V.; Storeygard, A.; Weil, D.N. Measuring Economic Growth from Outer Space. *Am. Econ. Rev.* **2012**, *102*. [CrossRef]
26. Michalopoulos, S.; Papaioannou, E. Pre-colonial Ethnic Institutions and Contemporary African Development. *Econom. J. Econom. Soc.* **2013**, *81*, 113–152. [CrossRef]
27. Wu, J.S.; Wang, Z.; Li, W.F.; Peng, J. Exploring factors affecting the relationship between light consumption and GDP based on DMSP/OLS nighttime satellite imagery. *Remote Sens. Environ.* **2013**, *134*, 111–119. [CrossRef]
28. Jiang, W.; He, G.J.; Liu, H.C. Modelling Regional Socio-economic Parameters Based on Comparison of NPP/VIIRS and DMSP/OLS Nighttime Light Imagery. *Remote Sens. Inf.* **2016**, *31*, 28–34.
29. Zhu, H.; Zhang, Q.L.; Zhang, S. Analysis of the temporal and spatial characteristics of social and economic development in Central Asia based on night-time light remote sensing from 1992 to 2017. *J. Geo-Inf. Sci.* **2020**, *22*, 1449–1462. [CrossRef]
30. Li, G.E.; Cai, Z.L.; Liu, X.J.; Liu, J.; Su, S.L. A comparison of machine learning approaches for identifying high-poverty counties: Robust features of DMSP/OLS night-time light imagery. *Int. J. Remote Sens.* **2019**, *40*, 5716–5736. [CrossRef]

31. Andreano, M.S.; Benedetti, R.; Piersimoni, F.; Savio, G. Mapping Poverty of Latin American and Caribbean Countries from Heaven through Night-Light Satellite Images. *Soc. Indic. Res.* **2020**, 1–30. [CrossRef]
32. Pokhriyal, N.; Jacques, D.C. Combining disparate data sources for improved poverty prediction and mapping. *Proc. Natl. Acad. Sci. USA* **2017**, *114*. [CrossRef] [PubMed]
33. Li, G.E.; Cai, Z.L.; Liu, J.; Liu, X.J.; Su, S.L.; Huang, X.R.; Li, B.Z. Multidimensional Poverty in Rural China: Indicators, Spatiotemporal Patterns and Applications. *Soc. Indic. Res.* **2019**, *144*. [CrossRef]
34. Donald, H.; Thomas, K.; Thomas, L.; Daniel, L.; Curtis, S.; Steven, M.; Jeremy, S.; Stanley, K.; Scott, B.; Tommy, J.; et al. First-light imagery from Suomi NPP VIIRS. *Bull. Am. Meteorol. Soc.* **2013**, *94*, 1019–1029. [CrossRef]
35. Jing, X.; Shao, X.; Cao, C.Y.; Fu, X.D.; Yan, L. Comparison between the Suomi-NPP Day-Night Band and DMSP-OLS for Correlating Socio-Economic Variables at the Provincial Level in China. *Remote Sens.* **2015**, *8*, 17. [CrossRef]
36. Elvidge, C.D.; Zhizhin, M.; Baugh, K.; Hsu, F.C. Automatic boat identification system for VIIRS low light imaging data. *Remote Sens.* **2015**, *7*, 3020–3036. [CrossRef]
37. Fox, H.S.A.; Gelling, M. "book-review" Signposts to the past. Place-Names and the History of England. *Geogr. Rev.* **1981**, *71*, 231. [CrossRef]
38. Chu, Y.; Yin, J.; Sun, D. *Basic Tutorial of Toponyms*; Surveying and Mapping Press: Beijing, China, 2009.
39. Conedera, M.; Vassere, S.; Neff, C.; Meurer, M.; Krebs, P. Using toponymy to reconstruct past land use: Acase study of 'brusada' (burn) in southern Switzerland. *J. Hist. Geogr.* **2007**, *33*, 729–748. [CrossRef]
40. Jett, S.C. Place-Naming, Environment, and Perception among the Canyon De Chelly Navajo of Arizona. *Prof. Geogr.* **1997**, *49*, 481–493. [CrossRef]
41. Evereet-Health, J. *Place Names of the World—Europe*; Palgrave Macmillan: London, UK, 2000.
42. Adrian, R. *Placenames of the World: Origins and Meanings of the Names for 6600 Countries, Cities, Territories, Natural Features and Historic Sites*, 2nd ed.; Choice Reviews Online; McFarland: Jefferson, NC, USA, 2006; Volume 43. [CrossRef]
43. Ray, I. Linguistic Geography of Chinese Dialects. *Cah. Linguist. Asie Orient.* **1995**, *24*, 195–227. [CrossRef]
44. Hartmann, J.F. The Power to Name Places: Ban, Muang, Chiang, Viang, Nakon, Krung. In *Studies in Tai and Southeast Asian Linguistics*; Harris, J., Burusphat, S., Eds.; Ekphimthai Ltd.: Bangkok, Thailand, 2007; pp. 139–154.
45. Xu, Z.N.; Gao, X.L. A novel method for identifying the boundary of urban build-up areas with POI data. *Acta Geogr. Sin.* **2016**, *71*, 928–939. [CrossRef]
46. Liu, Q.Y.; Zhan, Q.M.; Li, J.S.; Yang, C.; Liu, W. Extracting Built-up Areas Using Luojia-1A Nighttime Light Imageries in Wuhan, China. *Geomat. Inf. Sci. Wuhan Univ.* **2021**, *46*, 30–39. [CrossRef]
47. Heidenreich, N.B.; Schindler, A.; Sperlich, S. Bandwidth selection for kernel density estimation:A review of fully automatic selectors. *Adv. Stat. Anal.* **2013**, *97*, 403–433. [CrossRef]
48. Li, F.; Yan, Q.W.; Zou, Y.J.; Liu, B.L. Improving the Accuracy of Built-up Area Extraction Based on Nighttime Light Data and POI: A Case Study of NPP/VIIRS and Luojia 1-01 Nighttime Light Images. *J. Wuhan Univ. Inf. Sci. Ed.* **2019**, *46*, 1–14.
49. Zhao, F.; Ding, J.Y.; Zhang, S.J.; Luan, G.Z.; Song, L.; Peng, Z.Y.; Du, Q.Y.; Xie, Z.Q. Estimating Rural Electric Power Consumption Using NPP-VIIRS Night-Time Light, Toponym and POI Data in Ethnic Minority Areas of China. *Remote Sens.* **2020**, *12*, 2836. [CrossRef]
50. Zhou, Z.H.; You, R.J. *Dialects and Chinese Culture*; Shanghai People's Publishing House: Shanghai, China, 1986; pp. 7–9.
51. Liu, Y.H.; Xu, Y. Geographical Identification and Type Classification of Multidimensional Poverty in Rural China. *Acta Geogr. Sin.* **2015**, *70*, 993–1007. [CrossRef]
52. Zhang, E.M.; Deng, J.; Song, X.J.; Dai, K.R.; Shi, X.L. Multi-dimensional poverty identification and evolution analysis based on night-time light data. *Hubei Agric. Sci.* **2019**, *58*, 126–133. [CrossRef]
53. Shen, D.; Zhou, L.; Wang, P.A. Analysis of Poverty Spatial Differentiation Characteristics in Liupanshan Contiguous Destitute Area Based on Night Light Data. *Remote Sens. Inf.* **2018**, *33*, 42–48. [CrossRef]
54. Zhang, J.Q.; Wu, Y.J.; Ge, Y.; Wang, C.H.; Kung, H. Comprehensive Evaluation of Ecological Security in Poor Areas Based on Grey Relational Model—Taking Enshi Poor Area as an Example. *Geogr. Res.* **2014**, *33*, 1457–1466. [CrossRef]
55. Wen, K.Z. Thoughts on the Protection and Development of Ethnic Minority Culture. *J. Guangdong Inst. Social.* **2007**, *28*, 49–54.

remote sensing

MDPI

Article

A Global Assessment of Night Lights as an Indicator for Shipping Activity in Anchorage Areas

Semion Polinov [1,2,*], Revital Bookman [1] and Noam Levin [3,4]

[1] Department of Marine Geosciences, Charney School of Marine Sciences, University of Haifa, Mt. Carmel, Haifa 3498838, Israel; rbookman@univ.haifa.ac.il
[2] The Chaikin Chair for Geostrategy and Maritime Policy & Strategy Research Center, University of Haifa, Mt. Carmel, Haifa 3498838, Israel
[3] Department of Geography, Hebrew University of Jerusalem, Mount Scopus, Jerusalem 9190501, Israel; noamlevin@mail.huji.ac.il
[4] Remote Sensing Research Center, School of Earth and Environmental Sciences, University of Queensland, St Lucia, QLD 4072, Australia
* Correspondence: semion.polinov@gmail.com; Tel.: +972-546569687

Abstract: Accurate information on port shipping activities is critical for monitoring global and local traffic flows and assessing the state of development of the maritime industry. Such information is necessary for managers and analysts to make strategic decisions and monitor the maritime industry in achieving management goals. In this study, we used monthly night light (NTL) images of the Suomi National Polar-Orbiting Partnership (Suomi NPP) Visible Infrared Imaging Radiometer Suite (VIIRS) Day/Night Band, between 2012 and 2020, to study the night lights emitted by ships in ports' anchorage areas, as an indicator for shipping activity in anchorage areas and ports. Using a dataset covering 601 anchorage areas from 97 countries, we found a strong correspondence between NTL data and shipping metrics at the country level ($n = 97$), such as container port throughput ($R_s = 0.84$, $p < 0.01$) and maximum cargo carried by ships ($R_s = 0.66$, $p < 0.01$), as well as a strong correlation between the number of anchorage points and the NTL values in anchorage areas across the world ($R_s = 0.69$, $p < 0.01$; $n = 601$). The high correspondence levels of the VIIRS NTL data with various shipping indicators show the potential of using NTL data to analyze the spatio-temporal dynamic changes of the shipping activity in anchorage areas, providing convenient open access and a normalized assessment method for shipping industry parameters that are often lacking.

Keywords: VIIRS/DNB; nighttime lights; shipping; spatial analysis; anchorage area

Citation: Polinov, S.; Bookman, R.; Levin, N. A Global Assessment of Night Lights as an Indicator for Shipping Activity in Anchorage Areas. *Remote Sens.* **2022**, *14*, 1079. https://doi.org/10.3390/rs14051079

Academic Editors: Ran Goldblatt, Steven Louis Rubinyi and Hogeun Park

Received: 28 December 2021
Accepted: 14 February 2022
Published: 22 February 2022

Publisher's Note: MDPI stays neutral with regard to jurisdictional claims in published maps and institutional affiliations.

1. Introduction

The global rise in the standards of living, consumption volumes, as well as the development and use of marine resources, are leading to an increase in global shipping, despite a temporary slowdown in maritime trade growth in 2018 as a result of trade tensions, protectionism, Brexit [1], and shipping restrictions following the COVID-19 outbreak [2]. According to estimates of the United Nations Conference on Trade and Development (UNCTAD), the volume of international trade by sea accounts for approximately 80% of the volume of world trade [3]. The general trend points to a steadily increasing and developing rate of maritime transport over the past two decades [1]. Ports are important centers of trade between countries, and their cargo handling capacities (loading and unloading of goods) are one of the most basic and important indicators for measuring the development status of ports [4–6]. The port anchorage area (PAA) is a place where ships wait for their turn to enter the port outside the port areas. The PAA is an important part of the shipping and port management segment [5]. Accurate statistical information on the number of ships and cargo loaded and unloaded over a certain period in the port is of decisive importance for monitoring the movement of ships, assessing the state of development of the port, the country's economy, as well as assessing global trends [6–8]. To date, periodical statistics on port activity are published by various global organizations such as UNCTAD [9], as well as by the ports authorities themselves. Statistical data in frequent comparisons are difficult to access (Lloyd's List Intelligence) [10], varied, and rarely published, making it difficult to analyze data consistently or in near real-time. The recent development of automatic identification systems (AIS), which show the exact location of ships in almost real-time and with an average update of one minute, as well as additional parameters (ship dimensions, etc.) and dynamics (location, direction, speed, etc.), are expensive and complex in processing [11]. While the effects of some anthropogenic activities in marine areas are examined in detail, for example, oil spills from shipping [12,13], our understanding of the spatial and temporal trends of artificial night-time light (NTL) in port anchoring areas remains limited [14].

Conventional methods for monitoring ships at sea from space include optical and synthetic aperture radar (SAR) images obtained using remote sensing. Daytime optical sensors allow the detection of ships; however, their sensors are usually not sensitive enough for detecting low light levels as emitted at night-time [15]. While SAR images have all-weather and day and night capabilities, this approach for detecting ships at sea requires the processing of large amounts of data, and at the moment, there is no operational product offering vessel detection from SAR data [16–18]. Thus, there are still many gaps in the monitoring of ships at daily, monthly and annual time scales. Recent studies have promoted the use of VIIRS low light imaging data for monitoring ships that are using artificial lights within the fisheries industry [19,20]. In this study, we propose the use of night light data to monitor shipping activity in anchorage areas, thereby filling the research gap with a new method for assessing statistical parameters of shipping activities at the port level. This study provides a new approach based on monthly/annual NTL values at different geographic levels (port and country). The challenges we faced included variability in NTL emitted from different ships, the low sensitivity of satellite sensors (VIIRS) to small and weak light sources, and the need for correcting NTL data to minimize the influence of natural light, light from cities, and persistent cloud cover in certain areas [21,22], to measure the magnitude of NTL emitted by ships in the anchorage area.

Research Question and Objectives

Our main objective in this paper was, therefore, to determine to what degree can we use NTL as a proxy for shipping activities in anchorage areas, with the following two specific aims:

1. To what extent can night-time lights in anchorage areas serve as an indicator of shipping activity in port anchorage areas? To do this, we will examine the correspondence between the night-time lights and shipping data at the port level and the country level.
2. Which variables at the country level can explain the intensity of lighting in anchorage areas? To do this, we will examine various variables that represent economic activities such as GDP, exports, etc. and their correspondence with NTL.

2. Materials and Methods

An anchorage area is a place where boats and ships can safely drop anchor. Anchorage areas vary by the types of anchoring which are allowed (size of boat, type of anchor, vessel size, and type) and the authority in charge of the anchorage area (local government, county government, or state government) [23].

2.1. Process of Creating the Anchorage Polygons

To examine port activities, we created polygons of anchorage areas using the anchorage points (the offshore location where ships can lower anchors while waiting to be allowed in the ports) dataset of the Global Fishing Watch (GFW) (https://globalfishingwatch.org/data-download/datasets/public-anchorages:v20200316) (v1_20191205) (accessed on 2 August 2020). The anchorage points represent the centers of the anchorage circle (Figure 1). In different weather conditions, the ship moves radially around the anchor center, so as not to interfere with other ships and the movement of other ships in the port.

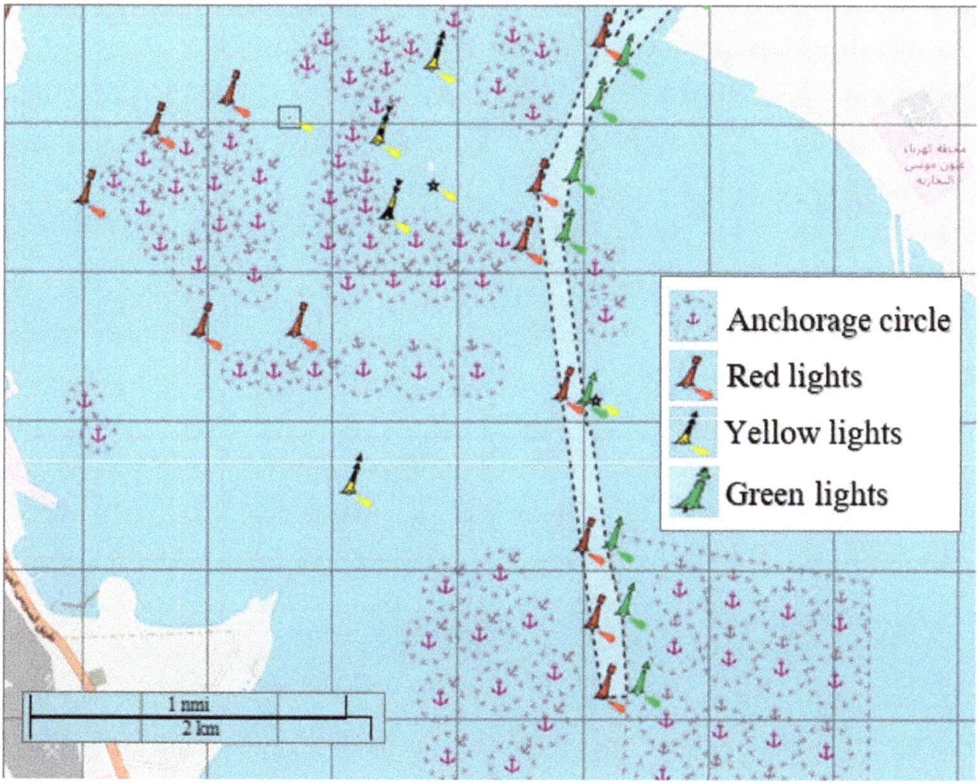

Figure 1. An anchorage area at the southern entrance to the Suez Canal. Circles provide anchoring places for different types of vessels, following their size and type (Tanker, Cargo, etc.). In the center of each circle is marked the "anchor", where the ship drops the anchor (source: openseamap.com, (accessed on 10 November 2021)). Colored symbols represent navigation buoys.

To extract the NTL values from VIIRS over the anchorage areas globally, we created a polygonal database of anchorage areas (Figure 2, showing as an example the anchorage areas of Fujairah, United Arab Emirates) covering most of the world's ports. To extract the polygons from the point layer of GFW, first, we transformed the point data to raster format with a cell size of 1 km, using an equal-area projection. We filled the gaps inside each polygon using spatial closing filters with a 7 × 7 moving window. We included in our analysis only anchorage areas with at least 10 anchorage points (Figure 2) and which were not close to the coastline so that night-time lights within them will be less affected by coastal urban lights. The night lights emitted by a coastal city can reach offshore areas via scattering by the atmosphere [24]. Moreover, this glow distance varies not only as a result of the density of light sources and the type of lighting but also via atmospheric scattering conditions. The purpose of the following section is to demonstrate the glow of urban coastal night lights into offshore areas.

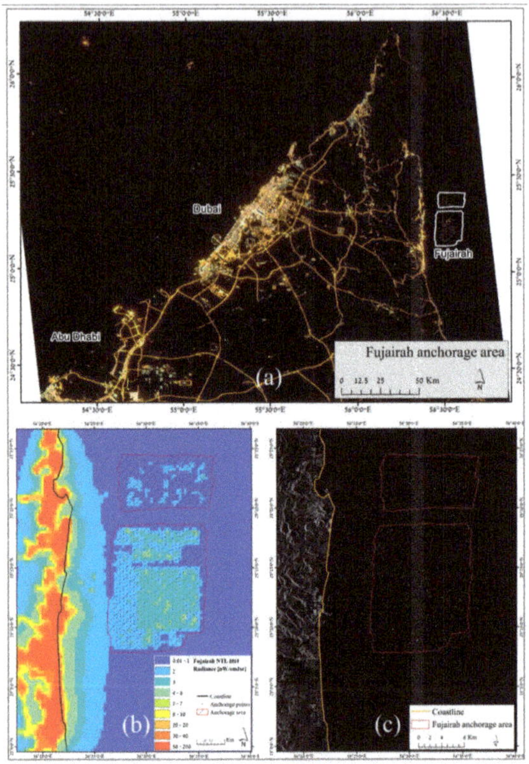

Figure 2. Astronaut photography (ISS038–E–16344) of the Eastern part of the Persian Gulf, 2013 (**a**). The average NTL values for 2019 at the vicinity of Fujairah port (**b**), UAE (black dots represents centers of anchorage circles for vessels, based on the dataset of Global Fishing Watch), whereas the (**c**) image shows an image of the same area acquired by Sentinel 1 (9 June 2019). The anchorage area of the port is about ten kilometers from the coastline. Coastal NTL does not affect the anchorage area of the port. The main type of vessel operating in the Fujairah port is tankers.

As can be seen from Figure 2, the urban and coastal areas of Fujairah were much more illuminated by artificial night-time lights (more than 50 nW cm^{-2} sr^{-1}) compared with the anchorage area in which the radiance values of night-time lights ranged between 2–7 nW cm^{-2} sr^{-1}. Despite the strong night illumination of the city, it did not affect the amount of light in the area of the anchorage area since the influence of city lights from Fujairah ended about 6–7 km from the coastline.

The resulting anchorage polygons which we created included 601 anchorage polygons belonging to 97 countries. Finally, these anchorage areas included 44,570 anchorage points out of a total of 119,478 anchorage points in the original dataset of GFW.

2.2. VIIRS Night–Time Light Data (Response Variables)

The VIIRS DNB data are more sensitive to low light levels than the DMSP/OLS and have a higher spatial resolution of 742 m × 742 m footprint from nadir out to the edge of scan [25]. The monthly VIIRS products are gridded to a 15 arc-second grid, which is slightly finer than the original pixel footprints [26]. In this study, we used VIIRS DNB monthly cloud–free average data products for the period between April 2012 and March 2020 provided by the NOAA service (https://ngdc.noaa.gov/eog/viirs/download_dnb_composites.html, (accessed on 20 June 2021)). The dataset includes radiance data and cloud-free coverages (the number of cloud-free acquisitions available for a given month for calculating night-time brightness in each pixel). The DNB radiance data excluded data impacted by stray light, lightning, lunar illumination, and cloud cover before averaging, while some temporal lights from auroras, fires, boats, etc., are reserved [27]. Two configurations of the VIIRS composites are available: "vcmcfg" excludes any data contaminated by stray light (typically solar illumination) and "vcmslcfg" excludes data impacted by stray light are corrected but not removed. We selected the "vcmslcfg" products as they offer greater temporal and spatial coverage [28]. Moreover, we calculated the annual sum of lights within each anchorage area by multiplying the polygon area with the average values of NTL. We used the equal–area Mollweide projection to calculate anchorage areas and to calculate the sum of lights (SOL) of anchorage areas. To extract NTL values of anchorage areas data, we used the Google Earth Engine (GEE) platform [29]. We extracted and merged both data configurations: 'VCMSLCFG' for the period of January 2014–March 2020 and 'VCMCFG' for the period of April 2012–December 2013, of two available bands:

- "Avg_rad"—value represents the monthly average value of NTL.
- "Cf_cfg"—cloud–free days (this was important to interpolate monthly radiance values for months that were too cloudy, as detailed below).

We used the correction coefficients provided by Coestfield [22] for correcting the temporal variation of natural light sources such as airglow and thus corrected the monthly time series of NTL using their published coefficients, which are especially important in areas with low levels of NTL.

In various anchorage areas, NTL values were underestimated in certain months due to high cloud cover for most of the month. We used a threshold of an average of at least one day without cloud cover per month within the anchorage area as the minimum threshold in which we accepted NTL values for a specific month that would be "valid" for our analysis of the anchorage areas. According to our dataset of NTL for 96 months and 601 anchors, almost 95% of the months in all ports had sufficient cloud–free data (Table 1). In the majority of cases, cloudy months (with cloud–free days below 1) were usually isolated (3.1% of all months; Table 1). More than 50% of all anchorage areas experienced at least one event of two consecutive months with persistent cloud coverage (cloud–free days below 1; Table 1).

Table 1. Distribution of the number of months without sufficient NTL data measurements due to high cloud cover. The left–hand column shows the number of consecutive months without sufficient cloud–free measurements (average cloud–free days < 1). Our dataset included 96 months and 601 anchorage areas (96 × 601 = 57,696). The first group, "0 = Zero", indicates months with no missing data.

Number of Consecutive Missing Months with Cloud Free Days (CFD) Value < 1	Sum of Months in Each Group across all Anchorage Areas	% of the Sum of Months in Each Group across all Anchorage Areas (out of 57,696)	Number of Anchorage Areas in Each Group of Missing Months	% Number of Anchorage Areas in Each Group of Missing Months (out of 601)
0	54,589	94.6%	113	18.8%
1	1772	3.1%	488	81.2%
2	731	1.3%	344	57.2%
3	375	0.6%	235	39.1%
4	149	0.3%	126	21.0%
5	56	0.1%	53	8.8%
6	24	0.04%	24	4.0%

To fill the monthly gaps in NTL values, we interpolated NTL values for months with average cloud–free days of less than one day, based on the six months before and after the missing month. Based on this method, the missing values were filled in, and the original values for months with CFD ≥ 1 were not changed (see example in Figure 3 for the anchorage area of Punta Arenas, Chile).

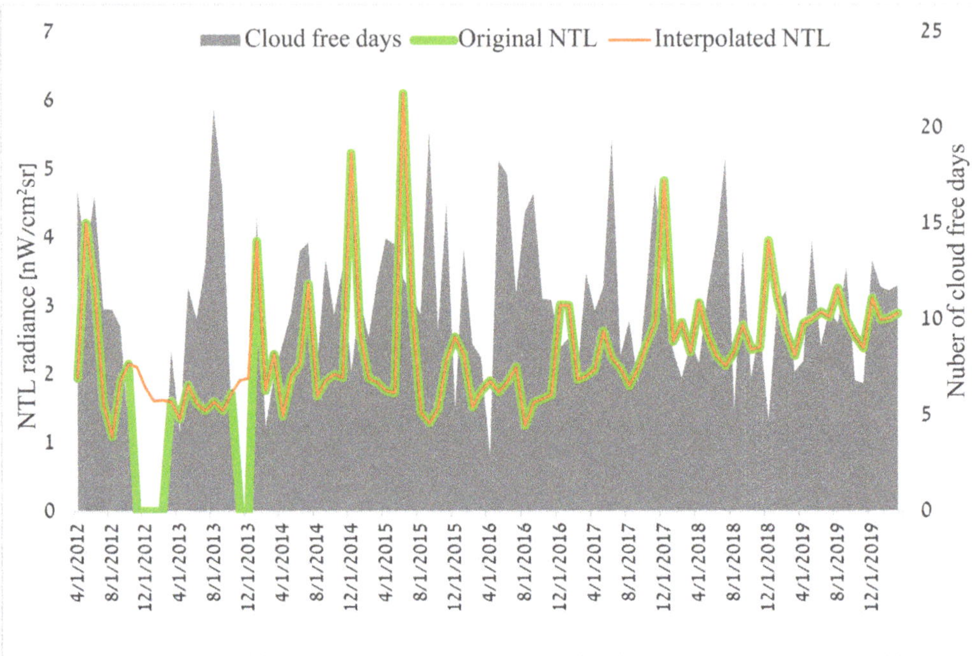

Figure 3. Interpolation of NTL values for cloudy months (for the anchorage area of Punta Arenas, Chile).

Finally, to estimate the impact of the interpolation method of "Interpolating NTL for cloudy months" on the average NTL values in all anchorage areas, we performed a 2-tails *t*-test. The test was used to determine if there was a significant difference between the means of the original NTL values vs. the interpolated values for each of the anchorage areas. Running a *t*-test between the NTL time series of each anchorage before and after the interpolation of months in which the number of the cloud-free day was less than one, we found that the null hypothesis (H_0—no difference between the original monthly values and the interpolated ones) could not be rejected (see example Figure 3 for Punta Arenas anchorage area) for any of the anchorage areas (i.e., the average NTL values before and after the interpolation were not different).

2.3. Explanatory Variables

We collected a range of quantitative information on maritime activities at the port level, as well as various parameters (economic, environmental, etc.) at the country level, using data sourced from the United Nations Conference on Trade and Development (UNCTAD) (https://unctad.org/statistics, accessed on 12 October 2021), the World Bank (https://data.worldbank.org, accessed on 12 October 2021), and www.trademap.com (accessed on 12 October 2021) at the port and country-level (Table 2). The UNCTAD "Port calls/Port performance" parameters group is one of the main sources of maritime shipping data and is part of a suite of port call and performance statistics that provide an overview of the characteristics of ships and the time they spent in a country's ports over a given period [30,31]. Together, the "Port calls/Port performance" includes statistics for up to eight parameters, covering: Median time in port—the median time vessels spent within port limits (in days); Average size of the vessel—the average gross tonnage of the vessels that have called in the country's ports during the year, as well as the "Maximum" parameters of the largest ships that have called during the period, the maximum cargo carrying capacity and the maximum container carrying capacity of container ships; Average cargo carrying capacity—the average deadweight tonnage of the vessels that have called in the country's ports during the year; Average container carrying capacity per container ship—the average capacity to carry the 20–foot equivalent units (TEU) of the container ships. Another data source from the UNCTAD is the container port throughput (CPT)—measurements of the container flow from land to sea transport modes, and vice versa, in 20–foot equivalent units (TEUs), a standard–size container [32]. Data refer to coastal shipping as well as international shipping. Trans–shipment traffic is counted as two lifts at the intermediate port (once to off–load and again as an outbound lift) and includes empty units. The liner shipping connectivity index (LSCI) represents the country's integration level into global liner shipping networks [33,34]. The LSCI is an index set at 100 for the country with the maximum value of country/port connectivity in the first quarter (Q1) as of 2006, which was then China. From the Wordlbank, we have downloaded the following data at the state level: Electric power consumption—electric power consumption per capita (kWh), which are the main indicators of the size and level of development of the country's economy; Fossil fuel consumption (% of total)—comprises coal, oil, petroleum, and natural gas products by country; Population—a country population based on national population censuses. Most of the explanatory variables used in the study represent the average values over a certain period (Table 2). Since the NTL data we used covered a wider time range (2012–2020), we calculated, for each of the explanatory variables, the mean over the period corresponding to the response variable (NTL data) or the widest time range that could be extracted, for example, TSC data were available for 2016–2020. Most of the explanatory variables were only available at the country level and only a few at the port level. These data describe various aspects directly or indirectly related to maritime ship activity in the ports at the country level and thus may explain the number of ships in the anchorage areas.

Table 2. List of explanatory (independent) variables at the country and port level.

Parameter	C—Country P—port	Years	Data Source
Number of anchorage points	C/P	2019	[35]
Average cargo carrying capacity	C	2018–2020	[31]
Average container carrying capacity	C	2018–2020	[31]
Average size of vessel	C	2018–2020	[31]
Average CO_2 emissions	C	2016	[36]
Container port throughput (CPT)	C	2016–2019	[37]
Electric power consumption	C	2013–2014	[38]
Fossil fuel consumption	C	2013–2015	[39]
Gross domestic product (GDP)	C	2016–2020	[40]
GDP growth (annual %)	C	2016–2020	[41]
Import	C	2016–2020	[42]
LSCI	C	2016–2020	[31]
Maximum cargo carrying capacity of vessels	C	2018–2020	[31]
Maximum container carrying capacity of vessel	C	2018–2020	[31]
Maximum size of vessels	C	2018–2020	[31]
Median time in port (days)	C	2018–2020	[31]
Population growth (%)	C	2016–2020	[43]
Population total	C	2016–2020	[44]
Monthly average number of vessels in the PAA of Santos	P	2016–2020	Sentinel 1
Santos port statistics (Import/Export, ship numbers by class and by waiting time)	P	2016–2020	[45]

Data collection on Vessel Numbers in Anchorage Areas from Sentinel 1

To further examine the potential of using NTL as a measure of temporal changes in ship activity in ports, we collected data on the number of ships anchored in the port anchorage area. Using Sentinel 1 (SAR) satellite images (which offer all–weather capabilities and enable ship detection; [46]) from January 2016 to March 2020 in the port of Santos (Brazil's largest port), the average number of vessels sighted in the anchorage area was calculated based on 4–5 images per month. The port of Santos was chosen because it is a relatively large port with a remote anchorage area. Such a large anchorage area can accommodate more than 50 ships at any given time, which can create a large amount of night–time light that will not be affected by city lights. Thus, a port such as Santos is a good place to test the hypothesis that night lights may serve as an indicator of temporal changes in shipping activity in the anchorage area.

To perform a detailed analysis at the port level for Brazil's largest port (the port of Santos), the following statistical data from the port's website (http://www.portodesantos.com.br/informacoes-operacionais/estatisticas/mensario-estatistico/ (accessed on 1 February 2022)) were also used: monthly volumes of imports and exports, the monthly number of ships waiting (total and those waiting >72 h) in the anchorage area, and the number of ships from different segments (general cargo, bulk solids, bulk liquids [tankers], passengers, etc.) that visited the port of Santos.

2.4. Analysis

We conducted a correlation analysis at two levels: at the port level and the country level. We chose to include a country–level analysis as well, as it is useful in our view to conduct between–countries comparisons, which is also common practice in other studies of economic activities using night lights [28,47,48]. For each spatial level, various explanatory and response variables were prepared (average NTL and Sum of Lights). For the country–level analysis, we calculated the mean values of the explanatory parameters and the mean value of the NTL of all anchorage areas of that country. At the port level, a comparison was made between the monthly average NTL and the average number of vessels, using the example of the port of Santos, Brazil. Spearman's rank correlation coefficient (denoted here by Rs) was calculated using SPSS to examine the correspondence between the average annual NTL 2012–2020 and the explanatory variables (Table 2). The use of NTL was proven to be effective for the study of large areas, such as at the country, state, county, or city level [49].

3. Results

Overall, we identified and analyzed 601 anchorage areas representing 97 countries with a temporal coverage of 96 months (April 2012–March 2020). In this study, an assessment was made of the level of the correlation between the average values of VIIRS (monthly, annual) sum of lights (SOL) both at the level of a single anchorage and at the country level.

The map of the distribution of the average annual NTL value for 2012–2020 (Figure 4) indicates high concentrations of anchorage points per port along the coasts of China (Figure 4c), the Persian Gulf (Figure 4b), the Mediterranean Sea (Figure 4a), Gulf of Guinea, the Southern Coast of Brazil, and the Southern Caribbean, which indicates high shipping activity in these regions during the study period. The spatial distribution of ports by their number of anchorage points varies greatly, with the highest number of anchorage points found along the coast of China. The anchorage area of the port of Xingang (China) had the highest number of anchorage points (1197), followed by the port of Fujairah (United Arab Emirates) with 725 anchorage points. The port of Shanghai had several anchorage areas, which combined included more than 1300 anchorage points; The port of Lome (Togo), with 620 anchorages, was the largest of any African port. Of the European ports, Malta had the largest number of anchorage points (498). The Brazilian port of Santos was the leading port in the number of anchorage points (449) within the Americas.

Figure 5 show a map of the anchorage area and their average annual sum of lights (SOL) values for 2012–2020. On overage, the ports of the Persian Gulf (Figure 5b) were among the most brightly lit (112 nW/cm^{2}cr). Anchorage areas located in Asia (especially in the East China Sea, Figure 5c) had significantly lower average values (43 nW/cm^{2}cr). The anchorage areas of the American continent, especially the northern part, were much less lit, with an average of 39 nW/cm^{2}cr. Of the seas of the European continent, the anchorage areas of the Mediterranean Sea (Figure 5a) were the most lit with an average of 48 nW/cm^{2}cr, with Malta's being the most lit with 443 nW/cm^{2}cr. The average NTL values of the anchorage areas of European ports in the Atlantic Ocean and the Northern Sea had lower SOL values (23 nW/cm^{2}cr), while the most lit anchorage area (Rotterdam) was ranked globally as 23rd most lit anchorage area with an average SOL of 187 nW/cm^{2}cr. China's anchorage areas had relatively low SOL values (average of 45 nW/cm^{2}cr based on 87 areas, with only 10 anchorage areas with SOL values higher than 100 nW/cm^{2}cr), despite a high average number of anchorage points of 155 per anchorage area (n = 87). The anchorage areas of Xingang and Ningbo (China), with an average of 354 and 328 nW/cm^{2}cr, respectively, were the most lit among Chinese ports regarding anchorage areas. Anchorage areas along the East Coast of Africa were less lit (37 nW/cm^{2}cr) than those of the West Coast of Africa (59 nW/cm^{2}cr) and the Southern Coast of the Mediterranean Sea (57 nW/cm^{2}cr). The port areas of Luanda (Angola) and Lome (Togo) were the most lit on the African continent, 252 nW/cm^{2}cr and 240 nW/cm^{2}cr, respectively. The anchorage area of the port of Kandla was the most lit (523 nW/cm^{2}cr) among all the investigated anchorage areas in India. The Port of Kandla had the largest number of anchorage points among all ports in India (178 anchorage points) and was ranked 53rd out of all anchorage areas in terms of its number of anchorage points.

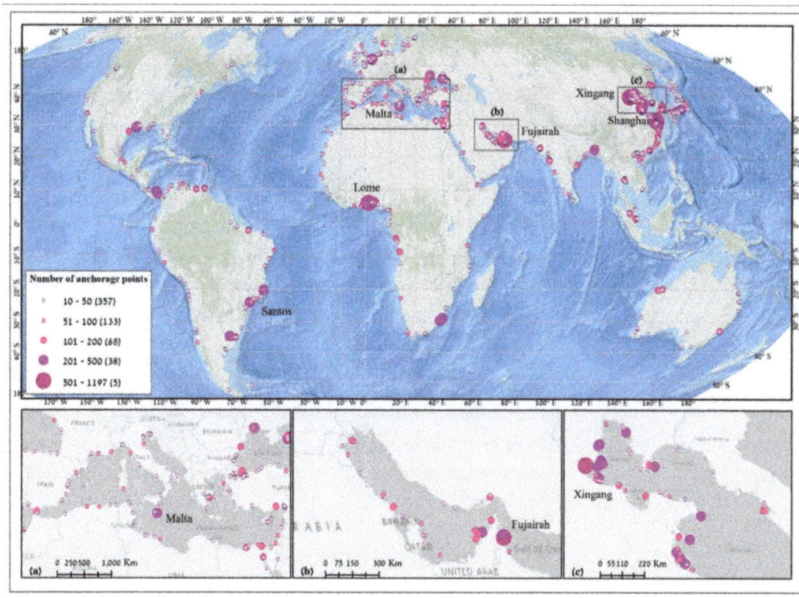

Figure 4. Map of anchorage areas (ports) symbolized by their number of anchorage points (source of anchorage points: www.globalfishingwatch.org, accessed on 5 October 2020). Enlarged maps show the following areas: (**a**) The Mediterranean Sea; (**b**) The Persian Gulf; (**c**) The Yellow Sea region.

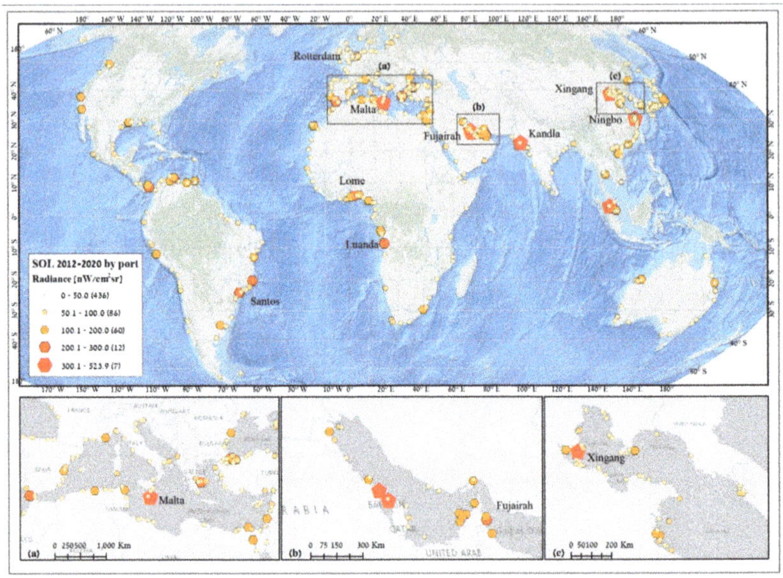

Figure 5. Map of anchorage areas and their average annual sum of lights (SOL) intensity for the period of April 2012–March 2020. Enlarged maps show the following areas: (**a**) The Mediterranean Sea; (**b**) The Persian Gulf; (**c**) The Yellow Sea region.

3.1. General Patterns of NTL as an Indicator of Shipping Activity

In this section, we present the results of the correlation analysis obtained at the global level. The purpose of this section is to assess the correlation between two fundamental parameters, NTL and the number of anchorage points at different geographical levels

Figure 6 show a statistically strong and significant ($R_s = 0.69$, $p < 0.01$) correlation between the number of anchorage points per area and the average annual value of NTL sum of light (SOL) for 2012–2020 based on 601 anchorage zones. This result indicates a strong relationship between the total number of anchorage points for ships within the anchorage area and the average SOL reflected from anchorages at sea. A higher number of anchorage points increases the amount of NTL lights measured by the satellite sensors.

Figure 7 present the correlation between the total number of anchorage points in the ports of each country (97 in total) and the average NTL value for 2012–2020. Despite the similarity of the two graphs (6 and 7), a higher correlation was obtained at the country level ($R_s = 0.84$, $p < 0.01$) between anchorage points and NTL than at the port level. In countries with a developed maritime industry such as China, Japan, Turkey and the USA, a large number of anchorage areas exist, and as a result, they emit high levels of NTL.

Figure 8 show a scatter plot of the correlation analysis between the annual average SOL and the signal to noise ratio (SNR = Average/Standard Deviation) for the period 2012–2020. We obtained a significantly strong correlation ($R_s = 0.49$, $p < 0.01$) between the annual average SOL value and the SNR over the investigated period of 2012–2020. Anchorage areas with SNR values below one indicate high variability in monthly NTL values. Out of 601 anchorage areas, 33 (5%) had SNR values below 1, while China had 12 (13% of all Chinese anchorage areas) and Germany had 5 and 7 anchorage areas. Anchorage areas with SNR values above 2 represent are areas with more stable monthly/annual NTL. The most stable NTL values were obtained in the anchorage areas of the port of Tenerife, Spain (Average/Stdv 258.37/19.9 nW/cm^{2}sr, SNR = 13), Valencia, Spain (166.45/13.7 nW/cm^{2}sr, SNR = 12.2), and Kuwait (164.56/15.5 nW/cm^{2}sr, SNR = 10.6), while the most unstable of the highly active ports (with average NTL above 150 nW/cm^{2}sr) was the anchorage area of Khalifa Bin Salman, Bahrain (153.9/484.8 nW/cm^{2}sr, SNR = 0.3).

Figure 6. Spearmen correlation between the number of anchorage points per anchorage area and averages of annual SOL from 2012 to 2020 ($R_s = 0.69$, $p < 0.01$), N = 601.

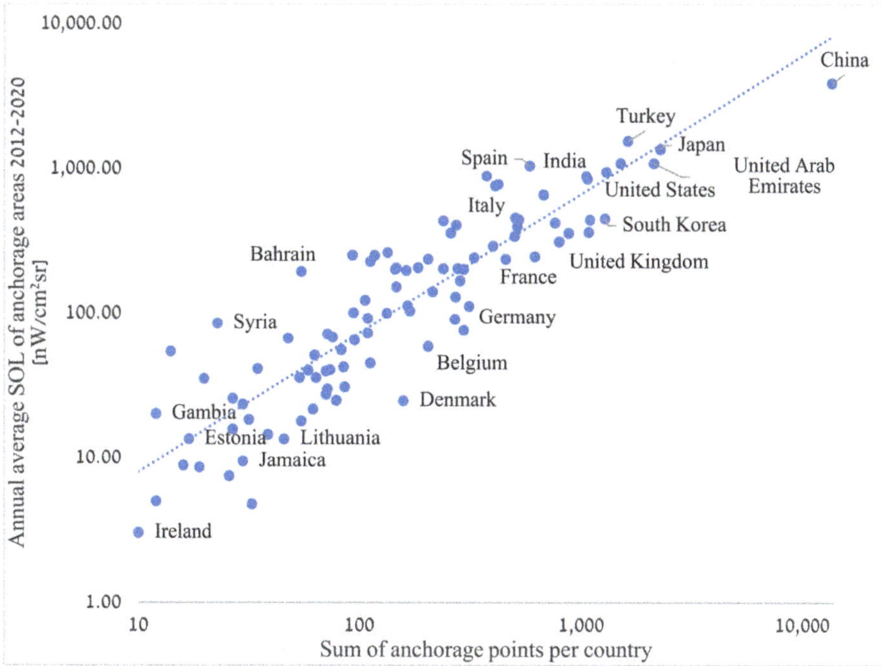

Figure 7. Logarithmic scatter plot of Spearmen correlation between the total number of anchorage points by country and averages of annual NTL sums, from 2012 to 2020 ($R_s = 0.84$, $p < 0.01$), based on N = 97.

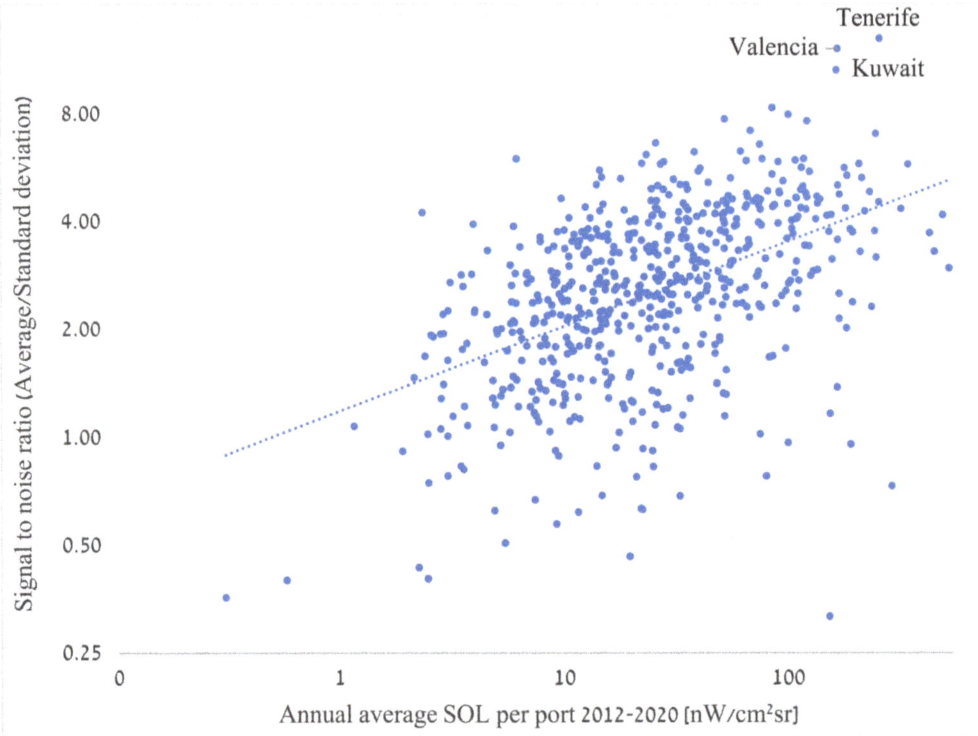

Figure 8. Scatter plot of all anchorage areas' average SOL and the signal to noise ratio (SNR = Average/Standard deviation) for the period of 2012–2020. (R_s = 0.49, $p < 0.01$), based on N = 601.

3.2. *Temporal Analysis of NTL Values within Anchorage Areas*

Figure 9 and Table 3 present the results of a correlation analysis between the NTL values within the port of Santos, Brazil and different statistical parameters of the port and the number of ships counted from Sentinel 1 images. The purpose of this section is to assess the feasibility of using the monthly average of NTL values as an indicator of the port shipping activity by estimating the number of ships in an in anchorage area on a monthly basis. The port of Santos is a relatively large port with a large anchorage area in which, during the study period, the average daily number of ships in the anchorage area was 60 with a standard deviation of 16 ships per month. The average monthly number of ships waiting for more than 72 h was 110 with a standard deviation of 23.

Over most of the study period, the five parameters in Figure 9 show statistically significant correlations, with corresponding peaks and lows. In the first half of 2019, the NTL values had a local peak which was not present in the number of ships as counted from Sentinel 1; however, this peak was found in the variables of total exports and ships waiting for more than 72 h.

VIIRS monthly sum of light values were moderately correlated with the number of ships counted from Sentinel 1 images (R_s = 0.51), the number of ships carrying bulk solids (R_s = 0.41), and the number of ships that waited for more than 72 h in the anchorage area (R_s = 0.41) (Table 3). Strong correlations ($Rs \geq 0.72$; Table 3) were found between all combinations of the following pairs of variables: the number of ships waiting, the number of ships waiting for more than 72 h (being on average 29% of all ships waiting), bulk solids ships (being on average 26% of all ships), and exports. Moreover, the number of bulk solids ships correlated with several other parameters: ships counted from Sentinel 1 ($R_s = 0.5$) as well as with exports ($R_s = 0.87$), while for the imports, a weaker correlation was found ($R_s = 0.39$). We also observed good correlations between the total monthly number of ships (based on the port's official statistics) and monthly exports ($R_s = 0.74$), the monthly number of bulk solid ships ($R_s = 0.72$), the number of ships counted from Sentinel 1 ($R_s = 0.48$), and with the monthly number of general cargo ships ($R_s = 0.47$) (Table 3).

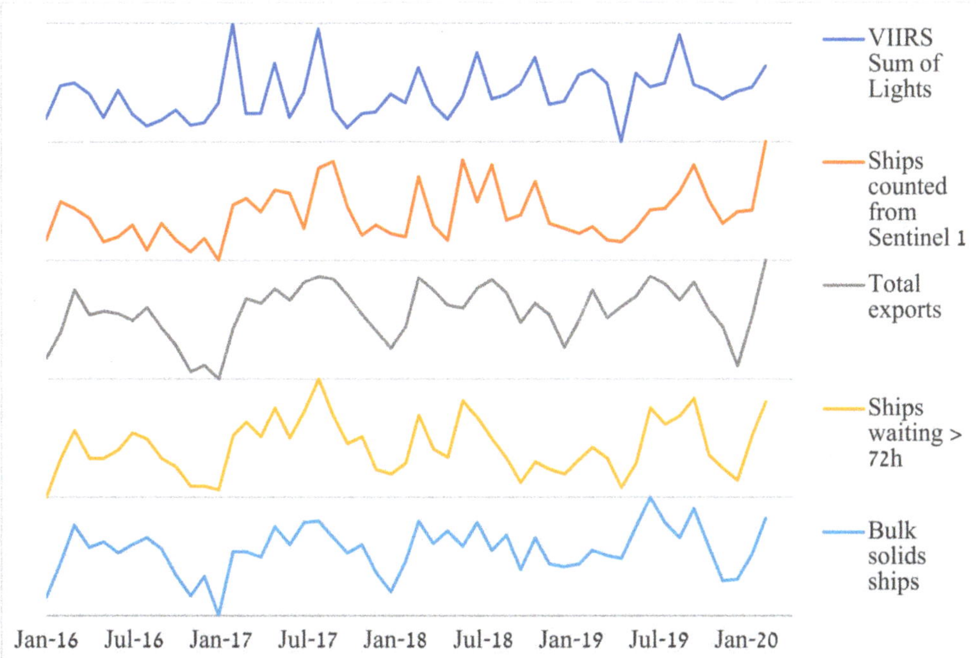

Figure 9. Time series presenting the correspondence between the average number of vessels counted from Sentinel 1, total exported goods, number of vessels waiting for more than 72 h in anchorage area, number of ships carrying bulk solids and average NTL over the anchorage area of Santos port, Brazil. All variables were normalized between their respective minimum and maximum values from Jan 2016 to Mar 2020 to ease the visual comparison between them.

Extending the temporal analysis to all anchorage areas, Figure 10 show a global map of temporal trends in NTL within anchorage areas for the period 2012–2020. In 219 (36%) of the anchorage areas, temporal trends were not significant (p-value > 0.05), with a relatively even distribution across continents. Of the 382 anchorage areas in which we found statistically significant results, in 112 (18%) of the anchorage areas, there was a decrease in NTL, while in 270 of the anchorage areas, there was an increase in NTL values. In the Mediterranean region (Figure 10a), the temporal trend of the NTL value of anchorage areas showed an increase (avg. $R_s = 0.31$, $p < 0.01$, Stdv = 0.38) based on 67 observations, while only 13 anchorage areas recorded a decrease. The NTL values growth in the Mediterranean region was mainly due to Turkey's anchorage areas, where 12 anchorage areas had a statistically significant and strong temporary trend of increase (avg. $R_s = 0.6$, $p < 0.01$). On the African continent, a significant difference in temporal trends was obtained between the Eastern part (avg. $R_s = 0.34$, $p < 0.01$, Stdv = 0.33) and the Western part (avg. $R_s = 0.8$, $p < 0.01$, Stdv = 0.53). The anchorage area of Mogadishu (Somalia) presented the highest rate of increase in NTL during the study period 2012–2020 with ($R_s = 0.94$, $p < 0.01$), probably reflecting its slow recovery from the long civil war. On the South American continent, on average, there was a downward trend in NTL with ($R_s = -0.18$, $p < 0.01$, Stdv = 0.41). In Northern America, NTL values increased with an average value of $R_s = 0.27$, $p < 0.01$, Stdv = 0.27. The NTL values in the Persian Gulf (Figure 10b) increased similarly to those in North America ($R_s = 0.25$, $p < 0.01$) but with a wider deviation between anchorage areas (Stdv = 0.47), mainly due to a decrease in the Qatar anchorage areas. The NTL values in China (Figure 10c) increased on average by $R_s = 0.32$, $p < 0.01$ and Stdv = 0.32, with only Turkey surpassing China. In Australia, there was a relatively stable neutral trend in NTL values. Among the countries, the decrease in NTL values occurred mainly in the anchorage areas of Japan (nine anchorage areas), Brazil (seven), Chile (seven), India (seven), while in the ports of Venterminals and Guanta (Venezuala), Capetown (South Africa) the most significant decrease occurred within the investigated period with a negative correlation below than $R_s = -0.72$, $p < 0.01$. A significant increase in NTL values also occurred in the anchorage areas of the ports of Huanghua (China) $R_s = 0.91$, $p < 0.01$, Izmir (Turkey) $R_s = 0.88$, $p < 0.01$, Taman (Ukraine) $R_s = 0.86$, $p < 0.01$, Poti (Georgia) $R_s = 0.83$, $p < 0.01$, Lagos (Nigeria) $R_s = 0.82$, $p < 0.01$, and Basrah (Iran) $R_s = 0.78$, $p < 0.01$.

Table 3. Matrix of Spearmen correlation coefficients of monthly statistical parameters for the port of Santos (Brazil) and VIIRS monthly values for the period of January 2016–March 2020 (n = 51). Positive correlation coefficients greater than 0.4 are highlighted.

	Remote Sensing		**Official Statistics from the Port of Santos**										
	VIIRS monthly sum	Ships counted Sentinel 1	Exports	Imports	Ships waiting	Ships waiting > 72 h	General cargo ships	Bulk solid ships	Tankers	Passenger ships	Roll-on/roll-off ships	Others ships	Total number of ships
VIIRS monthly sum	1												
Ships counted Sentinel 1	0.51	1											
Exports	0.37	0.59	1										
Imports	0.24	0.34	0.42	1									
Ships waiting	0.17	0.44	0.82	0.45	1								
Ships waiting > 72 h	0.41	0.68	0.78	0.25	0.72	1							
General cargo ships	−0.20	0.03	0.23	0.20	0.54	0.06	1						
Bulk solid ships	0.41	0.50	0.87	0.39	0.85	0.80	0.18	1					
Tankers	−0.07	0.15	0.21	0.38	0.34	0.19	0.05	0.12	1				
Passenger ships	0.09	−0.12	−0.48	−0.29	−0.62	−0.38	−0.34	−0.54	−0.31	1			
Roll-on/roll-off ships	−0.19	−0.06	0.20	−0.15	0.23	0.26	−0.10	0.13	0.28	−0.36	1		
Others ships	0.03	0.03	−0.10	0.29	−0.11	−0.13	0.05	−0.14	0.05	0.12	−0.16	1	
Total number of ships	0.22	0.48	0.74	0.38	0.85	0.67	0.47	0.72	0.33	−0.22	0.10	−0.02	1

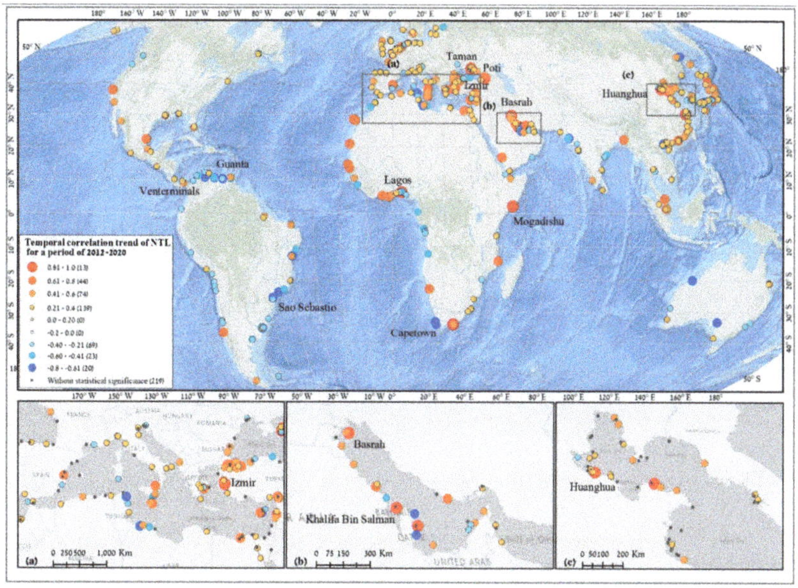

Figure 10. Global map of anchorage areas and their temporal trends over the 2012–2020 period. The temporal trends were assessed using the Spearmen correlation. The number in brackets represents the number of anchorage areas in each group of temporal trends. Enlarged maps show the following areas: (**a**) The Mediterranean Sea; (**b**) The Persian Gulf; (**c**) The Yellow Sea region.

3.3. Statistical Analysis at the Country–Level

At the country level, we found statistically significant correlations for 10 of the explanatory variables (Figure 11). The annual average values of NTL were very strongly correlated with the CPT ($R_s = 0.84$, $p < 0.01$), and strongly correlated with the country's population ($R_s = 0.68$, $p < 0.01$), maximum cargo capacity of the vessels ($R_s = 0.66$, $p < 0.01$), average import of the country ($R_s = 0.62$, $p < 0.01$), GDP (Rs = 0.61, $p < 0.01$), and LSCI ($R_s = 0.6$, $p < 0.01$). Moderate correlations were found for maximum container carrying capacity ($R_s = 0.55$, $p < 0.01$), maximum vessel size ($R_s = 0.51$, $p < 0.01$), and port calls $R_s = (0.42$, $p < 0.01)$.

Figure 12 provide a scatter plot of the correlation between CPT and NTL values at the country level. As shown in Figure 11, NTL data were most strongly and significantly correlated with "CPT" at the state level ($R_s = 0.84$, $p < 0.01$). China ranked first among the countries in terms of CPT (223,809,105 TEU), was four times higher than the USA CPT (52,716,134 TEU), and its average annual SOL was also four times higher (3876 and 943 TEU).

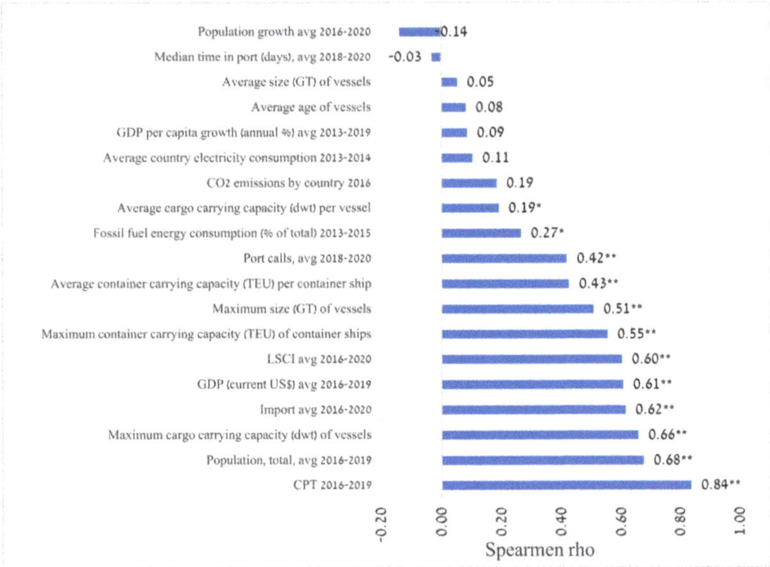

Figure 11. Spearmen correlation coefficients at the country level (* $p < 0.05$, ** $p < 0.01$; $n = 97$) for the response variable of average annual SOL values. Variables are ordered by the magnitude of their correlation coefficient with the SOL at the country level.

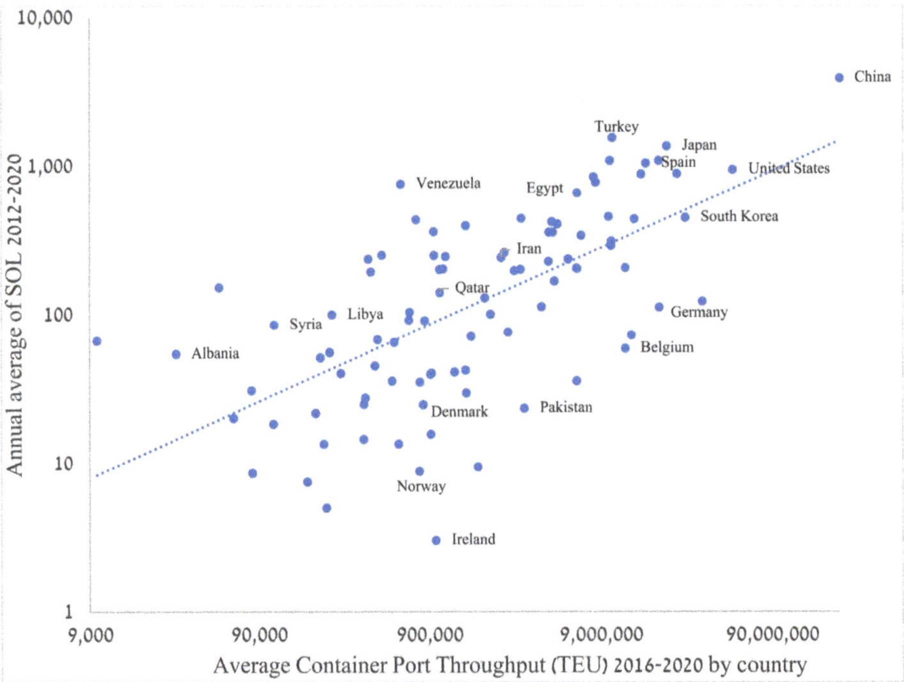

Figure 12. Scatter diagram of the correspondence ($R_s = 0.84$, $p < 0.01$) between the average Container Port Throughput (CPT) 2016–2020 by country and averages annual SOL for the period of 2012–2020, N = 97.

4. Discussion

To track global and local traffic flows and assess the state of development of the maritime industry, accurate data on port shipping activities is essential. Information of this nature is crucial for managers and analysts to make strategic decisions and monitor the maritime industry's progress towards its management goals.

Sources of artificial lights in the marine realm vary and include coastal cities, oil rigs, harbor lighting, fisheries, and anchoring ships [50]. The use of night–time remote sensing data for ship surveillance is a relatively new research field. Night light remote sensing technology has a unique night vision capability that can overcome the problem that traditional optical daytime remote sensing images cannot track targets at night. In existing studies, the VIIRS data on night illumination is widely used in the socio–economic sphere [28,51,52], the development of algorithms and applications for monitoring fishing vessels [19,53,54], assessment of the airport's throughput of traffic flow [51], assessment port economics scores [55], as well as assessment of countries' economic situation during the crisis [56]. Using VIIRS night light data, we showed that it is possible to assess shipping activities at the port level for which there is little or no consistent information at the global level. The NTL data on port anchorage areas are therefore a vital source of data for calibration and validation of, for example, the port collaborative decision making (PortCDM) concept [57], UNCTAD maritime and shipping parameters [33,58], and country economic assessments. Shipping activity in the water area near the port (anchorage area) proceeds without respite around the clock. To ensure normal operation at night–time, similarly to daytime, electricity is used on the ship, including for lighting the ship. This study is based on large data sets (601 anchorage areas) covering most of the anchorage areas of the world's major ports over the period from April 2012 to March 2020, thus providing a method for estimating both the average monthly number of ships in the anchorage area and various shipping and trade variables. Eight of the explanatory variables presented in Table 2 were significantly correlated ($R_s > 0.50$, $p < 0.01$) with the annual average (2012–2020) SOL values of VIIRS at the country level. The CPT index, which provides information on the number of unloaded and loaded containers by the port, was strongly correlated ($R_s = 0.84$, $p < 0.01$) with the NTL values. A possible reason might be that in most of the ports there is infrastructure for loading and unloading containers from ships as well as from container trucks (IMO, 2021). Moreover, based on other independent UNCTAD parameters (Table 2), a high correlation was obtained for the annual shipping parameters related to the aspects of "maximum" (instead of average): the maximum cargo carrying capacity (dwt) of vessels ($R_s = 0.66$, $p < 0.01$), the maximum container carrying capacity (TEU) of container ships ($R_s = 0.55$, $p < 0.01$), and the maximum size (GT) of vessels ($R_s = 0.51$, $p < 0.01$). A similar assessment of the use of night–time lighting data as an indicator has also been provided in studies of evaluating port economics comprehensive scores (PECS), based on UNCTAD, 1987 [59], with a correlation of ($R^2 > 0.85$) in the case of Shanghai port [55] and even an assessment of airport throughput (represent the annual number of aircraft movements or passengers [60]) with a significant correlation of ($R^2 > 0.85$) [51]. The domination of "Maximum metrics" and their higher correlation with NTL could be the result of port prioritization and commercial considerations by operating companies [61] in the entrance of large ships carrying high amounts of cargo [62]. Such prioritization leads to the long waiting time of smaller ships that wait for their queue to enter the port in the anchorage area [63,64]. Moreover, we also found significant correlations of NTL with countries' socio–economic parameters such as: country's average population 2016–2020 ($R_s = 0.68$, $p < 0.01$), average import 2016–2020 ($R_s = 0.62$, $p < 0.01$), and average GDP ($R_s = 0.61$, $p < 0.01$) [65], despite the relative decline in the importance of cities with large populations in global traffic [66]. Countries with large populations and high GDP often require high levels of imports and a large number of different ship segments that serve a high standard of living a large country's population. This conclusion is supported by the strong correlation of the LCSI parameter with NTL ($R_s = 0.60$, $p < 0.05$), which indicates the variability level of the country's integration into the global liner transportation networks [33,58]. The use of fossil fuels for energy ($R_s = 0.27$, $p < 0.05$) and CO_2 emission ($R_s = 0.19$) were not found to be strongly correlated with NTL, probably because the ships are not a significant factor [67] or are not counted in the overall balance of the country for each of these parameters. Thus, NTL data is an indicator of shipping that can be used to accurately estimate a wide range of activity (port, maritime, country, economy) parameters of ports and country, which is also confirmed by the results Liu (2019), where the researchers show that NTL data is a proxy indicator of economic assessment ports of China, despite high variability of port pixels and light interference around the port area [68]. The results obtained are of applied importance for assessing the dynamics of the development of seaports and the socio–economic parameters of the country, similar to how Bennet, 2017 [69] shows the correlation of NTL with socio–economic parameters on various spatio–temporal scales, thereby compensating for the lack of statistical data on the activities of ports and country parameters. On a local level, the observed negative temporal trend in the Venezuelan anchorage areas (five out of six anchorage areas having a temporal decrease of $R_s < -0.3$, and three with a decrease of $R_s < -0.6$) is confirmed by the results of a study conducted by Zhang et al. 2020 [56]; they assessed the economic crisis in Venezuela using NTL data for April 2012–December 2018, based on 12 cities, finding high correlations ($R^2 > 0.8$) between the sum of urban lights and several economic parameters (crude oil production, USD exchange rate and the number of asylum seekers), thus demonstrating the use of NTL data as an indicator of the economic state of Venezuela during the crisis. At the port level, results of statistical analysis showed lower correlations than at the country level, but the results of a detailed analysis we conducted for the Port of Santos, Brazil, (Table 3) shed light on some of the correspondence between NTL and the number of ships counted in Sentinel 1. In 2019, when we observed a discrepancy between the NTL value and the number of ships counted on Sentinel 1 images, only the parameter "number of ships waiting for more than 72 h" showed a similar trend. The bulk ship segment constitutes the majority of those waiting in the anchorage area of the Santos port ($R_s = 0.85$) and those that are over 72 h ($R_s = 0.67$), which are mainly dedicated to exports ($R_s = 0.87$). Hence, when there is a decrease in exports via bulk ships, fewer ships will be waiting in the anchoring area. Perhaps this is the reason for the discrepancy between the NTL value and the number of ships counted on Sentinel 1 flights in 2019. Although Sentinel 1 SAR data is not affected by cloud cover (whereas cloud cover hampers the detection of night lights by VIIRS), VIIRS acquires night–time imagery every night, whereas the revisit time of Sentinel 1 is lower (six days; [70]). Hence, the two sensors did not acquire their images at the same dates and time of day, which may explain some of the discrepancies between the data from these two sensors. Therefore, the proposed method using night lights as an indicator of ship activity is particularly suitable for assessing spatial and temporal trends in the maritime industry, complementing other methods of tracking ships (AIS, SAR images), especially where official statistics are not available.

A detailed analysis of the causes of temporal variations makes it possible to improve the sensitivity of DNB to changes in illumination [68]. In this study, we developed and implemented several correction methods for VIIRS data to better analyze the light emitted from anchorage areas, although VIIRS significantly improved quality over DMSP/OLS in terms of spatial resolution, dynamic range, quantization, calibrations, and spectral range availability over DMSP–OLS [25]. First, to minimize the influence of temporal variation of natural light such as airglow, we used the method proposed by Coesfeld, 2020 [22]. Thus, fluctuations in natural light sources that limit the ability of night light sensors to detect changes in small artificial light sources have been minimized, increasing the ability to analyze the light emitted by ships. Moreover, city lights also represent an important factor that influences scattering over the marine environment, as light from brightly lit coastal cities can reach considerable distances at sea (Figure 2) by scattering through the atmosphere [24]. As of today, there is no method of amendment developed to minimize the influence of city lights over adjacent coastal waters (to assess the amount of NTL emitted from the coastal waters themselves), and thus we focused on anchorage areas that were too close to the coast. Secondly, due to the high cloudiness in some areas of the earth, monthly data from VIIRS are often underestimated. To fill in the gaps, we applied an interpolation method for calculating the monthly values for the underestimated months based on the VIIRS values of 12 adjacent months. Results of the *t*-test showed that the interpolation method filled the data in cloudy months and did not affect the original data.

The use of VIIRS data has certain limitations. Among the limitations is the coarse spatial resolution of 742 m [25], which does not allow a more detailed analysis of small dense areas and objects. Cloudiness, which is in principle a frequent occurrence near coastal areas, and in some regions (e.g., the tropics) a frequent occurrence, does not allow information to be collected on a large percentage of the days of the year. Moreover, the coastal light emitted by cities and ports themselves is much stronger than the light emitted by ships, which in some cases makes this approach ineffective [71] if an anchorage area is located too close to a brightly lit city. Additional sources of variation in emissions of NTL from ships are associated with the types of ships in anchorage areas and their night lights used. In this study, we did not have data on ship types (e.g., oil tankers, cargo ships, etc.), which lowered our ability to explain the variability in NTL between anchorage areas. AIS data can be a suitable source for replenishing knowledge about the types of ships in future studies. In the case of comparison between the anchorage areas, there may be variations in the volumes and powers of lighting permitted by local port authorities. Moreover, this difference can also exist between the ports of the same country, and even the policy of the port about the use of night light can change over time, which can lead to temporal changes. The VIIRS sensor is panchromatic and does not measure night light in the blue channel, thereby losing the night light emitted by ships in blue wavelength (which is a significant component of light emitted by LED lightings) [72]. Finally, due to the wide coverage on the ground by a single image, VIIRS imagery is mostly not acquired at nadir [22,73], and changes in the zenith and azimuth view angles may affect the amount of light received by the sensor from the ships, as was documented for light emissions from cities [74].

The ability of NTL data to serve as a proxy for shipping activity also depends on the number of anchorage points and their density, the type of ships entering the port, and waiting times in the anchorage area. The world's standards for the construction of ships as well as the rules on movement at sea, require ships to be equipped with a variety of lights, most of which are standardized [75,76]. Consequently, more sensitive sensors will enable measurement and distinguish the NTL values for different types of shipping segments [54]. For example, fishing vessels are the most illuminated ships at night–time. Depending on the type of fishing, fishery vessels have different types and directions of night–time lights, both for fishing and for working onboard [77]. Generally, fishing is prohibited in the harbor area and at most anchorages; therefore, their night–time light should not affect the light emitted from anchorage areas. Moreover, in ports with long queues, a ship that waits for more than one night will increase its night signature compared with ships or anchorage areas with short anchorage time (without anchorage during night–time hours). All these factors affect the ability of satellite sensors to capture night light produced by ships and to be used as an indicator for shipping activity. For example, the ports of Fujairah and Malta are mostly hosting tankers [78]. Their anchorage areas are located a few kilometers from the city coast and the light they emit, which makes it possible to measure mainly the light emitted by one sector of the maritime industry, while such an assessment is almost impossible inside the port due to the strong light emitted by the port infrastructure [68]. Thus, for remote ports that are not exposed to coastal lights, it is maybe easier to use night light values as a proxy for shipping activities, considering additional parameters such as ship size, anchorage point density, etc. In ports with a small number of anchorage points and with a small number of vessels in the anchorage area, the proposed method of using night lights may be limited in its ability to assess shipping activities, as not enough lights will be emitted that can be captured by the VIIRS sensor, as also found for other small scale economic activities [79].

5. Conclusions

In this study, we carried out a global assessment of shipping activity using the VIIRS satellite data by measuring the night lights emitted from ships at anchorage areas. The analysis was carried out at three geographic levels: overtime at the anchorage level using the example of the port of Santos (Brazil), overall anchorage areas at the country level (n = 97), and across all anchorage areas globally (n = 601).

The main conclusion of this study is that monthly/annual VIIRS data can serve as a good proxy for estimating the number of vessels as well as various shipping metrics (such as CPT, LSCI) in anchorage areas at the port and country levels. The estimation of the number of ships in anchorage areas with a small number of ships is probably limited due to the low energy of night light emitted by a small number of ships, and in such cases, VIIRS data cannot be used as an indicator. VIIRS NTL data can be implemented in a wide global range of studies of shipping and for assessing the economic development of ports and country parameters. Moreover, this method allows analyzing shipping, ports, and countries parameters, for which we have obtained a significant correlation with NTL data, for example, container port throughput. As a result, we conclude that NTL data can be used as an indicator for a wide range of assessments of ports, countries, and the shipping industry in general, and is applicable for ports and countries that do not share information, as well as for tracking spatial and temporal trends.

Finally, the results should be useful to international maritime organizations, governments, policy–makers, and stakeholders in formulating effective strategies for developing tools to assess shipping activities in the anchorage area and their use in overall port operations.

Author Contributions: Conceptualization S.P., N.L. and R.B.; methodology S.P., N.L. and R.B.; software S.P. and N.L.; validation S.P. and N.L.; formal analysis S.P. and N.L.; investigation S.P., N.L. and R.B.; resources S.P. and N.L.; data curation S.P. and N.L.; writing—original draft preparation S.P.; writing—review and editing S.P., N.L. and R.B.; visualization S.P. and N.L.; supervision N.L. and R.B.; project administration S.P., N.L. and R.B.; funding acquisition S.P. All authors have read and agreed to the published version of the manuscript.

Funding: S.P. is grateful to the Department of Marine Geoscience, the Maritime Policy and Strategy Research Center (HMS), and the Wydra Institute of Shipping and Aviation Research of the University of Haifa in Israel for their financial support.

Conflicts of Interest: The authors declare no conflict of interest.

References

1. UNCTAD. *2019 Review of Maritime Transport*; UNCTAD: Geneva, Switzerland, 2019; ISBN 9789211129588.
2. March, D.; Metcalfe, K.; Tintoré, J.; Godley, B. Tracking the global reduction of marine traffic during the COVID-19 pandemic. *Nat. Commun.* **2020**, *12*, 2415. [CrossRef] [PubMed]
3. UNCTAD. *Reveiw of Maritime Transport 2018*; UNCTAD: Geneva, Switzerland, 2018; ISBN 9789211129281.
4. Guerrero, D.; Rodrigue, J.P. The waves of containerization: Shifts in global maritime transportation. *J. Transp. Geogr.* **2014**, *34*, 151–164. [CrossRef]
5. Hanson, S.E.; Nicholls, R.J. Demand for Ports to 2050: Climate Policy, Growing Trade and the Impacts of Sea-Level Rise. *Earth's Future* **2020**, *8*, e2020EF001543. [CrossRef]
6. Liu, A.; Wei, Y.; Yu, B.; Song, W. Estimation of cargo handling capacity of coastal ports in China based on panel model and DMSP-OLS nighttime light data. *Remote Sens.* **2019**, *11*, 582. [CrossRef]
7. Eskafi, M.; Kowsari, M.; Dastgheib, A.; Ulfarsson, G.F.; Taneja, P.; Thorarinsdottir, R.I. Mutual information analysis of the factors influencing port throughput. *Marit. Bus. Rev.* **2020**, *6*, 129–146. [CrossRef]
8. Milenković, M.; Milosavljevic, N.; Bojović, N.; Val, S. Container flow forecasting through neural networks based on metaheuristics. *Oper. Res.* **2021**, *21*, 965–997. [CrossRef]
9. UNCTAD. *Review of Maritime Transport 2020*; UNCTAD: Geneva, Switzerland, 2020.
10. Stefan, G. The Lloyd's List-A Global Intelligence Unit? In *Transfer between Sea and Land*; Sidestone Press: Leiden, The Netherlands, 2018; pp. 119–130.
11. Mazzarella, F.; Vespe, M.; Mazzarella, F.; Vespe, M.; Damalas, D.; Osio, G. Discovering vessel activities at sea using AIS data: Mapping of fishing footprints Discovering Vessel Activities at Sea using AIS Data: Mapping of Fishing Footprints. In Proceedings of the 17th International Conference on Information Fusion (FUSION), Salamanca, Spain, 7–10 July 2014; pp. 1–7.

12. Polinov, S.; Bookman, R.; Levin, N. Spatial and temporal assessment of oil spills in the Mediterranean Sea. *Mar. Pollut. Bull.* **2021**, *167*, 112338. [CrossRef]
13. Halpern, B.S.; Lbridge, S.; Selkoe, K.A.; Kappel, C.V.; Micheli, F.; D'Agrosa, C.; Bruno, J.F.; Casey, K.S.; Ebert, C.; Fox, H.E.; et al. A global map of human impact on marine ecosystems. *Science* **2008**, *319*, 948–952. [CrossRef]
14. Deter, J.; Lozupone, X.; Inacio, A.; Boissery, P.; Holon, F. Boat anchoring pressure on coastal seabed: Quantification and bias estimation using AIS data. *Mar. Pollut. Bull.* **2017**, *123*, 175–181. [CrossRef]
15. Levin, N.; Phinn, S. Illuminating the capabilities of Landsat 8 for mapping night lights. *Remote Sens. Environ.* **2016**, *182*, 27–38. [CrossRef]
16. Kang, M.; Ji, K.; Leng, X.; Lin, Z. Contextual region-based convolutional neural network with multilayer fusion for SAR ship detection. *Remote Sens.* **2017**, *9*, 860. [CrossRef]
17. Tello, M.; López-Martínez, C.; Mallorquí, J.J.; Greidanus, H. A novel algorithm for ship detection in ENVISAT SAR imagery based on the wavelet transform. In Proceedings of the 2004 ENVISAT and ERS Symposium, Salzburg, Austria, 6–10 September 2004; pp. 1557–1562.
18. Wackerman, C.C.; Friedman, K.S.; Pichel, W.G.; Clemente-Colon, P.; Li, X. Automatic detection of ships in RADARSAT-1 SAR imagery. *Can. J. Remote Sens.* **2001**, *27*, 568–577. [CrossRef]
19. Elvidge, C.D.; Zhizhin, M.; Baugh, K.; Hsu, F.C. Automatic boat identification system for VIIRS low light imaging data. *Remote Sens.* **2015**, *7*, 3020–3036. [CrossRef]
20. Hsu, F.C.; Elvidge, C.D.; Baugh, K.; Zhizhin, M.; Ghosh, T.; Kroodsma, D.; Susanto, A.; Budy, W.; Riyanto, M.; Nurzeha, R.; et al. Cross-matching VIIRS boat detections with vessel monitoring system tracks in Indonesia. *Remote Sens.* **2019**, *11*, 995. [CrossRef]
21. Ges, X.; Bará, S.; García-Gil, M.; Zamorano, J.; Ribas, S.J.; Masana, E. Light pollution offshore: Zenithal sky glow measurements in the mediterranean coastal waters. *J. Quant. Spectrosc. Radiat. Transf.* **2018**, *210*, 91–100. [CrossRef]
22. Coesfeld, J.; Kuester, T.; Kuechly, H.U.; Kyba, C.C.M. Reducing variability and removing natural light from nighttime satellite imagery: A case study using the VIIRS DNB. *Sensors* **2020**, *20*, 3287. [CrossRef] [PubMed]
23. Marine Cadaster Anchorage Areas. Available online: https://www.marinecadastre.gov/news/load.php?url=posts/anchorage-areas.html (accessed on 23 September 2021).
24. Sanchez de Miguel, A.; Kyba, C.C.M.; Zamorano, J.; Gallego, J.; Gaston, K.J. The nature of the diffuse light near cities detected in nighttime satellite imagery. *Sci. Rep.* **2020**, *10*, 7829. [CrossRef]
25. Elvidge, C.D.; Baugh, K.E.; Zhizhin, M.; Hsu, F.-C. Why VIIRS data are superior to DMSP for mapping nighttime lights. *Proc. Asia-Pac. Adv. Netw.* **2013**, *35*, 62. [CrossRef]
26. Elvidge, C.D.; Baugh, K.; Zhizhin, M.; Hsu, F.C.; Ghosh, T. VIIRS night-time lights. *Int. J. Remote Sens.* **2017**, *38*, 5860–5879. [CrossRef]
27. Zhao, X.; Li, D.; Li, X.; Zhao, L.; Wu, C. Spatial and seasonal patterns of night-time lights in global ocean derived from VIIRS DNB images. *Int. J. Remote Sens.* **2018**, *39*, 8151–8181. [CrossRef]
28. Levin, N.; Zhang, Q. A global analysis of factors controlling VIIRS nighttime light levels from densely populated areas. *Remote Sens. Environ.* **2017**, *190*, 366–382. [CrossRef]
29. Gorelick, N.; Hancher, M.; Dixon, M.; Ilyushchenko, S.; Thau, D.; Moore, R. Google Earth Engine: Planetary-scale geospatial analysis for everyone. *Remote Sens. Environ.* **2017**, *202*, 18–27. [CrossRef]
30. Pesquera, M.Á.; Ruiz, J.R. *Sustainable Development Strategies for Cities and Ports*; UNCTAD: Geneva, Switzerland, 1996.
31. UNCTAD. Port Call and Performance Statistics: Time Spent in Ports, Vessel Age and Size, Annual. Available online: https://unctadstat.unctad.org/wds/TableViewer/tableView.aspx?ReportId=194889 (accessed on 21 November 2021).
32. Ding, D.; Teo, C.P. World container port throughput follows lognormal distribution. *Marit. Policy Manag.* **2010**, *37*, 401–426. [CrossRef]
33. Fugazza, M.; Hoffmann, J. Liner shipping connectivity as determinant of trade. *J. Shipp. Trade* **2017**, *2*, 1. [CrossRef]
34. UNCTAD. *Review of Maritime Transport 2016*; UNCTAD: Geneva, Switzerland, 2016; ISBN 9789211128789.
35. GFW Global Fishing Watch | Data Download Portal. Available online: https://globalfishingwatch.org/data-download/datasets/public-anchorages:v20200316 (accessed on 24 August 2021).
36. The World Bank CO_2 Emissions. Available online: https://data.worldbank.org/indicator/EN.ATM.CO2E.PC?view=chart (accessed on 23 November 2021).
37. UNCTAD. Stat Annual Container Port Throughput. Available online: https://unctadstat.unctad.org/wds/TableViewer/tableView.aspx?ReportId=13321 (accessed on 23 November 2021).
38. The World Bank. Electric Power Consumption (kWh per Capita). Available online: https://data.worldbank.org/indicator/EG.USE.ELEC.KH.PC?view=chart (accessed on 23 November 2021).
39. The World Bank. Fossil Fuel Energy Consumption (% of Total). Available online: https://data.worldbank.org/indicator/EG.USE.COMM.FO.ZS?view=chart (accessed on 23 November 2021).
40. The World Bank. GDP per Capita. Available online: https://data.worldbank.org/indicator/NY.GDP.MKTP.CD (accessed on 23 November 2021).
41. The World Bank. GDP, Annual Growth. Available online: https://data.worldbank.org/indicator/NY.GDP.PCAP.KD.ZG?view=chart (accessed on 23 November 2021).

42. Trade Map Country Import. Available online: https://www.trademap.org/Country_SelProduct_TS.aspx?nvpm=1%7C%7C%7C%7C%7C%7CTOTAL%7C%7C%7C2%7C1%7C1%7C1%7C2%7C1%7C2%7C1%7C1%7C1 (accessed on 23 November 2021).
43. The World Bank. Population Growth (Annual %). Available online: https://data.worldbank.org/indicator/SP.POP.GROW (accessed on 23 November 2021).
44. The World Bank. Country Population, Total. Available online: https://data.worldbank.org/indicator/SP.POP.TOTL (accessed on 23 November 2021).
45. Santos Port Authority. Santos Port Authority Mensário Estatístico. Available online: http://www.portodesantos.com.br/informacoes-operacionais/estatisticas/mensario-estatistico/ (accessed on 1 February 2022).
46. Grover, A.; Kumar, S.; Kumar, A. Ship Detection Using Sentinel-1 SAR Data. In *Proceedings of the ISPRS Annals of the Photogrammetry, Remote Sensing and Spatial Information Sciences*; Copernicus GmbH: Göttingen, Germany, 2018; Volume 4, pp. 317–324.
47. Elvidge, C.D.; Baugh, K.E.; Kihn, E.A.; Kroehl, H.W.; Davis, E.R.; Davis, C.W. Relation between satellite observed visible-near infrared emissions, population, economic activity and electric power consumption. *Int. J. Remote Sens.* **1997**, *18*, 1373–1379. [CrossRef]
48. Henderson, J.V.; Storeygard, A.; Weil, D.N. Measuring economic growth from outer space. *Am. Econ. Rev.* **2012**, *102*, 994–1028. [CrossRef]
49. Kyba, C.C.M.; Garz, S.; Kuechly, H.; de Miguel, A.S.; Zamorano, J.; Fischer, J.; Hölker, F. High-resolution imagery of earth at night: New sources, opportunities and challenges. *Remote Sens.* **2015**, *7*, 1–23. [CrossRef]
50. Davies, T.W.; Duffy, J.P.; Bennie, J.; Gaston, K.J. The nature, extent, and ecological implications of marine light pollution. *Front. Ecol. Environ.* **2014**, *12*, 347–355. [CrossRef]
51. Ma, M.; Ge, W.; Shi, K. Airport's Throughput Estimation Using Nighttime Light Data in China Mainland. *IEEE Geosci. Remote Sens. Lett.* **2020**, *18*, 1357–1360. [CrossRef]
52. Mellander, C.; Lobo, J.; Stolarick, K.; Matheson, Z. Night-time light data: A good proxy measure for economic activity? *PLoS ONE* **2015**, *10*, e0139779. [CrossRef] [PubMed]
53. Li, H.; Liu, Y.; Sun, C.; Dong, Y.; Zhang, S. Satellite Observation of the Marine Light-Fishing and Its Dynamics in the South China Sea. *J. Mar. Sci. Eng.* **2021**, *9*, 1394. [CrossRef]
54. Zhong, L.; Liu, X.; Yang, P.; Lin, R. Explore the application of high-resolution nighttime light remote sensing images in nighttime marine ship detection: A case study of LJ1-01 data. *Open Geosci.* **2020**, *12*, 1169–1184. [CrossRef]
55. Li, C.; Chen, G.; Luo, J.; Li, S.; Ye, J. Port economics comprehensive scores for major cities in the Yangtze Valley, China using the DMSP-OLS night-time light imagery. *Int. J. Remote Sens.* **2017**, *38*, 6007–6029. [CrossRef]
56. Zhang, L.; Li, X.; Chen, F. Spatiotemporal Analysis of Venezuela's Nighttime Light during the Socioeconomic Crisis. *IEEE J. Sel. Top. Appl. Earth Obs. Remote Sens.* **2020**, *13*, 2396–2408. [CrossRef]
57. Lind, M.; Ward, R.; Bergmann, M.; Haraldson, S.; Zerem, A.; Hoffmann, J.; Eklund, E. Maritime Informatics for Increased Collaboration. In *Maritime Informatics for Increased Collaboration*; Springer: Cham, Switzerland, 2021; pp. 113–136.
58. Hoffmann, J.; Wilmsmeier, G.; Lun, Y.H.V. Connecting the world through global shipping networks. *J. Shipp. Trade* **2017**, *2*, 2. [CrossRef]
59. UNCTAD. *Manual on a Uniform System of Port Statitstics and Performance Indicators*; UNCTAD: Geneva, Switzerland; p. 1987.
60. Jacquillat, A.; Odoni, A.R. A roadmap toward airport demand and capacity management. *Transp. Res. Part A Policy Pract.* **2018**, *114*, 168–185. [CrossRef]
61. Poulsen, R.T.; Sampson, H. 'Swinging on the anchor': The difficulties in achieving greenhouse gas abatement in shipping via virtual arrival. *Transp. Res. Part D Transp. Environ.* **2019**, *73*, 230–244. [CrossRef]
62. IMO. *Ship-Port Interface Guide Practical Measures to Reduce GHG Emissions*; International Maritime Organization: London, UK, 2021.
63. Shahpanah, A.; Hashemi, A.; Nouredin, G.; Zahraee, S.M.; Helmi, S.A. Reduction of ship waiting time at port container terminal through enhancement of the tug/pilot machine operation. *J. Teknol. Sci. Eng.* **2014**, *68*, 63–66. [CrossRef]
64. Stergiopoulos, G.; Valvis, E.; Mitrodimas, D.; Lekkas, D.; Gritzalis, D. Analyzing Congestion Interdependencies of Ports and Container Ship Routes in the Maritime Network Infrastructure. *IEEE Access* **2018**, *6*, 63823–63832. [CrossRef]
65. Ducruet, C.; Cuyala, S.; El Hosni, A. Maritime networks as systems of cities: The long-term interdependencies between global shipping flows and urban development (1890–2010). *J. Transp. Geogr.* **2018**, *66*, 340–355. [CrossRef]
66. Ducruet, C.; Cuyala, S.; EL Hosni, A. The changing influence of city-systems on global shipping networks: An empirical analysis. *J. Shipp. Trade* **2016**, *1*, 4. [CrossRef]
67. Wang, R.; Tao, S.; Ciais, P.; Shen, H.Z.; Huang, Y.; Chen, H.; Shen, G.F.; Wang, B.; Li, W.; Zhang, Y.Y.; et al. High-resolution mapping of combustion processes and implications for CO_2 emissions. *Atmos. Chem. Phys.* **2013**, *13*, 5189–5203. [CrossRef]
68. Coesfeld, J.; Anderson, S.J.; Baugh, K.; Elvidge, C.D.; Schernthanner, H.; Kyba, C.C.M. Variation of individual location radiance in VIIRS DNB monthly composite images. *Remote Sens.* **2018**, *10*, 1964. [CrossRef]
69. Bennett, M.M.; Smith, L.C. Advances in using multitemporal night-time lights satellite imagery to detect, estimate, and monitor socioeconomic dynamics. *Remote Sens. Environ.* **2017**, *192*, 176–197. [CrossRef]
70. Torres, R.; Snoeij, P.; Geudtner, D.; Bibby, D.; Davidson, M.; Attema, E.; Potin, P.; Rommen, B.Ö.; Floury, N.; Brown, M.; et al. GMES Sentinel-1 mission. *Remote Sens. Environ.* **2012**, *120*, 9–24. [CrossRef]

71. Geronimo, R.C.; Franklin, E.C.; Brainard, R.E.; Elvidge, C.D.; Santos, M.D.; Venegas, R.; Mora, C. Mapping fishing activities and suitable fishing grounds using nighttime satellite images and maximum entropy modelling. *Remote Sens.* **2018**, *10*, 1604. [CrossRef]
72. Levin, N.; Kyba, C.C.M.; Zhang, Q.; Sánchez de Miguel, A.; Román, M.O.; Li, X.; Portnov, B.A.; Molthan, A.L.; Jechow, A.; Miller, S.D.; et al. Remote sensing of night lights: A review and an outlook for the future. *Remote Sens. Environ.* **2020**, *237*, 111443. [CrossRef]
73. Mills, S.; Weiss, S.; Liang, C. VIIRS day/night band (DNB) stray light characterization and correction. In Proceedings of the Earth Observing Systems XVIII, San Diego, CA, USA, 23 September 2013; Volume 8866, pp. 549–566.
74. Li, X.; Ma, R.; Zhang, Q.; Li, D.; Liu, S.; He, T.; Zhao, L. Anisotropic characteristic of artificial light at night–Systematic investigation with VIIRS DNB multi-temporal observations. *Remote Sens. Environ.* **2019**, *233*, 111357. [CrossRef]
75. Bandara, D.; Woodward, M.; Chin, C.; Jiang, D. Augmented reality lights for compromised visibility navigation. *J. Mar. Sci. Eng.* **2020**, *8*, 1014. [CrossRef]
76. Han, G.-Y. Ship Design Rules and Regulations-An overview of major themes. *Rogue Waves Work.* **2004**, *2004*, 20–22.
77. Nguyen, K.Q.; Winger, P.D. Artificial Light in Commercial Industrialized Fishing Applications: A Review. *Rev. Fish. Sci. Aquac.* **2019**, *27*, 106–126. [CrossRef]
78. Peng, P.; Yang, Y.; Cheng, S.; Lu, F.; Yuan, Z. Hub-and-spoke structure: Characterizing the global crude oil transport network with mass vessel trajectories. *Energy* **2019**, *168*, 966–974. [CrossRef]
79. Kyba, C.C.M.; Giuliani, G.; Franziskakis, F.; Tockner, K.; Lacroix, P. Artisanal and Small-Scale Mining Sites in the Democratic Republic of the Congo Are Not Associated with Nighttime Light Emissions. *J.* **2019**, *2*, 152–161. [CrossRef]

remote sensing

MDPI

Article

Spatio-Temporal Dynamics and Driving Forces of Multi-Scale CO_2 Emissions by Integrating DMSP-OLS and NPP-VIIRS Data: A Case Study in Beijing-Tianjin-Hebei, China

Shiyu Xia [1,2], Huaiyong Shao [1,2,*], Hao Wang [3], Wei Xian [4], Qiufang Shao [5], Ziqiang Yin [6] and Jiaguo Qi [7]

1 College of Earth Sciences, Chengdu University of Technology, Chengdu 610059, China
2 Key Laboratory of Earth Exploration and Information Technology, Ministry of Education, Chengdu 610059, China
3 Piesat Information Technology Company Limited, Beijing 100195, China
4 College of Resources and Environment, Chengdu University of Information Technology, Chengdu 610225, China
5 Teaching Steering Committee, Sichuan Tourism University, Chengdu 610100, China
6 College of Ecology and Environment, Chengdu University of Technology, Chengdu 610059, China
7 Center for Global Change and Earth Observations, Michigan State University, East Lansing, MI 48824, USA
* Correspondence: shaohuaiyong@cdut.edu.cn

Abstract: The emission of greenhouse gases, especially CO_2, is the main factor causing global warming. Due to incomplete statistical data on energy consumption at and below the urban scale of Beijing-Tianjin-Hebei (BTH), in this study, Defense Meteorological Satellite Program Operational Linescan System (DMSP-OLS) and Suomi National Polar-orbiting Partnership Visible Infrared Imaging Radiometer Suite (NPP-VIIRS) data were combined, and a neural network model and weighted average method based on DN (Digital Number) value were used to obtain CO_2 emissions at the municipal and county scales with a resolution of 1 km $\times$ 1 km from 2000–2019. Next, a spatial-temporal analysis model and spatial econometric model were used to study the CO_2 emissions at different scales of BTH. This study also solved the problem that STIRPAT analysis cannot be carried out due to insufficient urban statistical CO_2 emissions data. The results show that the energy CO_2 emissions in BTH present a distribution pattern of "East greater than West", with a trend of first rising and then slowing down. Moreover, the rapid growth areas are mainly located in Chengde and Tianjin. The degree of regional spatial aggregation decreased year by year from 2000–2019. Population, affluence and technology factors were positively correlated with CO_2 emissions in Tianjin and Hebei. For Beijing, in addition to foreign investment, factors such as urbanization rate, energy intensity, construction and transportation factors all contributed to the increase in CO_2 emissions. Among them, the growth of population is the main reason for the increase of CO_2 at the urban scale in BTH. Finally, based on the research results and the specific situation of the cities, corresponding policies and measures are proposed for the future low-carbon development of the cities.

Keywords: CO_2 emission; integration of two nighttime light datasets; spatial-temporal dynamics; STIRPAT; policy

Citation: Xia, S.; Shao, H.; Wang, H.; Xian, W.; Shao, Q.; Yin, Z.; Qi, J. Spatio-Temporal Dynamics and Driving Forces of Multi-Scale CO_2 Emissions by Integrating DMSP-OLS and NPP-VIIRS Data: A Case Study in Beijing-Tianjin-Hebei, China. *Remote Sens.* **2022**, *14*, 4799. https://doi.org/10.3390/rs14194799

Academic Editor: Anthony Brazel

Received: 9 August 2022
Accepted: 21 September 2022
Published: 26 September 2022

1. Introduction

In recent years, global warming has become an indisputable fact. Greenhouse gas emissions, especially CO_2 emissions, have been proven to be one of the main causes of global warming [1]. However, for densely populated urban areas, CO_2 from the burning of fossil fuels from human activities is the main source of global CO_2 emissions [2]. Urban regions utilized 75% of the world's energy and emitted 80% of global greenhouse gas emissions, even though they cover less than 3% of the Earth's surface [3]. Over the past century, CO_2 emissions have significantly increased due to the ongoing rise in energy consumption brought on by the process of global economic development [4]. China, a significant developing nation, since 2006, has ranked first in the world for CO_2 emissions [5]. China's CO_2 emissions have increased significantly due to the country's increasing industrialization and urbanization. According to the energy statistics provided by the International Energy Agency (IEA), China accounted for 28.13% of global CO_2 emissions in 2015, exceeding the combined emissions of the United States and the European Union [6]. China is facing serious environmental problems and pressure to deal with climate change. Thus, understanding the spatial and temporal distribution of CO_2 emissions at different administrative levels is crucial for China's future development [7].

In September 2020, China proposed the "dual carbon" goal. That is, China will work to peak its CO_2 emissions before 2030 and to become carbon neutral before 2060 [8]. In order to achieve this goal, China has clearly put forward the main implementation paths and policy proposals in many aspects. Due to the vast territory and uneven economic development level, the BTH region, as the key area of the Bohai economic circle, is the most energy-intensive region in China. Despite accounting for only 2.3% of China's total land area, the BTH contributed almost 15% of China's CO_2 emissions in 2018 [9]. As a result, the region must transition to low-carbon development to meet China's carbon reduction targets. However, there are huge differences amongst cities in terms of economic and social growth. Therefore, the analysis of CO_2 distribution and driving forces in different scale regions can contribute significantly to the formulation of urban development strategies and the future planning of BTH.

A detailed spatio-temporal analysis of carbon emissions can provide a basis for CO_2 emission strategies in different regions. Many previous studies have assessed energy consumption in different countries and regions at different scales [10–12]. However, most of the studies on CO_2 emissions in the BTH region concentrated on the municipalities (Beijing, Tianjin) and provinces (Hebei), and most of them were based on statistics from the national and provincial energy administrations, which only provided a digital record of CO_2 emissions by administrative units, but did not reveal their spatial distribution characteristics. According to numerous earlier studies, nighttime lighting data is a valuable indicator for assessing human activities and social economy, such as estimating regional populations [13,14], urban expansion [15], economic development [16], power consumption [17], gas combustion [18], etc. Based on DMSP-OLS data, many scholars have conducted regression analyses on the nighttime stable light (NSL) and CO_2 emission index, and confirmed that nighttime lighting data at the national, provincial and municipal levels and CO_2 emissions are significantly correlated [19,20]. Therefore, using nighttime light data opens a new perspective for dynamic monitoring of energy consumption. The studies of Zhang [20] and Zhao [21] showed that two types of nighttime light data both have benefits at different scales, and their combination can be used to investigate the spatial patterns of CO_2 emissions in long time series. Lv [22] integrated and analyzed the spatio-temporal dynamics of CO_2 emissions in China by using two nighttime lighting datasets, and her research results proved that DMSP-OLS and NPP-VIIRS data could be effectively combined to study CO_2 emissions. Meng [23] used linear equations to fit night light data and CO_2 data based on DMSP-OLS data. Considering that there is a nonlinear relationship between CO_2 emission and light value, conventional econometric methods may produce large errors [24]. The majority of the studies mentioned above were conducted on large-scale regions, such as national and provincial levels, lacking detailed studies on city and county scales, and are based on a single satellite. It is more necessary to meet the high accuracy of fitting the CO_2 emission and nighttime light values for the investigation of small-scale areas.

Previously conducted research on the influencing factors of CO_2 emissions primarily included logarithmic average weight, the Divisia decomposition method, the LMDI decomposition method, IPAT theory and the STIRPAT model, the factor decomposition method, Kuznets curve theory and the spatial econometric model and other methods and theories [25–28]. The IPAT model is widely used as a classical model to study CO_2 emissions [29,30]. Since this model cannot make assumptions on variables, York [31] proposed an improved model, the STIRPAT model, based on the IPAT model, which removed the influence of the same proportion change. The model comprehensively considers the impact of population, affluence and technology on the environment, and is widely used in various fields. But they are mostly used at the national and provincial levels, and few studies have been conducted at the city level, where data on CO_2 emissions are lacking. Therefore, in order to overcome the drawback that STIRPAT cannot be utilized to examine urban carbon emissions due to a lack of data, this study combines the two kinds of nighttime lighting data (DMSP-OLS and NPP-VIIRS) to get municipal carbon emission data.

The remainder of this paper is organized as follows. Section 2 describes the study area and data sources. Section 3 outlines the integration of DMSP-OLS with NPP-VIIRS data, as well as the estimation of CO_2 emissions at the city and county scales and the assessment of the characteristics of spatial and temporal changes in carbon emissions and the analysis of driving forces. Section 4 analyzes the main results obtained in this paper. Section 5 discusses the results. Finally, Section 6 lists the main conclusions and puts forward some strategies for the future low-carbon development of BTH.

2. Study Areas and Data Sources

The study area includes Beijing, Tianjin and the Hebei Province, which are located in the North China Plain, accounting for 2.27% of the total land area of China. The area is the center of Chinese politics, culture, economy, science and technology, and it is also the area where Chinese steel production is most concentrated. The analysis process involved 13 cities, 199 counties and other administrative units (Figure 1).

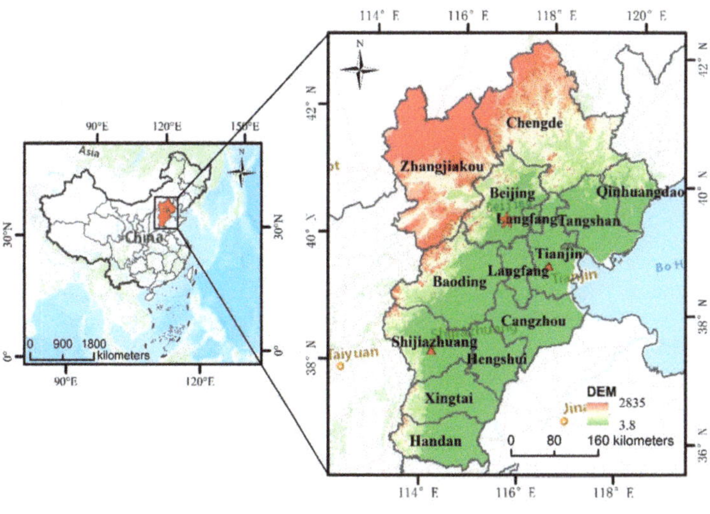

Figure 1. Geographical location of Beijing–Tianjin–Hebei.

DMSP-OLS nighttime lighting data was obtained from the National Oceanic and Atmospheric Administration's National Geophysical Data Center (NOAA/NGDC) website (https://ngdc.noaa.gov/eog/dmsp/downloadV4composites.html (accessed on 1 March 2022)). The product spans from 1992–2013, and the image is generated by six separate sensors, these being F10 (1992–1994), F12 (1994–1999), F14 (1997–2003), F15 (2000–2007), F16 (2004–2009), and F18 (2010–2013). Four sensor data, F14, F15, F16 and F18, were used in this study. The illumination area is large, but there is an obvious saturation problem, and the DN value ranges from 0–63. Annual data from version four of DMSP-OLS from 2000–2013 was used in this study. NPP-VIIRS, which was launched in 2011, is an Earth observation satellite NPP equipped with a visible/infrared radiation imager (VIIRS) which obtains remote sensing images of ground light at night in the DNB (Day/Night

Band) band. This study selected monthly data from 2012–2019, obtained from the website (https://eogdata.mines.edu/products/vnl/ (accessed on 5 March 2022)).

All statistical data were obtained from the China Energy Statistical Yearbook, the China Urban Statistical Yearbook and local statistical yearbooks, the Beijing Statistical Yearbook, the Tianjin Statistical Yearbook, and the Hebei Statistical Yearbook.

3. Methodology

3.1. Integration of Two Kinds of Night Light Data

The integration of the two kinds of night light data is mainly divided into three steps (Figure 2):

1. DMSP-OLS data were corrected by sensor, within year and between years;
2. Annual synthesis and denoising of NPP-VIIRS data;
3. Using a model to integrate the former two datasets to obtain the stable nighttime lighting data from 2000–2019.

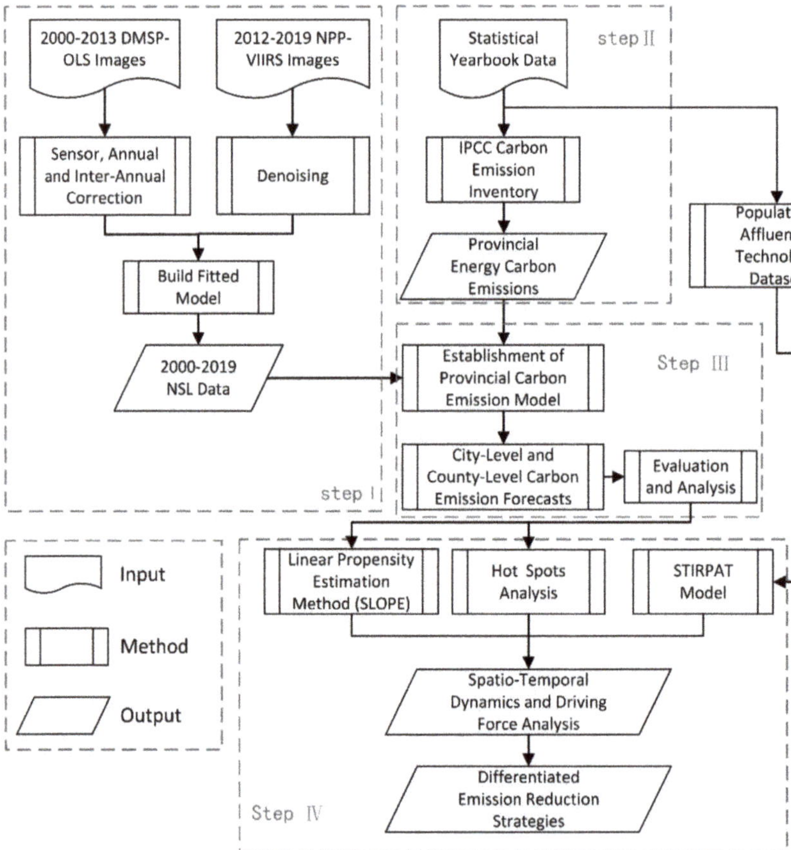

Figure 2. Flowchart of Beijing–Tianjin–Hebei carbon emission technology.

Because NPP-VIIRS are the monthly average data and are polluted by stray light, the light data from May to July in the middle and high latitude areas of summer in China are seriously distorted, and the pixel value is 0. Therefore, the missing months were eliminated and the other months were synthesized to obtain the annual comprehensive data. The images and DMSP-OLS images were then resampled to 1 km × 1 km and projected onto the Albers equal area projection. The invariant region method was selected to carry out sensor correction, annual correction and inter-annual correction for DMSP-OLS data, and then the continuity was improved. For the specific method, we referred to the study of Elvidge [18]. After viewing the distribution of the data, it was found that approximately 99% of the pixel values were low (DN < 0.3). In order to remove the weak light in the NPP-VIIRS data (usually regarded as noise), the threshold method was used to correct it, and 0.3×10^{-9} W·cm^{-2}·sr^{-1} was used as the empirical threshold, while values less than 0.3 in the NPP-VIIRS data was assigned as 0 [32]. Because there were abnormally high light values in NPP-VIIRS images, the method of Zhao [33] was adopted in this paper. The pixel values higher than those of China's three megacities (Beijing, Shanghai and Guangzhou) (DN_{MAX}) in other regions were considered as outliers caused by oil, gas or fire. Each outlier was iterated to the maximum value in its respective 3 × 3 neighborhood raster DN until no value exceeded DN_{MAX}.

Because of the dispersion of the two groups of data, it is difficult to fit the data directly, so the coefficient of variation (*cv*) method was used to deal with this problem [34]. The NPP-VIIRS data in 2012 was only available for six months, so the *cv* values of all pixel 3 × 3 windows in NPP-VIIRS and DMSP-OLS data of 2013 were chosen to be calculated. The *cv* value is calculated using the following formula:

$$cv = \frac{S}{x} \tag{1}$$

where S and x are the standard deviation and average value of the image radiation value, respectively.

Subsequently, the NPP-VIIRS data were log-transformed to reduce the difference between the two datasets and make the brightness range consistent with the DMSP-OLS data.

$$Log\_V_i = Ln(V_i + 1) \tag{2}$$

where V_i is the original radiation value of NPP-VIIRS, and Log_V_i is the result after logarithmic transformation. In order to make the logarithmic transformation valid, we added the constant 1 to the equation.

According to previous studies, the two groups of *cv* values show an S-shape after fitting (Figure 3), and ordinary function curves are difficult to fit accurately. In this study, the S-type function model BiDoseResp was adopted to integrate the two kinds of night lights. Meanwhile, in order to ensure its accurate application in the Beijing-Tianjin-Hebei region, unary linear regression, binary linear regression and the DoseResp function model were used for comparison. Finally, the results showed that the BiDoseResp model worked best among the four models. The order of R^2 was BiDoseResp (0.8836) > DoseResp (0.8833) > binomial (0.8378) > unary linear (0.8359), and the RSS of the BiDoseResp model was the smallest, which was 3.4687×10^5. Its model parameters are shown in Table 1. The BiDoseResp formula can be expressed as follows:

$$y = A1 + (A2 - A1)\left[\frac{p}{1 + 10^{(LOGx01 - x)h1}} + \frac{1 - p}{1 + 10^{(LOGx02 - x)h2}}\right] \tag{3}$$

where $A1$ and $A2$ are the minimum and maximum of the function, respectively, p represents the weight of Logistic, $LOGx01$ and $LOGx02$ respectively represent the average value of the independent variables of the two curves, and $h1$ and $h2$ represent the curve slope 1 and slope 2, respectively. Finally, the obtained NPP-VIIRS data was processed by Gaussian low-pass filtering to obtain a smooth image.

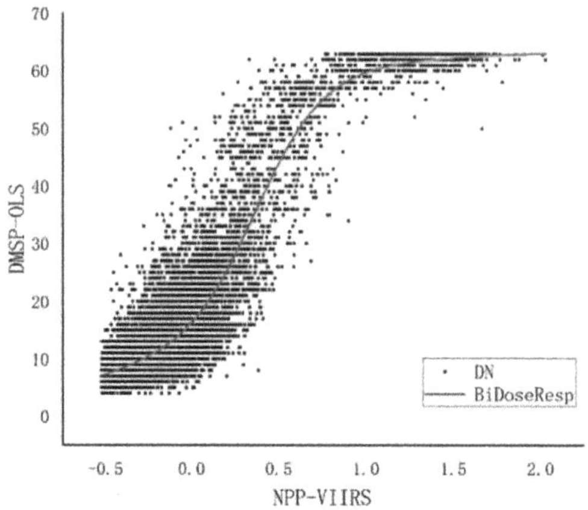

Figure 3. Image of the fitted Bidoseresp model in 2013.

Table 1. BiDoseResp model coefficients.

*A*1	*A*2	*LOGx*01	*LOGx*02	*h*1	*h*2	*P*
−35.4269	64.19793	0.36666	−1.98149	2.24954	0.41862	0.47464

3.2. Estimating CO_2 Emissions According to IPCC

The United Nations' Intergovernmental Panel on Climate Change (IPCC) provided the internationally agreed standard for assessing emissions of the greenhouse gas CO_2. This study used the IPCC method to calculate the CO_2 emissions of each province and city from 2000–2019. The formula is as follows:

$$Y_{CO_2} = \frac{44}{12} \times \sum_{i=1}^{n} K_i E_i \quad (4)$$

where i is the type of energy, n represents the quantity of energy, E_i represents the consumption of energy type i (calculated by standard coal), and K_i represents the carbon emission coefficient of energy i, the original data unit is J, and in order to keep the data unit consistent, it is converted into standard coal (SCE); the conversion coefficient is: 1×10^4 t standard coal is equal to 2.93×10^5 GJ. 44/12 is the amount of carbon dioxide produced by the complete combustion of one ton of carbon in oxygen. Table 2 lists the conversion of carbon emission coefficient and standard coal of different fossil fuels. These energy sources account for about 85% of China's total emissions.

Table 2. Fossil energy emission factors.

Energy Type	SCE Conversion Factor (tSCE/t)	CO_2 Emission Factor (t/SCE)
Raw coal	0.7143	0.7559
Coke	0.9714	0.855
Crude oil	1.4286	0.5857
Gasoline	1.4714	0.5538
Kerosene	1.4714	0.5714
Diesel fuel	1.4571	0.5921
Fuel oil	1.4286	0.6185
Natural gas	13.3	0.4483
Electricity	1.229	0.272

Note: The conversion of natural gas into standard coal is standard coal/10,000 m^3, and the conversion of electricity into standard coal is standard coal/10,000 kWh.

3.3. Estimating CO_2 Emissions at Municipal and County Levels

Regression modeling can be used to determine the association between CO_2 emissions and NSL. Due to the nonlinear relationship between CO_2 emissions and NSL, the traditional methods may lead to relatively high errors, especially for Beijing. Artificial neural networks can effectively solve this problem. It is a kind of machine learning, which is a nonlinear, dynamic and complex information processing system connected by a large number of neurons. Bp-neural networks are a kind of multi-layer feedforward neural network which is trained according to the error back propagation method (Figure 4), which generally includes three layers: the input layer, the hidden layer and the output layer, and each layer has multiple nodes. Therefore, we chose the BP-neural network method to fit NSL and CO_2 emission data. $x = [x_1, x_2, \ldots, x_n]^T$ was selected as the DN value of NSL, $y = [y_1, y_2, \ldots, y_m]^T$ was selected as the CO_2 emission data as the output value, and W_{ij} was the weight parameter. Considering the accuracy of the BP-neural network, 70% of the samples were used as the training set, and 30% of the samples were used as the test set and validation set. Since the types of economic development in Hebei Province, Beijing and Tianjin are quite different, Hebei Province was chosen to be fitted separately.

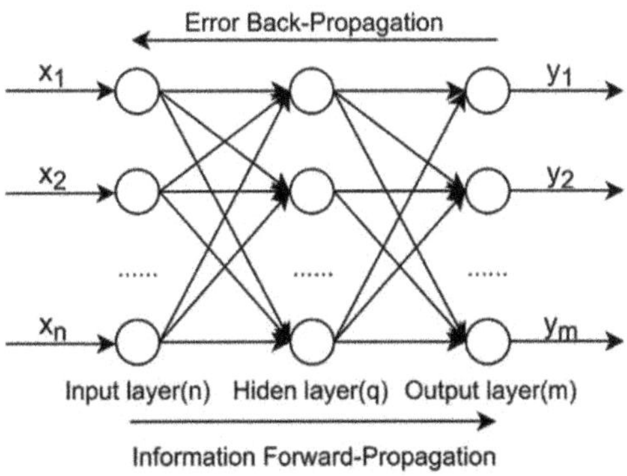

Figure 4. Three-layer BP neural network model.

The relationship between input and output during forward propagation is as follows:

$$y_j = \sum_{i=1}^{m} \omega_{ij} x_i + b_j \tag{5}$$

$$x_j = f(y_j) = \frac{1}{1 + e^{-y_j}} \tag{6}$$

The gradient descent method is adopted for reverse error propagation, and the loss function is:

$$E = \frac{1}{2} \sum_{j=1}^{p} (x_j - y_j)^2 \tag{7}$$

The revised weight value and threshold calculation formula are:

$$W_{ij} = w_{ij} + \Delta w_{ij} \tag{8}$$

$$B_j = b_j + \Delta b_j \tag{9}$$

$$\Delta w_{ij} = -\eta_1 \frac{\partial E}{\partial w_{ij}} \tag{10}$$

$$\Delta b_j = -\eta_2 \frac{\partial E}{\partial b_j} \tag{11}$$

where x_i is the input value of node i, x_j is the output value of node j, y_j is the net output value of node j, ω_{ij} represents the weight between nodes i and j, and b_j represents the threshold of node j, f is the transfer function of the node, p is the number of iterations; W_{ij} and B_j are the weight correction amount and the threshold between the i node and the j node, respectively.

In order to obtain CO_2 emission data at the city, district and county level, the weighted average method based on DN value was used in this study. The formula is as follows:

$$C_i = DN_i \times \frac{y_m}{\sum_{i=1}^{n} DN_i} \tag{12}$$

where i stands for different cities (i= 1,2,3...n), m represents the province where city i is located, C_i represents the CO_2 emissions of city i, y_m represents the CO_2 emissions of province m.

3.4. *Analysis of Spatio-Temporal Pattern*

3.4.1. Linear Propensity Estimation (Slope)

The linear propensity estimation method was used to obtain the propensity values of each county in different years from 2000–2019 so as to analyze the changing trends and types of carbon emissions. With the change of time, energy emissions will show a significant increase, a significant decline, or no obvious change. The area with a tendency value less than 0 is classified as having a downward trend, while the area with a tendency value greater than 0 is classified as having one of four growth speeds: slow, medium, faster, and rapid. It is calculated as follows:

$$SLOPE = \frac{n \times \sum_{i=1}^{n} x_i C_i - \sum_{i=1}^{n} x_i \sum_{i=1}^{n} C_i}{n \times \sum_{i=1}^{n} C_i^2 - (\sum_{i=1}^{n} C_i)^2} \tag{13}$$

where n is the total number of years, which is 20; C_i is the CO_2 emissions in the year i, x_i is the year i.

3.4.2. Global Autocorrelation

The global Moran's I index can judge the spatial distribution pattern of carbon emissions in Beijing-Tianjin-Hebei by reflecting the similarity or correlation degree of carbon emissions in the spatially adjacent regions. The specific formula is as follows:

$$I = \frac{n \sum_{i=1}^{n} \sum_{j=1}^{n} w_{ij}(x_i - \overline{x})(x_j - \overline{x})}{\sum_{i=1}^{n} \sum_{j=1}^{n} w_{ij} \sum_{i=1}^{n} (x_i - \overline{x})^2} \tag{14}$$

where I is the global Moran's I index, x_i and x_j represent the CO_2 emission values of county i and county j, n is the total number of units in the districts and counties, and w_{ij} is the standardized spatial weight matrix.

When the Moran index is greater than 0, it indicates that there is a positive correlation between regions; otherwise, there is a negative correlation. The Z-value size can judge the significance level; if $|Z| \leq 1.96$ and $p \geq 0.05$, it means that there is no correlation in the region; if $|Z| \geq 1.96, p < 0.05$, it means that there is significant spatial autocorrelation in the region.

3.4.3. Hot Spots Analysis

The "Getis-Ord G*" indicator can be used to determine the spatially clustered locations of carbon emissions, which obey the standard normal distribution. It identifies statistically significant hot and cold spots using local General G index statistics given a set of weighted elements. When the "Getis-Ord G*" statistic passes the significance test and is positive, it means that carbon emissions are clustered in space, that is, hot spots; otherwise, it means discrete, that is, cold spots. The "Getis-Ord G*" can be expressed as follows:

$$G_i^* = \frac{\sum_{j=1}^{n} W_{i,j}x_j - \overline{X}\sum_{j=1}^{n} w_{i,j}}{s\sqrt{\left[n\sum_{j=1}^{n} w_{i,j}{}^2 - \left(\sum_{j=1}^{n} w_{i,j}\right)^2\right]/(n-1)}} \quad (15)$$

where G_i^* is the Z-score, $W_{i,j}$ represents the spatial weight between space units i and j (adjacent is 1, non-adjacent is 0); n is the number of space units, x_j is the carbon emission value of the space unit; $\overline{X}$ represents the Mean, S the standard deviation.

3.5. Spatial Econometrics Models

The modified STIRPAT stochastic extensible model based on IPAT proposed by York [31] was selected. Because of its flexibility, the model has been widely used in various fields [35,36].

CO_2 emission from energy consumption can reflect regional environmental pressure, so STIRPAT can better explore the driving factors of CO_2 emission. The formula is as follows:

$$I = aP^bA^cT^de \quad (16)$$

where I is an environmental factor; P is the demographic factor, A is the wealth factor, and T is the technology factor; a is a constant term; e is an error; and b, c, and d are indices that need to be evaluated. Logarithm transformation was performed on both sides of the model (16) at the same time, and it is expressed as:

$$(LnI)_i = Lna_i + b_i(LnP)_i + c_i(LnA)_i + d_i(LnT)_i + Lne_i \quad (17)$$

where i is the city i. The model was then extended based on the actual situation and existing data. b_i, c_i, and d_i are elastic coefficients, which means that when P, A, and T change by 1%, b_i%, c_i%, and d_i% of I will change respectively. Table 3 shows the selected influencing factors. Since data on energy consumption per unit GDP and coal share were not available for cities in Hebei Province, these two terms were removed from the driving force analysis.

Table 3. STIRPAT model variable selection.

Variable	Factor		Symbol	Indicator	Unit
dependent variable	Environmental	CO_2 emissions	CE	Urban carbon emissions	10^4 tons
independent variable	population factor	total population	p	Year-end total population	10^4 people
		Urbanization rate	UR	The proportion of urban population to total population	%
	Wealth factor	GDP per capita	PP	GDP per capita	Yuan
		Foreign investment	FAI	foreign investment	Ten thousand dollars
	technical factors	Secondary industry	TI2	Proportion of added value of secondary industry in GDP	%
		Industrial added value	LAV	Annual industrial added value	billion

4. Results

4.1. NSL Mutual Correction Results

The radiation characteristics of a certain area in Beijing-Tianjin-Hebei (including rural areas, urban areas and suburbs) in the nighttime light data and Google Earth image were selected for cross-sectional examination, as shown in Figure 5. It can be seen that the distribution of BDR (BiDoseResp) NPP (adjusted NPP-VIIRS) data is consistent with that of

DMSP-OLS data. Figure 6 displays the nighttime stable light data (NSL) of the time series from 2000–2019 after the final continuous correction.

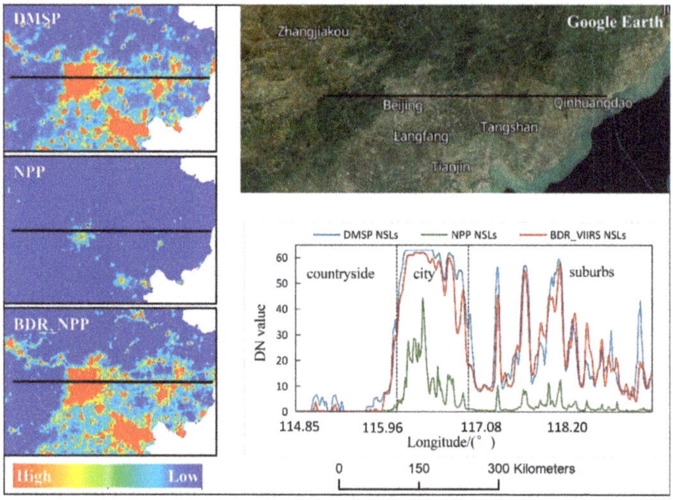

Figure 5. Profile distribution of radiation value of light images at different nights in 2013.

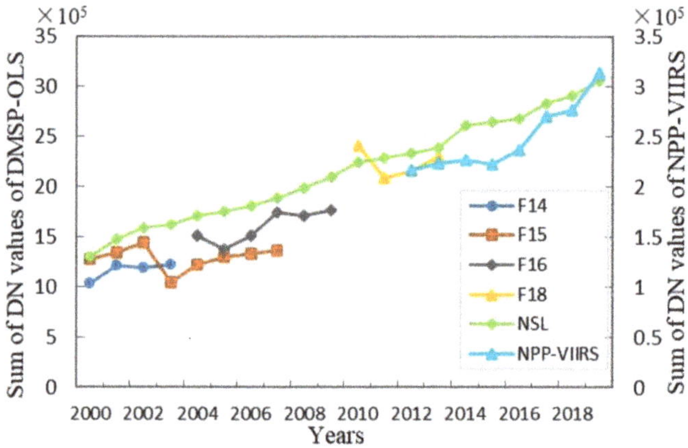

Figure 6. Night light values before and after correction.

4.2. Accuracy Evaluation of CO_2 Emission Estimation

The bp-neural network was used as a tool to fit the relationship between CO_2 emissions and nighttime lighting data. The predicted value was then compared with the actual statistical value, and other functions were also used for comparison in order to verify the accuracy. Table 4 shows the fitting results of different functional models on the DN value and CO_2 emissions. In comparison to other functional models, the BP-neural network was superior in every way. RMSE is the root mean square error, which measures the error rate of the regression model. The measurement's accuracy is indicated by the average relative error (RE), which is the difference between the absolute error and the real value. It can be seen that the accuracy of fitting night light and CO_2 emissions by the BP-neural network was good, so the model can be used to fit CO_2 emissions in BTH.

Table 4. Comparison of different function models.

	Model	R^2	RMSE	RE
Beijing	Linear	0.5600	1186.443	0.0753
	quadratic fit	0.6129	1387.492	0.0902
	Exponential function	0.5231	2094.901	0.1294
	Power function	0.5934	1169.759	0.0754
	bp-neural network	0.9799	260.061	0.0158
Tianjin	Linear	0.8674	1741.121	0.0793
	quadratic fit	0.9327	2799.401	0.1281
	Exponential function	0.7828	2503.933	0.1081
	Power function	0.8483	1846.445	0.0768
	bp-neural network	0.9978	314.669	0.0146
Hebei	Linear	0.8497	9870.159	0.1334
	quadratic fit	0.9574	10281.500	0.1341
	Exponential function	0.7357	14073.120	0.1645
	Power function	0.8179	11682.530	0.1306
	bp-neural network	0.9963	2455.139	0.0269

4.3. Spatio-Temporal Dynamics of CO_2 Emissions

4.3.1. Temporal Variations

The total CO_2 emissions in BTH changed significantly, increasing from 525.26 million tons in 2000 to 1381.01 million tons in 2019, with an average growth rate of 4.952%. The upward trend from 2000–2012 gradually stabilized after 2012 and showed a downward trend in 2014, 2016 and 2017 (Figure 7). The CO_2 emissions of Hebei Province were basically consistent with the total CO_2 emission trend of Beijing-Tianjin-Hebei, from 61.5% in 2000 to 75.31% in 2017, and the growth rate has slowed since 2012.

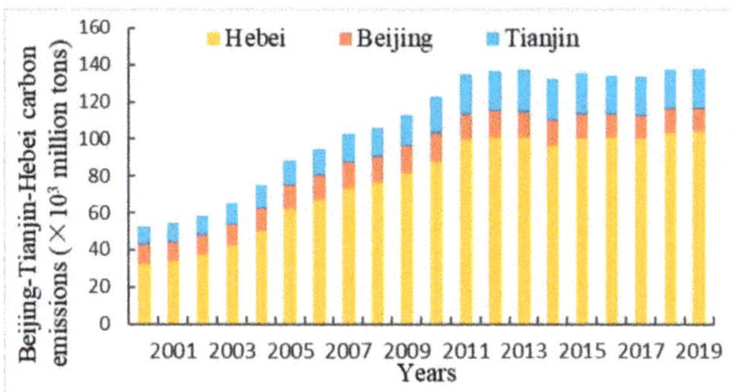

Figure 7. Beijing–Tianjin–Hebei carbon emissions from 2000–2019.

At the municipal level, as shown in Figure 8, the CO_2 emissions of Tianjin have been significantly higher than those of other cities since 2010. Since 2004, the CO_2 emissions of Tangshan have steadily ranked first in Hebei Province, followed by Baoding. At the county level, the faster developing areas in the eastern portion of the Beijing-Tianjin-Hebei region were where CO_2 emissions were highest (Figure 9). Over the past 20 years, the CO_2 emissions in BTH cities have changed significantly. Since 2000, Tianjin's Binhai New Area has consistently ranked first among all districts in terms of CO_2 emissions.

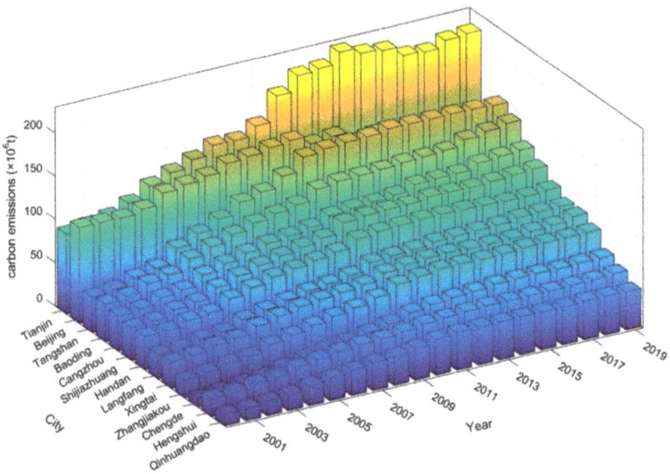

Figure 8. Beijing–Tianjin–Hebei municipal-level energy CO_2 emissions from 2000–2019.

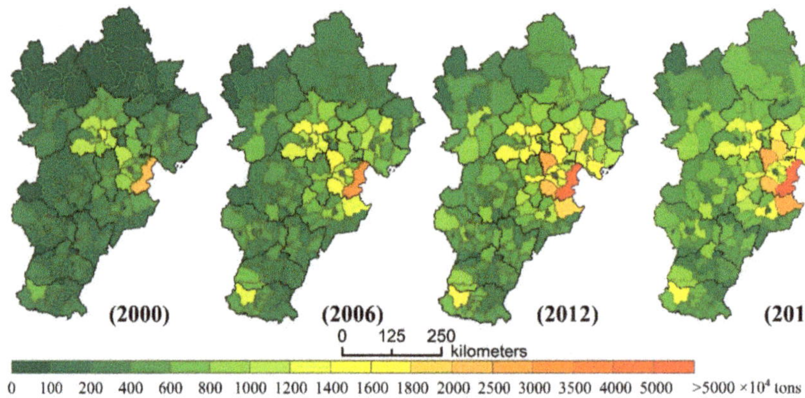

Figure 9. County-level CO_2 emissions in the Beijing–Tianjin–Hebei region in 2000, 2006, 2012 and 2019.

As shown in Figure 10, according to the county-level SLOPE results, the southern BTH region had the majority of the regions with significant tendencies. According to the classification standard, there were seven counties with a downtrend, concentrated in the center of Beijing and Tianjin, accounting for 0.89% of the total area; 96 slow growth areas; 69 medium growth areas; 24 faster growth areas; and three rapid growth areas, accounting for 71.79%, 22.95%, 4.28% and 0.084% of the total area, respectively. The places with faster growth were primarily spread out southeast of Beijing-Tianjin-Hebei, and the rapid growth areas were distributed in the Yingshouyingzi mining area of Chengde, the Hongqiao District, and the Heping District of Tianjin.

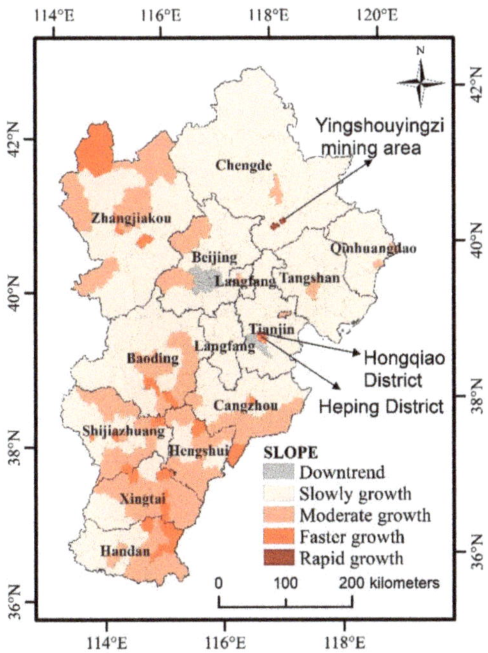

Figure 10. 2000–2019 Beijing–Tianjin–Hebei energy CO_2 emission trends.

4.3.2. Spatial Variations

The global Moran's I index was calculated in 2000, 2006, 2012 and 2019, which was 0.4501, 0.4023, 0.3938 and 0.3763 respectively ($p < 0.01$). This indicated that there was a positive correlation between county-level energy CO_2 emissions distribution in BTH during the past 20 years, and the correlation was gradually weakening. Over a 20-year period, CO_2 emissions became spatially dispersed.

Figure 11 shows that the southern and western counties of Hebei Province are the areas where the cold spot region mainly gathers. In 2019, the area of cold spot areas decreased, and the reduced cold spot areas were all converted into sub-cold spot areas. Beijing, Tianjin and surrounding areas were the main hot spots, and with the passage of time, the scope of hot spots gradually decreased and turned into sub-hot spots. The transformation of hot spots in Beijing was the most obvious, from the hot spots (16) in 2000 to the sub-hot spots (12) and sub-cold spots (4) in 2019, which indicates that the carbon reduction measures in Beijing have been very effective in recent years. Tianjin, as the economic center around the Bohai Sea, was also an international transportation hub, and has always remained in the hot spot.

4.4. Analysis of Driving Force

First, a Pearson analysis was performed on the data to remove insignificant factors and determine the influencing factors to be used. The least squares method was then used for regression to calculate the variance inflation factor (VIF) of the respective variables. The results demonstrated that most variables' VIF values were above 10, which confirmed the existence of multicollinearity. To eliminate the effects of multicollinearity, the data was re-regressed using ridge regression analysis with biased estimation. It is an improved method based on the least square method. In order to obtain more realistic and reliable regression coefficients, it loses part of the information and reduces the accuracy [37]. When K = 0, it is the ordinary least squares estimate.

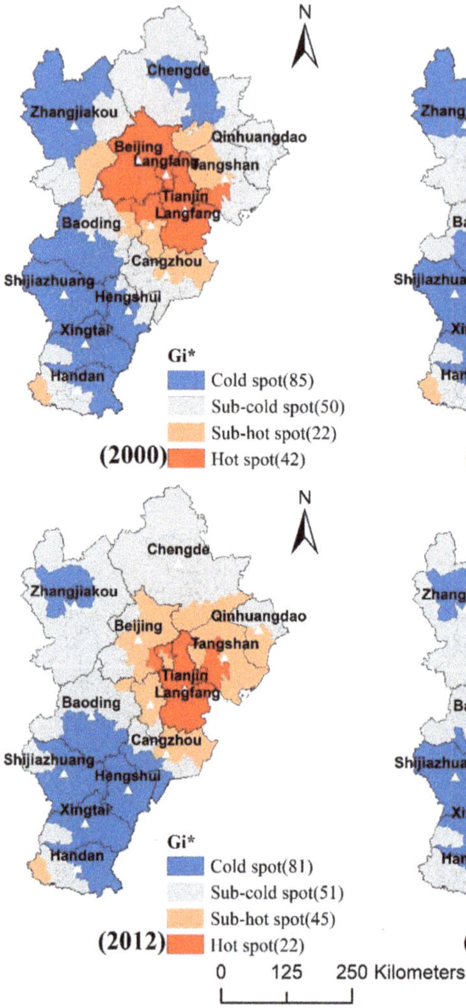

Figure 11. Aggregation of energy CO_2 emissions in Beijing-Tianjin-Hebei in 2000, 2006, 2012 and 2019.

Because energy CO_2 emissions were normally distributed in Beijing from 2000–2019, only population and wealth factors were significantly correlated with energy CO_2 emissions in the Pearson test. To explore the causes of energy CO_2 emissions in Beijing, it was necessary to decompose the technical factors. Therefore, on the original basis, three independent variables of building, transportation and energy intensity were added. The formula can be expressed as:

$$LnI = Lna + bLnP + cLnA + dLnT + mLnU + nLnR + tLnE + Lne \quad (18)$$

where U is the construction factor (CI), expressed by the number of construction enterprises; R is the transportation factor (PC), expressed by the number of private cars (10,000 vehicles); E is the energy intensity (GC), expressed by the unit GDP carbon dioxide consumption Quantity representation. Ridge regression analysis with biased estimation was conducted on the STIRPAT model, and the final regression results of Beijing are shown in Table 5, and the results of other cities are shown in Table 6.

Table 5. Ridge Regression Estimation Results of Beijing.

Variables	Ln(UR)	Ln(FAI)	Ln(GC)	Ln(CI)	Ln(PC)	Constant	Sig F	R-Squared	K
coefficient	1.090 *** (0.258)	−0.029 * (0.016)	0.037 ** (0.013)	0.246 *** (0.044)	0.025 *** (0.007)	2.938 * (1.069)	0.0000	0.858	0.15

Note: "()" is the standard error; ***, **, * indicate significance at the 1%, 5% and 10% levels, respectively.

Table 6. Ridge regression estimation results of Tianjin-Hebei.

	Tianjin	Shijiazhuang	Tangshan	Qinhuangdao	Handan	Xingtai
Ln(P)	0.249 *** (0.038)		1.711 *** (0.529)	1.941 *** (0.218)	0.323 *** (0.106)	0.436 ** (0.181)
Ln(UR)	1.418 *** (0.088)	0.572 *** (0.154)	0.406 *** (0.097)	0.258 ** (0.115)	0.282 *** (0.046)	0.339 *** (0.044)
Ln(PP)	0.132 *** (0.012)	0.092 ** (0.042)	0.149 *** (0.013)	0.151 *** (0.019)	0.129 *** (0.010)	0.134 *** (0.009)
Ln(FAI)	0.037 ** (0.017)		0.106 *** (0.017)	0.081 *** (0.019)	0.027 ** (0.012)	
Ln(TI2)	0.265 *** (0.086)					
Ln(LAV)	0.142 *** (0.012)	0.297 *** (0.055)	0.177 *** (0.020)	0.235 *** (0.029)	0.142 *** (0.015)	0.216 *** (0.021)
Constant	−2.32 *** (0.682)	3.731 *** (0.401)	−7.706 ** (3.447)	−7.598 *** (1.485)	3.136 *** (0.757)	1.999 * (1.125)
Sig F	0.0000	0.0000	0.0000	0.0000	0.0000	0.0000
R-squared	0.980	0.9541	0.963	0.939	0.943	0.961
K	0.03	0.05	0.26	0.24	0.4	0.25
	Baoding	**Zhangjiakou**	**Chengde**	**Cangzhou**	**Langfang**	**Hengshui**
Ln(P)	1.501 *** (0.223)	4.195 *** (0806)	2.425 *** (0.820)			1.409 *** (0.317)
Ln(UR)	0.258 *** (0.050)	0.432 *** (0.063)	0.639 *** (0.085)	0.634 *** (0.085)	0.741 *** (0.083)	0.330 *** (0.055)
Ln(PP)	0.115 *** (0.015)	0.142 *** (0.010)	0.166 *** (0.016)	0.199 *** (0.022)	0.114 *** (0.019)	0.100 *** (0.018)
Ln(FAI)	0.053 ** (0.025)	0.028 ** (0.013)			0.100 *** (0.026)	
Ln(TI2)					0.412 *** (0.076)	
Ln(LAV)	0.163 *** (0.030)	0.190 *** (0.019)	0.251 *** (0.029)	0.069 ** (0.029)	0.153 *** (0.022)	0.176 *** (0.034)
Constant	−5.056 *** (1.622)	−21.436 *** (4.898)	−11.487 ** (4.676)	4.236 *** (0.185)	1.089 ** (0.433)	−3.430 * (1.945)
Sig F	0.0000	0.0000	0.0000	0.0000	0.0000	0.0000
R-squared	0.933	0.943	0.988	0.985	0.990	0.900
K	0.33	0.48	0.08	0.04	0.05	0.37

Note: "()" is the standard error; ***, **, * indicate significance at the 1%, 5% and 10% levels, respectively.

Tables 5 and 6 show that the significance test of 10% was passed by all of the chosen independent variables. Ridge regression results showed that R^2 was greater than 85%, F significance test Sig F = 0.0000 < 0.01, passed the significance level test. Therefore, it can be concluded that the relationship between CO_2 emissions and their respective variables in Beijing-Tianjin-Hebei cities can be well explained by models (17) and (18).

The urbanization rate (UR) promoted CO_2 emissions in all of the 13 cities, and the total population (P) of each city had a promoting effect on energy CO_2 emissions to a certain extent. The increase in the population and the migration of rural people to cities have led to the increase in urbanization rate and the change of industrial form, which are among the reasons for the increase in energy consumption. For Tianjin and some cities in Hebei, foreign investment (FAI) played a promoting role in CO_2 emission. Foreign investment promoted the level of industrialization and economic growth and improved the energy intensity of infrastructure construction and the mechanized production process. Therefore, a large amount of investment can produce higher CO_2 emissions [38].

The ratio of secondary industry to GDP (TI2) was only significantly correlated with CO_2 emissions in Tianjin and Langfang, indicating that the industrial structure had a stimulating effect on CO_2 emissions in these two cities during 2000–2019, and the effective adjustment of industrial structure could restrain CO_2 emissions. Except for Beijing, the increase of industrial added value (LAV) will lead to the increase of CO_2 emissions in the other 12 cities because the continuous increase of industrial added value will stimulate social demand and increase production intensity, resulting in more CO_2 emissions. For Beijing, urbanization rate (UR) was the most critical positive factor affecting its energy CO_2

emissions, and its elasticity value was the highest. For every 1% increase in the urbanization rate of Beijing, its energy CO_2 emissions will increase by 1.09%. Increased foreign investment (FAI) will curb CO_2 emissions in Beijing, as a large amount of investment is often accompanied by advanced technologies that use clean energy in the production process.

5. Discussion

5.1. Accuracy Assessment of CO_2 Emissions

The lack of energy statistics and inconsistent statistical standards in prefecture-level cities bring some difficulties to the analysis of the spatio-temporal CO_2 emissions dynamics. In this study, the method of using night light (NSL) to assess carbon dioxide proved to be effective. In order to assess the reliability of the findings, linear equations, quadratic terms, power functions and exponential functions were also used for comparison. Coefficient of determination (R^2), relative error (RE), and root mean square error (RMSE) were used to assess the precision of the spatial distribution of CO_2 emissions. Among them, the coefficient of the neural network model is better than that of other function models, as shown in Table 4. However, this method is relatively dependent on the accuracy of NSL data. Due to some shortcomings of night light data, such as supersaturation and low accuracy, how to simulate CO_2 emissions more accurately through night light needs to be studied further.

5.2. The Role of Influencing Factors

The results of the spatial econometric model show that all the factors selected in this study have an effect on CO_2 emissions, especially the population factor, which is in line with the research findings of Wang [39]. Since 2010, there seems to be no direct correlation between Beijing's carbon emission curve and its economic development curve, because the rapid economic growth accompanied by the use of advanced technology and clean energy has slowed the growth of CO_2 emissions [40]. The impact of foreign investment (FAI) on CO_2 emissions in the BTH region is different, with Beijing being negative and the other 12 cities being positive. For Beijing, with high GDP, advanced technologies can be transferred to domestic enterprises through foreign investment, and technology transfer can help domestic enterprises adopt new technologies and improve the energy efficiency of energy use [41]. For Tianjin and the cities in Hebei Province, a large amount of foreign capital inflow is responsible for the regional economic growth, but it inevitably leads to environmental pollution, thereby increasing carbon emissions [42]. Many scholars have analyzed different areas of the environmental stress factors, but most of the studies focused on large scale regions and ignore the regional internal differences, and this research solved this problem. However, the selection of influencing factors needs to be discussed further. Different scholars choose different influencing factors, which might have an effect on the study's findings. At the same time, due to the high scalability of the STIRPAT model, the factors affecting energy consumption are complex. This study was conducted on the basis of available urban statistics, which did not include all of the influencing factors, especially the technical factor indicators. The establishment and improvement of the statistical index system of different scales will make the driving force research results more accurate. The detailed decomposition of the influence of each variable from a small-scale perspective is a direction for future research on regional environmental stress.

5.3. Compared with Previous Studies

Research on carbon emissions has become a hot topic in recent years. Previous studies have had a large scope and rarely involve city and county levels. For example, Shi [19] and Chen [43] chose China as their research area and studied China's carbon emissions through single satellite data. The method proposed in this study has obvious advantages. By integrating two kinds of nighttime light data, long-term carbon emission data at different administrative levels in different cities can be obtained with certain accuracy. This study focused on the Beijing-Tianjin-Hebei region, which can more specifically analyze the driving principles behind the carbon emissions of cities and counties. Compared with previous studies, most of the carbon emissions and NSL fitting studies adopted the traditional linear method [22,23], in which Lv obtained the fitting RE of Beijing, Tianjin and Hebei to be 6.64%, 0.45% and 38.86%, respectively, while the fitting RE of this study was 1.58%, 1.46% and 2.69%, respectively. The errors of Beijing and Hebei are much smaller than the results of Lv. Therefore, the neural network method proposed in this study is superior to some other studies to a certain extent, and the final results are more accurate. Finally, in the analysis of driving factors, this study added some factors on the basis of previous studies. For example, traffic and building factors are added to the study of Beijing. Compared with the study of Wen [44], this research has a smaller scale and is accurate to each city, so it has more advantages.

6. Conclusions and Policy Implications

To some extent, the low-carbon development strategies formulated by the BTH can provide some references for other cities. However, the accurate analysis of small-scale regional carbon emissions is limited by the lack of statistical data in some cities. In order to resolve this issue, this work employed the integration of two kinds of nighttime lighting data to fit the CO_2 emissions of Beijing, Tianjin, and Hebei, and obtained the CO_2 emissions at different scales. Then, spatial analysis technology was used to examine spatial and temporal changes, and the driving force of CO_2 emissions was analyzed by the STIRPAT model with biased estimation ridge regression. The results show that the distribution pattern of energy CO_2 emission in Beijing-Tianjin-Hebei is "greater in the east than in the west" and is mainly concentrated in the eastern coastal cities and the cities with higher population density and higher GDP. There were variances in CO_2 emission levels across cities, and most displayed a pattern of first growing and then slowing down. With the exception of the Hongqiao District and Heping District, carbon emissions in central urban areas of Beijing and Tianjin showed a general decreasing trend. Tangshan, Baoding, Cangzhou and Shijiazhuang, which are the industrial, economic and political centers, are the areas with high carbon emissions in Hebei Province, and the CO_2 emissions of some counties in the south and northwest of Hebei Province show an increasing trend. The aggregation degree of carbon emissions in each district was weakened, and most of them showed a slowing trend.

According to the spatial econometric model's findings, carbon emissions in Tianjin and Hebei are positively affected by population, wealth, and technology. In addition to the negative impact of foreign investment in Beijing, other factors are also positive. In the Beijing-Tianjin-Hebei region, population is the most significant factor determining CO_2 emissions. The weight of the same influence factor is different in different cities, and the difference is obvious within cities.

This study offers some policy suggestions for controlling CO_2 emissions based on the above discussion and conclusions. First, we should control the size of the population, optimize the population structure and alleviate the problem of high population density. At the same time, we should improve the quality of the population, and guide people's environmental awareness and consumption concepts. Second, strictly controlling the approval of foreign investment and energy-intensive projects and raising the investment threshold to develop low-carbon industries can effectively control CO_2 emissions. Third, we should adjust the industrial structure and reduce the proportion of industries that consume large amounts of fossil energy. We need to encourage high-tech industries and establish a green and ecological industrial development model, especially in Tianjin, Shijiazhuang, Chengde and Langfang. For regions with underdeveloped economic development, such as Zhangjiakou and Hengshui, CO_2 emissions should be reduced by saving energy, improving energy utilization efficiency and actively cultivating low-carbon and green concepts. In addition, for Beijing, besides controlling the population size, the proportion of clean energy in total energy consumption can be increased, the consumption of fossil energy can be reduced, and public transportation can be advocated. Local governments should improve regional industrial guidance, encourage foreign investment in high-tech industries, and introduce resource-saving investments. At the same time, we need to develop low-carbon buildings and green buildings, and improve the relevant technical norms and standards.

Author Contributions: Conceptualization, S.X. and H.S.; methodology, S.X. and H.S.; software, H.W.; validation, W.X. and Q.S.; formal analysis, Z.Y.; investigation, J.Q.; writing—original draft preparation, S.X. and H.W.; writing—review and editing, H.S. and W.X.; All authors have read and agreed to the published version of the manuscript.

Funding: This research was funded by National Natural Science Foundation of China (grant number 41401659) and Natural Resources Science and Technology Project of Sichuan Province (grant number AZ20210002).

Data Availability Statement: Not applicable.

Acknowledgments: The authors are grateful for the support offered by National Natural Science Foundation of China and Natural Resources Science and Technology Project of Sichuan Province.

Conflicts of Interest: The authors declare that they have no conflict of interest.

References

1. Soytas, U.; Sari, R.; Ewing, B.T. Energy consumption, income, and carbon emissions in the United States. *Ecol. Econ.* **2007**, *62*, 482–489. [CrossRef]
2. Ciais, P.; Wang, Y.; Andrew, R.; Bréon, F.M.; Chevallier, F.; Broquet, G.; Nabuurs, G.J.; Peters, G.; McGrath, M.; Meng, W. Biofuel burning and human respiration bias on satellite estimates of fossil fuel CO_2 emissions. *Environ. Res. Lett.* **2020**, *15*, 074036. [CrossRef]
3. Fang, C.; Wang, S.; Li, G. Changing urban forms and carbon dioxide emissions in China: A case study of 30 provincial capital cities. *Appl. Energy* **2015**, *158*, 519–531. [CrossRef]
4. Su, Y.; Chen, X.; Li, Y.; Liao, J.; Ye, Y.; Zhang, H.; Huang, N.; Kuang, Y. China's 19-year city-level carbon emissions of energy consumptions, driving forces and regionalized mitigation guidelines. *Renew. Sustain. Energy Rev.* **2014**, *35*, 231–243. [CrossRef]
5. Apergis, N.; Payne, J.E. The causal dynamics between coal consumption and growth: Evidence from emerging market economies. *Appl. Energy* **2010**, *87*, 1972–1977. [CrossRef]
6. Guo, D.; Chen, H.; Long, R. Can China fulfill its commitment to reducing carbon dioxide emissions in the Paris Agreement? Analysis based on a back-propagation neural network. *Environ. Sci. Pollut. Res.* **2018**, *25*, 27451–27462. [CrossRef]
7. Wang, M.; Cai, B. A two-level comparison of CO_2 emission data in China: Evidence from three gridded data sources. *J. Clean. Prod.* **2017**, *148*, 194–201. [CrossRef]
8. Mallapaty, S. How China could be carbon neutral by mid-century. *Nature* **2020**, *586*, 482–484. [CrossRef]
9. Yu, X.; Liang, Z.; Fan, J.; Zhang, J.; Luo, Y.; Zhu, X. Spatial decomposition of city-level CO_2 emission changes in Beijing-Tianjin-Hebei. *J. Clean. Prod.* **2021**, *296*, 126613. [CrossRef]
10. Geng, Y.; Tian, M.; Zhu, Q.; Zhang, J.; Peng, C. Quantification of provincial-level carbon emissions from energy consumption in China. *Renew. Sustain. Energy Rev.* **2011**, *15*, 3658–3668. [CrossRef]
11. Müller, D.B.; Liu, G.; Løvik, A.N.; Modaresi, R.; Pauliuk, S.; Steinhoff, F.S.; Brattebø, H. Carbon emissions of infrastructure development. *Environ. Sci. Technol.* **2013**, *47*, 11739–11746. [CrossRef] [PubMed]
12. O'neill, B.C.; Dalton, M.; Fuchs, R.; Jiang, L.; Pachauri, S.; Zigova, K. Global demographic trends and future carbon emissions. *Proc. Natl. Acad. Sci. USA* **2010**, *107*, 17521–17526. [CrossRef] [PubMed]
13. Amaral, S.; Câmara, G.; Monteiro, A.M.V.; Quintanilha, J.A.; Elvidge, C.D. Estimating population and energy consumption in Brazilian Amazonia using DMSP night-time satellite data. *Comput. Environ. Urban Syst.* **2005**, *29*, 179–195. [CrossRef]
14. Sutton, P.C.; Taylor, M.J.; Elvidge, C.D. Using DMSP OLS imagery to characterize urban populations in developed and developing countries. In *Remote Sensing of Urban and Suburban Areas*; Springer: Berlin/Heidelberg, Germany, 2010; pp. 329–348.
15. Alahmadi, M.; Atkinson, P.M. Three-fold urban expansion in Saudi Arabia from 1992 to 2013 observed using calibrated DMSP-OLS night-time lights imagery. *Remote Sens.* **2019**, *11*, 2266. [CrossRef]
16. Gibson, J.; Olivia, S.; Boe-Gibson, G.; Li, C. Which night lights data should we use in economics, and where? *J. Dev. Econ.* **2021**, *149*, 102602. [CrossRef]
17. Chand, T.K.; Badarinath, K.; Elvidge, C.; Tuttle, B. Spatial characterization of electrical power consumption patterns over India using temporal DMSP-OLS night-time satellite data. *Int. J. Remote Sens.* **2009**, *30*, 647–661. [CrossRef]
18. Elvidge, C.D.; Ziskin, D.; Baugh, K.E.; Tuttle, B.T.; Ghosh, T.; Pack, D.W.; Erwin, E.H.; Zhizhin, M. A fifteen year record of global natural gas flaring derived from satellite data. *Energies* **2009**, *2*, 595–622. [CrossRef]
19. Shi, K.; Chen, Y.; Yu, B.; Xu, T.; Chen, Z.; Liu, R.; Li, L.; Wu, J. Modeling spatiotemporal CO_2 (carbon dioxide) emission dynamics in China from DMSP-OLS nighttime stable light data using panel data analysis. *Appl. Energy* **2016**, *168*, 523–533. [CrossRef]
20. Zhang, X.; Wu, J.; Peng, J.; Cao, Q. The uncertainty of nighttime light data in estimating carbon dioxide emissions in China: A comparison between DMSP-OLS and NPP-VIIRS. *Remote Sens.* **2017**, *9*, 797. [CrossRef]

21. Zhao, J.; Chen, Y.; Ji, G.; Wang, Z. Residential carbon dioxide emissions at the urban scale for county-level cities in China: A comparative study of nighttime light data. *J. Clean. Prod.* **2018**, *180*, 198–209. [CrossRef]
22. Lv, Q.; Liu, H.; Wang, J.; Liu, H.; Shang, Y. Multiscale analysis on spatiotemporal dynamics of energy consumption CO_2 emissions in China: Utilizing the integrated of DMSP-OLS and NPP-VIIRS nighttime light datasets. *Sci. Total Environ.* **2020**, *703*, 134394. [CrossRef] [PubMed]
23. Meng, L.; Graus, W.; Worrell, E.; Huang, B. Estimating CO_2 (carbon dioxide) emissions at urban scales by DMSP/OLS (Defense Meteorological Satellite Program's Operational Linescan System) nighttime light imagery: Methodological challenges and a case study for China. *Energy* **2014**, *71*, 468–478. [CrossRef]
24. Chen, J.; Gao, M.; Cheng, S.; Hou, W.; Song, M.; Liu, X.; Liu, Y.; Shan, Y. County-level CO_2 emissions and sequestration in China during 1997–2017. *Sci. Data* **2020**, *7*, 1–12. [CrossRef]
25. Ang, B.W. The LMDI approach to decomposition analysis: A practical guide. *Energy Policy* **2005**, *33*, 867–871. [CrossRef]
26. Fan, Y.; Liu, L.-C.; Wu, G.; Wei, Y.-M. Analyzing impact factors of CO_2 emissions using the STIRPAT model. *Environ. Impact Assess. Rev.* **2006**, *26*, 377–395. [CrossRef]
27. Kaika, D.; Zervas, E. The Environmental Kuznets Curve (EKC) theory—Part A: Concept, causes and the CO_2 emissions case. *Energy Policy* **2013**, *62*, 1392–1402. [CrossRef]
28. Vaninsky, A. Factorial decomposition of CO_2 emissions: A generalized Divisia index approach. *Energy Econ.* **2014**, *45*, 389–400. [CrossRef]
29. Brizga, J.; Feng, K.; Hubacek, K. Drivers of CO_2 emissions in the former Soviet Union: A country level IPAT analysis from 1990 to 2010. *Energy* **2013**, *59*, 743–753. [CrossRef]
30. Ozcan, B.; Ulucak, R. An empirical investigation of nuclear energy consumption and carbon dioxide (CO_2) emission in India: Bridging IPAT and EKC hypotheses. *Nucl. Eng. Technol.* **2021**, *53*, 2056–2065. [CrossRef]
31. York, R.; Rosa, E.A.; Dietz, T. STIRPAT, IPAT and ImPACT: Analytic tools for unpacking the driving forces of environmental impacts. *Ecol. Econ.* **2003**, *46*, 351–365. [CrossRef]
32. Ma, T.; Zhou, C.; Pei, T.; Haynie, S.; Fan, J. Responses of Suomi-NPP VIIRS-derived nighttime lights to socioeconomic activity in China's cities. *Remote Sens. Lett.* **2014**, *5*, 165–174. [CrossRef]
33. Zhao, M.; Zhou, Y.; Li, X.; Zhou, C.; Cheng, W.; Li, M.; Huang, K. Building a series of consistent night-time light data (1992–2018) in Southeast Asia by integrating DMSP-OLS and NPP-VIIRS. *IEEE Trans. Geosci. Remote Sens.* **2019**, *58*, 1843–1856. [CrossRef]
34. Jeswani, R.; Kulshrestha, A.; Gupta, P.K.; Srivastav, S. Evaluation of the consistency of DMSP-OLS and SNPP-VIIRS Night-time Light Datasets. *J. Geomat* **2019**, *13*, 98–105.
35. Liddle, B.; Lung, S. Age-structure, urbanization, and climate change in developed countries: Revisiting STIRPAT for disaggregated population and consumption-related environmental impacts. *Popul. Environ.* **2010**, *31*, 317–343. [CrossRef]
36. Shahbaz, M.; Chaudhary, A.; Ozturk, I. Does urbanization cause increasing energy demand in Pakistan? Empirical evidence from STIRPAT model. *Energy* **2017**, *122*, 83–93. [CrossRef]
37. Meng, M.; Zhou, J. Has air pollution emission level in the Beijing–Tianjin–Hebei region peaked? A panel data analysis. *Ecol. Indic.* **2020**, *119*, 106875. [CrossRef]
38. Zhao, X.; Zhang, X.; Shao, S. Decoupling CO_2 emissions and industrial growth in China over 1993–2013: The role of investment. *Energy Econ.* **2016**, *60*, 275–292. [CrossRef]
39. Wang, Z.; Yin, F.; Zhang, Y.; Zhang, X. An empirical research on the influencing factors of regional CO_2 emissions: Evidence from Beijing city, China. *Appl. Energy* **2012**, *100*, 277–284. [CrossRef]
40. Liu, Z.; Guan, D.; Moore, S.; Lee, H.; Su, J.; Zhang, Q. Climate policy: Steps to China's carbon peak. *Nature* **2015**, *522*, 279–281. [CrossRef]
41. Lee, J.W. The contribution of foreign direct investment to clean energy use, carbon emissions and economic growth. *Energy Policy* **2013**, *55*, 483–489. [CrossRef]
42. Peng, H.; Tan, X.; Li, Y.; Hu, L. Economic growth, foreign direct investment and CO_2 emissions in China: A panel granger causality analysis. *Sustainability* **2016**, *8*, 233. [CrossRef]
43. Chen, H.; Zhang, X.; Wu, R.; Cai, T. Revisiting the environmental Kuznets curve for city-level CO2 emissions: Based on corrected NPP-VIIRS nighttime light data in China. *J. Clean. Prod.* **2020**, *268*, 121575. [CrossRef]
44. Wen, L.; Liu, Y. The Peak Value of Carbon Emissions in the Beijing-Tianjin-Hebei Region Based on the STIRPAT Model and Scenario Design. *Pol. J. Environ. Stud.* **2016**, *25*. [CrossRef]

remote sensing

Article

Differential Spatiotemporal Patterns of CO_2 Emissions in Eastern China's Urban Agglomerations from NPP/VIIRS Nighttime Light Data Based on a Neural Network Algorithm

Lei Zhou [1,2], Jun Song [1], Yonggang Chi [1,*] and Quanzhou Yu [3]

1 College of Geography and Environmental Sciences, Zhejiang Normal University, Jinhua 321004, China
2 Key Laboratory of Ecosystem Network Observation and Modelling, Institute of Geographic Sciences and Natural Resources Research, Chinese Academy of Sciences, Beijing 100101, China
3 School of Geography and Environment, Liaocheng University, Liaocheng 252059, China
* Correspondence: chiyonggang@zjnu.cn; Tel.: +86-579-8228-2273

Abstract: Urban agglomerations, such as Beijing-Tianjin-Hebei Region, Yangtze River Delta and Pearl River Delta, are the key regions for energy conservation, carbon emission reduction and low-carbon development in China. However, spatiotemporal patterns of CO_2 emissions at fine scale in these major urban agglomerations are not well documented. In this study, a back propagation neural network based on genetic algorithm optimization (GABP) coupled with NPP/VIIRS nighttime light datasets was established to estimate the CO_2 emissions of China's three major urban agglomerations at 500 m resolution from 2014 to 2019. The results showed that spatial patterns of CO_2 emissions presented three-core distribution in the Beijing-Tianjin-Hebei Region, multiple-core distribution in the Yangtze River Delta, and null-core distribution in the Pearl River Delta. Temporal patterns of CO_2 emissions showed upward trends in 28.74–43.99% of the total areas while downward trends were shown in 13.47–15.43% of the total areas in three urban agglomerations. The total amount of CO_2 emissions in urban areas was largest among urban circles, followed by first-level urban circles and second-level urban circles. The profiles of CO_2 emissions along urbanization gradients featured high peaks and wide ranges in large cities, and low peaks and narrow ranges in small cities. Population density primarily impacted the spatial pattern of CO_2 emissions among urban agglomerations, followed by terrain slope. These findings suggested that differences in urban agglomerations should be taken into consideration in formulating emission reduction policies.

Keywords: carbon emission; energy conservation; genetic algorithm; nighttime light data; urban agglomerations

Citation: Zhou, L.; Song, J.; Chi, Y.; Yu, Q. Differential Spatiotemporal Patterns of CO_2 Emissions in Eastern China's Urban Agglomerations from NPP/VIIRS Nighttime Light Data Based on a Neural Network Algorithm. *Remote Sens.* **2023**, *15*, 404. https://doi.org/10.3390/rs15020404

Academic Editors: Ran Goldblatt, Steven Louis Rubinyi and Hogeun Park

Received: 21 November 2022
Revised: 26 December 2022
Accepted: 30 December 2022
Published: 9 January 2023

1. Introduction

Global cities only account for 2% of the Earth's area, but produce more than 70% of the world's anthropogenic CO_2 emissions [1,2]. The extent and intensity of global climate change will be strengthened in the next few decades, according to the sixth assessment report of the Intergovernmental Panel on Climate Change (IPCC). CO_2 emissions are increasing at an unprecedented rate due to the energy consumption related to human activities [3–6]. The urbanization rate of China increased to 59.6% in 2018 [7]. Rising energy consumption due to the rapid development of urbanization and industrialization has increased China's CO_2 emissions significantly [8,9]. A study reported that China's 35 largest cities accounted for 40% of the country's CO_2 emissions [10]. The huge amount and rapid growth rate of CO_2 emissions have made China face huge pressure to balance economic growth and sustainable development [11,12]. Therefore, exploring the patterns of carbon emissions in urban agglomerations are crucial for formulating carbon reduction policies.

Satellite nighttime light information at high resolution may provide accurate estimation of spatial and temporal patterns of CO_2 emissions [13]. The relationships between statistical energy consumption and nighttime light data can be obtained by a variety of regression models [14–16]. However, the conventional regression method is not effective due to the fixed model parameters [17]. Back propagation neural networks allow a high variance model without suffering from overfitting, which can provide more accurate regression models than traditional regression models. Jasiński (2019) used artificial neural networks to model electricity consumption based on nighttime light images [18]. Yang et al. (2020) proposed a structure-based neural network ensemble to analyze the nonlinear relationship between nighttime light data and CO_2 emissions [17]. Genetic neural networks integrated with the satellite nighttime light data at high resolution and local statistical CO_2 emissions data may improve estimation accuracy of CO_2 emissions.

Fine-scale information of national and regional scale CO_2 emissions is essential for the design of emissions mitigation policies [19–21]. Nighttime light data from satellite information is highly associated with the footprint of human activities, and can provide an effective proxy for estimating energy consumption and CO_2 emissions [14,15]. The Defense Meteorological Satellite Program's Operational Linescan System (DMSP/OLS) data have been commonly used for the estimation of carbon emissions [16]. For example, Doll et al. (2000) first found a strong relationship between DMSP/OLS nighttime light data and CO_2 emissions [22]. Wang and Liu (2017) utilized DMSP/OLS data from 1992 to 2013 to analyze regional inequalities and spatial agglomeration of urban CO_2 emissions [23]. Shi et al. (2018) combined DMSP/OLS images and statistical energy consumption data to explore spatiotemporal variations of CO_2 emissions from urban agglomeration to national scales [13]. Soon afterwards, the Suomi National Polar-Orbiting Partnership (NPP) Visible Infrared Imaging Radiometer Suite (VIIRS) using nighttime light data with high spatial resolution and wide radiometric detection proved to be better than DMSP/OLS data in simulating CO_2 emissions [24,25]. Shi et al. (2014) analyzed the correlation between nighttime light, gross domestic product (GDP) and electric power consumption, and found that NPP/VIIRS data were powerful tools to model socioeconomic indicators [26]. Zhao et al. (2018) modelled CO_2 emissions in residential sectors at the urban scale and found that the performance of NPP/VIIRS data in simulating residential carbon emissions were better than DMSP/OLS data [25].

Beijing-Tianjin-Hebei, the Yangtze River Delta and the Pearl River Delta are the most populated urban agglomerations with the highest comprehensive strength in China. By 2019, the GDP of the three urban agglomerations accounted for 39.65% of the national GDP, while their population accounted for only 23.79% of the national population. Here, spatiotemporal variations of CO_2 emissions among urban agglomerations and the profiles of CO_2 emissions along urbanization gradients were explored using NPP/VIIRS nighttime light datasets at 500 m resolution from 2014 to 2019 based on a back propagation neural network with genetic algorithm optimization (GABP). Our aims were to: (1) integrate multi-source data to estimate fine-scale carbon emissions of three urban agglomerations; (2) analyze the spatial and temporal dynamics of carbon emissions among urban agglomerations and the profiles of CO_2 emissions along urbanization gradients within cities; and (3) reveal the main influencing factors of carbon emissions.

2. Materials and Methods

2.1. Study Area

Beijing-Tianjin-Hebei, the Yangtze River Delta, and the Pearl River Delta regions are China's three major economic growth poles. These urban agglomerations, located on the eastern coast of Mainland China, are focus areas for energy conservation and emission reduction (Figure 1). Beijing-Tianjin-Hebei consists of two municipalities, Beijing and Tianjin, as well as 11 cities in Hebei province. The Yangtze River Delta consists of 26 cities (i.e., Shanghai, Hangzhou, Nanjing, etc.). The Pearl River Delta is composed of nine cities in Guangdong province. Furthermore, Beijing-Tianjin-Hebei is dominated by temperate continental monsoon climate [27], while the Pearl River Delta and Yangtze River Delta are dominated by subtropical/tropical monsoon climates [28] and maritime monsoon subtropical climates [29], respectively.

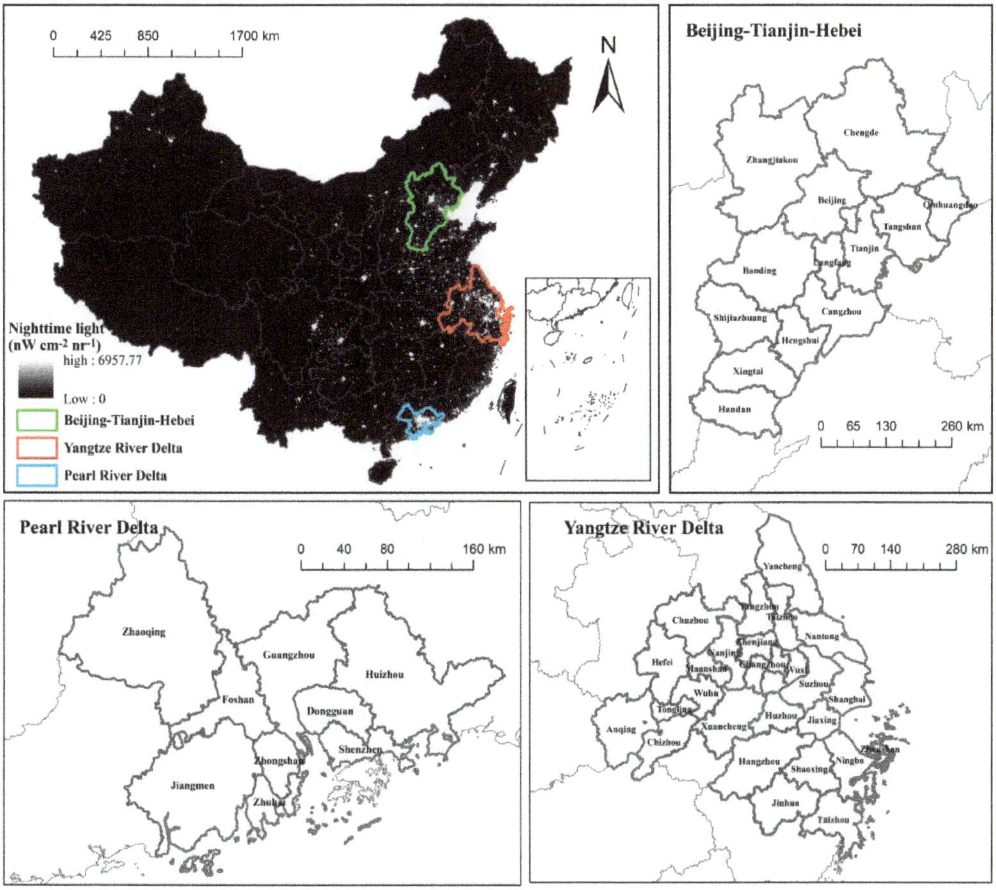

Figure 1. Locations of three urban agglomerations such as Beijing-Tianjin-Hebei Region, Yangtze River Delta, and Pearl River Delta.

2.2. Data Sources

Datasets consisted of NPP/VIIRS nighttime light data, fossil fuel combustion data, socioeconomic data, and basic geographic information data. Monthly NPP/VIIRS nighttime light data (vcmsl version) at 500 m resolution during 2014–2019 were obtained from Colorado School of Mines (https://payneinstitute.mines.edu/eog, accessed on 11 October 2019). The radiation value of pixels in this data represents the intensity of the light. The fossil fuel combustion data and socioeconomic data were obtained from China Statistical Yearbook, China Regional Statistical Yearbook, and statistical yearbooks of each city. The details of these data are listed in Table 1.

2.3. Methods

The flowchart of this study is summarized in Figure 2. Three key steps are included: (1) estimating the statistical carbon emissions using the IPCC method; (2) estimating CO_2 emissions using GABP neural networks coupled with satellite data and statistical carbon emissions; (3) exploring the spatiotemporal dynamics of CO_2 emissions at 500 m resolution and their influencing factors.

Table 1. Data used in this study.

Data Name	Data Description	Source
Nighttime light	NPP/VIIRS nighttime light at a spatial resolution of 500 m for 2014–2019	Colorado School of Mines (https://payneinstitute.mines.edu/eog/, accessed on 11 October 2019)
Fossil fuel combustion data	Annual total data of ten energy types, such as raw coal, coke, crude oil, gasoline, kerosene, diesel oil, fuel oil, natural gas, heat and electricity, during 2014–2019	China Statistical Yearbook, China Regional Statistical Yearbook, China Energy Statistical Yearbook and statistical yearbooks of each city
Socioeconomic data	Annual statistical data of six types, such as permanent population, GDP, per capita GDP, primary industry GDP, secondary industry GDP and tertiary industry GDP, during 2014–2019	China Statistical Yearbook, China City Statistical Yearbook, China Regional Statistical Yearbook, and statistical yearbooks of each city
Administrative boundaries	Vector files of provinces, prefectures in China	National Catalogue Service For Geographic Information
Population density	Annual data with a spatial resolution of 30 arc-seconds (approximately 1km at the equator)	WorldPop (https://www.worldpop.org/, accessed on 13 August 2021)
Terrain slope	Spatial resolution of 90 m	Geospatial Data Cloud (http://www.gscloud.cn/, accessed on 16 July 2021)
Temperature	Annual mean temperature unit with a spatial resolution of 1 km	Resource and Environment Science and Data Center (https://www.resdc.cn/, accessed on 3 September 2021)
Urban area	Redefined data in 2016	Beijing City Lab Database [30]

2.3.1. Correction of NPP/VIIRS Nighttime Light Data

The NPP/VIIRS nighttime lights were not filtered to remove light detections associated with fires, gas flares and background noise [8,26]. First, a few additional outliers caused by lights from the flaring of oil and gas wells should be eliminated [31]. Because Beijing, Shanghai and Guangzhou are major cities in each urban agglomeration, nighttime light digital number (DN) values in these cities should be the largest compared to that in other cities. The largest DN value in Beijing, Shanghai and Guangzhou was used as a threshold to correct the outliers. The 8-neighbor denoising method was used to remove high value noise in the image [32]. These pixels where the DN value was larger than the threshold were assigned a new value, which equaled to the maximal DN value within the eight direct neighbors of the pixel [33]. Secondly, we removed the background noise of nighttime light data. Referring to Google Earth images, large-scale water areas were selected to set as the sample area, and the radiation value was averaged to be used as the minimum threshold. The pixels where the DN value was smaller than this minimum threshold were changed to zero [34]. Finally, monthly NPP/VIIRS nighttime light data were averaged to annual nighttime light images for the periods 2014–2019.

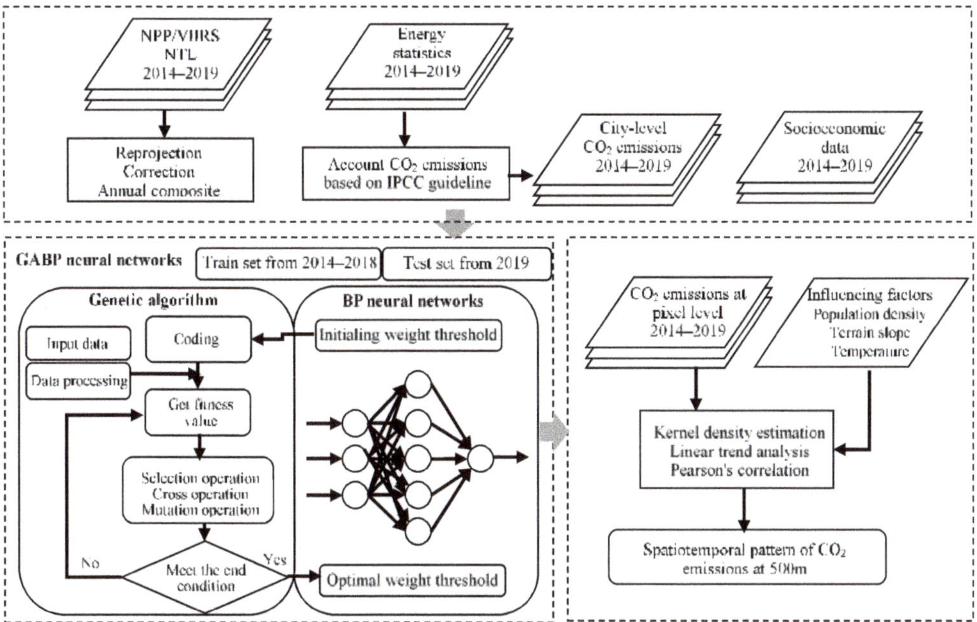

Figure 2. The flowchart of this study.

2.3.2. Estimating the Statistical Carbon Emissions Using the IPCC Method

The IPCC guidelines recommended a unified standard method to evaluate CO_2 emissions from greenhouse gases [35]. This study used the IPCC method to calculate the statistical CO_2 emissions of energy consumptions [36]. The energy sources selected in this study were raw coal, coke, crude oil, gasoline, kerosene, diesel oil, fuel oil, natural gas, heat, and electricity. The conversion formula is:

$$CO_2 = \frac{44}{12} \times \sum_{i=1}^{10} K_i E_i \quad (1)$$

While i represents the types of energy; E_i represents the standard coal consumption of energy type i; and K_i represents the effective CO_2 emission factor of each energy type. According to the IPCC guidelines, various types of fuel consumptions should be converted to a standard coal consumption based on the calorific value of each fuel type.

2.3.3. Estimating CO_2 Emissions by GABP Neural Networks and Nighttime Light Data

GABP neural networks have a strong ability to construct nonlinear relationships and overcomes the shortcomings of the traditional neural networks. A BP neural network is multi-layer feedforward neural network based on the error back propagation algorithm, composed of an input layer, hidden layer and output layer [37]. Genetic algorithm optimization of BP neural network mainly obtains the optimal weights and thresholds to substitute into the BP neural network for prediction [38]. Previous studies have shown that there were positive correlations between CO_2 emissions, GDP and population, indicating that GDP and population have large impacts on CO_2 emissions [13,39]. As result, major socioeconomic factors related to human activities were selected as the input layer of GABP model to explore the effect of human activity on environment. Therefore, we used a GABP to establish the relationships between NPP/VIIRS nighttime light data, socioeconomic data, and statistical carbon emissions in three urban agglomerations. Firstly, we used DN, permanent population, GDP, per capita GDP, primary industry GDP, secondary industry

GDP, and tertiary industry GDP to determine the input layers of GABP. Variables such as DN, permanent population and primary industry GDP with the strongest ties to statistical carbon emissions were finally selected as the input layer. In detail, statistical carbon emissions have the highest correlation with DN (r = 0.893, $p < 0.05$), followed by primary industry GDP (r = 0.796, $p < 0.05$) and permanent population (r = 0.758, $p < 0.05$). In terms of parameter settings, we then reduced the model training error through multiple trainings by the GABP neural networks, and finally determined the optimal parameters for constructing the carbon emission model estimation. The datasets during 2014–2018 were utilized for training and the data in 2019 was used for validation. The GABP neural networks in this study were divided into three layers. In detail, the input layer node was 3, the hidden layer node was 5, and the output layer node was 1.

CO_2 emissions estimated from GABP neural networks were validated with the city-level statistical CO_2 emissions (Figure 3). Three indicators of root mean square error (RSME), mean absolute percentage error (MAPE) and determination coefficient (R^2) were used to verify the fitting performance. The RMSE, MAPE and R^2 of CO_2 emissions estimated from GABP neural networks and the statistical data in this study were 1386.23 $\times$ 10^4 t, 11.01% and 0.977, respectively, indicating that the GABP neural networks method showed a good performance in estimating regional energy carbon emissions [17].

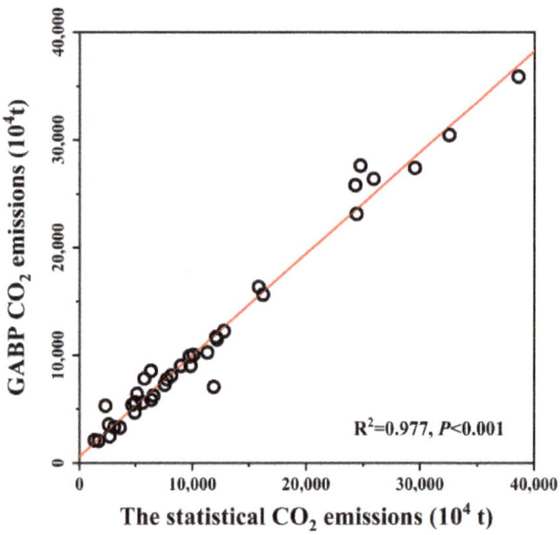

Figure 3. Scatter plots of total CO_2 emissions calculated from statistical energy consumption data and simulated CO_2 emissions by GABP neural networks of three agglomerations in 2019.

The NPP/VIIRS nighttime light was used to allocate the city-level carbon emissions to 500 m pixel. Radiance values of all pixels belong to a region were summed, and the original value at each pixel was normalized by the regional sum. The CO_2 emission intensity at a pixel was scaled by multiplying the normalized radiance with the annual total emissions of a country or a region [40]. Therefore, we obtained energy carbon emissions at 500 m resolution for three urban agglomerations in China from 2014 to 2019.

2.3.4. Spatiotemporal Dynamics of CO_2 Emissions Based on GIS-Based Buffer Analysis, Kernel Density Estimation and Linear Regression Analysis

Kernel density estimation (KDE) is a nonparametric density estimation method to estimate the probability density of random variables. In this study, we used KDE to estimate the continuous probability density curves of carbon emissions to capture the kernel density maps of carbon emissions in three urban agglomerations [41].

The city areas in urban agglomerations adopted for the GIS-based buffer analysis were divided into three urbanization gradients. We designed two buffer zones around urban areas. Two kinds of buffer zones were established: first-level urban circle was the buffer zone with a width of 500 m outside the urban area, while second-level urban circle was the buffer zone with a width of 1 km outside the first-level urban circle [42]. Two circular buffer systems (first-level urban circle and second-level urban circle) were built by creating buffer zones to compare the characteristics of CO_2 emissions between urban areas and sub-urban areas in urban agglomerations.

We calculated the variation slope (S_{slope} index) of energy consumption CO_2 emissions from 2014 to 2019 by establishing linear regression model between CO_2 emissions and years [43].

$$S_{slope} = \frac{\sum_{i=1}^{n} x_i t_i - \frac{1}{n}\sum_{i=1}^{n} x_i \sum_{i=1}^{n} t_i}{\sum_{i=1}^{n} t_i^2 - \frac{1}{n}(\sum_{i=1}^{n} t_i)^2} \quad (2)$$

where n is the total number of years; x_i is the serial number of year i; t_i is the CO_2 emission amount in the i year. When S_{slope} index was larger than 0, it meant that CO_2 emissions showed an increasing trend. Significant trends of CO_2 emissions in three urban agglomerations were divided into four levels. The first level was a high-decline region where S_{slope} was less than -0.16 and the second level was named as a low-decline region where S_{slope} was greater than -0.16 and less than 0. The third levels ($0 < S_{slope} < 0.16$) were named low-growth regions, and the last one ($S_{slope} > 0.16$) was called a high-growth region.

Profiles of CO_2 emissions along urbanization gradients were investigated using the section line that passed through the city center. Cities with high CO_2 emissions and low CO_2 emissions were selected, respectively, in each urban agglomeration. We chose Beijing and Langfang in Beijing-Tianjin-Hebei, Shanghai and Chizhou in Yangtze River Delta, Guangzhou and Zhaoqing in Pearl River Delta as case studies.

2.3.5. Statistical Analysis on the Influencing Factors of CO_2 Emissions among Urban Agglomerations

Population density, annual mean temperature and terrain slope excluding the predictors of the carbon emission model were selected as potential influencing factors of spatial pattern of CO_2 emissions among urban agglomerations. First, population density was an important socioeconomic factor, which was often found to positively correlate with energy consumptions and CO_2 emissions [44]. Second, climate was considered as a determinant of energy consumption and CO_2 emissions due to heating and cooling [45,46]. Meanwhile, annual mean temperature was different among three urban agglomerations. Third, topography, for example, terrain slope, was a crucial natural factor because CO_2 emissions of the car increased with the increase of terrain slope. In order to analyze the realistic factors affecting CO_2 emissions, the strength of correlation between CO_2 emissions and population density, annual mean temperature and terrain slope were estimated using the Pearson's correlation coefficient. All influencing factors, such as population density, annual mean temperature and terrain slope, were resampled to 500 m resolution.

3. Results

3.1. Spatiotemporal Variations of CO_2 Emissions

The spatial distribution of CO_2 emissions appeared as different patterns in three urban agglomerations (Figures 4 and 5). CO_2 emissions in Beijing-Tianjin-Hebei Region presented a three-core distribution, such as Beijing, Tianjin and Tangshan. CO_2 emissions in the Yangtze River Delta showed a multiple-core distribution, such as Nanjing, Zhenjiang, Wuxi, Suzhou, Shanghai and Ningbo. CO_2 emissions in Pearl River Delta appeared as a null-core distribution (Figure 5).

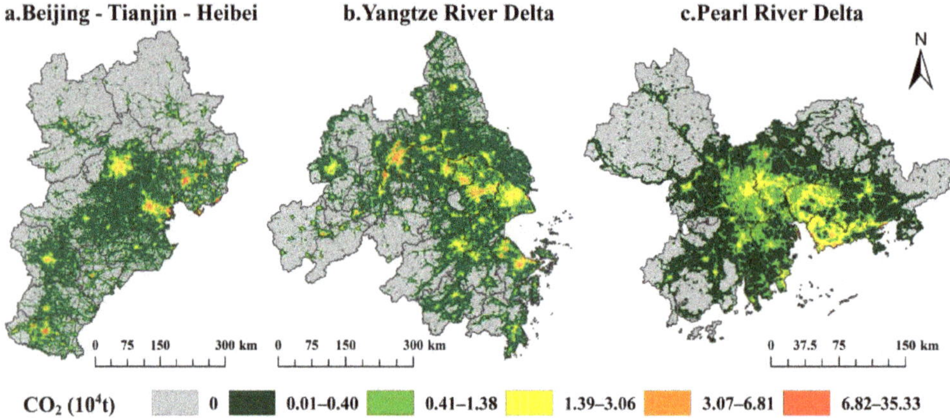

Figure 4. Spatial distribution of annual CO_2 emissions in three urban agglomerations averaged from 2014 to 2019.

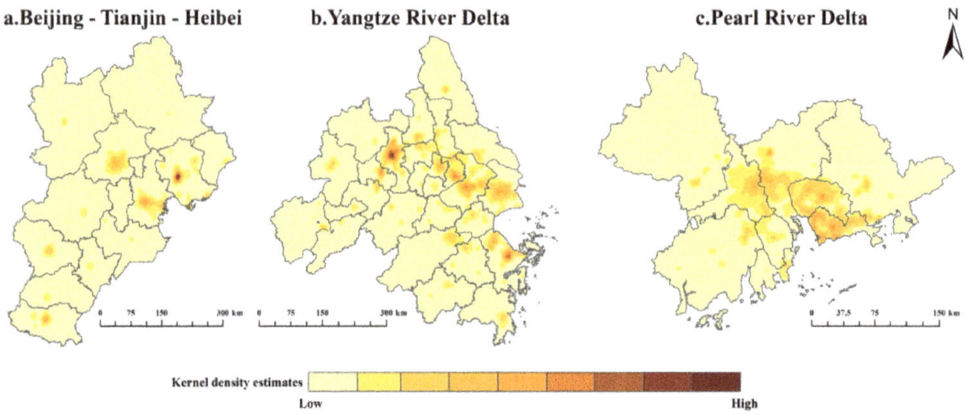

Figure 5. Kernel density estimation maps of CO_2 emissions of three urban agglomerations in 2019.

The Yangtze River Delta has the largest total CO_2 emissions, followed by Beijing-Tianjin-Hebei region and the Pearl River Delta (Figure 6). CO_2 emissions per unit in cities of the three agglomerations showed that the city with the lowest carbon emissions per unit in Beijing-Tianjin-Hebei was Langfang, while the highest was Tangshan. CO_2 emissions per unit of Tangshan were 8.29 times that of Langfang. In the Yangtze River Delta, the city with the lowest carbon emissions per unit was Jinhua, while the highest was Maanshan. CO_2 emissions per unit in Maanshan were 4.39 times that of Jinhua. In the Pearl River Delta, the city with the lowest carbon emissions per unit was Zhongshan, and the highest was Shenzhen. CO_2 emissions per unit of Shenzhen was 2.46 times that of Zhongshan. Consequently, Beijing-Tianjin-Hebei Region have the largest variations in carbon emissions at city level, while the Pearl River Delta have the smallest variations.

Temporal patterns of CO_2 emissions showed that th 28.74%, 43.99% and 43.45% of the areas appeared to have significant upward trends of CO_2 emissions in Beijing-Tianjin-Hebei Region, Yangtze River Delta, and Pearl River Delta during 2014–2019, respectively (Figure 7). Meanwhile, the percentages of regions with significant downward trends of CO_2 emissions were 15.43%, 13.47% and 15.31% in Beijing-Tianjin-Hebei, Yangtze River Delta, and Pearl River Delta during study period, respectively. In general, the areas with downward trends of carbon emissions were concentrated in southern Beijing, southern

Tianjin, northern Tangshan, and western Handan in Beijing-Tianjin-Hebei, while these were dispersed in the Yangtze River Delta and Pearl River Delta. Moreover, the percentages of high-growth regions and low-growth regions in the upward trend regions were 2.88%, 5.46%, 3.44% and 4.75%, 3.97%, 2.22% in Beijing-Tianjin-Hebei, the Yangtze River Delta, and Pearl River Delta, respectively.

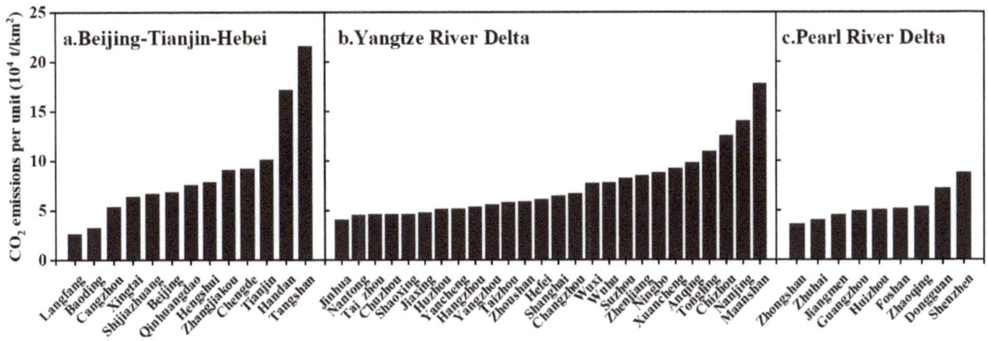

Figure 6. CO_2 emissions per unit in cities of three urban agglomerations.

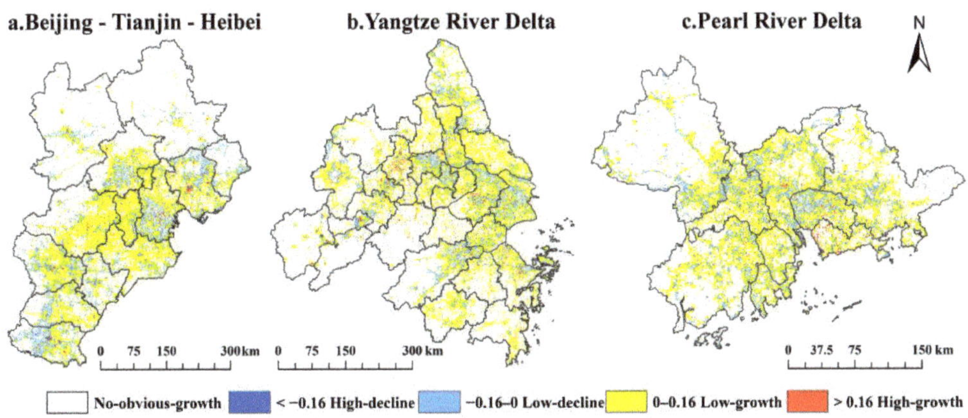

Figure 7. Temporal trends of CO_2 emissions in three urban agglomerations from 2014 to 2019. The white area was not passed the significance test ($p > 0.05$). High-decline: $S_{slope} < -0.16$; Low-decline: $-0.16 < S_{slope} < 0$; Low-growth: $0 < S_{slope} < 0.16$; High-growth: $0.16 < S_{slope}$.

3.2. Carbon Emissions within Cities

Urban areas have a larger amount of CO_2 emissions than first-level urban circle and second-level urban circle (Figure 8a–c). In detail, the S_{slope} of carbon emissions in urban areas of the Yangtze River Delta was the largest, reaching 0.270, while that of Beijing-Tianjin-Hebei Region was the smallest, at only 0.018. The slopes of CO_2 emissions in urban areas of Yangtze River Delta and Pearl River Delta were both larger than those in the other two urban circles.

On the contrary, CO_2 emissions per unit in urban areas, first-level urban circles and second-level urban circles in Beijing-Tianjin-Hebei were the highest among three urban agglomerations (Figure 8d–f). The temporal trends of CO_2 emissions per unit in three urban circles of Yangtze River Delta and Pearl River Delta have increased significantly. The temporal slopes of CO_2 emissions per unit in first-level urban circle and second-level urban circle of Yangtze River Delta were larger than that of Pearl River Delta, respectively (Figure 8d–f).

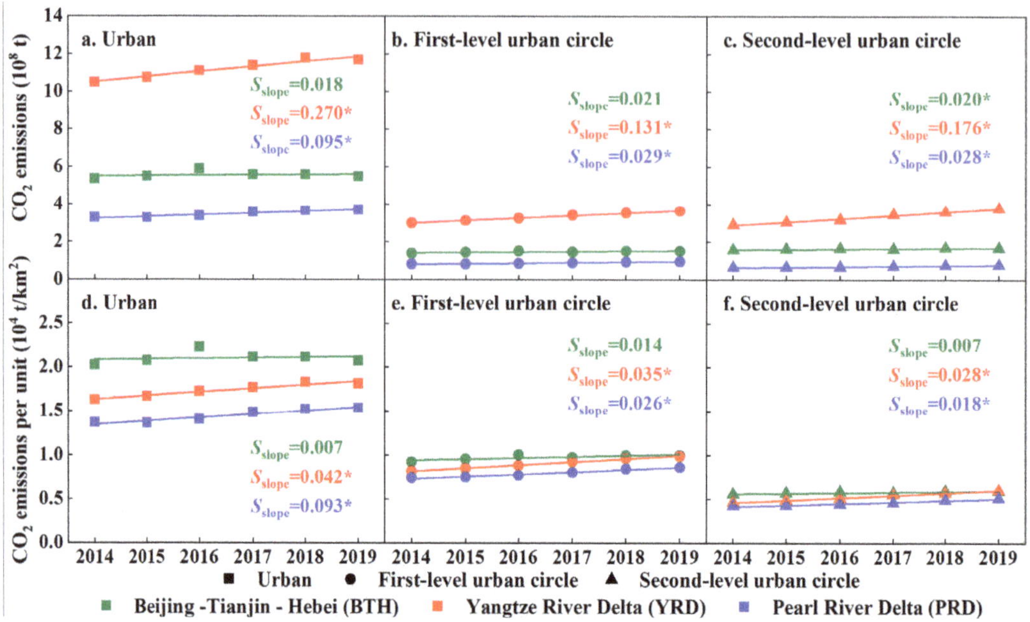

Figure 8. CO_2 emissions and CO_2 emissions per unit within cities. The fitting curve indicated the temporal trend of CO_2 emissions in three urban circles for each urban agglomeration and S_{slope} indicated the slope of energy CO_2 emissions from 2014 to 2019. The star behind S_{slope} value mean the significance test of trend was passed ($p < 0.05$).

In order to further investigate the pattern of CO_2 emissions along urbanization gradients, the section line that passed through the city center was used to obtain profile lines, which represented the structure of CO_2 emissions along the distance away from urban centers. In this study, cities with high CO_2 emissions and low CO_2 emissions were selected respectively in each urban agglomeration (Figure 9). Profiles of CO_2 emissions along urbanization gradients behaved as a radiation peak pattern with a high peak and a wide range in big cities (Figure 9a–c). Correspondingly, profiles of CO_2 emissions showed an independent peak pattern with a low peak and a narrower range in small cities (Figure 9d–f).

3.3. *Influencing Factors of Spatial Pattern of CO_2 Emissions*

Significant positive relationships between CO_2 emissions and population density can be observed in each urban agglomeration ($r_{BTH} = 0.486$; $r_{YRD} = 0.470$; $r_{PRD} = 0.535$; all $p < 0.05$) (Figure 10). Meanwhile, there were significant negative correlations between terrain slope and CO_2 emissions in three agglomerations ($r_{BTH} = -0.178$; $r_{YRD} = -0.156$; $r_{PRD} = -0.335$; all $p < 0.05$), which meant that CO_2 emissions were generally decreasing when terrain slope increased. Furthermore, CO_2 emissions have weak correlations with annual mean temperature in each agglomeration ($r_{BTH} = 0.172$; $r_{YRD} = 0.004$; $r_{PRD} = -0.124$; all $p < 0.05$). It indicated that population density was the primary influencing factor of CO_2 emissions in all urban agglomerations. Meanwhile, the Pearl River Delta was one of the three urban agglomerations which CO_2 emissions most affected by terrain slope.

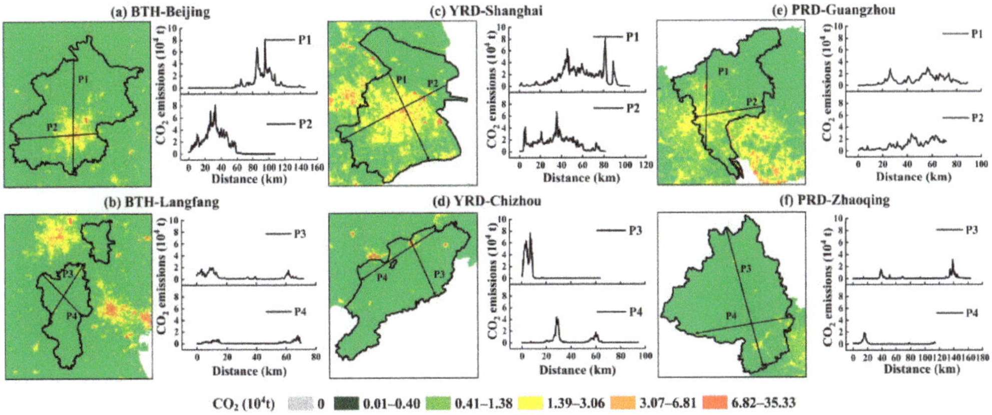

Figure 9. Profiles of CO_2 emissions along urbanization gradients in: (**a**) Beijing; (**b**) Langfang of Beijing-Tianjin-Hebei (BTH); (**c**) Shanghai; (**d**) Chizhou of Yangtze River Delta (YRD); (**e**) Guangzhou; and (**f**) Zhaoqing of Pearl River Delta (PRD).

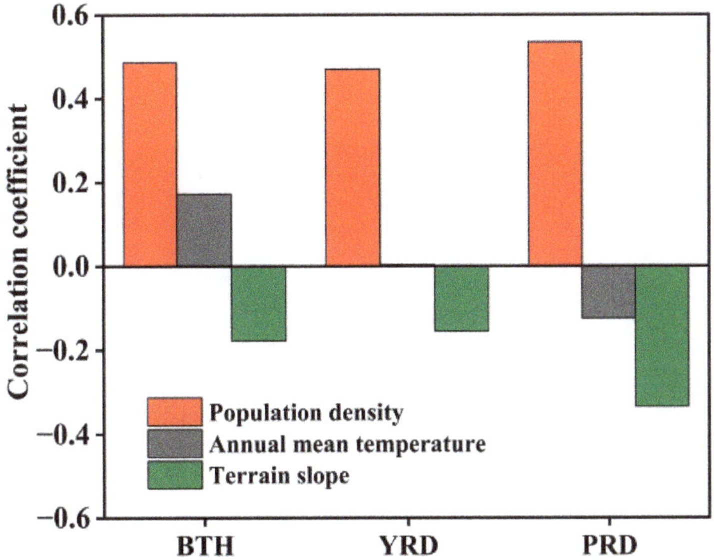

Figure 10. Correlations between CO_2 emissions and influencing factors, such as population density, annual mean temperature and terrain slope, in three major urban agglomerations. BTH: Beijing-Tianjin-Hebei; YRD: Yangtze River Delta; PRD: Pearl River Delta.

In general, there were significant positive correlations between carbon emissions and population density among three urban circles (Table 2), which indicated that CO_2 emissions were increasing when population density increased. In detail, the correlations between CO_2 emissions and population density in urban areas were more robust than other two urban circles, expect for Pearl River Delta, where CO_2 emissions in second-level urban circle has largest correlations with population density. Meanwhile, CO_2 emissions in urban areas of three urban agglomerations have the smallest correlation with the terrain slope, compared to other two urban circles. Additionally, temperature has the smallest impact on CO_2 emissions (Table 2).

Table 2. Relationships between CO_2 emissions and population density, annual mean temperature and terrain slope among urban circles, such as urban area, first-level urban circle and second-level urban circle.

Factor	Urban Agglomeration	Urban	First-Level Urban Circle	Second-Level Urban Circle
Population density	Beijing-Tianjin-Hebei	0.209 *	0.184 *	0.152 *
	Yangtze River Delta	0.257 *	0.087 *	0.188 *
	Pearl River Delta	0.333 *	0.172 *	0.384 *
Annual mean temperature	Beijing-Tianjin-Hebei	−0.013	−0.062	−0.012
	Yangtze River Delta	−0.020 *	−0.047	−0.016
	Pearl River Delta	−0.283 *	−0.122 *	0.025
Terrain slope	Beijing-Tianjin-Hebei	−0.026 *	−0.070 *	−0.091 *
	Yangtze River Delta	−0.060 *	−0.101 *	−0.113 *
	Pearl River Delta	−0.062 *	−0.178 *	−0.175 *

* Means significant at confidence level 5%.

4. Discussion

4.1. NPP/VIIRS Nighttime Light Integrated Genetic Neural Network Showed a Good Performance in Estimating CO_2 Emissions

Remote sensing nighttime light data can distinguish human urban areas with artificial lights from the dark background at night [14], and have a high correlation with the energy consumptions and carbon emissions [20]. One the one hand, the county-level statistical carbon emissions based on published energy use data were limited to coarse spatial/temporal resolution and short period [47]. Remote sensing nighttime light data were often selected as a good indicator to downscale the county-level statistical CO_2 emissions [13,26]. Because the nighttime light data have wide time-span and coverage [48]. On the other hand, a range of methods, including linear regression model [26], power regression model [24] and log-log regression model [32] have been used in modeling spatiotemporal patterns of CO_2 emissions based on nighttime light data and statistical energy consumption. However, traditional regression methods were not effective to track the nonlinear relationship between statistical CO_2 emissions and nighttime light data due to fixed parameters [17]. Genetic neural networks have higher feasibility and reliability than traditional regression models. Jasiński (2019) modeled total electricity consumption based on nighttime light images with artificial neural networks, and found that the results achieved by artificial neural networks has higher precision than that using linear regressions [18]. For three northeastern provinces in China, Yang et al. (2020) found that the performance of a neural network model was superior to traditional regression models in analyzing the nonlinear relationship between nighttime light data and statistical carbon emissions [17]. For central and western regions of China, Lin et al. (2022) confirmed that a deep neural network ensemble model was the best method to establish the relationship between the multi-dimensional data characteristics and carbon emissions [49]. Our study demonstrated that genetic neural network can estimate the spatiotemporal patterns of CO_2 emissions by integrating multiple datasets.

However, some limitations of nighttime light data will lead to uncertainty of CO_2 emissions regardless of the linear function, power function or even complex methods. To a certain extent, over-estimation of CO_2 emissions may exist in urban areas due to the limitation of nighttime light data. For example, some factories without any nighttime lights were likely to have intensive human activities and CO_2 emissions during the day [50]. In general, the statistical carbon emissions consisted of whole urban CO_2 emissions during the day and night. Many studies have reported that nighttime light data have a high correlation with energy consumptions and carbon emissions [17,35]. It indicated that brighter the nighttime lights, more daytime human activities in urban areas. For example, nighttime light data was commonly used as an indicator of overall energy use in urban, though electricity was often considered as the primary source of energy used for producing artificial light [51]. Therefore, remote sensing nighttime light data can still be considered as a good proxy of CO_2 emissions in large-scale areas.

4.2. Three Urban Agglomerations Exhibited Diverse Spatial Patterns of CO_2 Emissions

Beijing-Tianjin-Hebei Region behaved as three-core, Yangtze River Delta showed multiple-core and Pearl River Delta presented null-core structure in CO_2 emissions, respectively (Figures 4 and 5). The differential spatial patterns of CO_2 emissions may be ascribed to various development modes in three urban agglomerations. Beijing-Tianjin-Hebei Region is one of China's heavy industrial bases, which consists of a large number of high-energy-consuming enterprises, such as thermal power, steel and cement manufacturing. A Study indicated that energy was wasted in most cities in the Beijing-Tianjin-Hebei region [52]. Yangtze River Delta has low variations in urban development levels within agglomerations [53]. Shanghai, Zhejiang and Jiangsu Provinces in Yangtze River Delta could be classified as high development level regions, where CO_2 emissions are relatively high and concentrated. As a hotspot for investment in the manufacturing industry, Pearl River Delta is making efforts on promoting green energy and building the national green development [54]. For example, cities like Shenzhen has witnessed great progress in cleaner production [55]. Therefore, diverse spatial patterns of carbon emissions among urban agglomerations reflected differences in socioeconomic structure.

4.3. Population Density Versus Terrain Slope Featured Opposite Effects on Spatial Pattern of CO_2 Emissions

Positive relationships between population density and CO_2 emissions while negative correlations between terrain slope and CO_2 emissions were observed in three urban agglomerations (Figure 10). A study using STIRPAT model with panel data of China's 30 provinces from 1997 to 2012 indicated that population size has a strong explanatory power on CO_2 emissions [56]. The expansion of population scale directly leads to the extrusion of individual living space, posing a huge threat to population and environment [39]. Meanwhile, previous studies found that topographical factors were important limiting factors that influence population distribution and economic development [57]. Generally, the terrain of urban areas which less restricted the development of human activities is flat. But, in areas with complex terrain, such as mountainous areas and forest areas, human activities are greatly affected by slope, and the paving of roads and railways is also affected by natural environment. Therefore, inclination of terrain slope may decrease CO_2 emissions. In addition, CO_2 emissions decreased gradually with the increase of distance away from urban center in each urban agglomeration (Figures 8 and 9). Decrease in effects of population density while increase in effects of terrain slope on CO_2 emissions were found along the gradients from urban areas to first-level urban circles then to second-level urban circles (Table 2). So, differences in urban circles should be taken into consideration in formulating emission reduction policies.

5. Conclusions

Urban agglomerations, such as Beijing-Tianjin-Hebei, Yangtze River Delta and Pearl River Delta are the most developed regions in China, which are playing leadership roles in low-carbon development. We proposed a neural networks model based on nighttime light data to estimate CO_2 emissions of three major urban agglomerations from 2014 to 2019. The results indicated that spatial distribution of CO_2 emissions exhibited diversity among urban agglomerations. The areas appearing upward temporal trends of CO_2 emissions were larger than that with downward temporal trends in each urban agglomeration. The total amount of CO_2 emissions in the urban areas was largest among three urban circles, followed by first-level urban circle and second-level urban circle. In addition, population density had the greatest impact on spatial pattern of CO_2 emissions, while temperature had the least impact. Urban agglomerations should coordinate development strategies in economic growth versus carbon reduction.

Author Contributions: Conceptualization, L.Z., J.S. and Y.C.; Methodology, L.Z., J.S. and Y.C.; Formal analysis, L.Z. and J.S; Writing-original draft, L.Z. and J.S.; Writing–review & editing, L.Z., Y.C. and Q.Y.; Supervision, Y.C.; funding acquisition, Y.C. All authors have read and agreed to the published version of the manuscript.

Funding: This research was funded by National Natural Science Foundation of China (41871084), National Key Research and Development Program of China (2017YFB0504000), Soft Science Research Program of Zhejiang Provincial Department of Science and Technology (2022C35095), Jinhua Science and Technology Research Program (2021-4-340 and 2020-4-184), and Self-Design Project in Zhejiang Normal University (2021ZS07).

Data Availability Statement: The multi-source data used in this paper are publicly available, and the dataset information is shown in Table 1.

Conflicts of Interest: The authors declare no conflict of interest.

References

1. Pablo-Romero, M.d.P.; Pozo-Barajas, R.; Yñiguez, R. Global changes in residential energy consumption. *Energy Policy* **2017**, *101*, 342–352. [CrossRef]
2. Zheng, J.; Dong, S.; Hu, Y.; Li, Y. Comparative analysis of the CO_2 emissions of expressway and arterial road traffic: A case in Beijing. *PLoS ONE* **2020**, *15*, e0231536. [CrossRef] [PubMed]
3. Shi, K.; Yu, B.; Zhou, Y.; Chen, Y.; Yang, C.; Chen, Z.; Wu, J. Spatiotemporal variations of CO_2 emissions and their impact factors in China: A comparative analysis between the provincial and prefectural levels. *Appl. Energy* **2019**, *233–234*, 170–181. [CrossRef]
4. Zhou, D.; Zhao, S.; Liu, S.; Zhang, L.; Zhu, C. Surface urban heat island in China's 32 major cities: Spatial patterns and drivers. *Remote Sens. Environ.* **2014**, *152*, 51–61. [CrossRef]
5. Grimm, N.B.; Faeth, S.H.; Golubiewski, N.E.; Redman, C.L.; Wu, J.; Bai, X.; Briggs, J.M. Global change and the ecology of cities. *Science* **2008**, *319*, 756–760. [CrossRef]
6. Dong, F.; Yu, B.; Hadachin, T.; Dai, Y.; Wang, Y.; Zhang, S.; Long, R. Drivers of carbon emission intensity change in China. *Resour. Conserv. Recycl.* **2018**, *129*, 187–201. [CrossRef]
7. Yu, X.; Wu, Z.Y.; Zheng, H.R.; Li, M.Q.; Tan, T.L. How urban agglomeration improve the emission efficiency? A spatial econometric analysis of the Yangtze river delta urban agglomeration in China. *J. Environ. Manag.* **2020**, *263*, 110061. [CrossRef]
8. Chen, J.; Gao, M.; Cheng, S.; Liu, X.; Hou, W.; Song, M.; Li, D.; Fan, W. China's city-level carbon emissions during 1992–2017 based on the inter-calibration of nighttime light data. *Sci. Rep.* **2021**, *11*, 3323. [CrossRef]
9. Cai, B.; Zhang, L. Urban CO_2 emissions in China: Spatial boundary and performance comparison. *Energy Policy* **2014**, *66*, 557–567. [CrossRef]
10. Dhakal, S. Urban energy use and carbon emissions from cities in China and policy implications. *Energy Policy* **2009**, *37*, 4208–4219. [CrossRef]
11. Sheng, Y.; Miao, Y.; Song, J.; Shen, H. The Moderating Effect of innovation on the relationship between urbanization and CO_2 emissions: Evidence from three major urban agglomerations in China. *Sustainability* **2019**, *11*, 1633. [CrossRef]
12. Sheng, P.; Guo, X. The long-run and short-run impacts of urbanization on carbon dioxide emissions. *Econ. Model.* **2016**, *53*, 208–215. [CrossRef]
13. Shi, K.; Chen, Y.; Li, L.; Huang, C. Spatiotemporal variations of urban CO_2 emissions in China: A multiscale perspective. *Appl. Energy* **2018**, *211*, 218–229. [CrossRef]
14. Elvidge, C.D.; Baugh, K.E.; Kihn, E.A.; Kroehl, H.W.; Davis, E.R.; Davis, C.W. Relation between satellite observed visible-near infrared emissions, population, economic activity and electric power consumption. *Int. J. Remote Sens.* **1997**, *18*, 1373–1379. [CrossRef]
15. Oda, T.; Maksyutov, S.; Elvidge, C.D. Disaggregation of national fossil fuel CO_2 emissions using a global power plant database and DMSP nightlight data. *Proc. Asia-Pac. Adv. Netw.* **2010**, *30*, 219–228. [CrossRef]
16. Ou, J.; Liu, X.; Li, X.; Li, M.; Li, W. Evaluation of NPP-VIIRS Nighttime Light Data for Mapping Global Fossil Fuel Combustion CO_2 Emissions: A Comparison with DMSP-OLS Nighttime Light Data. *PLoS ONE* **2015**, *10*, e0138310. [CrossRef] [PubMed]
17. Yang, D.; Luan, W.; Qiao, L.; Pratama, M. Modeling and spatio-temporal analysis of city-level carbon emissions based on nighttime light satellite imagery. *Appl. Energy* **2020**, *268*, 114696. [CrossRef]
18. Jasiński, T. Modeling electricity consumption using nighttime light images and artificial neural networks. *Energy* **2019**, *179*, 831–842. [CrossRef]

19. Li, H.; Zhao, Y.; Qiao, X.; Liu, Y.; Cao, Y.; Li, Y.; Wang, S.; Zhang, Z.; Zhang, Y.; Weng, J. Identifying the driving forces of national and regional CO_2 emissions in China: Based on temporal and spatial decomposition analysis models. *Energy Econ.* **2017**, *68*, 522–538. [CrossRef]
20. Mi, Z.; Zheng, J.; Meng, J.; Zheng, H.; Li, X.; Coffman, D.M.; Woltjer, J.; Wang, S.; Guan, D. Carbon emissions of cities from a consumption-based perspective. *Appl. Energy* **2019**, *235*, 509–518. [CrossRef]
21. Chen, L.; Xu, L.; Cai, Y.; Yang, Z. Spatiotemporal patterns of industrial carbon emissions at the city level. *Resour. Conserv. Recycl.* **2021**, *169*, 105499. [CrossRef]
22. Doll, C.H.; Muller, J.P.; Elvidge, C.D. Night-time imagery as a tool for global mapping of socioeconomic parameters and greenhouse gas emissions. *AMBIO* **2000**, *29*, 157–162. [CrossRef]
23. Wang, S.; Liu, X. China's city-level energy-related CO_2 emissions: Spatiotemporal patterns and driving forces. *Appl. Energy* **2017**, *200*, 204–214. [CrossRef]
24. Shi, K.; Chen, Z.; Cui, Y.; Wu, J.; Yu, B. NPP-VIIRS nighttime light data have different correlated relationships with fossil fuel combustion carbon emissions from different sectors. *IEEE Geosci. Remote Sens. Lett.* **2020**, *18*, 2062–2066. [CrossRef]
25. Zhao, J.; Chen, Y.; Ji, G.; Wang, Z. Residential carbon dioxide emissions at the urban scale for county-level cities in China: A comparative study of nighttime light data. *J. Clean. Prod.* **2018**, *180*, 198–209. [CrossRef]
26. Shi, K.; Yu, B.; Huang, Y.; Hu, Y.; Yin, B.; Chen, Z.; Chen, L.; Wu, J. Evaluating the ability of NPP-VIIRS nighttime light data to estimate the gross domestic product and the electric power consumption of China at multiple scales: A comparison with DMSP-OLS data. *Remote Sens.* **2014**, *6*, 1705–1724. [CrossRef]
27. Sun, T.; Sun, R.; Khan, M.S.; Chen, L. Urbanization increased annual precipitation in temperate climate zone: A case in Beijing-Tianjin-Hebei region of North China. *Ecol. Indic.* **2021**, *126*, 107621. [CrossRef]
28. Chen, L.; Liu, M.; Xu, Z.; Fan, R.; Tao, J.; Chen, D.; Zhang, D.; Xie, D.; Sun, J. Variation trends and influencing factors of total gaseous mercury in the Pearl River Delta—A highly industrialised region in South China influenced by seasonal monsoons. *Atmos. Environ.* **2013**, *77*, 757–766. [CrossRef]
29. Ye, Y.; Zhang, H.; Liu, K.; Wu, Q. Research on the influence of site factors on the expansion of construction land in the Pearl River Delta, China: By using GIS and remote sensing. *Int. J. Appl. Earth Obs. Geoinf.* **2013**, *21*, 366–373. [CrossRef]
30. Song, Y.; Long, Y.; Wu, P.; Wang, X. Are all cities with similar urban form or not? Redefining cities with ubiquitous points of interest and evaluating them with indicators at city and block levels in China. *Int. J. Geogr. Inf. Sci.* **2018**, *32*, 2447–2476. [CrossRef]
31. Zhang, W.; Cui, Y.; Wang, J.; Wang, C.; Streets, D.G. How does urbanization affect CO_2 emissions of central heating systems in China? An assessment of natural gas transition policy based on nighttime light data. *J. Clean. Prod.* **2020**, *276*, 123188. [CrossRef]
32. Shi, K.; Yu, B.; Hu, Y.; Huang, C.; Chen, Y.; Huang, Y.; Chen, Z.; Wu, J. Modeling and mapping total freight traffic in China using NPP-VIIRS nighttime light composite data. *GISci. Remote Sens.* **2015**, *52*, 274–289. [CrossRef]
33. Liu, H.; Ma, L.; Xu, L. Estimating spatiotemporal dynamics of county-level fossil fuel consumption based on integrated nighttime light data. *J. Clean. Prod.* **2021**, *278*, 123427. [CrossRef]
34. Bennett, M.M.; Smith, L.C. Advances in using multitemporal night-time lights satellite imagery to detect, estimate, and monitor socioeconomic dynamics. *Remote Sens. Environ.* **2017**, *192*, 176–197. [CrossRef]
35. Su, Y.; Chen, X.; Li, Y.; Liao, J.; Ye, Y.; Zhang, H.; Huang, N.; Kuang, Y. China's 19-year city-level carbon emissions of energy consumptions, driving forces and regionalized mitigation guidelines. *Renew. Sustain. Energy Rev.* **2014**, *35*, 231–243. [CrossRef]
36. Fang, C.; Wang, S.; Li, G. Changing urban forms and carbon dioxide emissions in China: A case study of 30 provincial capital cities. *Appl. Energy* **2015**, *158*, 519–531. [CrossRef]
37. Rumelhart, D.E.; Hinton, G.E.; Williams, R.J. *Learning Internal Representations by Error Propagation*; California University San Diego, Institute for Cognitive Science: La Jolla, CA, USA, 1985.
38. Yu, L.; He, Y. Evaluation of sports training effect based on GABP neural network and artificial intelligence. *J. Ambient. Intell. Humaniz. Comput.* **2021**, 1–11. [CrossRef]
39. Wang, P.; Wu, W.; Zhu, B.; Wei, Y. Examining the impact factors of energy-related CO_2 emissions using the STIRPAT model in Guangdong Province, China. *Appl Energy* **2013**, *106*, 65–71. [CrossRef]
40. Oda, T.; Maksyutov, S. A very high-resolution (1 km × 1 km) global fossil fuel CO_2 emission inventory derived using a point source database and satellite observations of nighttime lights. *Atmos. Chem. Phys.* **2011**, *11*, 543–556. [CrossRef]
41. Ke, N.; Lu, X.; Zhang, X.; Kuang, B.; Zhang, Y. Urban land use carbon emission intensity in China under the "double carbon" targets: Spatiotemporal patterns and evolution trend. *Environ. Sci. Pollut. Res. Int.* **2022**, 1–14. [CrossRef]
42. Li, X.; Zhang, L.; Liang, C. A GIS-based buffer gradient analysis on spatiotemporal dynamics of urban expansion in Shanghai and its major satellite cities. *Procedia Environ. Sci.* **2010**, *2*, 1139–1156. [CrossRef]
43. He, C.; Ma, Q.; Li, T.; Yang, Y.; Liu, Z. Spatiotemporal dynamics of electric power consumption in Chinese Mainland from 1995 to 2008 modeled using DMSP/OLS stable nighttime lights data. *J. Geogr. Sci.* **2012**, *22*, 125–136. [CrossRef]
44. Puliafito, S.E.; Puliafito, J.L.; Grand, M.C. Modeling population dynamics and economic growth as competing species: An application to CO_2 global emissions. *Ecol. Econ.* **2008**, *65*, 602–615. [CrossRef]
45. Nie, H.G.; Kemp, R.; Xu, J.H.; Vasseur, V.; Fan, Y. Drivers of urban and rural residential energy consumption in China from the perspectives of climate and economic effects. *J. Clean. Prod.* **2018**, *172*, 2954–2963. [CrossRef]
46. Zhu, D.; Tao, S.; Wang, R.; Shen, H.; Huang, Y.; Shen, G.; Wang, B.; Li, W.; Zhang, Y.; Chen, H.; et al. Temporal and spatial trends of residential energy consumption and air pollutant emissions in China. *Appl. Energy* **2013**, *106*, 17–24. [CrossRef]

47. Guan, D.; Liu, Z.; Geng, Y.; Lindner, S.; Hubacek, K. The gigatonne gap in China's carbon dioxide inventories. *Nat. Clim. Chang.* **2012**, *2*, 672–675. [CrossRef]
48. Chen, J.; Gao, M.; Cheng, S.; Hou, W.; Song, M.; Liu, X.; Liu, Y.; Shan, Y. County-level CO_2 emissions and sequestration in China during 1997–2017. *Sci. Data* **2020**, *7*, 391. [CrossRef]
49. Lin, X.; Ma, J.; Chen, H.; Shen, F.; Ahmad, S.; Li, Z. Carbon emissions estimation and spatiotemporal analysis of china at city level based on multi-dimensional data and machine learning. *Remote Sens.* **2022**, *14*, 3014. [CrossRef]
50. Zhao, N.; Samson, E.L.; Currit, N.A. Nighttime-lights-derived fossil fuel carbon dioxide emission maps and their limitations. *Photogramm. Eng. Remote Sens.* **2015**, *81*, 935–943. [CrossRef]
51. Mellander, C.; Lobo, J.; Stolarick, K.; Matheson, Z. Night-time light data: A good proxy measure for economic activity? *PloS ONE* **2015**, *10*, e0139779. [CrossRef]
52. Sun, J.; Wang, Z.; Zhu, Q. Analysis of resource allocation and environmental performance in China's three major urban agglomerations. *Environ. Sci. Pollut. Res. Int.* **2020**, *27*, 34289–34299. [CrossRef] [PubMed]
53. Zhou, Y.; Chen, M.X.; Tang, Z.P.; Mei, Z.A. Urbanization, land use change, and carbon emissions: Quantitative assessments for city-level carbon emissions in Beijing-Tianjin-Hebei region. *Sustain. Cities Soc.* **2021**, *66*, 102701. [CrossRef]
54. Wang, M.X.; Zhao, H.H.; Cui, J.X.; Fan, D.; Lv, B.; Wang, G.; Li, Z.H.; Zhou, G.J. Evaluating green development level of nine cities within the Pearl River Delta, China. *J. Clean. Prod.* **2018**, *174*, 315–323. [CrossRef]
55. Wang, S.J.; Zeng, J.Y.; Huang, Y.Y.; Shi, C.Y.; Zhan, P.Y. The effects of urbanization on CO_2 emissions in the Pearl River Delta: A comprehensive assessment and panel data analysis. *Appl. Energy* **2018**, *228*, 1693–1706. [CrossRef]
56. Wang, Y.A.; Kang, Y.Q.; Wang, J.; Xu, L.N. Panel estimation for the impacts of population-related factors on CO_2 emissions: A regional analysis in China. *Ecol. Indic.* **2017**, *78*, 322–330. [CrossRef]
57. Zhang, J.J.; Zhu, W.B.; Zhu, L.Q.; Cui, Y.P.; He, S.S.; Ren, H. Topographical relief characteristics and its impact on population and economy: A case study of the mountainous area in western Henan, China. *J. Geogr. Sci.* **2019**, *29*, 598–612. [CrossRef]

remote sensing

MDPI

Article

Nightlight Intensity Change Surrounding Nature Reserves: A Case Study in Orbroicher Bruch Nature Reserve, Germany

Jillian LaRoe [1,*], Christopher M. Holmes [1] and Thorsten Schad [2]

[1] Applied Analysis Solutions, 535 McDonald Rd, Winchester, VA 22602, USA
[2] Bayer AG, Alfred-Nobel-Str. 50, 40789 Monheim, Germany
* Correspondence: jillianlaroe@appliedanalysis.solutions

Abstract: Persistent global urbanization has a direct relationship to measurable artificial light at night (ALAN), and the Defense Meteorological Satellite Program has served an important role in monitoring this relationship over time. Recent studies have observed significant declines in insect abundance and populations, and ALAN has been recognized as a contributing factor. We investigated changes in nightlight intensity at various spatial scales surrounding insect traps located in Orbroicher Bruch Nature Reserve, Germany. Using a time series of global nighttime light imagery (1992–2010), we evaluated pixel-level trends through linear regressions and the Mann–Kendall test. Paired with urban land cover delineation, we compared nightlight trends across rural and urban areas. We utilized high-resolution satellite imagery to identify landscape features potentially related to pixel-level trends within areas containing notable change. Approximately 96% of the pixel-level trends had a positive slope, and 22% of pixels experienced statistically significant increases in nightlight intensity. We observed that 80% of the region experienced nightlight intensity increases >1%, concurrent with the observed decline in insect biomass. While it is unclear if these trends extend to other geographic regions, our results highlight the need for future studies to concurrently investigate long-term trends in ALAN and insect population decline across multiple scales, and consider the spatial and temporal overlaps between these patterns.

Keywords: nighttime lights; DMSP-OLS; data fusion; urbanization processes; insect decline; artificial light at night

Citation: LaRoe, J.; Holmes, C.M.; Schad, T. Nightlight Intensity Change Surrounding Nature Reserves: A Case Study in Orbroicher Bruch Nature Reserve, Germany. *Remote Sens.* **2022**, *14*, 3876. https://doi.org/10.3390/rs14163876

Academic Editors: Ran Goldblatt, Steven Louis Rubinyi, Hogeun Park and Tilottama Ghosh

Received: 8 June 2022
Accepted: 29 July 2022
Published: 10 August 2022

1. Introduction

A decline in insect abundance and populations has been observed across many regions in recent decades [1,2]. Previous studies measuring insect biomass through insect traps have found significant declines in biomass through long-term, multiyear observations [2–4]. Owens et al. [1] suggest that the combination of climate change, habitat loss, chemical pollutants, invasive species, and artificial light at night (ALAN) are key factors driving insect decline. Approximately 30% of vertebrates and 60% of invertebrates across the world are nocturnal [5]. Light is an important factor for many key behavioral and biological processes of insects. They rely on nocturnal light for navigation, avoiding predation, foraging, reproduction, and regulating their biological clocks [1,2], and ALAN has been observed to affect these biological processes (Owens et al., 2020). ALAN has been identified as a factor directly contributing to declines in moth populations and other insects with strong phototactic responses [5,6]. Additionally, ALAN has been observed to influence the community composition of aquatic primary producers [2].

The increases in ALAN have been primarily attributed to human settlements and the process of urbanization. ALAN is widespread and has been increasing over the last decades worldwide at an annual rate of 2–6%, imposing an unprecedented alteration of natural light regimes [2,5]. The impacts of ALAN are not limited to the immediate surroundings of the light source. More remote areas can be affected by light pollution [1,2] through skyglow [7,8], which can reach dozens of kilometers beyond the original source [9,10] and the effects are amplified by overcast skies [11]; therefore, even nature reserves and other protected areas are not always shielded from ALAN pollution [12]. Given the average increasing rate of ALAN worldwide, the effects of ALAN on declining insect populations should be further investigated at local and regional spatial scales [7].

The Defense Meteorological Satellite Program Operational Linescan System (DMSP-OLS) satellites detect low levels of electromagnetic radiation emitted at night at wavelengths within the visible and near-infrared (VNIR) portion of the electromagnetic spectrum, spanning 0.5 nm–0.9 nm [13–15]. Documented sources of nightlight emission include city lights, gas flares, and wildfires [13–15]. Previous studies have found that nightlight imagery can be used to map urban areas since spatiotemporal patterns in nightlight intensity correspond to patterns in modern human settlements [13,15,16]. Artificial lights at night have also been used to develop indicators of economic productivity [17]. In Europe, trends in ALAN over time captured by DMSP-OLS satellites have been related to specific landscape changes [18]; increases in ALAN have generally been attributed to suburban and industrial development, and decreases related to energy efficiency improvements and periods of economic decline [18].

Increases in ALAN at different scales can impact trophic interactions [19], ecological functions [2,20], and alter biodiversity [20,21]. Some studies have emphasized the need to consider and manage ALAN within protected areas [12,22]. The impacts of ALAN on populations, communities, and ecosystems vary across regions and spatial scales. Previous studies utilizing remotely sensed imagery and landscape ecology concepts have found that ALAN has significantly reduced the areas considered to be suitable habitat (lacking nighttime light pollution) and has negatively impacted biodiversity within natural reserves [12]. ALAN impacting protected nature reserves is a globally observed phenomenon. Fan et al. [23] suggest that biodiversity data should be paired with nightlight satellite imagery to investigate the effects of light pollution and optimize buffer distances surrounding protected areas. The impacts of lighting on ecosystems vary based on the economic and social contexts, suggesting that the optimal buffer distance around protected areas differs for each unique scenario [23,24]. There is not a one-size-fits-all solution for optimizing the configuration of light pollution, anthropogenic structures, and their proximity to nature reserves, presenting the need to consider all relevant factors in landscape planning.

In this study, we investigated changes in nightlight intensity captured by satellite imagery at varying radii (1, 2, and 10 km) surrounding the two insect trap locations in the Orbroicher Bruch Nature Reserve in Germany utilized by Hallmann et al. [4]. First, we tested trends in ALAN between 1992 and 2010, using global nightlight intensity satellite imagery to investigate pixel-level changes using linear regressions and the Mann–Kendall test. Paired with an urban land cover designation from CORINE (2012) and high-resolution satellite imagery, we compared trends of nightlight intensity through time across rural and urban areas. Within areas where we identified significant changes in nightlight intensity, we attributed the specific alterations on the land surface as potential causes or contributing factors to the change using a combination of aerial imagery, historical records, and expert knowledge. By pairing the timeframe of change with specific surface features, we outline a methodology that highlights specific landscape changes to consider as potential contributors to insect decline in future investigations. We present a novel investigation of observed insect decline and nightlight satellite imagery corresponding to the time period and geographic region [4]. These methods have the potential to be further developed and refined to assess the efficacy of nightlight mitigation efforts, identify structures with significant impacts on nightlight pollution, consider the effects of changes from smaller intense light and wider area diffuse light, and further support landscape management decisions in the context of insect conservation.

2. Materials and Methods

2.1. Study Area and Time Period

Within North Rhine-Westfalia, Germany we examined the landscape within a 10 km radius surrounding two insect trap locations (Figure 1) to investigate changes in nightlight intensity between 1992 and 2010. The total area spans roughly 318 km^2, and overlaps with 11 municipalities, although the majority of the study area is comprised of Krefeld, Kempen, Rheurdt, Tönisvorst, Kerken, and Neukirchen-Vluyn. The traps are located near Krefeld within Orbroicher Bruch, which is a nature reserve. Nature reserves are a class of protected areas under Germany's Federal Nature Conservation Act primarily used to facilitate species conservation. Orbroicher Bruch covers roughly 1 km^2 of protected forest, wet grasslands, and marshes. The surrounding region is comprised of cropland, forest, grasslands, and urban areas. The population of North Rhine-Westphalia has grown from approximately 17.1 million in 1989 (source) to 17.9 million in 2010 (retrieved from data commons timelines based on Europa.eu, measurement method = Eurostat regional population data). Previous studies using satellite imagery to model land cover have estimated that impervious surfaces from urban development have increased by roughly 30% (1670 km^2) between 1985 and 2017 in North Rhine-Westphalia [25].

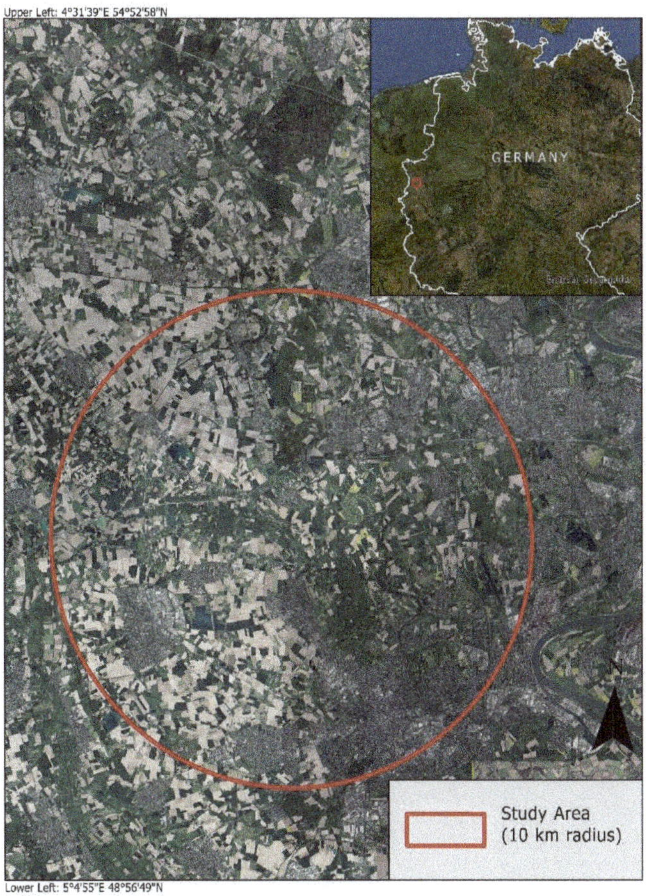

Figure 1. Study area spanning a 10 km radius around insect trap locations near Krefeld in Orbroicher Bruch Nature Reserve in North Rhine-Westphalia, Germany.

2.2. Remote Sensing Data (DMSP-OLS Nighttime Lights Time Series)

We gathered a time series of satellite images between 1992 and 2010 of Global Night-Time Light intensity [Version 4] captured by the DMSP-OLS at a 30 arc-second spatial resolution (roughly 1.5 km × 0.9 km at 51.4° N) (Figure 2) (data retrieved from https://ngdc.noaa.gov/eog/dmsp/downloadV4composites.html, accessed on 20 April 2022). We selected the DMSP-OLS satellite imagery since it provided global data measuring nightlight intensity corresponding to our time period of interest. Images covering the 10 km radius region surrounding the Orbroicher Bruch Nature Reserve were acquired from seven DMSP-OLS satellites (F10, F11, F12, F14, F15, F16, and F18) to capture the full 19-year period. This dataset contains annual, cloud-free composites of light intensity at night and has been processed to remove sunlit images, moonlit images, images containing glare, as well as any light features from the aurora in the Northern hemisphere [15,26]. In years where two annual composites were available because two satellites collected overlapping data, we selected a single composite collected by one satellite for the given year to maintain consistency with the years with only one composite. We selected the stable lights average band within this dataset, which covers cities, towns, and other regions with perpetual light emission; these data are provided in digital numbers and have been filtered for background noise from other sources of VNIR emissions (e.g., gas flares) [15,26]. Since the DMSP-OLS does not have on-board calibration, it is not feasible to compare digital numbers across multiple years without additional processing [27]. Therefore, we performed an intercalibration of DMSP-OLS images using the software GRASS GIS (Version 7.4) based on the coefficients and approach of Wu et al. [27].

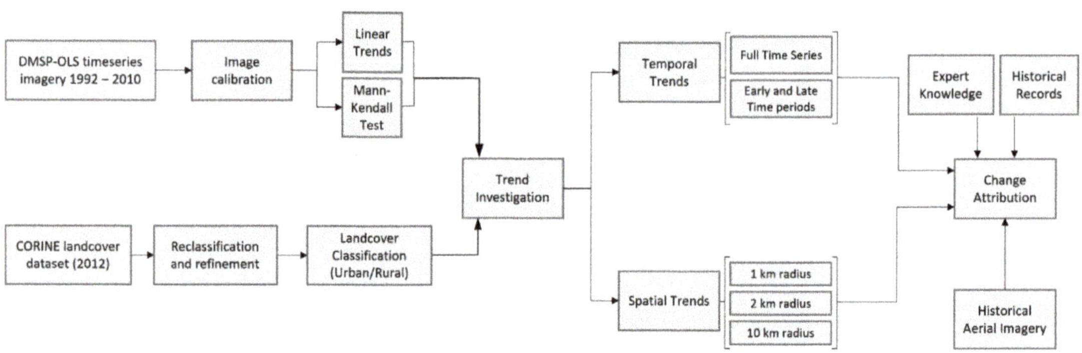

Figure 2. Methodological workflow for processing DMSP-OLS data, analyzing pixel-level trends using linear regressions and the Mann–Kendall test, and refining the CORINE dataset to reflect urban and rural land cover. Combining the trend analyses with the urban/rural land cover classes served as the basis for investigating temporal and spatial trends at various scales. Further integrating aerial imagery, expert knowledge, and historical records provided a framework for attributing local changes in nightlight intensity to specific landscape features.

2.3. Remote Sensing Trend Analysis

We assessed spatial and temporal trends in nightlight intensity values from the calibrated DMSP-OLS time series at the individual pixel scale to identify significant changes in nightlight intensity over time (Figure 2). To identify trends, we estimated linear regressions on a pixel-level basis and examined the regression slope within each pixel. We selected the Mann–Kendall statistical test for use in this analysis as a non-parametric test for environmental time series data, to measure the direction and strength of monotonic, pixel-level trends. We used R [V 3.4.3] to estimate linear regressions and apply the Mann–Kendall test using the "Kendall" package [V 2.2] [28–31]. The results from the Mann–Kendall test include Kendall's tau statistic, which indicates trend strength and direction, as well as a two-sided p-value to determine the statistical significance of the monotonic trends.

Kendall's tau statistic ranges from –1 to 1, where positive or negative values closer to zero are less significant, and values closer to −1 or 1 exhibit stronger trends, with the sign indicating the directionality of the trend.

We applied this methodology to the full time period (1992–2010) as well as the early (1992–1999) and late (2003–2010) time periods we defined to analyze the differences between the beginning and end of the investigation period. For the subset of annual images captured within both the early and late time periods, the median nightlight intensity value was calculated for each pixel. Additionally, we calculated the relative difference between the early and late time periods for each pixel to identify substantial changes in nightlight intensity in areas where values did not fall on the extreme ends of the spectrum. The full time period of interest captured the overall trend, whereas the comparisons between the early and late time periods allowed us to compare regional changes in nightlight intensity.

2.4. Urbanization Analysis

In order to explore potential causes of change in nightlight intensity, we first characterized land use and land cover (LULC) across our study area, since urbanization coincides with increasing nightlight intensity (Figure 2). We utilized the CORINE land cover dataset from 2012 (data retrieved from https://land.copernicus.eu/pan-european/corine-land-cover/clc-2012, accessed on 4 May 2022) to delineate urban and rural areas within the full study area. We selected all areas defined as continuous (1.1.1) and discontinuous urban fabric (1.1.2), construction sites (1.3.3), dump sites (1.3.2), as well as industrial or commercial units (1.2.1) to represent urban land cover. Port areas (1.2.3) and airports (1.2.4) were not present within the study region, while green urban areas (1.4.1) and sport and leisure facilities (1.4.2) were not included in the urban areas we defined since they contain vegetation cover. All areas that did not meet these criteria were classified as rural (i.e., not urban); rural areas included farmland, water, forests, and other land cover classes, although many of these areas may contain sparse or patchy developed areas (human features).

To connect this classification with the nightlights time series imagery, a grid representing the individual pixels of the nightlight's dataset was intersected with the rural/urban land cover classification, and pixels were designated as urban or rural based on the majority type (≥50% of the given pixel). We summarized the total area of pixels classified as rural and urban across the full study area for 2012. In addition, we compared the pixel-level trends in nightlight intensity (from linear regressions based on the early and late time periods and the Mann-Kendall test across the full study period) between the pixels classified as urban and rural to characterize patterns occurring within these contrasting land cover types.

2.5. Change Attribution

We utilized pixel-level statistics to identify temporal and spatial trends in nightlight intensity indicative of landscape feature changes, and to describe potential agents or features contributing to the pixel-level trend (Figure 2). Within the full time period (1992–2010), we used the Mann–Kendall test to identify pixel-level instances of relative, significant local and regional changes in nightlight intensity within 1, 2, and 10 km radii around the insect trap locations. Within the 1 and 2 km radii, we considered all pixels at each scale, and identified all potentially relevant features that had changed; we chose to examine all surrounding features at these scales due to the effects that direct lighting can have on insect habitats and nature preserves. We also examined the 10 km radius (full study area) to identify regional changes in nightlight intensity; it is important to consider this scale based on the range of insect movement, insect metapopulation dynamics, and because nightlight pollution (indirect effects) can span many kilometers, especially with overcast skies. Within the 10 km radius, we selected two subsets of pixels that met specific criteria to further explore local, feature-based changes. The first group of pixels (ID's 1–10) had a slope > 0 and $R^2 > 0.3$ based on the linear trend analysis; they also had a relative early to late difference (between 1992–1999 and 2003–2010) > 0 and statistically significant change ($\alpha < 0.1$). The second group of pixels that was further investigated (ID's 11–17) had high slope values >0.86. These pixels were not selected based on any level of alpha significance or any particular R^2 value.

At each spatial scale considered, we utilized historical aerial imagery to investigate sub-pixel features (Figure 2). Historical imagery collected through the North Rhine-Westphalia aerial photography catalog was provided by the District Government of Cologne, and captured in 1989 (start of the study period) at a 40 × 40 cm spatial resolution and in 2013 (end of the study period) at a 10 × 10 cm spatial resolution. Chronological aerial imagery from Google Earth was also used to supplement the identification of changes that took place during the study period. Features of interest identified via aerial imagery were further verified by incorporating expert knowledge and relevant, supporting documentation (Figure 2). External expert knowledge included a local landscape photographer, Paul Maaßen, and community officials in Krefeld. Identifying specific features that changed based on historical aerial imagery, and supporting these findings with external knowledge and documentation, provided the framework for relating these landscape features to observed trends in nightlight intensity.

While the spatial resolution of the DMSP-OLS time series data is coarser than the landscape features investigated, we also considered that a variety of sub-pixel spatial configurations of landscape features could contribute to the pixel-level trend. For example, a pixel with a one percent increase in nightlight intensity has many potential spatial configurations, as displayed in Figure A1. A single hectare that experienced a 100 percent increase in nightlight intensity, or 20 hectares that experienced a five percent increase in nightlight intensity, are both potential spatial configurations of the generalized pixel-level trend (one percent increase). The variety of sub-pixel spatial configurations responsible for pixel-level trends is an influential factor that was considered throughout the process of evaluating and attributing changes to specific landscape features.

3. Results

3.1. Image (Pre-)Processing

Nightlight intensity values from DMSP-OLS V4 image collections are stored as digital numbers which are the raw values captured at the sensor corresponding to the intensity of reflectance. The original digital numbers range from 1 to 63, from lowest to highest intensity; the stable lights average composites were processed to remove the temporary events causing background noise, and these values were replaced with 0. Areas completely covered by clouds are represented by the value 255. After intercalibration of the images using the invariant region method [27], the digital numbers of the nightlight intensity values are rescaled, as shown in Figure 3 (ranging from 1 to 100+). In both the uncalibrated and calibrated DMSP-OLS time series images, pixel regions with the greatest nightlight intensity (highest values, shown in yellow) generally correspond with urban areas. The areas with the brightest nightlight can be identified in the southeastern portion of the study area in Figure 3b, which geographically corresponds to the areas classified as urban in Figures 1 and 4.

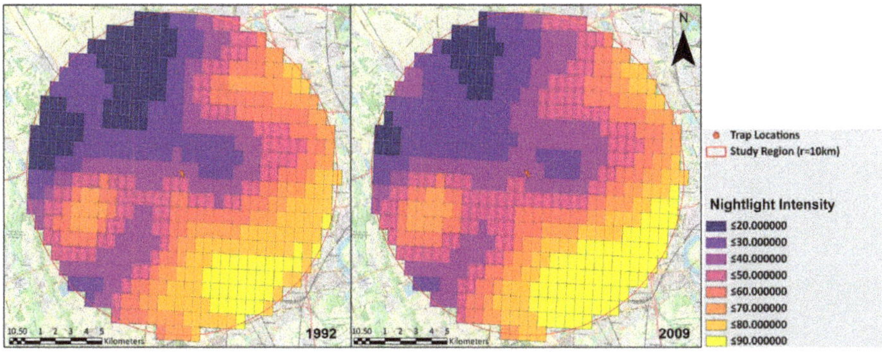

(a) (b)

Figure 3. Example of intercalibrated DMSP-OLS intensity data over the study area in the first (**a**, 1992) and second to last (**b**, 2009) years of the time series. The southeast portion of the study area (highlighted most in **b**) corresponds to the highest density urban areas.

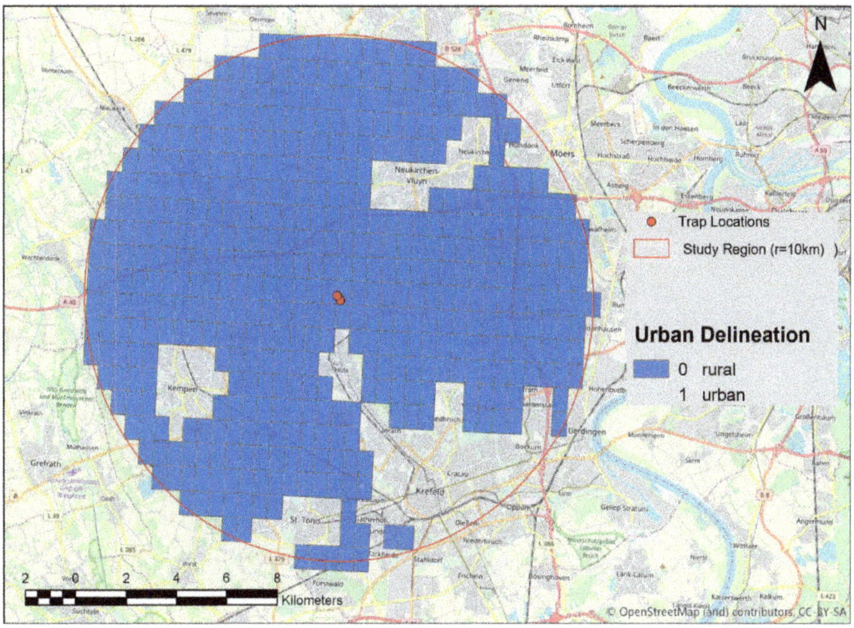

Figure 4. Classification of rural (blue) and urban (transparent) areas based on the 2012 CORINE land cover data within a 10 km buffer around the insect trap locations. Urban classification is based on the reassignment of urban classes from the original CORINE dataset and their spatial relationships to city administrative units. The majority of the study area was classified as rural.

3.2. Nightlight Trend Analysis

3.2.1. Overall Trends

The linear regression estimation and Mann–Kendall test revealed comparable relationships between increasing urbanization and increasing intensity values. Across the full time series (1992–2010, DMSP-OLS), we observed a general increase in nightlight intensity that was more prominent within localized clusters (Figure 5), likely reflecting development and urbanization patterns. Approximately 22% of the region demonstrated statistically significant ($\alpha < 0.1$) trends in nightlight intensity based on the Mann–Kendall test, and 13.4% of the region had statistically significant trends of increasing or decreasing nightlight intensity over the 19-year period at $\alpha < 0.05$ (Table A1). Based on the Mann–Kendall tau statistic, many of the statistically significant changes in nightlight intensity were clustered in the northwest of the study area (Figure 5). Decreases in nightlight intensity were only observed in 13 of the 590 (3%) pixels covering the region (also indicated by the Mann–Kendal tau statistic) (Figure 5).

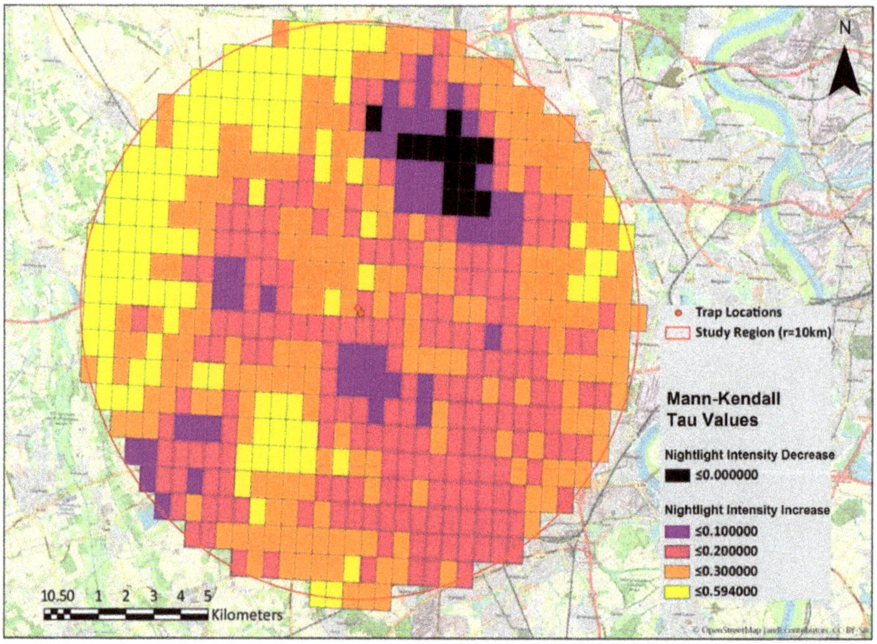

Figure 5. Mann–Kendall tau statistic across the region for the full time period; values closer to −1 and 1 represent more significant trends and values closer to 0 are less significant. Positive and negative values represent the directionality of monotonic trends.

Median nightlight intensity in the early and late time periods are shown in Figures 6 and A2. Between the early and late time periods, the mean pixel-level change in nightlight intensity across the region was an increase of 7%, with a maximum increase of 47% and decrease of −17% (Figure 7). We also evaluated the number of pixels that exceeded varying thresholds (1, 5, 10, and 25%) of nightlight intensity increase between the two time periods. We generally observed that 80% of pixels exceeded the 1% threshold in nightlight intensity between the early and late time periods (Figure 8). A general trend of increasing lighting was observed across the majority of the study area. The pixels that exceeded the 5% and 10% nightlight intensity increase thresholds corresponded with urbanization and varying degrees of new settlement or development (Figure 8). However, only 4% of the study area experienced greater than a 25% increase in nightlight intensity, and this was attributed to intensive land use changes in these areas, with high lighting requirements related to industry and greenhouses (Figure 8). The nightlight intensity thresholds exceeded by a given pixel between the early and late time periods depend on the amount of existing urban development present and the initial observation of intensity in the early time period, implying that change is relative to each pixel (Figure 7). The magnitude of change observed is multifactorial, and we further explored the types of infrastructure and spatial configuration of land use changes corresponding to the observed trends in Section 3.3.

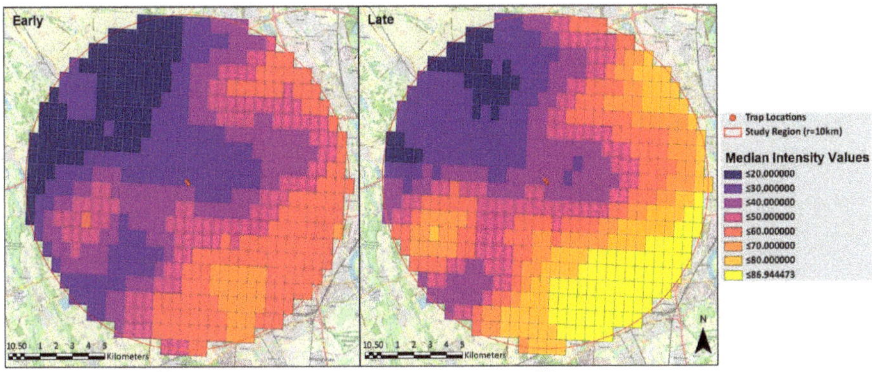

(a) (b)

Figure 6. Median nightlight intensity from the early time period (**a**, 1992–1999) and late time period (**b**, 2003–2010).

(a)

(b)

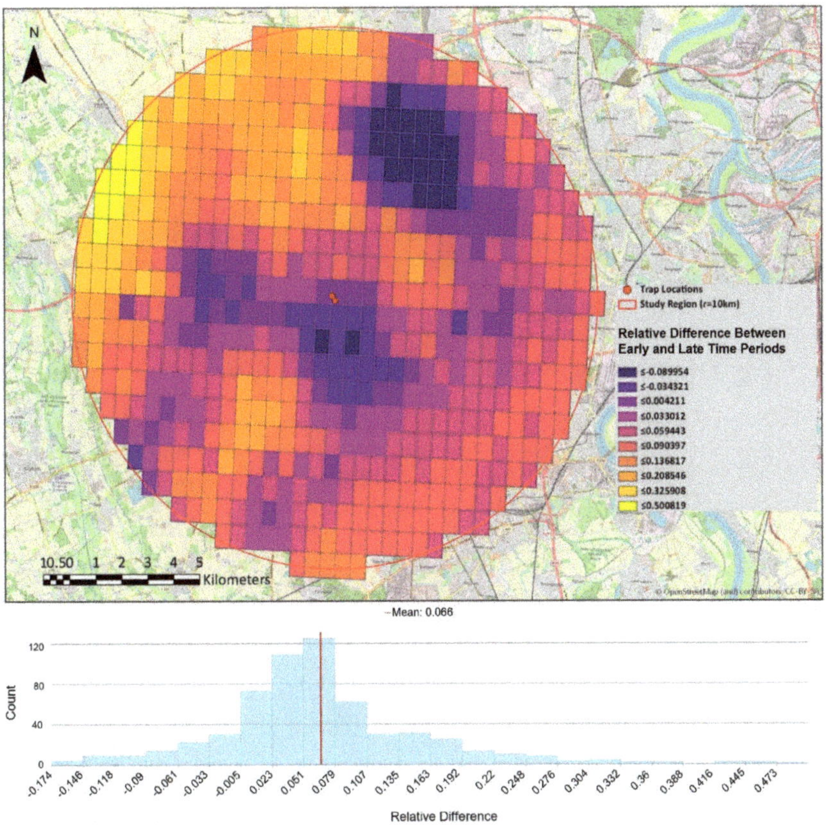

Figure 7. The spatial distribution of the relative difference in nightlight intensity for each pixel between the early time period (1992–1999) and late time period (2003–2010) derived through linear trend analysis (**a**). The distribution (frequency) of the relative difference in nightlight intensity for all pixels is shown below (**b**), where the mean relative difference (red line) was approximately a 7% increase in nightlight intensity.

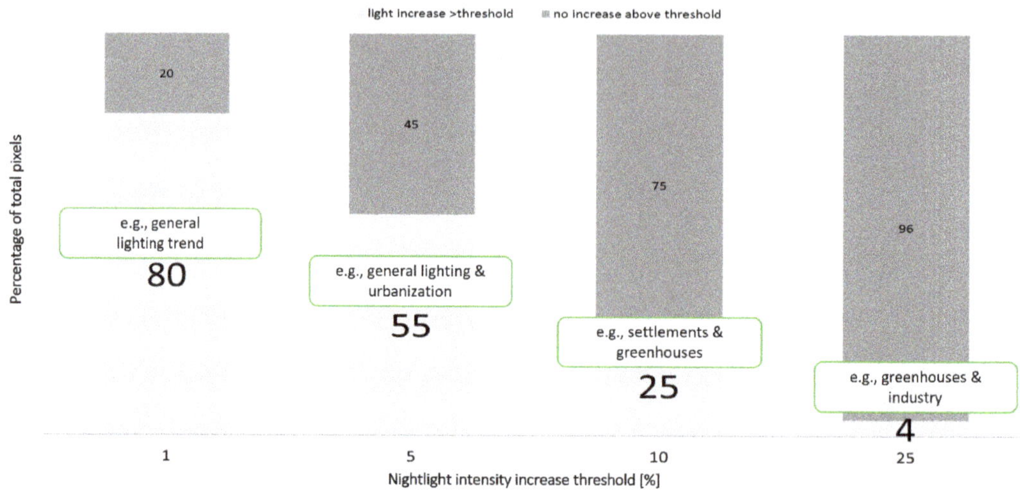

Figure 8. Proportion of pixel increases in nightlight intensity exceeding various percent change thresholds observed between the early (1992–1999) and late (2003–2010) time periods.

3.2.2. Urban/Rural

Through historical, fine-resolution aerial imagery, we visually identified an increase in urban land cover between 1989 and 2013. Approximately 82% of the region (out of the 590 pixels) was classified as rural, and inversely, 18% of the region was classified as urban in 2012 (Table A2). The predominant categorization across the region (10 km radius) was classified as rural, and overall, we observed a more prominent increase in nightlight intensity within these rural areas compared to urban areas (Figure 9). Most significant changes in nightlight intensity (Mann–Kendall *p*-values, $\alpha < 0.05$) were related to increases that occurred within rural areas (Figures 4 and 5). A larger proportion of the study area classified as rural had positive Mann–Kendall tau values and corresponded to statistically significant increases in nightlight intensity (*p*-values, $\alpha < 0.05$), while a smaller proportion of urban areas experienced statistically significant increases. Our results from the Mann–Kendall test revealed some pixels with negative tau values, indicating nightlight intensity decreases (3% of the region) corresponding to urban areas.

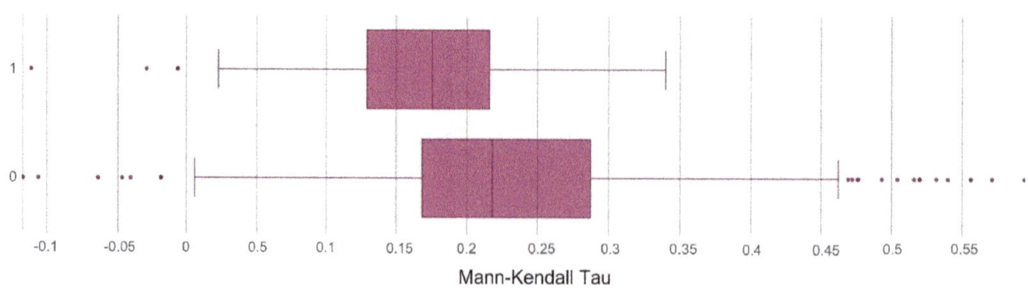

Figure 9. Increases of nightlight intensity across urban (top) and rural (bottom) areas between 1992 and 2010 using Mann–Kendall tau statistic.

3.3. Change Attribution

3.3.1. Nightlight Intensity Changes within 1 km and 2 km Radii

The change attribution findings were reported as part of a descriptive analysis, and features were visually identified using historical aerial imagery from 1989 and 2013. Within the 1 km radius and immediate surroundings of the trap locations, there was a minor increase in nightlight intensity; however, the single pixel overlapping the actual trap locations retained relatively low absolute nightlight intensity (Figure A3), as expected, since they are located within a nature reserve. We identified a new farm extension, settlement, and greenhouse within the periphery of the trap, as well as a farm reduction. Within the 2 km radius around the trap locations, the results demonstrate a moderate increase in ALAN during the observation period without any indications of decreases (Figure A4). The increases likely correspond with settlement extensions identified via historical aerial imagery. Within the 2 km radius, we identified five specific features likely contributing to significant increases detected with the Mann–Kendall tau statistic. A new greenhouse complex, three new greenhouse extensions, and a golf course were identified as potential contributors to peripheral increases in nightlight intensity within 2 km of the insect trap locations (Figure A4).

3.3.2. Nightlight Intensity Changes within 10 km Radius–Evaluation Points

Within the full study area (i.e., 10 km radius), in the northwest cluster of statistically significant increases in nightlight intensity (Figures 5 and 7), the primary driver of these significant trends is likely urbanization. Within the significant evaluation points (ID 1–10, criteria outlined in Section 2.5), some specific features identified that are likely responsible for the increases in nightlight intensity include courtyard extensions, nursery extensions, greenhouses, and campsites. Within the non-significant evaluation points that still had a trend slope value greater than 0.86 (ID 11–17), new settlements and growing urban infrastructure were identified via aerial imagery as changes that occurred within the study period that likely contributed to increased nightlight intensity. Additional pixels were investigated using this same framework and significant features or changes were not identified via aerial imagery; only the significant findings are reported here.

Within the 3% of the region that experienced a decrease in nightlight intensity, further investigation suggested that the decreases within some of these areas (urban areas) could be related to decreased consumption (see discussion; Figure A5). Within the city of Krefeld, efforts were made to reduce the number of lights utilized per streetlamp from two to one at night, resulting in a 30% decrease in intensity [32]. Data collected by the City of Krefeld Energy Conservation Program revealed decreasing trends in energy consumption in urban areas from 12.4 million kWh in 1998 to 8 million kWh in 2012 (Figure A5). The temporal patterns in regional nightlight intensity (Figure 10) likely correspond to changes occurring with the Niederberg coal mine, which was opened in 1912 and closed in 2001. Dismantling of the local briquette factory began in 1995 and corresponds to regional decreases in nightlight intensity (Figure 10). In the following years, we identified new residential developments and buildings that correlate with increasing nightlight intensity (Figure 10); the start and end values in the evaluation points time series are similar to the mean difference between the regional medians of the early and late time periods (Figure A2).

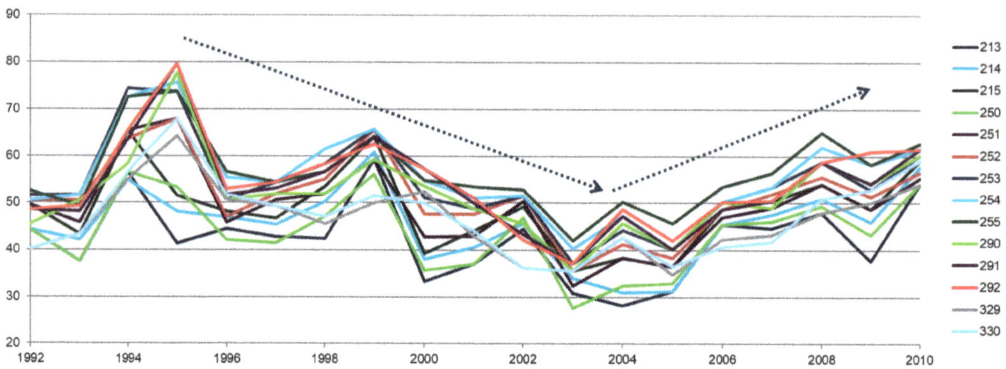

Figure 10. Trends in nightlight intensity examined for evaluation points (pixels) which reflect patterns in land use change, showing decreasing trends in the early time period (1992–1999), and increasing trends in the late time period (2003–2010). Values on the right correspond to individual raster cell IDs.

4. Discussion

The spatial distribution of nightlight intensity trends resulting from linear regression and the Mann–Kendall test reflect patterns of land use (Figures 4 and 5) where, across the region, nightlight intensity generally increased between 1992 and 2010. The overall increase in nightlight across the region and study period is likely driven by urbanization [5,15,16]. We identified a mean 7% increase in nightlight intensity (and a maximum 47% increase), which is comparable to the trends identified by Sánchez de Miguel et al. [33] across Germany.

Across the various spatial scales investigated (1, 2, and 10 km radii), we identified different trends in nightlight intensity, and a variety of features potentially contributing to those trends; the variance within observations at different spatial scales is a well-documented concept [34]. It is important to evaluate the nightscape of the area surrounding nature reserves since they can significantly degrade habitat quality and have a negative impact on the entire ecosystem [12]. Through the trends in nightlight intensity and historical aerial imagery, we identified a new greenhouse, greenhouse extensions, and golf course within a two-kilometer radius of the insect traps. Considering the significant insect biomass decline found by Hallmann et al. [4] within the Orbroicher Bruch Nature Reserve, it may be valuable to further investigate the contributions to light pollution from the identified features. Additionally, it may be important to consider how these entities may relate to declining insect abundance, and other ecological implications within the nature reserve near these features. Across the full study area (10 km radius), we identified a variety of new urban features constructed between 1992 and 2010 that have likely altered nightlight intensity. We found that approximately 96% of the pixel-level trends had a positive slope (>0, Figure 5), and 22% of the pixels within the region experienced statistically significant increases in nightlight intensity (Table A2). With the majority of the study area experiencing nightlight increases >1%, further investigation may reveal regional effects or implications. The broader trend of increasing nightlight intensity across this region may be important to consider in relation to local insect population decline.

Within a small proportion of the urban area (3% of the region), we identified decreases in nightlight intensity that may be related to improved energy efficiency and/or reduced energy consumption. While we cannot conclusively determine the cause of decreasing trends in nightlight intensity without including ground-level observations or auxiliary data, this observation points to the possible impact that city energy conservation programs could have on ALAN. Energy use could be reduced by introducing new lighting technology while maintaining or increasing the amount of light pollution. Alternatively, energy use could also be reduced by using fewer lights, thus decreasing light pollution. Nightlight intensity over time could potentially serve as one metric that could provide more information to local or regional energy programs to assess their effectiveness or plan future improvements. This could be useful for broader scale management or assessments of light energy efficiency for countries to monitor their collective progress through time. However, these results may be confounded by several factors, including overglow, which is more prominent in urban areas [35], and other limitations of the DMSP-OLS data discussed below.

Attributing specific anthropogenic changes to pixel-level trends in nightlight intensity can help with conservation and landscape planning efforts. This framework has the potential to direct research into which structures cause the most significant increases in nightlight intensity near key insect habitats. It may also be helpful in determining what methods are most effective at decreasing nightlight intensity or the radius that subsequent light pollution spans across surrounding areas. Additionally, this type of investigation may help expand understanding of which types of lights contribute the most pollution, or what energy reduction efforts and policies are most effective. In scenarios without a quantifiable change in nightlight intensity and observed insect decline, the potential effects of temperature [36], chemical/biological pollutants, habitat loss or degradation, invasive species, or other factors should be investigated. In regions with significant nightlight intensity increases and observed insect decline, the outlined methods could be refined and expanded upon to help identify which specific structure(s) may be responsible, and to evaluate the contributions of nightlight. Isolating specific features potentially causing increases in ALAN can guide in situ measurements to further investigate the impacts of luminous flux. Ultimately, this type of investigation may have implications for ALAN mitigation efforts specific to the identified structure(s). Furthermore, the outlined methodology could be refined to assess the effectiveness of nightlight mitigation efforts across broader spatial extents.

There are several factors that have introduced different degrees of uncertainty in this analysis, and it is important to consider their potential implications for the outlined results. The limited significance of pixel-level trends across the majority of the region is likely due to cloud cover, sensor calibration, spatial resolution, the precision of georegistration, variable sensor gain, overglow, and other known limitations of the DMSP-OLS data detailed by Huang et al. [37]. The coarser spatial resolution of the DMSP-OLS dataset (30 arc-second) is suited for regional analyses, but made it challenging to directly attribute a quantitative change in nightlight intensity to sub-pixel landscape features (identified via fine-resolution aerial imagery); the inferences that can be drawn at the 1 km and 2 km scales surrounding the insect trap locations are limited by the original 5 km spatial resolution of the DMSP-OLS sensors. We selected the DMSP-OLS dataset to investigate the time period corresponding to the study conducted by Hallman et al. [4]. However, it would be preferable to use finer-resolution satellite imagery, when possible. Future studies investigating ALAN during more recent time periods (2012 and onward) should consider utilizing imagery from the Visible Infrared Imaging Radiometer Suite (VIIRS) sensor as a finer-resolution option to navigate the uncertainty of attributing nightlight intensity trends to local features [38]. Specific to the nightlight time series data utilized, the data from 2010 may introduce noise and some uncertainty; however, excluding the 2010 data is unlikely to change the overall directionality of trends across the 19-year study period (Figures 10 and A2). Additionally, the DMSP-OLS dataset only captures the VNIR portion of the electromagnetic spectrum; there may be other light sources emitting in different portions of the spectrum, like blue light from LED or ultraviolet wavelengths, that may influence insect behavior [39–42]. This should be considered as there may be additional ALAN emissions not captured by the DMSP-OLS satellites that could have negative impacts on different insect species. Furthermore, static satellite images captured within a single timeframe may not represent the full dynamic variation in the light emitted by various structures [39]. The timing of important biological processes of different species implies that impact on individual species may vary based on the timing and composition of ALAN [39,43]. Previous studies investigating aquatic insects have noted that satellite imagery captures light emitted toward the satellite and does not necessarily capture light emitted into a specific habitat or other directions [11]. Luminous flux also varies based on the light composition. It is recommended to pair nightlight remote sensing imagery with ground-level light measurements to monitor insects [7]. Advancements in concepts surrounding calibration and correction of nighttime light imagery are still emerging, and future investigations should consider variability in airglow and additional factors that may influence light emitted upward toward the satellite, as described by Kyba & Coesfeld [44]. Further investigations into the Orbroicher Bruch Nature Reserve should consider using ground-level measurements of light to explore the impacts of ALAN.

5. Conclusions

Our detailed examination of nightlight time series data paired with aerial imagery revealed trends in nightlight intensity that reflected dynamic patterns of specific landscape features that have altered ALAN within the region. Within the 10 km radius surrounding insect traps in Orbroicher Bruch Nature Reserve, Germany, we observed a mean long-term median increase of 7% (and a maximum of 47%) between the early and late time periods we considered, and the majority of the study area experienced nightlight intensity increases greater than 1%. Numerous studies across the world have reported independently on either the patterns of increasing ALAN via satellite imager or observed insect population decline. While it is unclear if the identified trends extend to other geographic regions, our results further highlight the need for future studies to concurrently investigate long-term trends in ALAN and insect decline; ideally, future investigations should span multiple scales and consider the spatial and temporal overlaps between these patterns. This highlights an important gap between regional trends in ALAN and identifying specific local features contributing to these patterns. Ultimately, further development and refinement of these methods has the potential to guide in situ investigations and mitigations of ALAN, and targeting efforts toward the most impactful landscape features could provide an additional strategy to support insect conservation.

Author Contributions: Conceptualization, T.S.; methodology, T.S.; investigation, T.S.; resources, T.S.; data curation, T.S.; writing—original draft preparation, J.L. and C.M.H.; writing—review and editing, J.L., C.M.H. and T.S. All authors have read and agreed to the published version of the manuscript.

Funding: The research has been funded by Bayer AG, Alfred-Nobel-Str. 50, 40789 Monheim, Germany. Bayer Project MEAAL024.

Data Availability Statement: Data are available via corresponding author, including individual time series nightlight intensity measurements of all grid cells, reclassification of urban/rural areas, and codes for analysis. The basic geo information of the state survey of North Rhine-Westphalia is protected by law. Use is only permitted with the permission of the district government of Cologne, department 7 Geobasis NRW. In the case of presentations and copies, the following approval note must be clearly visible: Basic geodata of the municipalities and the state of North Rhine-Westphalia (c) Geobasis NRW, 2017. The right of use only applies to the internal use of the authorized user. The transfer of basic geo information to a contractor for earmarked order processing is permitted. Any use beyond this requires a special contractual arrangement with the district government of Cologne, Department 7 Geobasis NRW. Credit for DMSP-OLS: Image and data processing by NOAA's National Geophysical Data Center. DMSP data collected by US Air Force Weather Agency. Insect trap locations provided by Hallmann et al. (2017).

Acknowledgments: The authors would like to thank Markus Neteler (Mundialis), Frederik van der Stouwe (knoell Consult), and Fabian Löw (Mundialis) for their guidance and efforts for this work.

Conflicts of Interest: The authors declare the following financial interests/personal relationships which may be considered as potential competing interests: The research has been funded by Bayer AG, Alfred-Nobel-Str. 50, 40789 Monheim, Germany. T.S. is an employee of Bayer Crop Science, a manufacturer of agricultural products. J.L. and C.H. are employees of Applied Analysis Solutions and received compensation for the preparation of manuscript.

Appendix A

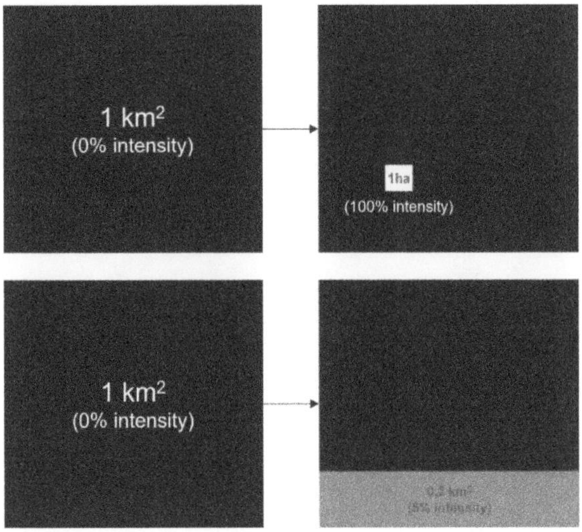

Figure A1. Examples of potential combinations of spatial patterns, area, and intensity changes that could cause a 1% increase in nightlight intensity within the time series trend of a single pixel.

Table A1. Significant Mann-Kendall pixel-level trends at varying levels of alpha significance.1.

MKtau p-Value	Number of Pixels (Out of 590)	Percent of Region of Interest
$\alpha < 0.01$	18	3.1%
$\alpha < 0.05$	79	13.4%
$\alpha < 0.10$	130	22%

Table A2. Urban land cover delineation across the full study area (10 km radius) based on CORINE 2012 dataset. After resampling the classification to match the DMSP-OLS dataset, "urban" was assigned to any pixel with $\geq$50% urban cover. The final number of pixels assigned to urban and rural were totaled to estimate the percentage of the study region comprised of urban and rural land cover.

Land Cover Type	Number of Pixels (Out of 590)	Percent of Region of Interest
Rural	484	82%
Urban	106	18%

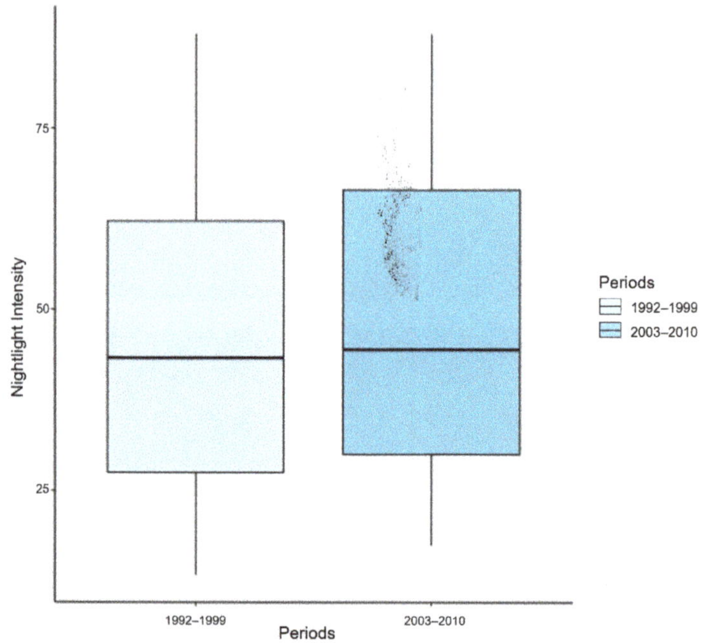

Figure A2. Median nightlight intensity between the early (left) and late (right) time periods across the region, with the late time period showing a general increase in nightlight intensity across the full study area.

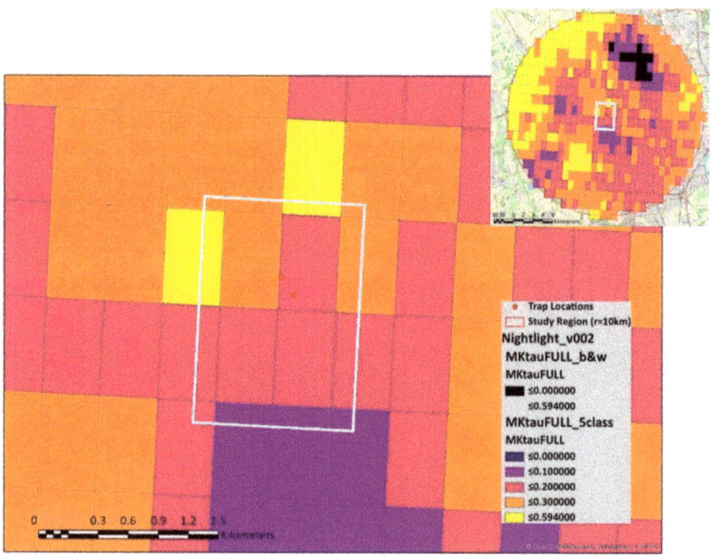

Figure A3. Mann-Kendall tau statistic values for pixels within a 1 km radius of the insect trap locations in Orbroicher Bruch Nature Preserve, with minimal changes in nightlight intensity occurring within the pixel containing the insect trap locations.

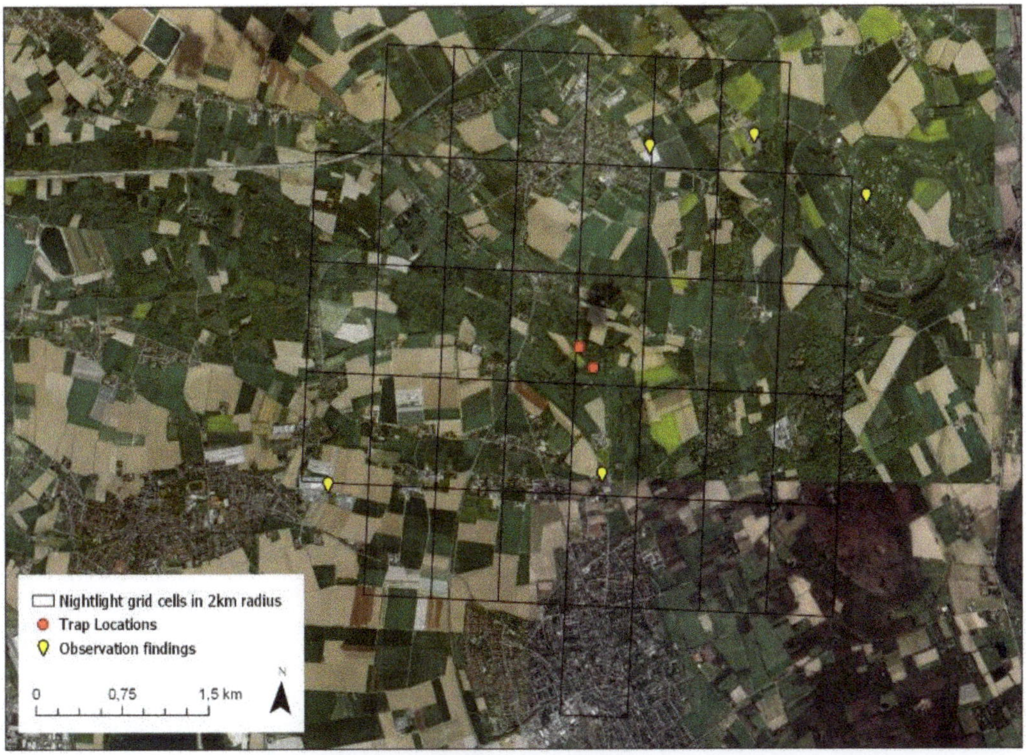

Figure A4. Grid cells (black lines) represent the pixels that overlapped the 2 km radius around the insect trap locations (red points). Within the 2 km radius, we identified five potential landscape features (yellow points) potentially contributing to increasing nightlight intensity within the pixels they are centered in/adjacent to. These features were not present in the 1989 aerial imagery and were identified as changes in the 2013 imagery (with 2013 imagery shown). These features include a greenhouse complex, three new greenhouse extensions, and a golf course.

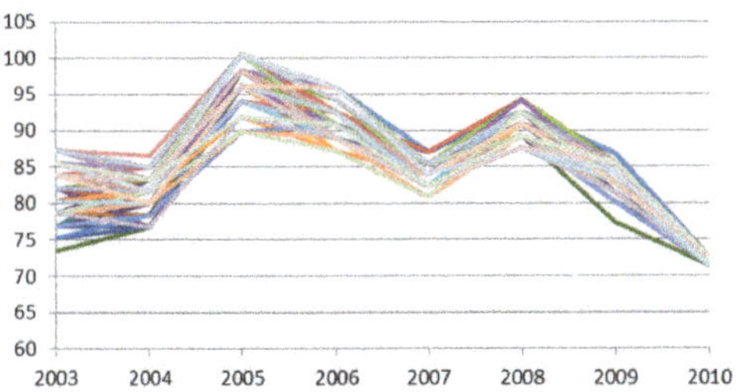

(**a**)

Year	Energy Consumption [kWh]	# Street lamps
1998	12,400,000	27,500
2008		29,000
2011		28,900
2012	8,000,000	29,000
2013		28,864
2015	7,500,000	

(**b**)

Figure A5. Decreases in nightlight intensity within urban areas where each different colored line represents an individual pixel (**a**) corresponding with improved energy efficiency based on data collected by the City of Krefeld Energy Conservation Program (**b**).

References

1. Owens, A.C.S.; Cochard, P.; Durrant, J.; Farnworth, B.; Perkin, E.K.; Seymoure, B. Light pollution is a driver of insect declines. *Biol. Conserv.* **2020**, *241*, 108259. [CrossRef]
2. Grubisic, M.; Van Grunsven, R.; Kyba, C.; Manfrin, A.; Hölker, F. Insect declines and agroecosystems: Does light pollution matter? *Ann. Appl. Biol.* **2018**, *173*, 180–189. [CrossRef]
3. Bell, J.R.; Blumgart, D.; Shortall, C.R. Are insects declining and at what rate? An analysis of standardised, systematic catches of aphid and moth abundances across Great Britain. *Insect Conserv. Divers.* **2020**, *13*, 115–126. [CrossRef]
4. Hallmann, C.A.; Sorg, M.; Jongejans, E.; Siepel, H.; Hofland, N.; Schwan, H.; Stenmans, W.; Müller, A.; Sumser, H.; Hörren, T.; et al. More than 75 percent decline over 27 years in total flying insect biomass in protected areas. *PLoS ONE* **2017**, *12*, e0185809. [CrossRef] [PubMed]
5. Hölker, F.; Moss, T.; Griefahn, B.; Kloas, W.; Voigt, C.C.; Henckel, D.; Hänel, A.; Kappeler, P.M.; Völker, S.; Schwope, A.; et al. The Dark Side of Light: A Transdisciplinary Research Agenda for Light Pollution Policy. *Ecol. Soc.* **2010**, *15*, 13. [CrossRef]
6. Langevelde, F.; Braamburg-Annegarn, M.; Huigens, M.E.; Groendijk, R.; Poitevin, O.; Deijk, J.R.; Ellis, W.N.; Grunsven, R.H.A.; Vos, R.; Vos, R.A.; et al. Declines in moth populations stress the need for conserving dark nights. *Glob. Chang. Biol.* **2018**, *24*, 925–932. [CrossRef]
7. Kalinkat, G.; Grubisic, M.; Jechow, A.; Grunsven, R.H.A.; Schroer, S.; Hölker, F. Assessing long-term effects of artificial light at night on insects: What is missing and how to get there. *Insect Conserv. Divers.* **2021**, *14*, 260–270. [CrossRef]
8. Falchi, F.; Cinzano, P.; Duriscoe, D.; Kyba, C.C.M.; Elvidge, C.D.; Baugh, K.; Portnov, B.A.; Rybnikova, N.A.; Furgoni, R. The new world atlas of artificial night sky brightness. *Am. Assoc. Adv. Sci.* **2016**, *2*, e1600377. [CrossRef] [PubMed]
9. Ges, X.; Bará, S.; García-Gil, M.; Zamorano, J.; Ribas, S.J.; Masana, E. Light pollution offshore: Zenithal sky glow measurements in the mediterranean coastal waters. *J. Quant. Spectrosc. Radiat. Transf.* **2018**, *210*, 91–100. [CrossRef]
10. Jechow, A.; Kolláth, Z.; Ribas, S.J.; Spoelstra, H.; Hölker, F.; Kyba, C.C.M. Imaging and mapping the impact of clouds on skyglow with all-sky photometry. In *Scientific Reports*; Springer Science and Business Media LLC: Berlin, Germany, 2017; Volume 7. [CrossRef]
11. Jechow, A.; Hölker, F. Snowglow—The Amplification of Skyglow by Snow and Clouds Can Exceed Full Moon Illuminance in Suburban Areas. *J. Imaging* **2019**, *5*, 69. [CrossRef] [PubMed]
12. Marcantonio, M.; Pareeth, S.; Rocchini, D.; Metz, M.; Garzon-Lopez, C.X.; Neteler, M. The integration of Artificial Night-Time Lights in landscape ecology: A remote sensing approach. *Ecol. Complex.* **2015**, *22*, 109–120. [CrossRef]
13. Wei, Y.; Liu, H.; Song, W.; Yu, B.; Xiu, C. Normalization of time series DMSP-OLS nighttime light images for urban growth analysis with Pseudo Invariant Features. *Landsc. Urban Plan.* **2014**, *128*, 1–13. [CrossRef]
14. Elvidge, C.D.; Baugh, K.E.; Dietz, J.B.; Bland, T.; Sutton, P.C.; Kroehl, H.W. Radiance Calibration of DMSP-OLS Low-Light Imaging Data of Human Settlements. *Remote Sens. Environ.* **1999**, *68*, 77–88. [CrossRef]
15. Elvidge, C.D.; Baugh, K.E.; Kihn, E.A.; Kroehl, H.W.; Davis, E.R. Mapping city lights with nighttime data from the DMSP Operational Linescan System. *Photogramm. Eng. Remote Sens.* **1997**, *63*, 727–734.
16. Zhang, Q.; Seto, K.C. Can Night-Time Light Data Identify Typologies of Urbanization? A Global Assessment of Successes and Failures. *Remote Sens.* **2013**, *5*, 3476–3494. [CrossRef]
17. Ghosh, T.; Powell, R.; Elvidge, C.D.; Baugh, K.E.; Sutton, P.C.; Anderson, S. Shedding light on the global distribution of economic activity. *Open Geogr. J.* **2010**, *3*, 148–161. Available online: https://ngdc.noaa.gov/eog/dmsp/download_gdp.html (accessed on 15 May 2022).
18. Bennie, J.; Davies, T.W.; Duffy, J.P.; Inger, R.; Gaston, K.J. Contrasting trends in light pollution across Europe based on satellite observed night time lights. *Sci. Rep.* **2014**, *4*, 3789. [CrossRef] [PubMed]
19. Miller, C.R.; Barton, B.T.; Zhu, L.; Radeloff, V.C.; Oliver, K.M.; Harmon, J.P.; Ives, A.R. Combined effects of night warming and light pollution on predator–prey interactions. *Proc. R. Soc. B Boil. Sci.* **2017**, *284*, 20171195. [CrossRef]
20. Barentine, J.C. Methods for Assessment and Monitoring of Light Pollution around Ecologically Sensitive Sites. *J. Imaging* **2019**, *5*, 54. [CrossRef]
21. Koen, E.L.; Minnaar, C.; Roever, C.L.; Boyles, J.G. Emerging threat of the 21st century lightscape to global biodiversity. *Glob. Chang. Biol.* **2018**, *24*, 2315–2324. [CrossRef] [PubMed]
22. Longcore, T.; Rich, C. Ecological light pollution. *Front. Ecol. Environ.* **2004**, *2*, 191–198. [CrossRef]
23. Fan, L.; Zhao, J.; Wang, Y.; Ren, Z.; Zhang, H.; Guo, X. Assessment of Night-Time Lighting for Global Terrestrial Protected and Wilderness Areas. *Remote Sens.* **2019**, *11*, 2699. [CrossRef]
24. Gaston, K.J.; Davies, T.W.; Bennie, J.; Hopkins, J. Reducing the ecological consequences of night-time light pollution: Options and developments. *J. Appl. Ecol.* **2012**, *49*, 1256–1266. [CrossRef] [PubMed]
25. Ghazaryan, G.; Rienow, A.; Oldenburg, C.; Thonfeld, F.; Trampnau, B.; Sticksel, S.; Jürgens, C. Monitoring of Urban Sprawl and Densification Processes in Western Germany in the Light of SDG Indicator 11.3.1 Based on an Automated Retrospective Classification Approach. *Remote Sens.* **2021**, *13*, 1694. [CrossRef]
26. Baugh, K.; Elvidge, C.D.; Ghosh, T.; Ziskin, D. Development of a 2009 Stable Lights Product using DMSP-OLS data. *Proc. Asia-Pac. Adv. Netw.* **2010**, *30*, 114. [CrossRef]
27. Wu, J.; He, S.; Peng, J.; Li, W.; Zhong, X. Intercalibration of DMSP-OLS night-time light data by the invariant region method. *Int. J. Remote Sens.* **2013**, *34*, 7356–7368. [CrossRef]

28. R Core Team. R: A Language and Environment for Statistical Computing. Available online: https://www.R-project.org/ (accessed on 18 April 2018).
29. McLeod, A.I. Kendall: Kendall Rank Correlation and Mann-Kendall Trend Test. R Package Version 2.2. 2011. Available online: https://CRAN.R-project.org/package=Kendall (accessed on 22 April 2018).
30. Kendall, M.G. *Rank Correlation Methods*, 4th ed.; Charles Griffin: London, UK, 1975.
31. Mann, H.B. Nonparametric tests against trend. *Econometrica* **1945**, *13*, 245. [CrossRef]
32. NRW. Regional Examination: Preliminary Report by the City of Krefeld in 2014. Available online: http://gpanrw.de/media/1432714191_stadt_krefeld_gesamtbericht_2014_internet.pdf (accessed on 7 June 2022).
33. Miguel, A.S.; Bennie, J.; Rosenfeld, E.; Dzurjak, S.; Gaston, K.J. First Estimation of Global Trends in Nocturnal Power Emissions Reveals Acceleration of Light Pollution. *Remote Sens.* **2021**, *13*, 3311. [CrossRef]
34. Levin, S. The Problem of Pattern and Scale in Ecology: The Robert H. MacArthur Award Lecture. *Ecology* **1992**, *73*, 1943–1967. [CrossRef]
35. Bluhm, R.; McCord, G.C. What Can We Learn from Nighttime Lights for Small Geographies? Measurement Errors and Heterogeneous Elasticities. *Remote Sens.* **2022**, *14*, 1190. [CrossRef]
36. Welti, E.A.R.; Zajicek, P.; Frenzel, M.; Ayasse, M.; Bornholdt, T.; Buse, J.; Classen, A.; Dziock, F.; Engelmann, R.A.; Englmeier, J.; et al. Temperature drives variation in flying insect biomass across a German malaise trap network. *Insect Conserv. Divers.* **2021**, *15*, 168–180. [CrossRef]
37. Huang, Q.; Yang, X.; Gao, B.; Yang, Y.; Zhao, Y. Application of DMSP/OLS nighttime light images: A meta-analysis and a systematic literature review. *Remote Sens.* **2014**, *6*, 6844–6866. [CrossRef]
38. Elvidge, C.D.; Baugh, K.E.; Zhizhin, M.; Hsu, F.-C. Why VIIRS Data Are Superior to DMSP for Mapping Nighttime Lights. *Proc. Asia-Pac. Adv. Netw.* **2013**, *35*, 62. [CrossRef]
39. Sordello, R.; Busson, S.; Cornuau, J.H.; Deverchère, P.; Faure, B.; Guetté, A.; Hölker, F.; Kerbiriou, C.; Lengagne, T.; Le Viol, I.; et al. A plea for a worldwide development of dark infrastructure for biodiversity—Practical examples and ways to go forward. *Landsc. Urban Plan.* **2022**, *219*, 104332. [CrossRef]
40. Levin, N.; Kyba, C.C.; Zhang, Q.; de Miguel, A.S.; Román, M.O.; Li, X.; Portnov, B.A.; Molthan, A.L.; Jechow, A.; Miller, S.D.; et al. Remote sensing of night lights: A review and an outlook for the future. *Remote Sens. Environ.* **2020**, *237*, 111443. [CrossRef]
41. Kyba, C.; Garz, S.; Kuechly, H.; de Miguel, A.S.; Zamorano, J.; Fischer, J.; Hölker, F. High-Resolution Imagery of Earth at Night: New Sources, Opportunities and Challenges. *Remote Sens.* **2014**, *7*, 1–23. [CrossRef]
42. Elvidge, C.D.; Keith, D.M.; Tuttle, B.T.; Baugh, K.E. Spectral Identification of Lighting Type and Character. *Sensors* **2010**, *10*, 3961–3988. [CrossRef]
43. Longcore, T.; Rodríguez, A.; Witherington, B.; Penniman, J.F.; Herf, L.; Herf, M. Rapid assessment of lamp spectrum to quantify ecological effects of light at night. *J. Exp. Zool. Part A Ecol. Integr. Physiol.* **2018**, *329*, 511–521. [CrossRef]
44. Kyba, C.C.M.; Coesfeld, J. Satellite Observations Show Reductions in Light Emissions at International Dark Sky Places During 2012–2020. *Int. J. Sustain. Light.* **2021**, *23*, 51–57. [CrossRef]

remote sensing

MDPI

Article

Spatiotemporal Patterns of Light Pollution on the Tibetan Plateau over Three Decades at Multiple Scales: Implications for Conservation of Natural Habitats

Yihang Wang [1,2,3], Caifeng Lv [1,2], Xinhao Pan [1,2], Ziwen Liu [1,2], Pei Xia [4], Chunna Zhang [5,6,7] and Zhifeng Liu [1,2,5,*]

1 State Key Laboratory of Earth Surface Processes and Resource Ecology (ESPRE), Beijing Normal University, Beijing 100875, China
2 School of Natural Resources, Faculty of Geographical Sciences, Beijing Normal University, Beijing 100875, China
3 Tsinghua University High School-Wangjing, Beijing 100102, China
4 College of Urban and Environmental Sciences, Peking University, Beijing 100871, China
5 Academy of Plateau Science and Sustainability, People's Government of Qinghai Province, Beijing Normal University, Beijing 100875, China
6 Qinghai Province Key Laboratory of Physical Geography and Environment Process, College of Geographical Science, Qinghai Normal University, Xining 810008, China
7 Key Laboratory of Tibetan Plateau Land Surface Processes and Ecological Conservation (Ministry of Education), Qinghai Normal University, Xining 810008, China
* Correspondence: zhifeng.liu@bnu.edu.cn

Abstract: Light pollution (LP), induced by human activities, has become a crucial threat to biodiversity on the Tibetan plateau (TP), but few studies have explored its coverage and dynamics. In this study, we intended to measure the spatiotemporal patterns of LP on the TP from 1992 to 2018. First, we extracted the annual extent of LP from time-series nighttime light data. After that, we analyzed its spatiotemporal patterns at multiple scales and identified the natural habitats and the species habitats affected by LP. Finally, we discussed the main influencing factors of LP expansion on the TP. We found that the LP area increased exponentially from 1.2 thousand km^2 to 82.8 thousand km^2, an increase of nearly 70 times. In 2018, LP accounted for 3.2% of the total area of the TP, mainly concentrated in the eastern and southern areas. Several national key ecological function zones (e.g., the Gannan Yellow river key water supply ecological function zone) and national nature reserves (e.g., the Lalu Wetland National Nature Reserve) had a large extent of LP. The proportion of LP area on natural habitats increased from 79.6% to 91.4%. The number of endangered species with habitats affected by LP increased from 89 to 228, and more than a quarter of the habitats of 18 endangered species were affected by LP. We also discovered that roadways as well as settlements in both urban and rural areas were the main sources of LP. Thus, to lessen LP's negative effects on biodiversity, effective measures should be taken during road construction and urbanization on the TP.

Keywords: Tibetan plateau; light pollution; urbanization; highway; railway; biodiversity; landscape sustainability

Citation: Wang, Y.; Lv, C.; Pan, X.; Liu, Z.; Xia, P.; Zhang, C.; Liu, Z. Spatiotemporal Patterns of Light Pollution on the Tibetan Plateau over Three Decades at Multiple Scales: Implications for Conservation of Natural Habitats. *Remote Sens.* **2022**, *14*, 5755. https://doi.org/10.3390/rs14225755

Academic Editors: Ran Goldblatt, Steven Louis Rubinyi and Hogeun Park

Received: 13 October 2022
Accepted: 10 November 2022
Published: 14 November 2022

1. Introduction

LP refers to the phenomenon of diffuse light, reflected light and glare from modern urban buildings and night lighting that causes interference or negative effects on people, animals and plants [1,2]. An increasing number of studies show that LP can affect species feeding, sleep, migration, reproduction, navigation, communication, habitat selection and social interaction, resulting in a series of ecological and environmental issues, such as disturbance of circadian rhythm, limited survival and reproduction of species, and destruction of ecosystem structure [3–16]. At present, social and economic activities such as global urbanization are accelerating the expansion of LP at an unprecedented speed and degree; LP is becoming a hot topic in the field of biodiversity conservation and sustainable development [17,18].

The Tibetan plateau (TP) is undergoing rapid urbanization and road construction, driven by the "Belt and Road" initiative, the "New Urbanization" plan and the "Western Development" strategy [19]. Just as the "Belt and Road" initiative drove the rapid development of Central Asia and other regions [20–23], the human footprint on the TP increased by 28.43% from 1990 to 2010, much higher than the world average (9%) [24]. Increased LP due to urbanization and road construction affects the quantity, quality and connectivity of natural habitats, thus posing a serious threat to biodiversity [25]. In addition, the ecosystems on the TP are fragile and have low resilience [26]. Once disturbed by human activities, they start to deteriorate rapidly [27]. To limit the influences of LP, it is essential to reveal the temporal and spatial patterns of LP and its impacts on ecological protection and sustainable development on the TP. However, there is still a lack of LP research focusing on the whole region of the TP.

Nighttime light (NTL) data provide an effective way to assess LP on the TP. There are two ways to measure LP using nighttime light data. One is to characterize LP directly from the spatial distribution of areas exposed to light at nighttime. Using this approach, Fan et al. [28] analyzed the impact of global LP on terrestrial nature reserves and wilderness areas based on Defense Meteorological Satellite Program-Operational Linescan System (DMSP-OLS) nighttime light data. Kumar et al. [29] analyzed the change in LP in India based on DMSP-OLS nighttime light data. Another is to build models to assess LP based on night light data and ecological indicators. For example, Koen et al. [18] analyzed the level of threat to biodiversity from LP during 1992–2012 based on DMSP-OLS nighttime light data, using the number of pixels and brightness value of night light in a 20 km × 20 km quadrate as an LP index. Cabrera-cruz et al. [30] used NPP-VIIRS nighttime light data to establish the relationship between nighttime light and bird migration routes and analyzed the impact of LP on migratory bird migration. The first method is simpler and more feasible and can fully reflect the characteristics of the spatial and temporal patterns of regional LP.

In existing studies, the two widely used sources of nighttime light data, i.e., DMSP-OLS and Suomi National Polar-Orbiting Partnership-Visible Infrared Imaging Radiometer Suite (NPP-VIIRS) have different archival times, which limits the time of most LP-related studies to a short period (before or after 2013), leading to a decrease in the spatiotemporal continuity of related studies [29,31]. To improve the temporal extent of nighttime light data, Li et al. [32] recently created a long time series nighttime light dataset GHNTL (Harmonization of DMSP and VIIRS nighttime light data from 1992–2018 at the global scale) by integrating the two types of nighttime light data, providing new data to support the study of the spatiotemporal pattern of LP and its impact on the TP in the last 30 years.

This study aimed to analyze the spatiotemporal patterns of LP and its impacts on the TP based on the GHNTL nighttime light data. First, we quantified the distribution and dynamics of LP on the TP during 1992–2018. We quantified LP at four scales: ecoregion, national key ecological function zone, national nature reserve and national park. Second, we analyzed the effects of LP on various natural habitats and habitats of different species on the TP. Then, we analyzed the reasons for LP and gave policy suggestions to deal with LP on the TP.

2. Study Area and Materials

2.1. Study Area

There are 27 ecoregions, 10 national key ecological function zones, 46 national nature reserves and 4 national parks on the TP (Figure 1). According to incomplete statistics by the International Union for the Conservation of Nature [33], there are at least 1714 species on the TP, including 388 species of mammals, 1050 species of birds, 117 species of reptiles and 159 species of amphibians.

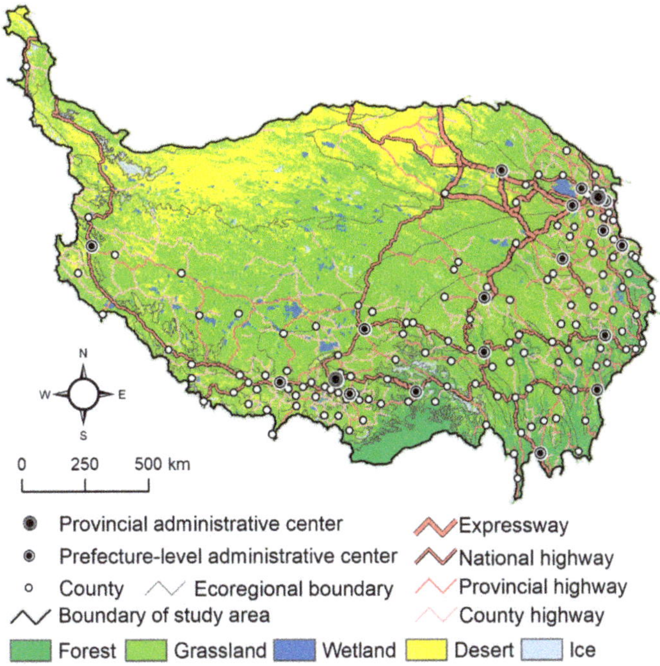

Figure 1. Study area. Note: the land cover information was obtained from 2018 version of ESACCI global land cover product.

The TP has undergone rapid social and economic development in recent years. The urbanization level of Qinghai province increased from 34.76% in 2000 to 55.52% in 2019, an increase of 21.8%. The urbanization level of the Tibet Autonomous Region increased from 18.93% in 2000 to 31.54% in 2019, an increase of 12.6% [34].

2.2. Data

NTL data obtained from the GHNTL [32] were used to analyze the LP over the TP. This dataset contains DMSP-OLS NTL time series data calibrated from 1992 to 2013 and DMSP-like NTL time series data transformed from NPP-VIIRS NTL from 2014 to 2018, with a spatial resolution of 1 km.

The land cover data from 1992 to 2018 used to assess the impact of LP on natural habitats were derived from the ESACCI global land cover product (http://maps.elie.ucl.ac.be/CCI/viewer/index.php) (accessed on 5 November 2022), with a spatial resolution of 300 m. The data include 22 land cover types. Referring to He et al. [35] and Zalles et al. [36], we adopted the habitat classification of the IUCN to extract natural habitats from ESACCI global land cover.

The species data used in this study were obtained from the IUCN Red List of Threatened Species (https://www.iucnredlist.org/resources/spatial-data-download) (accessed on 1 June 2021). Species data can be divided into four categories: mammals, birds, reptiles and amphibians. Each species was classified as Data Deficient (DD), Least Concern (LC), Near Threatened (NT), Vulnerable (VU), Endangered (EN), Critically Endangered (CR) and other categories according to the degree of endangered (Table 1). In addition, referring to Koen et al. [18], a separate division of nocturnal animals was summarized in this study.

Table 1. Number of species on the TP.

Category (Abbreviation)	Mammals	Birds	Reptiles	Amphibians	Nocturnal Species	Total
Data Deficient (DD)	24	2	8	26	7	60
Least Concern (LC)	285	931	100	91	103	1407
Near Threatened (NT)	29	52	1	16	6	98
Vulnerable (VU)	28	41	4	17	6	90
Endangered (EN)	20	15	3	7	0	45
Critically Endangered (CR)	2	8	1	2	0	13
Total	**388**	**1050**	**117**	**159**	**122**	**1714**

The ecoregion boundaries used in this study were derived from the Worldwide Fund for Nature (WWF) World Terrestrial ecoregion database [37] (http://worldwildlife.org/publications/terrestrial-ecoregions-of-the-world) (accessed on 1 June 2021). There are a total of 27 ecoregions on the TP. The boundary of national key ecological function zones was derived from the National and Regional Planning for Main Functional Zones [38]. There are 10 national key ecological function zones on the TP. The national nature reserve boundaries were derived from the Resources and Environment Science Data Center, and there are a total of 46 national nature reserves on the TP (http://www.resdc.cn/data.aspx?DATAID=272) (accessed on 1 June 2021). National park boundaries were derived from vectorization of national park planning maps [27]. There were four national parks on the TP: Giant Panda National Park, Sanjiangyuan National Park, Qilian Mountain National Park and Pudacuo National Park. The glacier dataset was provided by the National Cryosphere Desert Data Center [39] (http://www.ncdc.ac.cn) (accessed on 1 June 2021). The data of built-up areas used for LP attribution analysis came from the National Data Center for TP Science [40]. The basic geographic information came from the China National Basic Geographic Information Center, including 1:1,000,000 administrative boundaries, administrative centers, roads, settlements, rivers and lakes (http://ngcc.sbsm.gov.cn) (accessed on 1 June 2021).

3. Methods

3.1. Extraction of the LP Range

First, referring to Koen et al. [18], the night light data used in this study were comprehensively corrected to extract the LP range year by year from 1992 to 2018 (Figure 2). The value of night light from human activities is above 6.5 [32], and pixels with values lower than 6.5 are mostly abnormal values caused by the reflection of moonlight from glaciers or lakes. Therefore, the value of 6.5 was selected as the standard to binarize the original nighttime light data to remove abnormal values from the dataset.

Second, to further eliminate errors, the nighttime light range was corrected based on glacier and settlement data. As nighttime light patches should contain at least one residential area, nighttime light patches without residential areas were removed. In addition, patches of nighttime light containing glaciers were also removed.

Then, as the intensity of human activities on the TP is increasing year by year, we assumed that the range and intensity of nighttime light continuously increase. Referring to Liu et al. [41], based on this hypothesis, interannual correction was carried out from 1992 to 2018. After the above corrections, the areas exposed to nighttime light were extracted as the LP ranges [28,29].

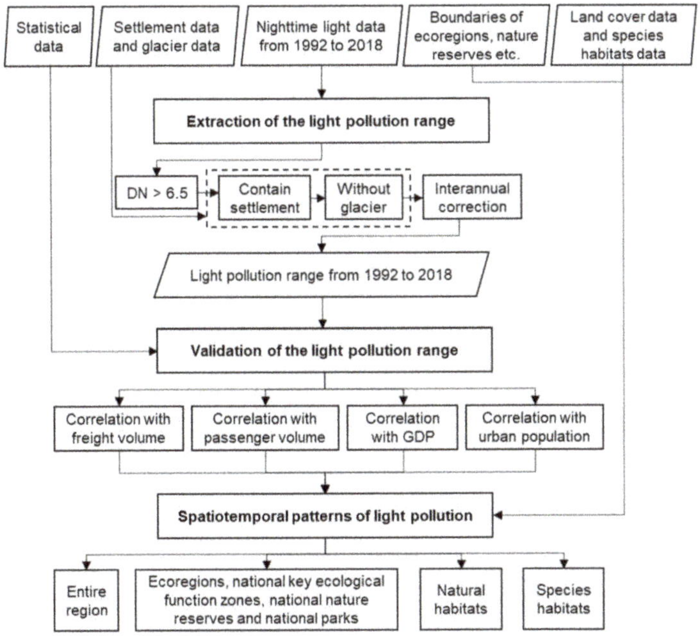

Figure 2. Flow chart.

3.2. *Validation of the LP Range*

Referring to Li et al. [42], provincial and prefecture-level socioeconomic indicators closely related to LP were selected. The accuracy of the annual LP ranges in Qinghai Province and the Tibet Autonomous Region during 1992–2018 were evaluated by the correlation analysis method. As the main sources of nighttime LP on the TP are roads and towns, four indices were selected: freight volume, passenger volume, GDP in secondary and tertiary industries and urban population. Freight volume and passenger volume represent the level of transportation development in this region, while GDP in secondary and tertiary industries and urban population represent the level of urbanization in this region.

3.3. *Analysis of Spatiotemporal Patterns of LP*

First, referring to Liu et al. [43], the landscape indices were used to quantify the spatial pattern of LP on the TP during 1992–2018. We selected total area (TA), patch density (PD), landscape shape index (LSI), aggregation index (AI) and largest patch index (LPI). PD refers to the number of patches per unit area, reflecting the degree of fragmentation or dispersion of light-polluted areas. The LSI is the normalized ratio of the patch perimeter to the patch area, which is an index used to measure the shape complexity of LP patches. The AI measures the concentration of LP patches. The LPI is the largest polluted patch on the entire TP. All indices were calculated by Fragstats software (V4.2) [44].

Second, referring to Fan et al. [28], Williams et al. [45] and Mu et al. [46], we analyzed the impact of LP expansion on the TP during 1992–2018 at the scale of ecoregions, national key ecological function zones, national nature reserves and national parks. We measured the degree of LP based on the proportion of LP area to the total area of each subregion. Since a threshold for defining the high-level LP at such large scales cannot be found, we used the approach adopted by Koen et al. [18] and He et al. [35], i.e., the regional average value, to extract the important areas facing relatively high-level LP.

Finally, referring to Koen et al. [18], He et al. [35] and McDonald et al. [47], we quantified the impacts of LP on biodiversity on the TP during 1992–2018 based on spatial overlay analysis and statistical analysis of the LP range with different types of natural habitats and species habitats. To understand the effects of LP on natural habitats, we calculated the area of LP that overlapped with each type of natural habitats and calculated the proportion of this area to the total area of LP range year by year. The effects of LP on the habitats of different species were quantified from two aspects: one is the number of species with potential habitats affected by LP (i.e., species whose habitats overlap with LP areas), and the other is the proportion of light-polluted habitat area of the species. Additionally, the endangered species with high proportion of light-polluted habitat area were identified. To highlight key messages, we selected the LP range at seven milestones (1992, 1995, 2000, 2005, 2010, 2015, and 2018) from annual LP range between 1992 and 2018 for the above analyses.

4. Results

4.1. Validation of the LP Range

LP area is significantly correlated with various socioeconomic indicators (Figure 3). Correlation coefficients were all above 0.8 and passed the significance test of 0.001. The correlation coefficient between LP and GDP in secondary and tertiary industries is the highest (R = 0.89), followed by freight volume (R = 0.88), passenger volume (R = 0.84) and urban population (R = 0.82).

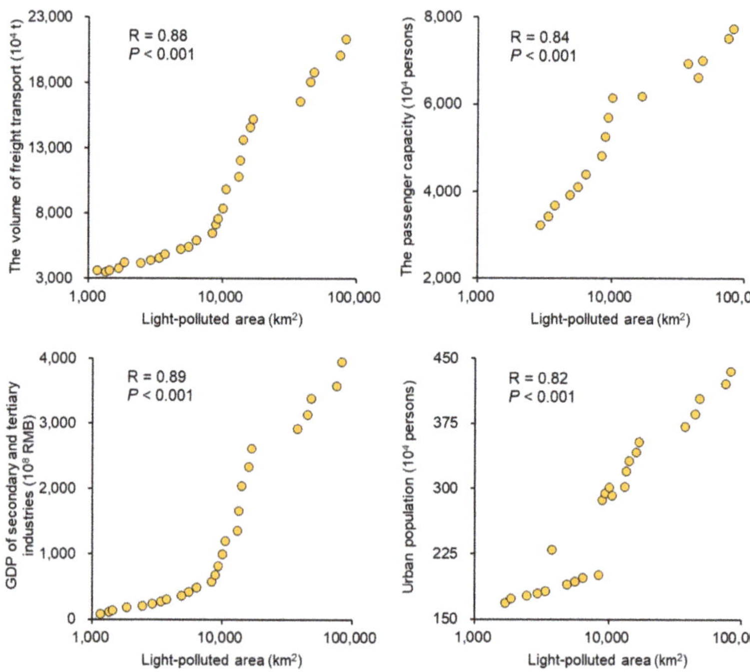

Figure 3. Accuracy evaluation of the light pollution range on the TP during 1992–2018.

4.2. Changes in LP on the TP from 1992 to 2018

From 1992–2018, the LP area on the TP increased exponentially. The LP area increased from 1.2 thousand km^2 to 82.8 thousand km^2, an increase of approximately 70 times, with an average annual growth rate of 30.6% (Figure 4a). In 2018, 3.2% of the total area of the TP was affected by LP. The LP in the eastern and southern TP was more serious and mainly occurred around urban and traffic routes. The LP in Xining and Lhasa was the most obvious (Figure 4). From 1992–2018, the degree of fragmentation in light-polluted areas on the TP gradually increased, and the PD increased from 1.6×10^{-5} patches/km^2 in 1992 to

3.8×10^{-4} patches/km^2 in 2018, an increase of nearly 24 times (Figure 4b). The complexity of LP patches on the TP also gradually increased, and the LSI increased from 6.6 to 36.5 (Figure 4c). The concentration degree of LP showed an overall increasing trend, and the AI increased from 82.8 to 87.6 (Figure 4d). The LPI increased from 9.2×10^{-5} to 4.1×10^{-3}, with an annual increase of 26.7% (Figure 4e).

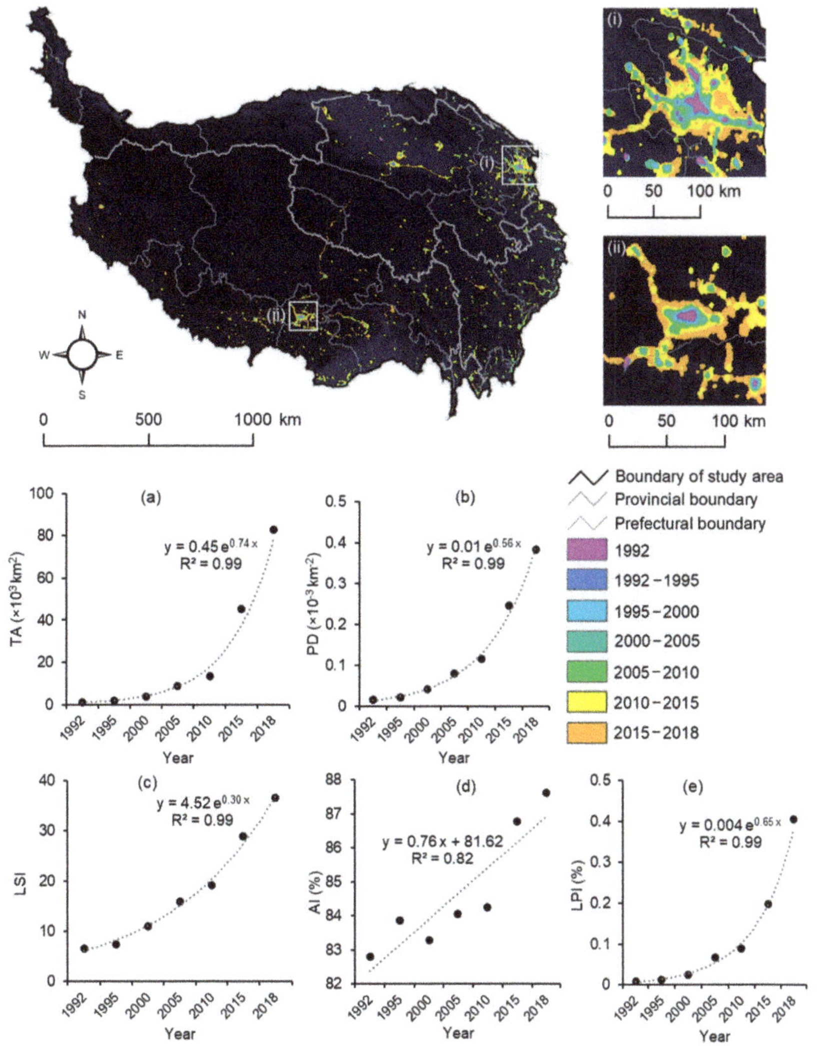

Figure 4. Changes in light pollution over the TP from 1992 to 2018: (**a**) total area (TA); (**b**) patch density (PD); (**c**) landscape shape index (LSI); (**d**) agglomeration index (AI); and (**e**) largest patch index (LPI) (Table S1).

4.3. Changes in LP at Different Scales

The results show that 26 of the 27 ecoregions on the TP were affected by LP, and 14 of them were affected by LP at a higher level than 3%, i.e., the average level of LP on the TP in 2018 (Figure 5a). The ecoregions in the southern TP were more affected, while the ecoregions in the northwest were less affected. In terms of time periods, the influence of LP on the ecoregions of the TP has increased significantly since 2010 (Figure S1). Specifically, the proportion of LP area in five ecoregions, including Brahmaputra Valley semievergreen forests (No. 1 in Figure 5a), Yarlung Tsangpo arid steppe (No. 2) and Qionglai-Minshan conifer forests (No. 3), accounted for 12% or more. It should be noted that the most affected ecoregion was the Brahmaputra Valley semievergreen forest, located on the southern edge of the TP, where 32.90% of the entire region was affected by LP. The proportion of LP area in Qilian Mountain coniferous forests (No. 6), Qin Ling Mountain deciduous forests (No. 7) and Eastern Himalayan broadleaf forests (No. 8) accounted for 9—12%. The proportion of LP area in Hengduan Mountains subalpine coniferous forests (No. 9) and northeastern Himalayan subalpine coniferous forests (No. 10) was between 6% and 9%. The proportion of LP area in four ecoregions was between 3% and 6%: Eastern Himalayan subalpine conifer forests (No. 11), Southeast Tibet shrublands and meadows (No. 12), Nujiang Langcang Gorge alpine conifers and mixed forests (No. 13) and Qaidam Basin semi-deserts (No. 14).

All 10 national key ecological function zones on the TP were affected by LP, and four of them were affected by LP of more than the average level of LP on the TP in 2018 (Figures 5b and S2). These were the ecological function zone of the Gannan Yellow River water supply (No. 1 in Figure 5b), the forest and biodiversity in Sichuan and Yunnan (No. 2), the forest at the margin of the plateau in southeastern Tibet (No. 3), and the Zoige Grassland Wetland ecological function zone (No. 4). These national key ecological function zones were all located in the southeast of the TP. The national key ecological zones in the northwest of the TP were less affected by LP.

Thirty-four of the 46 national nature reserves on the TP were affected by LP, and 17 of them were affected by LP, accounting for more than the average level of LP on the TP in 2018 (Figures 5c and S3). Specifically, LP in the Lalu Wetland (No. 1 in Figure 5c), Lianhua Mountains (No. 2) and Xunhua Mengda Nature Reserve in Gansu (No. 3) accounted for 12% or more of the area. The area of LP in Wolong (No. 6) and Four Girls Mountain National Nature Reserve in Sichuan (No. 7) is between 9% and 12%. LP accounts for between 6% and 9% of the area of five national nature reserves, including the Haloxylon forest in Qinghai-Chaidam, Qinghai Province (No. 8), Baihe in Sichuan Province (No. 9) and Taizi Mountains in Gansu Province (No. 10). Five national nature reserves, including Sichuan Gexigou (No. 13), Gansu Duoer (No. 14) and Sichuan Gongga Mountain (No. 15), accounted for between 3% and 6% of the area of LP. It is worth noting that the Lalu Wetland National Nature Reserve, which was adjacent to Lhasa city and is mainly protected for its alpine wetland ecosystem, has been continuously affected by LP since 1992. By 2000, the range of LP affected the whole range of the reserve.

The impact of LP on the TP in 2018 was relatively small in national parks (Figure 5d). The area of LP in Giant Panda National Park (No. 1 in Figure 5d) accounts for 6.75% of its own area. and 2.22% of the source regions of the Yangtze River, Yellow River and Lancang River National Parks (No. 2) were affected by LP. Only 0.68% of the area of Qilian Mountain National Park (No. 3) was affected by LP, and Pudacuo National Park (No. 4) was not affected by LP. In terms of the time period, the impact of LP in Giant Panda National Park from 2000 to 2005 was greater, and, after 2010, the impact of LP in other national parks except Pudacuo Park increased significantly (Figure S4).

4.4. *LP for Different Habitat Typess*

The area of LP on natural habitats increased from 938 km^2 (79.6% of total LP area) in 1992 to 74.2 thousand km^2 (91.4% of total LP area) in 2018 (Table 2). The largest area of LP was found on grassland, accounting for 46.9% of LP area in 1992 and 54.7% of LP area in 2018 (Figure 6). The area of LP on forest is growing the fastest, and increased from 7.6% of LP area in 1992 to 26.0% of LP area in 2018 (Figure 6). Desert, wetland and ice accounted for 9.1%, 1.2% and 0.3% of the total LP area in 2018, respectively.

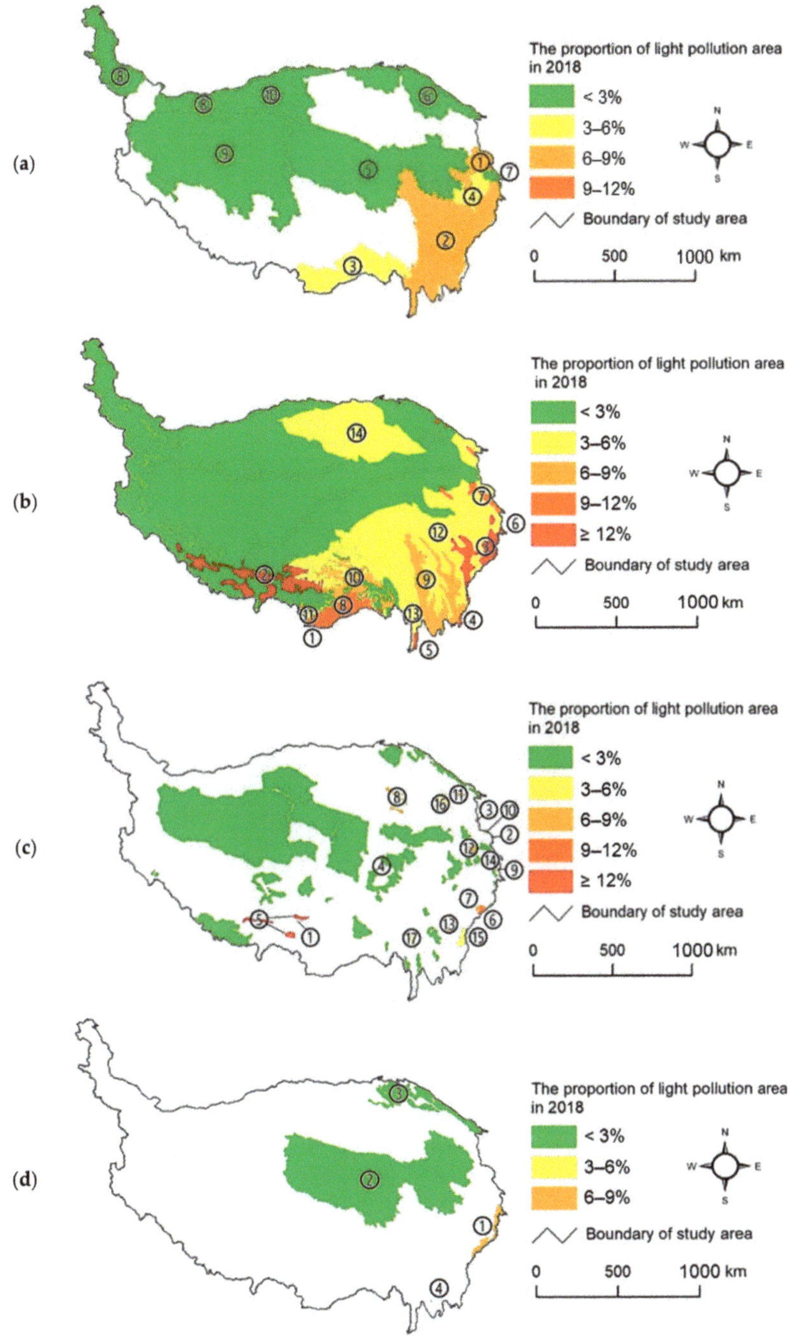

Figure 5. Light pollution in 2018 at the different scales. The names of labelled regions are shown in Figures S1–S4. (**a**) National key ecological function zone, (**b**) ecoregion, (**c**) national nature reserve, and (**d**) national park.

Table 2. Natural habitats affected by light pollution *.

Natural Habitat	Year						
	1992	**1995**	**2000**	**2005**	**2010**	**2015**	**2018**
Grassland	552 (46.9%)	855 (49.9%)	1903 (50.5%)	4583 (51.7%)	6860 (51.1%)	22,144 (49.9%)	44,424 (54.7%)
Forest	90 (7.6%)	152 (8.9%)	695 (18.4%)	1928 (21.8%)	3207 (23.9%)	13,574 (30.6%)	21,096 (26.0%)
Desert	265 (22.5%)	352 (20.6%)	488 (13.0%)	785 (8.9%)	1096 (8.2%)	3558 (8.0%)	7395 (9.1%)
Wetland	31 (2.6%)	35 (2.0%)	88 (2.3%)	163 (1.8%)	230 (1.7%)	611 (1.4%)	999 (1.2%)
Ice	0 (0.0%)	0 (0.0%)	0 (0.0%)	0 (0.0%)	0 (0.0%)	0 (0.0%)	254 (0.3%)
Sum	**938 (79.6%)**	**1394 (81.4%)**	**3174 (84.3%)**	**7459 (84.2%)**	**11,393 (84.9%)**	**39,887 (89.8%)**	**74,168 (91.4%)**

* The number denotes the area of natural habitat affected by light pollution, while the proportion in parentheses denotes the percentage of natural habitat affected by light pollution to total light pollution area.

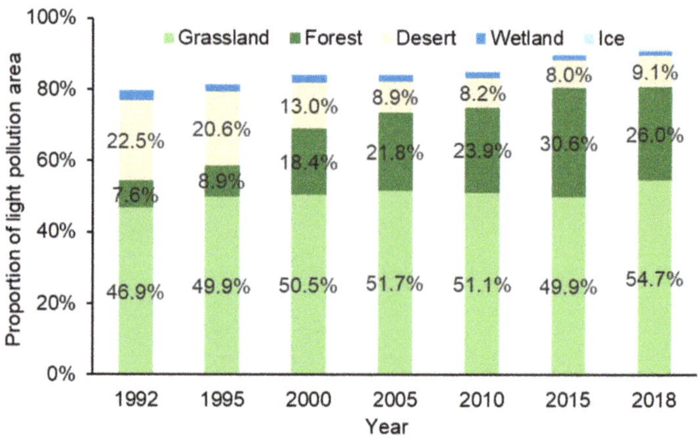

Figure 6. Natural habitats affected by light pollution.

4.5. LP for Different Species Habitats

The number of species with habitats affected by LP gradually increased on the TP during 1992–2018 (Figure 7a). Overall, the number of species with habitats affected by LP nearly doubled from 835 to 1619. The number of mammals with habitats affected by LP increased from 214 to 359, an increase of 0.7 times. The number of birds with habitats affected by LP increased from 532 to 1040, an increase of 0.9 times. The number of reptiles with habitats affected by LP increased from 52 to 110, more than doubling. The number of amphibians with habitats affected by LP increased by 2.7 times from 37 to 136. Among these, the number of nocturnal species with habitats affected by LP increased from 43 to 104, an increase of 1.4 times.

In addition, the number of endangered species with habitats affected by LP also increased year by year (Figure 7a). In total, the number of endangered species with habitats affected by LP increased from 89 to 228, an increase of approximately 1.5 times. The number of mammals affected increased 1.1 times, birds 1.6 times, reptiles 3.5 times, amphibians 2.8 times, and nocturnal animals approximately 2.7 times. Birds had the largest number of endangered species with habitats affected by LP (Figure 7a), while amphibians had the largest proportion of endangered species with habitats affected by LP (Figure 7b).

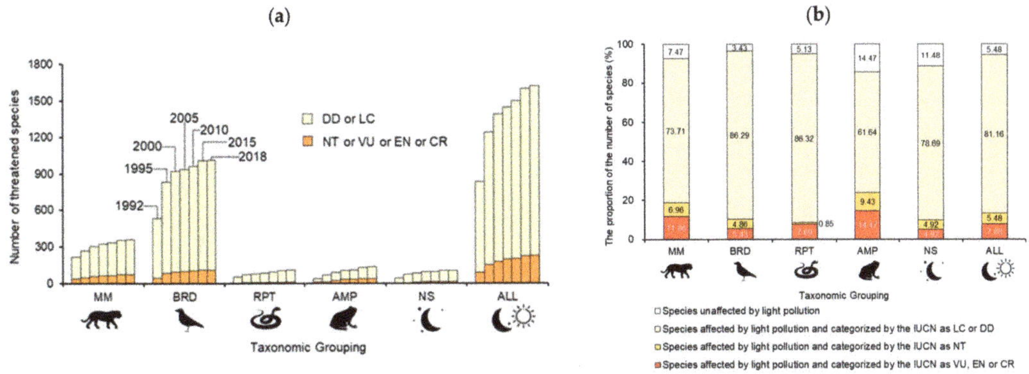

Figure 7. Different species habitats affected by light pollution on the TP. (**a**) Number of species with habitats affected by LP, 1992–2018 and (**b**) proportion of species with habitats affected by LP in 2018. Abbreviation: mammals (MM), birds (BRD), reptiles (RPT), amphibians (AMP), nocturnal species (NS).

The area of habitats affected by LP has gradually increased, especially since 2010 (Figure 8). The proportion of habitat area affected by LP for endangered species was generally larger than the average for all species. In 2018, a total of 18 endangered species, including 14 birds, 3 mammals and 1 reptile, were affected by LP, accounting for more than a quarter of their habitat area (Figure 8b; Tables S2 and S3).

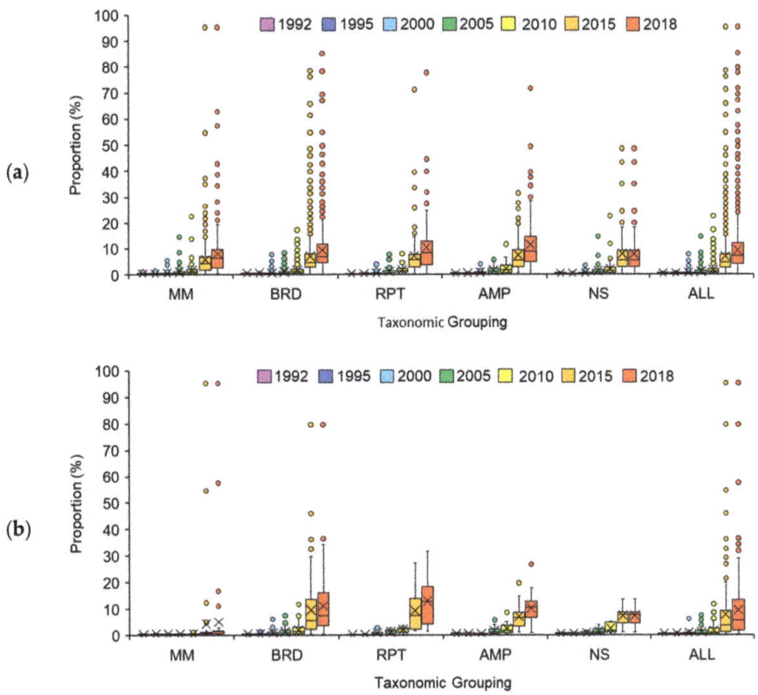

Figure 8. Proportion of habitat affected by light pollution on the TP. (**a**) Total species and (**b**) endangered species (i.e., NT, VU, EN and CR species).

Species richness increased from northwest to southeast, and endangered species also showed a similar pattern, which was consistent with the spatial distribution pattern of LP. According to the correlation analysis between the average level of species richness in different ecoregions and the percentage of increased area of LP from 1992 to 2018, the correlation coefficient between the percentage of increased area of LP and total species richness was 0.90 ($p < 0.001$) and that between the percentage of increased area of LP and endangered species richness was 0.99 ($p < 0.001$). The impact of LP on biodiversity, especially on endangered species, cannot be ignored.

5. Discussion

5.1. Main Reasons for the Increase in the LP Range

LP on the TP mainly occurs along traffic routes and around towns (Figures 1 and 4). We further conducted attribution analysis on the expanded range of LP on the TP from 1992 to 2018. Specifically, we analyzed the main sources of LP according to the distance of LP pixels from roads and settlements and LP pixels that had equal distances to roads and to settlements to the joint effects of both. The results show that the LP on the TP during 1992–2018 was mainly attributed to the influence of roads, accounting for 87.9% of the total area of new LP (Figure 9). In addition, 8.1% of the LP area was attributed to settlements. A total of 4.1% of the LP area was attributed to the combined influence of roads and settlements.

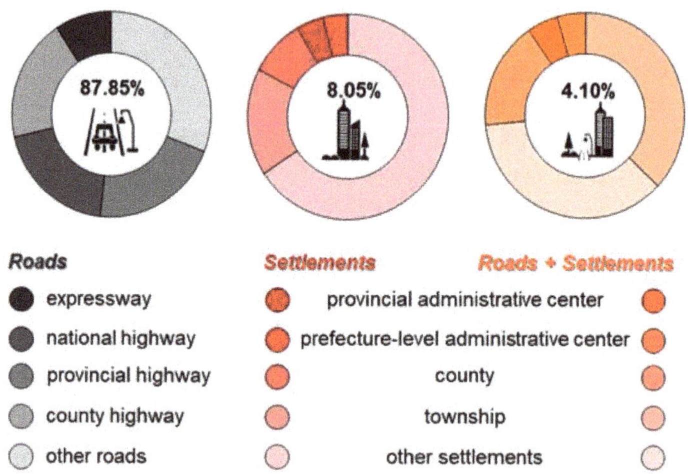

Figure 9. Attribution analysis of light pollution on the TP during 1992–2018.

5.2. Policy Implications

In recent years, the rapid increase in LP on the TP has posed a potential threat to nature reserves and endangered species. Therefore, it is suggested that LP should be reduced as much as possible in the future development process of the TP. Measures such as reducing lighting intensity, changing lighting spectral composition, and limiting lighting duration should be taken to effectively reduce LP [48]. For example, turning off an unnecessary artificial light source or adjusting the light source to the weakest brightness while still meeting the demand strictly limits the "blue" light that interferes with circadian rhythm and dark vision [49] and realizes adaptive lighting (intelligent switch) through technology. At the same time, the government should pay attention to the construction of "dark" ecological corridors, improve public awareness of LP, and strengthen relevant policy formulation and legislation (such as incorporating methods to avoid and mitigate LP in protected areas into management plans) [25,50,51]. Through the implementation of ecological compensation, the government can strengthen the protection and restoration of the ecological environment within the scope of LP and take targeted development measures for different regions to achieve the coordinated development of regional ecological and environmental benefits, economic benefits and social benefits [52,53].

Considering that the impact of roads is the main factor for the increase in LP on the TP in recent years, it is particularly important to make overall planning in the future construction and development of roads. On the one hand, the impact of LP on the environment can be controlled by optimizing the planning of traffic networks. On the other hand, road light sources should be rationally arranged. The intensity of light sources should be strictly limited to reduce LP caused by roads, avoid damage to ecological corridors and curb adverse effects on biodiversity [54,55]. Future urban construction should also focus on the conservation of endangered species as much as possible (Tables S2 and S3), strictly restricting human activity within the territory of these species.

To promote effective measures for mitigating effects of LP on biodiversity conservation, we visually identified the major sources of LP for national nature reserves with relatively high-level LP based on the approach used in attribution analysis. In specific, both human settlements (including cities, towns, and villages) and roads were identified (Table 3). Although these areas caused a small proportion of the total LP, they had influences on the key natural habitats for biodiversity conservation. Therefore, measures should first be taken to control LP in these areas.

Table 3. Major source of light pollution for national nature reserves.

ID in Figure 5c	Name	Abbreviation	Major Source of Light Pollution
1	Lalu Wetland	LW	Chengguan Distract in Lhasa city
2	Gansu Lianhua Mountain	GLM	Xiacheng town, Lianlu town
3	Xunhua Mengda	XM	Dahejia town
4	Longbao	Longbao	Longbao town
5	Black-necked crane of Brahmaputra River	BNCBR	Lhunzhub county, Maizhuokunggar county, Shigatse city, Road from Lhasa to Lhunzhub (G561), Road from Lhasa to Maizhuokunggar (G349)
6	Wolong	Wolong	Wolong town, Genda town
7	Four Girls Mountain	FGM	Siguniangshan town, Dawei town
8	Haloxylon forest in Qinghai-chaidam	HFIQC	Beijing-tibet highway (G6), Road from Beijing to Lhasa (G109)
9	Baihe	Baihe	Jiuhong road (S301)
10	Taizi mountains	TM	Songming town
11	The source region of rivers in the north of Datong	SRRND	Road from Xining to Zhangye (G227)
12	Gahai-zecha	GZ	Road from Gahai to Maqu (S204), Gaxiu village, Gahai village, Gongba village, Langmushi town
13	Gexigou	Gexigou	Yajing county, Xiangkezong
14	Duoer	Duoer	Road from Lianghekou to Maqu (S313)
15	Gongga Mountain	GM	Kangding city
16	Qinghai Lake	QL	Road from Beijing to Lhasa (G109)
17	Mangkang Yunnan snub-nosed monkey	MYSNM	Rumei town, Naxi nationality Town, Quzika Town

Note: only the national nature reserves whose total light pollution range is above the average level of light pollution on the TP are listed.

5.3. Future Perspectives

In this study, the spatial-temporal pattern of LP on the TP during 1992–2018 was determined based on remotely sensed nighttime light data at multiple scales. The impacts of LP changes on natural habitat and species habitat were analyzed, therefore helping to fill the gap in the field of LP research on the TP. In the previous studies, Koen et al. [18] assessed the global LP from 1992 to 2012, while Li et al. [11] evaluated the LP for natural reserves in China in the same period. They both found that TP had relatively low LP between 1992 and 2012. However, the previous studies did not measure LP on the TP after 2012 in light of the inconsistence in NTL data. This study provided an assessment of LP over the past three decades based on the harmonized NTL data, and found that the LP had increased widely since 2012, posing threats to natural habitats and species habitats on the TP.

The shortcoming of this study is that we only considered the range of LP, and the intensity of LP was not fully considered. In fact, the farther away from the center of the light source, the weaker its ecological impacts. The degree of LP influence also depends on the duration of light and the weather conditions at that time. In addition, we only quantified the species with habitats covered by LP, and did not assess the effects of LP on species (e.g., reproduction and growth) on the TP. Therefore, we will further use nighttime light intensity information to assess the impact of LP on biodiversity on the TP in the future, and the field investigations into the effects of LP on species are also needed in order to reveal thresholds of coverage and intensity for LP that affect habitats.

6. Conclusions

LP has been increasing rapidly in recent years, which has caused a serious impact on the natural and species habitats in the region. From 1992–2018, the LP area of the TP increased from 1.2 thousand km^2 to 82.8 thousand km^2, an increase of approximately 70 times, which was mainly distributed in the east and south of the TP. Since 2010, the speed of LP coverage has significantly increased, and by 2018, the coverage of LP on the TP accounted for 3.22% of the total area. LP has seriously affected some ecoregions, national key ecological function zones, national nature reserves and national parks. The area of LP on natural habitats increased from 79.6% of total LP area to 91.4% of total LP area. The number of endangered species with habitats affected by LP rose from 89 to 228. Roads were the main source of LP, followed by urban and rural settlements. In the process of urbanization and road construction in the future, effective measures should be taken to control the scope and intensity of LP, focusing on endangered species affected by LP to promote biodiversity conservation.

Supplementary Materials: The following are available online at https://www.mdpi.com/article/10.3390/rs14225755/s1. Figure S1: Light pollution at the ecoregional scale; Figure S2: Light pollution at the scale of national key ecological function zone; Figure S3: Light pollution at the scale of national nature reserve; Figure S4: Light pollution at the national park scale; Table S1: Landscape index of light pollution area on the TP from 1992 to 2018; Table S2: Threatened species whose habitats are affected by light pollution by more than 25%; Table S3: Near threatened species whose habitats are affected by light pollution by more than 25%.

Author Contributions: Y.W. drafted the manuscript. Z.L. (Zhifeng Liu) conceived and guided this study. Y.W., C.L., X.P., Z.L. (Ziwen Liu), P.X. and C.Z. gave important advice on methodology and providing suggestions on the revision of the manuscript. All authors have read and agreed to the published version of the manuscript.

Funding: This work was supported by the Second Tibetan Plateau Scientific Expedition and Research Program (Grant No.2019QZKK0405) and the National Natural Science Foundation of China (Grant No. 41871185 & 41971271). It was also supported by the project from the State Key Laboratory of Earth Surface Processes and Resource Ecology, China.

Institutional Review Board Statement: Not applicable.

Informed Consent Statement: Not applicable.

Data Availability Statement: Not applicable.

Conflicts of Interest: The authors declare no conflict of interest.

References

1. Longcore, T.; Rich, C. Ecological light pollution. *Front. Ecol. Environ.* **2004**, *2*, 191–198. [CrossRef]
2. Encyclopædia Britannica. Light Pollution. 2020. Available online: https://www.britannica.com/science/light-pollution (accessed on 10 May 2021).
3. Peregrym, M.; Pénzesné Kónya, E.; Falchi, F. Very important dark sky areas in Europe and the Caucasus region. *J. Environ. Manag.* **2020**, *274*, 111167. [CrossRef] [PubMed]
4. Svechkina, A.; Portnov, B.A.; Trop, T. The impact of artificial light at night on human and ecosystem health: A systematic literature review. *Landsc. Ecol.* **2020**, *35*, 1725–1742. [CrossRef]
5. Bennie, J.; Davies, T.W.; Cruse, D.; Inger, R.; Gaston, K.J. Artificial light at night causes top-down and bottom-up trophic effects on invertebrate populations. *J. Appl. Ecol.* **2018**, *55*, 2698–2706. [CrossRef]
6. Dimitriadis, C.; Fournari-Konstantinidou, I.; Sourbes, L.; Koutsoubas, D.; Mazaris, A.D. Reduction of sea turtle population recruitment caused by nightlight: Evidence from the Mediterranean region. *Ocean Coast. Manag.* **2018**, *153*, 108–115. [CrossRef]
7. Dominoni, D.; Smit, J.A.H.; Visser, M.E.; Halfwerk, W. Multisensory pollution: Artificial light at night and anthropogenic noise have interactive effects on activity patterns of great tits (Parus major). *Environ. Pollut.* **2020**, *256*, 113314. [CrossRef]
8. Ffrench-Constant, R.H.; Somers-Yeates, R.; Bennie, J.; Economou, T.; Hodgson, D.; Spalding, A.; McGregor, P.K. Light pollution is associated with earlier tree budburst across the United Kingdom. *Proc. R. Soc. B-Biol. Sci.* **2016**, *283*, 20160813. [CrossRef]
9. Grubisic, M.; van Grunsven, R.H.A.; Kyba, C.C.M.; Manfrin, A.; Holker, F. Insect declines and agroecosystems: Does light pollution matter? *Ann. Appl. Biol.* **2018**, *173*, 180–189. [CrossRef]
10. Haim, A.; Zubidat, A.E. Artificial light at night: Melatonin as a mediator between the environment and epigenome. *Philos. Trans. R. Soc. B-Biol. Sci.* **2015**, *370*, 20140121. [CrossRef]
11. Li, G.; Gao, J.; Li, L.; Hou, P. Human pressure dynamics in protected areas of China based on nighttime light. *Glob. Ecol. Conserv.* **2020**, *24*, e01222. [CrossRef]
12. Maggi, E.; Bongiorni, L.; Fontanini, D.; Capocchi, A.; Dal Bello, M.; Giacomelli, A.; Benedetti-Cecchi, L. Artificial light at night erases positive interactions across trophic levels. *Funct. Ecol.* **2020**, *34*, 694–706. [CrossRef]
13. Massetti, L. Assessing the impact of street lighting on Platanus x acerifolia phenology. *Urban For. Urban Green.* **2018**, *34*, 71–77. [CrossRef]
14. Skvareninova, J.; Tuharska, M.; Skvarenina, J.; Babalova, D.; Slobodnikova, L.; Slobodnik, B.; Stredova, H.; Mindas, J. Effects of light pollution on tree phenology in the urban environment. *Morav. Geogr. Rep.* **2017**, *25*, 282–290. [CrossRef]
15. Touitou, Y.; Reinberg, A.; Touitou, D. Association between light at night, melatonin secretion, sleep deprivation, and the internal clock: Health impacts and mechanisms of circadian disruption. *Life Sci.* **2017**, *173*, 94–106. [CrossRef] [PubMed]
16. Yang, Y.F.; Liu, Q.; Wang, T.; Pan, J.M. Wavelength-specific artificial light disrupts molecular clock in avian species: A power-calibrated statistical approach. *Environ. Pollut.* **2020**, *265*, 114206. [CrossRef] [PubMed]
17. Kyba, C.C.M.; Kuester, T.; Sánchez de Miguel, A.; Baugh, K.; Jechow, A.; Hölker, F.; Bennie, J.; Elvidge, C.D.; Gaston, K.J.; Guanter, L. Artificially lit surface of Earth at night increasing in radiance and extent. *Sci. Adv.* **2017**, *3*, e1701528. [CrossRef]
18. Koen, E.L.; Minnaar, C.; Roever, C.L.; Boyles, J.G. Emerging threat of the 21st century lightscape to global biodiversity. *Glob. Chang. Biol.* **2018**, *24*, 2315–2324. [CrossRef]
19. Zhang, Y.; Liu, L.; Wang, Z.; Bai, W.; Ding, M.; Wang, X.; Yan, J.; Xu, E.; Wu, X.; Zhang, B.; et al. Temporal and spatial characteristics of land use and cover change on the Tibetan Plateau. *Chin. Sci. Bull.* **2019**, *64*, 2865–2875.
20. Fan, D.; Ni, L.; Jiang, X.G.; Fang, S.F.; Wu, H.; Zhang, X.P. Spatiotemporal Analysis of Vegetation Changes Along the Belt and Road Initiative Region From 1982 to 2015. *IEEE Access* **2020**, *8*, 122579–122588. [CrossRef]
21. Jiang, Y.; Lin, W.P.; Wu, M.Q.; Liu, K.; Yu, X.M.; Gao, J. Remote Sensing Monitoring of Ecological-Economic Impacts in the Belt and Road Initiatives Mining Project: A Case Study in Sino Iron and Taldybulak Levoberezhny. *Remote Sens.* **2022**, *14*, 3308. [CrossRef]
22. Nguyen, C.T.; Chidthaisong, A.; Limsakul, A.; Varnakovida, P.; Ekkawatpanit, C.; Diem, P.K.; Diep, N.T.H. How do disparate urbanization and climate change imprint on urban thermal variations? A comparison between two dynamic cities in Southeast Asia. *Sustain. Cities Soc.* **2022**, *82*, 103882. [CrossRef]
23. Song, Y.; Aryal, J.; Tan, L.C.; Jin, L.; Gao, Z.H.; Wang, Y.Q. Comparison of changes in vegetation and land cover types between Shenzhen and Bangkok. *Land Degrad. Dev.* **2021**, *32*, 1192–1204. [CrossRef]
24. Li, S.; Zhang, Y.; Wang, Z.; Li, L. Mapping human influence intensity in the Tibetan Plateau for conservation of ecological service functions. *Ecosyst. Serv.* **2018**, *30*, 276–286. [CrossRef]
25. Challéat, S.; Barré, K.; Laforge, A.; Lapostolle, D.; Franchomme, M.; Sirami, C.; Le Viol, I.; Milian, J.; Kerbiriou, C. Grasping darkness: The dark ecological network as a social-ecological framework to limit the impacts of light pollution on biodiversity. *Ecol. Soc.* **2021**, *26*, 15. [CrossRef]

26. Piao, S.; Zhang, X.; Wang, T.; Liang, E.; Wang, S.; Zhu, J.; Niu, B. Response feedback of Tibetan Plateau ecosystem to climate change. *Chin. Sci. Bull.* **2019**, *64*, 2842–2855. [CrossRef]
27. Chen, W.; Huang, H.; Tian, Y.; Du, Y. Dynamic monitoring and analysis of ecological environment quality in Sanjiangyuan region based on Google Earth Engine platform. *J. Geo-Inf. Sci.* **2019**, *21*, 1382–1391.
28. Fan, L.; Zhao, J.; Wang, Y.; Ren, Z.; Zhang, H.; Guo, X. Assessment of Night-Time Lighting for Global Terrestrial Protected and Wilderness Areas. *Remote Sens.* **2019**, *11*, 2699. [CrossRef]
29. Kumar, P.; Rehman, S.; Sajjad, H.; Tripathy, B.R.; Rani, M.; Singh, S. Analyzing trend in artificial light pollution pattern in India using NTL sensor's data. *Urban Clim.* **2019**, *27*, 272–283. [CrossRef]
30. Cabrera-Cruz, S.A.; Smolinsky, J.A.; Buler, J.J. Light pollution is greatest within migration passage areas for nocturnally-migrating birds around the world. *Sci. Rep.* **2018**, *8*, 3261. [CrossRef]
31. Morshed, M.M.; Chakraborty, T.; Mazumder, T. Measuring Dhaka's Urban Transformation Using Nighttime Light Data. *J. Geovis. Spat. Anal.* **2022**, *6*, 25. [CrossRef]
32. Li, X.; Zhou, Y.; Zhao, M.; Zhao, X. A harmonized global nighttime light dataset 1992–2018. *Sci. Data* **2020**, *7*, 168. [CrossRef]
33. IUCN. *IUCN Red List of Threatened Species*; IUCN: Gland, Switzerland, 2013; Available online: www.iucnredlist.org (accessed on 20 October 2022).
34. *China Statistical Yearbook 2001–2020*; National Bureau of Statistics: Beijing, China, 2001–2020.
35. He, C.; Liu, Z.; Tian, J.; Ma, Q. Urban expansion dynamics and natural habitat loss in China: A multiscale landscape perspective. *Glob. Chang. Biol.* **2014**, *20*, 2886–2902. [CrossRef] [PubMed]
36. Zalles, V.; Hansen, M.C.; Potapov, P.V.; Parker, D.; Stehman, S.V.; Pickens, A.H.; Parente, L.L.; Ferreira, L.G.; Song, X.P.; Hernandez-Serna, A.; et al. Rapid expansion of human impact on natural land in South America since 1985. *Sci. Adv.* **2021**, *7*, eabg1620. [CrossRef] [PubMed]
37. Olson, D.M.; Dinerstein, E.; Wikramanayake, E.D.; Burgess, N.D.; Powell, G.V.N.; Underwood, E.C.; D'Amico, J.A.; Itoua, I.; Strand, H.E.; Morrison, J.C.; et al. Terrestrial Ecoregions of the World: A New Map of Life on Earth. *BioScience* **2001**, *51*, 933–938. [CrossRef]
38. National Development and Reform Commission. *National and Regional Functional Zones Are Planned*; People's Publishing House: Beijing, China, 2015.
39. Wu, L. Global 1:1 Million Glacier Dataset. National Cryophere Desert Data Center. 2013. Available online: http://www.ncdc.ac.cn (accessed on 1 June 2021).
40. He, C.; Liu, Z.; Wang, Y. *Urban Distribution and Urbanization Index Dataset on the Tibetan Plateau (2019)*; National Data Center for Tibetan Plateau Science: Beijing, China, 2019; Available online: http://data.tpdc.ac.cn/en/ (accessed on 20 October 2022).
41. Liu, Z.; He, C.; Zhang, Q.; Huang, Q.; Yang, Y. Extracting the dynamics of urban expansion in China using DMSP-OLS nighttime light data from 1992 to 2008. *Landsc. Urban Plan.* **2012**, *106*, 62–72. [CrossRef]
42. Li, X.; Xu, H.; Chen, X.; Li, C. Potential of NPP-VIIRS Nighttime Light Imagery for Modeling the Regional Economy of China. *Remote Sens.* **2013**, *5*, 3057–3081. [CrossRef]
43. Liu, Y.; Wu, J.; Yu, D. Characterizing spatiotemporal patterns of air pollution in China: A multiscale landscape approach. *Ecol. Indic.* **2017**, *76*, 344–356. [CrossRef]
44. McGarigal, K.; Cushman, S.A.; Ene, E. *FRAGSTATS v4: Spatial Pattern Analysis Program for Categorical and Continuous Maps*; University of Massachusetts: Amherst, MA, USA, 2012.
45. Williams, D.R.; Clark, M.; Buchanan, G.M.; Ficetola, G.F.; Rondinini, C.; Tilman, D. Proactive conservation to prevent habitat losses to agricultural expansion. *Nat. Sustain.* **2020**, *4*, 314–322. [CrossRef]
46. Mu, H.; Li, X.; Du, X.; Huang, J.; Su, W.; Hu, T.; Wen, Y.; Yin, P.; Han, Y.; Xue, F. Evaluation of Light Pollution in Global Protected Areas from 1992 to 2018. *Remote Sens.* **2021**, *13*, 1849. [CrossRef]
47. McDonald, R.I.; Kareiva, P.; Forman, R.T.T. The implications of current and future urbanization for global protected areas and biodiversity conservation. *Biol. Conserv.* **2008**, *141*, 1695–1703. [CrossRef]
48. Hu, J.; Liu, Y.; Fang, J. Ecological Corridor Construction Based on Least-Cost Modeling Using Visible Infrared Imaging Radiometer Suite (VIIRS) Nighttime Light Data and Normalized Difference Vegetation Index. *Land* **2021**, *10*, 782. [CrossRef]
49. Falchi, F.; Cinzano, P.; Duriscoe, D.; Kyba, C.C.M.; Elvidge, C.D.; Baugh, K.; Portnov, B.A.; Rybnikova, N.A.; Furgoni, R. The new world atlas of artificial night sky brightness. *Sci. Adv.* **2016**, *2*, e1600377. [CrossRef] [PubMed]
50. Gaston, K.J.; Davies, T.W.; Bennie, J.; Hopkins, J. REVIEW: Reducing the ecological consequences of night-time light pollution: Options and developments. *J. Appl. Ecol.* **2012**, *49*, 1256–1266. [CrossRef]
51. Kamrowski, R.L.; Sutton, S.G.; Tobin, R.C.; Hamann, M. Balancing artificial light at night with turtle conservation? Coastal community engagement with light-glow reduction. *Environ. Conserv.* **2014**, *42*, 171–181. [CrossRef]
52. Wei, W.; Li, W.; Song, Y.; Xu, J.; Wang, W.; Liu, C. The Dynamic Analysis and Comparison of Emergy Ecological Footprint for the Qinghai–Tibet Plateau: A Case Study of Qinghai Province and Tibet. *Sustainability* **2019**, *11*, 5587. [CrossRef]
53. Chai, L.; Tian, L.; Ao, Y.; Wang, X. Effects of human disturbance on vegetation cover change in Tibetan Plateau. *Res. Soil Water Conserv.* **2021**, *28*, 382–388.

54. Brooks, T.M.; Mittermeier, R.A.; da Fonseca, G.A.B.; Gerlach, J.; Hoffmann, M.; Lamoreux, J.F.; Mittermeier, C.G.; Pilgrim, J.D.; Rodrigues, A.S. Global Biodiversity Conservation Priorities. *Science* **2006**, *313*, 58–61. [CrossRef]
55. Leng, W.; He, G.; Jiang, W. Investigating the Spatiotemporal Variability and Driving Factors of Artificial Lighting in the Beijing-Tianjin-Hebei Region Using Remote Sensing Imagery and Socioeconomic Data. *Int. J. Environ. Res. Public Health* **2019**, *16*, 1950. [CrossRef]

remote sensing

Article

Examining Thresholding and Factors Impacting Snow Cover Detection Using Nighttime Images

Renato Stopic and Eduardo Dias *

SPINlab, Department of Spatial Economics, Vrije Universiteit Amsterdam, 1081 HV Amsterdam, The Netherlands
* Correspondence: ess580@vu.nl

Abstract: Nighttime remote sensing data from the Visible Infrared Imaging Radiometer suite day/night band (VIIRS DNB) enable snow cover detection from full moonlight reflection. Using nighttime data is particularly relevant in areas with limited daytime hours due to high latitudes. Previous studies demonstrated the potential of using thresholding methods in detecting snow, but more research studies are needed to understand the factors that influence their accuracy. This study explored seven thresholding algorithms in four case study areas with different characteristics and compared the classified snow results to the MODIS MOD10A1 snow cover product. The results found that Li thresholding delivers higher accuracies for most case studies, with an overall accuracy between 65% and 81%, while mean thresholding performed best in mountainous regions (70%) but struggled in other areas. Most false negatives are caused by forests, especially closed and evergreen forests. The analysis of NDVI data matches these findings, with the NDVI of false negatives being significantly higher than true positives. False positives appear to be primarily located in or around built-up areas. This study provides insights into where nighttime VIIRS DNB data can be used to increase snow cover data temporal and spatial coverage.

Keywords: remote sensing; snow extent; nighttime; moonlight; VIIRS; thresholding; land cover

Citation: Stopic, R.; Dias, E. Examining Thresholding and Factors Impacting Snow Cover Detection Using Nighttime Images. *Remote Sens.* **2023**, *15*, 868. https://doi.org/10.3390/rs15040868

Academic Editors: Ran Goldblatt, Steven Louis Rubinyi and Hogeun Park

Received: 24 December 2022
Revised: 28 January 2023
Accepted: 2 February 2023
Published: 4 February 2023

1. Introduction

Knowing which areas of the planet are covered by snow or ice is of major importance to many fields of earth sciences, including climatology, meteorology, and hydrology. Snow has an impact on both local environments and on global climate [1,2]. Since the beginning of remote sensing in the early 1970s, snow was one of the first to be investigated successfully because of its high albedo [3]. The introduction of MODIS (Moderate Resolution Imaging Spectroradiometer) enabled the creation of daily snow cover products at a global scale. Due to its high temporal resolution and global coverage, the MODIS snow cover product is widely used in many scientific fields, including climatology and hydrology [4]. The MODIS snow cover extent product continues to be actively released, and the most recent version of the product is version 6.1 of MOD10A1 [5], but the MODIS snow cover product is limited by light availability and cloud coverage. The limitation of low light is particularly impactful in high-latitude regions that receive few hours of daylight in winter months due to their position on the globe [6]. Even when there is sufficient light, the view of the ground may be limited by the presence of clouds. Beyond the limited winter availability of light, the impact of clouds is noticeable in both high-latitude regions and other potentially snow-covered areas in mid-latitude regions.

Satellite images captured during the night can circumvent both potential obstacles. The illumination from moonlight during certain lunar phases can be sufficient for detecting ground features with a high albedo, which also includes snow [7,8]. The ability to use illumination from moonlight means that data collection is not limited by the low number of daytime hours in winter in high-latitude regions.

The use of nighttime data would also increase the possibility of data being captured in cloud-free conditions anywhere on the planet. Regions covered by clouds during the day would have no data available. If clouds clear during the night, it is then possible to determine the snow cover extent. Combined with the assumption that there are fewer clouds at night due to the reduced convection, this means that more data would be available [9,10].

Snow detection with nighttime images has been around since the early days of nighttime satellites. The first attempts were made using the Defense Meteorological Satellite Program's Operational Line-Scan System (DMSP-OLS) in the 1980s and 1990s, but their usefulness was limited due to their coarse resolution. Shortly after the launch of the Suomi National Polar Partnership (S-NPP) satellite equipped with VIIRS, the first study detecting snow cover was conducted, and it showed that this approach is feasible [6]. Further research into this field has come in recent years, with the two most notable studies being Huang et al. [7] and Liu et al. [11].

Miller et al. [6] was the first study to demonstrate that the detection of snow was possible using VIIRS DNB and its potential to provide additional data about snow coverage. While this study demonstrated the viability of the approach, it was based on visual identification, which limited its practical applications.

Later, Liu et al. [11] studied the classification of VIIRS DNB images using a Random Forest algorithm. They classified the image into four categories: snow, farmland, river, and other. The overall accuracy achieved in this study was 79.80%, and the kappa value was 0.45. They tested the approach to only one case study area, and it is still unknown how this approach performs in areas with different land-use and topographic characteristics. In addition, this approach is semi-automatic, requiring the operators to provide training and test labels, meaning that the results also depend on the quality of the manually created training and test data.

To the best of the authors' knowledge, only one study has applied an automated algorithm to snow detection, and that is the minimum error thresholding algorithm by Huang et al. [7]. It demonstrated that automatic thresholding was possible for snow detection at night. They applied the minimum error thresholding algorithm to two case study areas. The overall accuracy reached 76.7% and 80.3% for each case study. The paper further analyzed the amount of data gaps that can be addressed using this approach and studied the impact of lunar phases on the results. However, it did not test the performance of different algorithms and only used two case study areas. While the variety of thresholding methods used for our specific purpose is limited, this is not the case in other areas of remote sensing. Moreover, for snow detection during the day, there have been studies that examined multiple different thresholding algorithms, including the study conducted by Yin et al. [12], which benchmarked nine different algorithms.

An additional challenge to nighttime snow detection ©s that other factors beyond snow cover can impact the brightness of the data collected and, therefore, the results. Huang et al. [7] looked at what an adequate lunar angle would be to determine snow cover extent and if there was a difference between different lunar phases. Land cover can also play a crucial role in VIIRS day/night band (DNB) data since different land covers have different albedos and, therefore, different pixel values when captured by the sensor. The impact of land cover on VIIRS data has already been documented in a study that showed that seasonal land cover change impacted the overall brightness of VIIRS DNB images [13].

Nighttime snow extent estimation is still in its early phases, and some questions still need to be answered before researchers can more widely adopt it. Two notable questions are as follows: (1) which thresholding methods perform best (under which conditions) and (2) what factors impact the accuracy of the methods.

This study examined seven automated thresholding algorithms to determine the best performance for snow detection using nighttime imagery captured by the S-NPP VIIRS DNB. After identifying the best automated threshold algorithm, the false estimates were analyzed to determine what factors influence the misclassifications. Improvements in this field can lead to a better understanding of what areas and under which conditions could be most suitable for snow cover detection at night. Furthermore, snow cover detection at night could further enable better spatial and temporal coverage of snow cover extent data, which could play a significant role in further understanding the impact snow coverage has on weather, climate, climate change, and hydrology.

2. Materials and Methods

2.1. Case Study Areas

For the purposes of this study, four different case study areas were selected. The selection was made to incorporate different latitudes, elevations, topographies, and land covers. The four case study areas are spread throughout North America, including both the United States of America and Canada. The case study areas are located in Colorado, USA (a); Ontario, Canada (b); Alaska, USA (c); and Saskatchewan, Canada (d). Their extents are depicted in Figure 1.

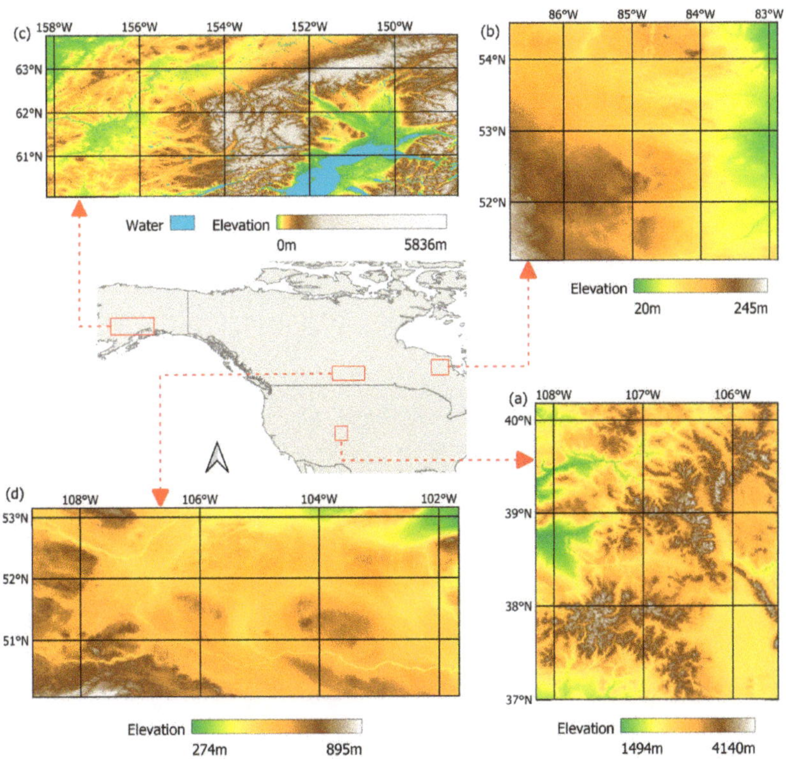

Figure 1. Locations of case study areas: Colorado, USA (**a**); Ontario, Canada (**b**); Alaska, USA (**c**); Saskatchewan, Canada (**d**).

The case study areas in Colorado (a) and Alaska (c) are primarily mountainous, with the area in Colorado encompassing a part of the Rocky Mountains, while on the other hand, the area in Alaska encompasses part of the Alaska Range. The major difference between the two areas lies in their latitude. The Colorado case study area is in a mid-latitude region, while the Alaska case study area is in a high-latitude region. Even though the Colorado case study area is not limited by the number of daytime hours due to it being located in a mid-latitude region, it can still experience cloud coverage during the day; therefore, it is included in this study as snow cover extent data can be added if nights are clear.

The other two case study areas are located in Canada and are mostly on flat terrain. The major difference between the two areas lies in their land cover, with the area in Saskatchewan (d) being primarily covered in cropland; in contrast, the area in Ontario (b) is covered mainly by several types of natural land cover, including several types of forests, shrubs, and herbaceous vegetation.

In addition to the geographic location of the case study areas, another important factor was selecting the exact date and time at which the data were captured. Three major factors determined the appropriate date for the data capture. The first factor was ensuring that the data were captured during a full moon. While Huang et al. [7] suggested that other lunar phases are also suitable for nighttime snow detection, only full moon phases were selected for this study to reduce the potential impact different lunar phases could have on the results. In addition to the data being captured during a full moon night, it was also essential to guarantee a sufficient number of day and night hours. Since this study uses both MODIS data captured during the day and VIIRS DNB data captured at night, only images from March and October were selected; during both of these months, there was snow cover in the selected case study areas, and both the day and night were sufficiently long for capturing data. Finally, the last factor was to ensure that there was no or little cloud coverage. Table 1 lists the exact dates of data collection for the four case study areas.

Table 1. Date and time of data collection by case study area.

Case Study	Location	VIIRS/MODIS Collection Date	VIIRS Time (Local)
a	Colorado	2 March 2018	01:30
b	Ontario	2 March 2018	03:24
c	Alaska	13 October 2019	05:06
d	Saskatchewan	21 March 2019	04:06

2.2. Data Collection

Data sources collected for the four case study areas included VIIRS DNB data for the nighttime snow extent estimates; the MODIS snow cover product for benchmarking; and land cover data to explore what factors could influence the false estimates. Table 2 lists the datasets used, including the acquisition date, spatial resolution, and access links.

Table 2. Study dataset names, spatial resolution, and links from which they can be accessed.

Dataset	Dates	Resolution
VIIRS DNB [1]	2018–2019	750 m
MODIS Snow Cover [2]	2018–2019	500 m
MODIS NDVI [3]	2018–2019	500 m
Copernicus Land Use [4]	2018–2019	10 m

[1] https://ladsweb.modaps.eosdis.nasa.gov/search/order/2/NPP_VDNES_L1--5000, (accessed on 6 May 2022). [2] https://nsidc.org/data/mod10a1/versions/61 (accessed on 5 May 2022). [3] https://lpdaac.usgs.gov/products/mod09gav061/ (accessed on 7 May 2022). [4] https://land.copernicus.eu/global/products/lc (accessed on 7 May 2022).

The VIIRS DNB data were captured by the Suomi National Polar Partnership (S-NPP) satellite. The captured data were accessed at the Level-1 and Atmosphere Archive and Distribution System Distributed Active Archive Center (LAADS DAAC), operated by the National Aeronautics and Space Administration (NASA). The data were part of the NPP_VDNES_L1-VIIRS/NPP Day/Night Band 6-Min L1 Swath SDR 750 m product, and for this study, raw radiance values from the DNB were used [14].

MODIS snow cover extent data were also used in this study, specifically, the latest version 6.1 of the MOD10A1 snow cover extent product [5]. This data source was selected because it is commonly used in many hydrology and climate studies and has relatively high accuracy [4].

For land cover data, since this study covers different countries, it was important to select a land cover dataset with global coverage. Beyond global coverage, the level of detail with respect to the data was also an important factor in the selection, as a more detailed breakdown of forest and herbaceous vegetation types could lead to a clearer understanding of how different land covers impact snow detection. For this reason, Copernicus Global Land Service data were used for both 2018 and 2019 [15,16].

To calculate the NDVI, another MODIS product was used to guarantee consistency and data availability: version 6.1 of MOD09GA [17].

2.3. Data Preparation

To maximize the potential of accurate threshold calculations, parts of the VIIRS DNB data were masked as they were either not considered in this study, such as bodies of water, or because they emitted light, which could skew the data, such as built-up areas [7]. Copernicus land cover data were used to determine which parts of the VIIRS data should be masked. Areas masked out from the analysis included urban/built-up, permanent water bodies, and open sea. In this step, any clouds present would also have been masked, but since the case study areas were explicitly chosen to avoid cloud coverage, this step was not necessary for this study.

In order to ensure consistency between this study and previous studies, the same preparation steps outlined in Huang et al.'s studies [7,18] were used: MODIS snow cover extent data were processed to create a binary snow cover map. Pixels were marked as snow-covered, snow-free, or as no data (in case of cloud coverage or other data gaps). While VIIRS data were completely cloud-free, some cloud coverage was present in the MODIS data.

The NDVI values per case study area were calculated using the MOD09GA data source. The standard NDVI was used, where MOD09GA Band 1 is the red band and MOD09GA Band 2 is the near-infrared (NIR) band. The resulting NDVI maps for the case study areas ranged between -1 and 1, with values closer to 1 indicating higher vegetation presence.

As noted in Table 2, the four datasets used have different spatial resolutions. To address this, all datasets were resampled using the nearest neighbor operator to match VIIRS data.

2.4. Methodology

Seven different thresholding algorithms were selected and applied to the preprocessed VIIRS data. During the calculations, values that were previously masked during the preprocessing phase were not taken into consideration. The seven thresholding algorithms used are as follows: (1) Otsu [19], (2) Li [20], (3) Yen [21], (4) triangle [22], (5) minimum [23], (6) mean [23], and (7) Isodata [24]. While thresholding algorithms were initially created for different purposes, similarly to general computer vision, text identification, or microbiology, they have since been used in many fields, including remote sensing [25]. The seven algorithms are nonparametric and unsupervised. Their calculations are based on the histogram of all values present in the image. While the specific ways in which they are calculated vary from algorithm to algorithm, the end result of all algorithms is a numerical value that splits the results into two discrete categories. A succinct overview of the different algorithms was proposed by Sekertekin [25], where they also benchmarked the different methods, however, in the classification of water.

Following this step, the binary snow cover maps made by thresholding (test datasets) were compared to MODIS binary snow cover data for the same case study area (reference dataset). All pixels where there were data available in both the test and the reference dataset were considered true positive (TP), true negative (TN), false positive (FP), or false negative (FN).

There are two distinct ways in which these results can be displayed, which serve different purposes. The first is to create a confusion matrix (calculating the percentage value for each category in the entire case study area). The confusion matrix is particularly useful in order to determine the overall accuracy of the thresholding results in comparison to the MODIS snow cover product. Specifically, the higher the sum of the true positive and true negative percentages, the more accurate the thresholding snow cover extent. The second way is to map the accuracy results. This is particularly important to determine which factors cause both false negatives and false positives. By mapping their position, it is possible to relate the performance per land cover category and relate it to the NDVI value.

Discriminating the results into discrete categories enables determining if any particular characteristic impacts the overall accuracy. This entails the creation of confusion matrices per land cover category as previously proposed in a study by Yang et al. [26]. If the land cover has no or a limited impact on the results, the accuracy should be relatively consistent between different land covers. Conversely, if specific land covers do impact the results, the accuracy within those should vary significantly.

Furthermore, version five of the MODIS snow cover product treated areas with high NDVI values separately from regions with low NDVI values [27]; therefore, this study also investigated the influence of NDVI. In version five of the MODIS snow cover product, areas with high NDVI values were given a lower threshold in order to be considered snow-covered compared to areas with low NDVI values [27]. If the impact of NDVI on VIIRS matches the impact it had on MOIDS, it is expected that the NDVI values of pixels marked as false negatives would be higher than those marked as true positives, indicating that the potential reason for the misclassification would be the NDVI. This factor was examined by looking at the distribution and mean values of the NDVI of true positives and false negatives per case study area.

Finally, some false positive and false negative values might not be explainable using the methods outlined above; therefore, visual analysis was also performed to explore possible hypotheses explaining some of the results and potentially create the basis for further research.

3. Results

3.1. Overall Thresholding Results

The confusion matrices of the seven different thresholding algorithms for the four case study areas are shown in Table 3. The results are expressed as percentages, with the number representing the total fraction of the classification relative to the total number of all classified pixels. Both the overall accuracy (the sum of both the true positive and true negative percentages) and Cohen's kappa [28,29] are included in Table 3.

Table 3. Confusion matrix, overall accuracy, errors, and Cohen's kappa for the seven thresholding algorithms and the four case study areas.

Case Study	Thresholding Algorithm	TP	TN	FP	FN	Overall Accuracy	Kappa
a. Colorado	Otsu	28.42	31.32	0.14	40.13	59.73	0.31
	Li	36.86	31.19	0.26	31.68	68.05	0.42
	Yen	0.03	31.44	0.01	68.51	31.47	0.00
	Triangle	7.98	31.42	0.04	60.57	39.40	0.08
	Minimum	0.01	31.45	0.00	68.54	31.45	0.00
	Mean	38.36	31.15	0.30	30.19	69.51	0.44
	Isodata	29.73	31.30	0.16	38.82	61.02	0.32
b. Ontario	Otsu	62.34	0.00	0.00	37.66	62.34	n/a
	Li	69.17	0.00	0.00	30.83	69.17	n/a
	Yen	77.91	0.00	0.00	22.09	77.91	n/a
	Triangle	0.07	0.00	0.00	99.93	0.07	n/a
	Minimum	0.00	0.00	0.00	100.00	0.00	n/a
	Mean	54.75	0.00	0.00	45.25	54.75	n/a
	Isodata	64.97	0.00	0.00	35.03	64.94	n/a

Table 3. *Cont.*

Case Study	Thresholding Algorithm	TP	TN	FP	FN	Overall Accuracy	Kappa
c. Alaska	Otsu	19.17	40.25	0.22	40.36	59.42	0.27
	Li	25.72	40.16	0.31	33.81	65.88	0.37
	Yen	0.01	40.42	0.05	59.53	40.42	0.00
	Triangle	0.20	40.39	0.08	59.33	40.59	0.00
	Minimum	0.00	40.47	0.00	59.53	40.47	0.00
	Mean	30.50	40.03	0.44	29.03	70.53	0.45
	Isodata	23.38	40.19	0.27	36.15	63.57	0.34
d. Saskatchewan	Otsu	55.10	22.26	0.17	22.47	77.36	0.52
	Li	59.37	22.21	0.22	18.20	81.58	0.59
	Yen	0.02	22.38	0.04	77.55	22.41	0.00
	Triangle	0.15	22.34	0.08	77.42	22.50	0.00
	Minimum	0.00	22.42	0.00	77.57	22.42	0.00
	Mean	52.68	22.27	0.16	24.89	74.95	0.48
	Isodata	55.10	22.26	0.17	22.47	77.36	0.52

In Figure 2, the thresholding method with the best overall accuracy was used to map the confusion matrix. For the Colorado and Alaska case study areas, mean thresholding was used; for the Ontario case study area, Yen thresholding was used; finally, Li thresholding results were used for the Saskatchewan case study area. The resulting four maps for the case study area show the geographic locations of TP, TN, FP and FN values.

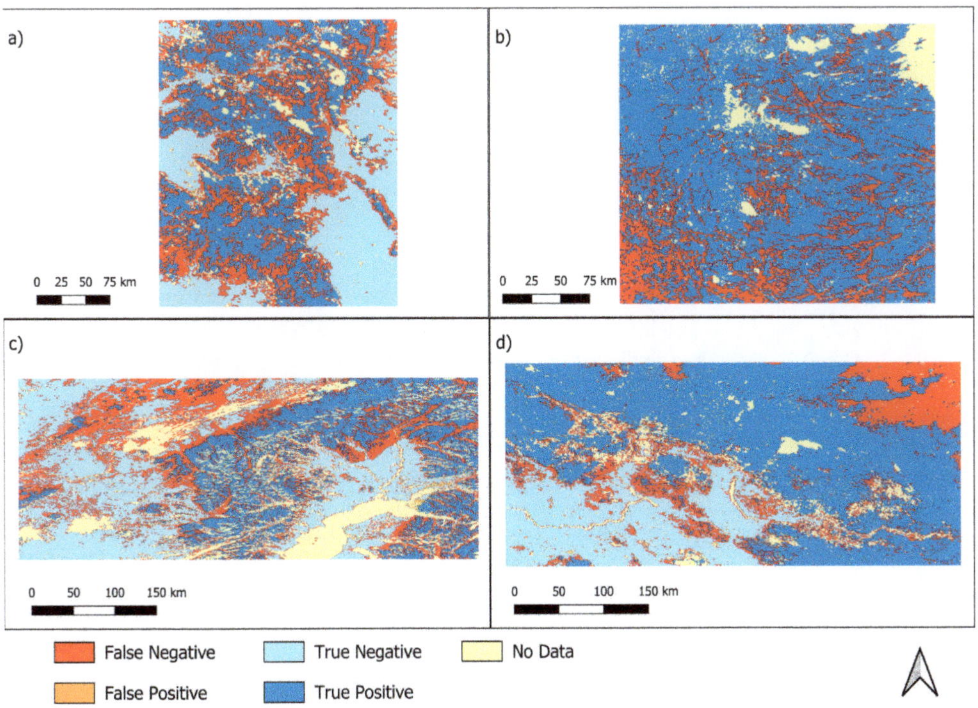

Figure 2. Maps of the confusion matrices: (**a**) Colorado—mean thresholding; (**b**) Ontario—Yen thresholding; (**c**) Alaska—mean thresholding; (**d**) Saskatchewan—Li thresholding.

3.2. Land Cover Discrimination

The mapping of the confusion matrix values onto their geographic location also enabled the creation of discrete results per land cover type. For this, the maps from Figure 2 were compared per land cover type of the Copernicus data for the appropriate year [15,16]. The resulting confusion matrices (Table 4) discriminate the per land cover type. As the four case study areas cover different geographic regions, there are differences between the land cover types present in them. Additionally, some land cover types are infrequent in some case study areas; therefore, land covers with under 500 appearances were removed from the results.

Table 4. Confusion matrix per land cover type for the four case study areas.

Case Study	Thresholding Algorithm	TP	TN	FP	FN	Overall Accuracy	Kappa
a. Colorado	Shrubs	25.45	58.43	0.73	15.39	83.88	0.65
	Herbaceous vegetation	46.45	41.22	0.46	11.88	87.67	0.76
	Cultivated and managed vegetation/agriculture	7.72	86.68	0.64	4.95	94.40	0.70
	Bare/sparse vegetation	73.37	24.09	0.13	2.37	97.50	0.93
	Closed forest, evergreen needle leaf	22.81	15.33	0.08	61.77	38.15	0.10
	Closed forest, deciduous broad leaf	70.13	0.11	0.03	29.74	70.24	0.00
	Closed forest, unknown	43.42	20.74	0.14	35.69	64.16	0.33
	Open forest, evergreen needle leaf	34.72	21.94	0.18	43.15	56.66	0.26
	Open forest, deciduous broad leaf	72.95	0.07	0.00	26.99	73.01	0.00
	Open forest, unknown	43.14	29.40	0.24	27.22	75.86	0.54
b. Ontario	Shrubs	85.81	0.00	0.00	14.19	85.81	n/a
	Herbaceous vegetation	94.17	0.00	0.00	5.83	94.17	n/a
	Herbaceous wetland	97.09	0.00	0.00	2.91	97.09	n/a
	Closed forest, evergreen needle leaf	36.94	0.00	0.00	63.06	36.94	n/a
	Open forest, evergreen needle leaf	63.92	0.00	0.00	36.08	63.92	n/a
	Open forest, unknown	74.73	0.00	0.00	25.27	74.73	n/a
c. Alaska	Shrubs	20.65	56.49	0.25	22.61	77.14	0.50
	Herbaceous vegetation	48.50	33.71	0.31	17.48	82.21	0.65
	Bare/sparse vegetation	84.51	1.14	0.28	14.06	85.66	0.12
	Snow and Ice	78.48	0.18	0.03	21.30	78.67	0.01
	Herbaceous wetland	3.37	82.83	2.28	11.52	86.20	0.27
	Closed forest, evergreen needle leaf	6.13	38.31	0.35	55.22	44.44	0.07
	Closed forest, mixed	0.95	69.95	2.47	26.63	70.90	0.00
	Closed forest, unknown	6.66	56.01	0.74	36.59	62.67	0.16
	Open forest, evergreen needle leaf	11.60	31.55	0.91	55.93	43.16	0.10
	Open forest, unknown	12.24	58.13	0.45	29.18	70.37	0.32
d. Saskatchewan	Herbaceous vegetation	48.16	37.93	0.59	13.33	86.08	0.72
	Cultivated and managed vegetation/agriculture	67.79	20.76	0.13	11.31	88.55	0.71
	Herbaceous wetland	62.41	20.66	0.18	16.75	83.07	0.60
	Closed forest, evergreen needle leaf	6.43	0.08	0.00	93.50	6.50	0.00
	Closed forest, deciduous broad leaf	29.14	0.22	0.00	70.63	29.37	0.00
	Closed forest, mixed	3.46	0.07	0.00	96.48	3.52	0.00
	Closed forest, unknown	35.92	5.77	0.00	58.31	41.69	0.07
	Open forest, evergreen needle leaf	57.12	0.92	0.00	41.96	58.04	0.02
	Open forest, deciduous broad leaf	58.94	0.26	0.00	40.80	59.20	0.01
	Open forest, unknown	60.25	7.63	0.08	32.03	67.88	0.22

3.3. NDVI Influence

NDVI values were also investigated in this study. If false negative areas have a higher NDVI than true positives, this may indicate that the thresholding value should be lower for the areas with higher NDVI. In order to determine if the NDVI values are higher, two boxplots per case study area were created: one for true positives and one for false negatives. The boxplots indicate mean values, the first and third quantiles of the data, and the minimum and maximum (see Figure 3).

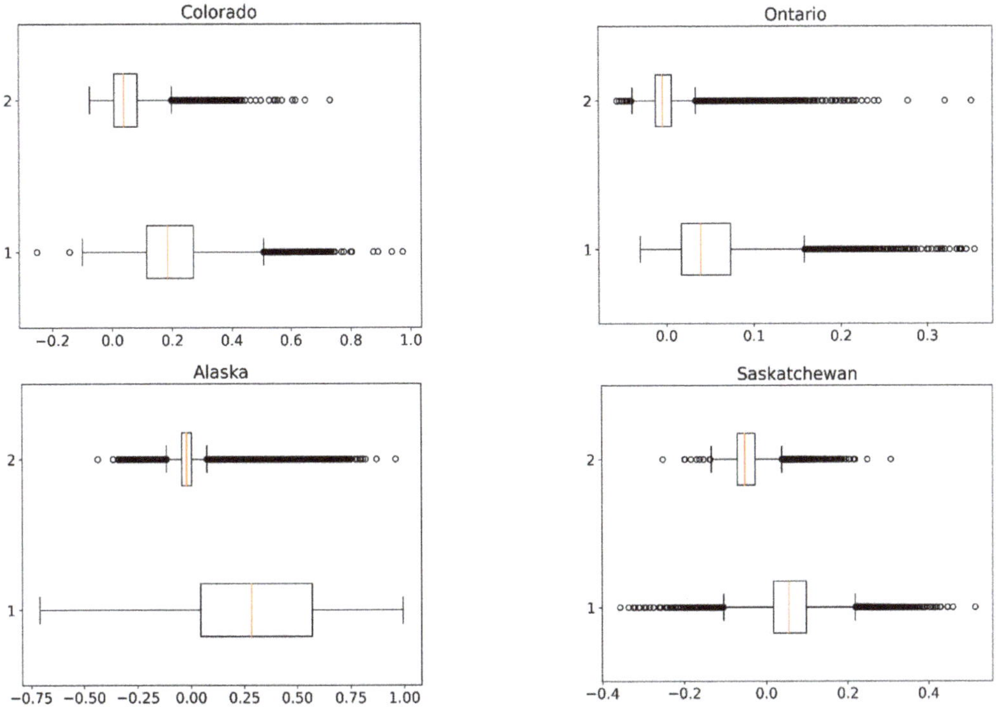

Figure 3. Boxplots of NDVI values per case study area. NDVI values for true positives (2) and false negatives (1).

3.4. Additional Factors

To explain false positives (which account for less than 1% of the pixels), visual analyses were performed. What is clearly noticeable in the spatial distribution of false positives is the proximity to urban/built-up areas. Examples of this phenomenon can be observed in Figure 4, where the built-up areas from the Copernicus data are presented, as well as false positives. The Ontario case study area had no false positives; therefore, it was not included in the visual analysis.

Even though most false negatives occurred in forested areas, other land covers also had false negatives. The distribution and cause of these cannot be explained by tree canopies; therefore, another factor must be their cause. The two different false negatives can be best observed in the Saskatchewan case study area, which is pictured in Figure 5. While the false negative values in the north and northeast can be explained by the forest present there, the false negatives in the center and southwest are present in a mixture between cropland and herbaceous vegetation.

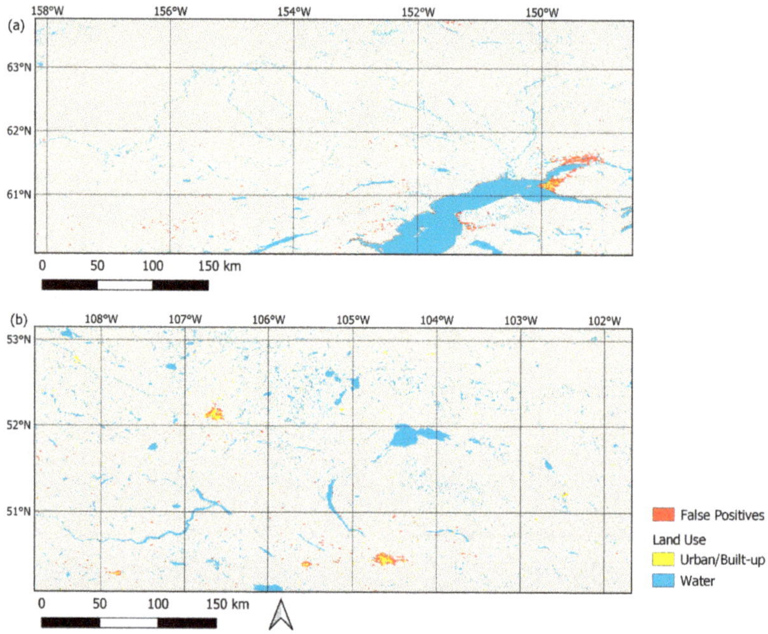

Figure 4. False positive values and urban/built-up areas in Alaska (**a**) and Saskatchewan (**b**) case study areas.

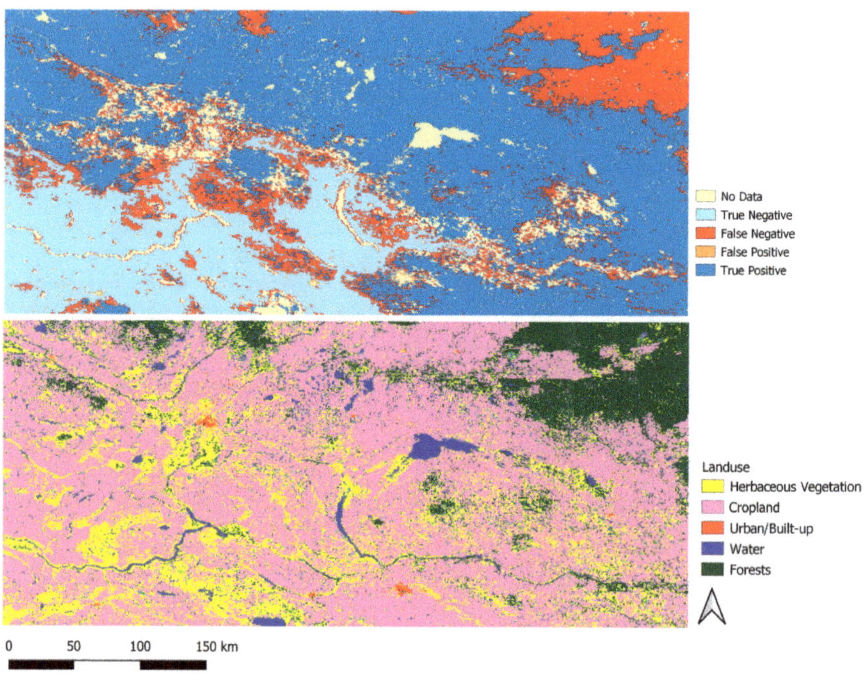

Figure 5. Map of the confusion matrix for the Saskatchewan case study area (**top**) and land use for the Saskatchewan case study area (**bottom**).

4. Discussion

The seven thresholding algorithms investigated in this study had notably different performances. The overall accuracy varied for each algorithm and between the four case study areas, but some overall conclusions can still be drawn from them. The overall accuracy of the algorithms had quite widespread values, with the lowest overall accuracy being 0%, while the highest overall accuracy was 81.58%.

The two thresholding algorithms that performed the worst were triangle thresholding and minimum thresholding, both of which had their highest overall score at only around 40%. Similarly, Yen thresholding also had results of less than or approximately 40%, with the exception of the Ontario case study area, where it performed the best with a 77.91% overall accuracy. Otsu and Isodata thresholding had middling performances in all four case study areas, with neither of them being the best-performing one in any case study area. Their overall accuracies ranged between 59% and 77%. Another algorithm, mean thresholding, had a similar range of overall accuracy, with its lowest result being 54.75% and the highest being 74.95%. However, what is notable is that mean thresholding was the best-performing algorithm in both the Colorado and Alaska case study areas. However, it underperformed in the Ontario case study area, where it achieved only a 54.75% overall accuracy. These results indicate that mean thresholding is the best-performing algorithm in mountainous regions, as both the Colorado and Alaska case study areas are predominantly mountainous.

While mean thresholding performed the best in mountainous areas, the algorithm with the best range of overall accuracy and the highest individual accuracy in the Saskatchewan case study area was Li thresholding. This algorithm always had an overall accuracy above 65% and was either the highest or second-highest result in every case study area. Therefore, even though mean thresholding did have the best result in two out of the four case study areas, the results indicate that its choice depends on context. The fact that mean thresholding measurably underperformed in the Ontario case study area means that in order to ensure the best overall and consistently high overall accuracy, the Li thresholding algorithm is the best choice out of all seven algorithms tested in this study. If the thresholding is applied exclusively to mountainous areas, using mean thresholding could yield better results. However, if various terrains are present, Li thresholding will provide fair overall and consistent results.

The highest overall accuracy results among all case studies range between 70% and 82%, which is comparable with the results achieved by Huang et al. [7], where the overall accuracy reached 77% and 80%. Moreover, similarly to the results achieved by Liu et al. [11], the overall accuracy reported was 80% and the Kappa coefficient was 0.45. Therefore, the results were also within the range of the maximums achieved by the thresholding algorithms in this study.

Determining the best thresholding algorithm is not the only conclusion that can be drawn from the results. What is especially noteworthy is the difference between false positives and false negatives. The percentage of false negatives is significantly larger than the false positives for all thresholding algorithms and in all case study areas. In fact, false positives in every case study area and thresholding algorithm make up less than 0.5%. This indicates that the vast majority of overall error comes from the underestimation of the snow extent by the VIIRS DNB snow extent model when compared to the MODIS snow cover product.

The causes of this underestimation and the locations where it occurs can be potentially explained by the land cover's presence. While the four case study areas had different land cover types present, some overall trends are still noticeable. The two land covers that had the lowest overall accuracy in all case study areas are open and closed evergreen needle leaf forests. Besides those two, all other types of forests, when present in the case study area, had lower overall accuracies than non-forest land covers such as shrubs or herbaceous vegetation. Considering forest types, evergreen needle leaves had the lowest overall accuracy, but when comparing open and closed forests, closed forests have lower overall accuracies. In all cases, the low overall accuracy can be entirely explained by false negatives, with false positives being rare or nonexistent for most forests. The remaining non-forested areas performed well with high true-positive and true-negative values, resulting in high overall accuracy. The false negatives were in the 10–25% range, and false positives were present but usually made up less than 1%.

The effect that forests have on the accuracy of snow cover extent is not unexpected. Forests have been a factor that limited the accuracy of snow cover extent estimates in many different products. The presence of tree canopies would effectively block the view of the ground and, therefore, any snow that is present, causing underestimations [30]. Furthermore, while it is possible that the canopy itself would capture some of the snow and therefore increase its albedo, studies suggest that even when that does occur, the overall albedo is still lower [31].

The differences between closed and open forests relate mainly to their canopies. In the Copernicus land cover data, closed forests were considered all areas that have more than 70% of the ground covered by canopies, while open forests have 15–70% canopy cover [32]. Since canopies cover more of the view of the ground and any snow present there in closed forests, they would have more false negative estimates than in open forests, where more of the ground is visible.

The impact of canopies would also explain the lower overall accuracy of evergreen forests compared to other forest types. In most cases, the evergreen forests performed roughly 20–30% worse than other types of forests. This relates to the fact that the images were captured during March and October, when evergreen trees still retain their full canopy while deciduous trees have lost most of their canopy. The lack of canopies in deciduous forests would mean that the ground—and therefore any potential snow on the ground—would be more visible and therefore easier to classify.

Another way in which this effect could possibly be measured is by investigating the NDVI. As tree canopies, and especially evergreen trees, should have a high NDVI value at the time the data were captured, it would be expected that the NDVI values of the false negatives would show consistently higher values than the NDVI in true positives. As observed from the box plots in Figure 3, the mean, quantiles, and maximum values of the NDVI are higher for false negatives than for true positives. This indicates that a higher NDVI is correlated with false-negative identification. This factor, combined with the previous discussion of the impact of forest canopies, suggests that high-density forests and especially evergreen forests have the most negative impact on identification.

The cause of the false negatives can be explained by their location at the periphery of the snow cover. Because these areas mark the transition between snow and no snow, they should have relatively shallow snow depth. Studies have shown that even during the day, the accuracy of remote sensing of snow decreases with lower snow depths [26]. Therefore, nighttime snow detection accuracy seems to also be susceptible to underestimations with low snow depth, explaining the underestimates in those periphery regions. The snow depth could have also slightly changed in the period between when MODIS and VIIRS DNB data were captured, which could explain some of the discrepancies between them. While in most non-forested areas, the VIIRS DNB data performed well, the accuracy of the boundaries will only become apparent with an in situ study, which could analyze the impact of snow depth on the accuracy.

The spatial distribution of false positives (e.g., in Figure 4) clearly shows that most false negatives are within the vicinity of urban/built-up areas. This is predominant near the larger built-up areas present in case study areas such as Anchorage in Alaska or Saskatoon and Regina in Saskatchewan. While the specific built-up areas were masked during the data preparation step, the surrounding areas were noticeably much brighter due to human activity. Our quantitative analysis of the accuracy within the different land covers is affected by the accuracy of the land cover reference dataset. If urban areas are underestimated in the reference dataset, we overestimate the false positives as not enough build-up is masked.

This phenomenon has been described in previous studies suggesting that suburban and rural areas do emit more artificial light toward the horizon, making the surrounding area brighter [33]. However, not all false positives are obviously near built-up areas (or at least near areas categorized as urban/built-up land cover). When exploring FPs distant from built-up land cover, we observed that a majority of these false positives still appear to be linked to human activity, but it is not enough to mark the area as built-up in the land cover dataset. This is true in the case of smaller communities, such as the town of Beechy in Saskatchewan, which is categorized as cropland due to its small size and population (the predominant land cover that completely surrounds the town). While being relatively small, there is still enough artificial light emitted to skew the results into false positives. Other areas in the three case study areas with false positive results follow the same pattern. While this approach seems to explain the false positives, it remains only a visual identification of the pattern, and to determine if it is the main reason behind it, a more rigorous and analytical study would be necessary. One way to (potentially) avoid these false positives is by creating a mask layer that is not based on land use but instead on light. Since the NASA Black Marble product suite releases monthly and annual data, it could be used to create a mask layer as all the brightest areas would be a result of human activity [34]. This approach would also remove uncertainty caused by inaccuracies of the land cover data.

Naturally, this study has certain limitations that can have some impact on the results. The main limitation is that the VIIRS snow cover extent is compared to another snow cover model and not to in situ measurements. The selected MODIS snow cover product has known limitations [35], but it is still a widely used model, and agreement between the models still indicates that it is acceptable for many studies in hydrology, climate science, or other areas where MODIS data are commonly used [4].

Exploring accuracies in evergreen or all forested areas is an avenue for further research. Daytime snow cover extent products have dealt with the forest issues [30], and improving the results in those areas could be possible by following the strategy of lowering the threshold for areas with high NDVI or by calculating a separate threshold for forested areas.

Furthermore, since future Joint Polar Satellite System (JPSS) missions are already planned and will be equipped with VIIRS, data availability in the future will only increase [36]. Hopefully, this will reduce the impact cloud coverage has on VIIRS DNB data as multiple satellites will be capturing the data, enabling even better coverage in the future.

Identifying snow cover extent using VIIRS DNB data is possible and achievable with relatively high overall accuracy. The false estimates are usually underestimations, with overestimations present but usually limited to areas near human activity. The false negatives are usually present in forested areas, especially in closed and evergreen forests. Some false negatives also occur on the border between areas covered and not covered by snow. While this could impact the usefulness of the VIIRS DNB data for some purposes, this is an area where the results could be improved with in situ research.

5. Conclusions

Of the seven tested thresholding algorithms, Li thresholding achieved the most consistently high overall accuracy in all case study areas, while the mean thresholding algorithm performed the best in mountainous regions, but it did not have high overall accuracies in other areas. Depending on the use case, either of the two thresholding algorithms could provide results that would be useful for climate or hydrology research. If the terrain of the case study area is diverse or if there are multiple different areas, Li thresholding seems to be the most effective option. If the case study is predominantly mountainous, then mean thresholding can potentially achieve better results. While some limitations are present, the high overall accuracy and the distribution of accurate identification should still prove useful in filling in some spatial or temporal data gaps for snow cover extent maps, especially in regions with few forests.

Author Contributions: Conceptualization, R.S. and E.D.; methodology, R.S. and E.D.; software, R.S.; validation, R.S. and E.D.; formal analysis, R.S.; investigation, R.S.; resources, R.S.; data curation, R.S.; writing—original draft preparation, R.S.; writing—review and editing, E.D. and R.S.; visualization, R.S.; supervision, E.D.; project administration, E.D.; funding acquisition, E.D. All authors have read and agreed to the published version of the manuscript.

Funding: This research received no external funding.

Data Availability Statement: See Section 2.2 for information about how to access the data used in this study.

Conflicts of Interest: The authors declare no conflict of interest.

References

1. Levis, S.; Bonan, G.B.; Lawrence, P.J. Present-Day Springtime High-Latitude Surface Albedo as a Predictor of Simulated Climate Sensitivity. *Geophys. Res. Lett.* **2007**, *34*, 1–4. [CrossRef]
2. Musselman, K.N.; Clark, M.P.; Liu, C.; Ikeda, K.; Rasmussen, R. Slower Snowmelt in a Warmer World. *Nat. Clim. Chang.* **2017**, *7*, 214–219. [CrossRef]
3. Foster, J.L.; Hall, D.K. Observations of Snow and Ice Features during the Polar Winter Using Moonlight as a Source of Illumination. *Remote Sens. Environ.* **1991**, *37*, 77–88. [CrossRef]
4. Da Ronco, P.; Avanzi, F.; De Michele, C.; Notarnicola, C.; Schaefli, B. Comparing MODIS Snow Products Collection 5 with Collection 6 over Italian Central Apennines. *Int. J. Remote Sens.* **2020**, *41*, 4174–4205. [CrossRef]
5. Hall, D.K.; Riggs, G.A. *MODIS/Aqua Snow Cover Daily L3 Global 500 m Grid, Version 61*; NASA National Snow and Ice Data Center Distributed Active Archive Center: Boulder, CO, USA. [CrossRef]
6. Miller, S.D.; Straka, W.; Mills, S.P.; Elvidge, C.D.; Lee, T.F.; Solbrig, J.; Walther, A.; Heidinger, A.K.; Weiss, S.C. Illuminating the Capabilities of the Suomi National Polar-Orbiting Partnership (NPP) Visible Infrared Imaging Radiometer Suite (VIIRS) Day/Night Band. *Remote Sens.* **2013**, *5*, 6717–6766. [CrossRef]
7. Huang, Y.; Song, Z.; Yang, H.; Yu, B.; Liu, H.; Che, T.; Chen, J.; Wu, J.; Shu, S.; Peng, X.; et al. Snow Cover Detection in Mid-Latitude Mountainous and Polar Regions Using Nighttime Light Data. *Remote Sens. Environ.* **2022**, *268*, 112766. [CrossRef]
8. Lee, T.E.; Miller, S.D.; Turk, F.J.; Schueler, C.; Julian, R.; Deyo, S.; Dills, P.; Wang, S. The NPOESS VIIRS Day/Night Visible Sensor. *Bull. Am. Meteorol. Soc.* **2006**, *87*, 191–200. [CrossRef]
9. Foster, J.L. *Observations of the Earth Using Nighttime Visible Imagery*; Aitken, G.W., Ed.; SPIE: Arlington, MA, USA, 1983; pp. 187–193.
10. Wiesnet, D.R.; Mcginnis, J.; Matson, M.; Pritchard, J.A. *Evaluation of Hcmm Satellite Data for Estuarine Tidal Circulation Patterns and Thermal Inertia Soil Moisture Measurements*; Interim; Final Report; National Oceanic and Atmospheric Administration: Washington, DC, USA, 1981.
11. Liu, D.; Zhang, Q.; Wang, J.; Wang, Y.; Shen, Y.; Shuai, Y. The Potential of Moonlight Remote Sensing: A Systematic Assessment with Multi-Source Nightlight Remote Sensing Data. *Remote Sens.* **2021**, *13*, 4639. [CrossRef]
12. Yin, D.; Cao, X.; Chen, X.; Shao, Y.; Chen, J. Comparison of Automatic Thresholding Methods for Snow-Cover Mapping Using Landsat TM Imagery. *Int. J. Remote Sens.* **2013**, *34*, 6529–6538. [CrossRef]
13. Levin, N. The Impact of Seasonal Changes on Observed Nighttime Brightness from 2014 to 2015 Monthly VIIRS DNB Composites. *Remote Sens. Environ.* **2017**, *193*, 150–164. [CrossRef]
14. LAADS DAAC VIIRS/NPP Day/Night Band 6-Min L1 Swath SDR 750m—LAADS DAAC. Available online: https://ladsweb.modaps.eosdis.nasa.gov/missions-and-measurements/products/NPP_VDNES_L1#overview (accessed on 6 May 2022).
15. Buchhorn, M.; Smets, B.; Bertels, L.; Roo, B.D.; Lesiv, M.; Tsendbazar, N.-E.; Herold, M.; Fritz, S. Copernicus Global Land Service: Land Cover 100 m: Collection 3: Epoch 2018: Globe 2020. Available online: https://zenodo.org/record/3518038 (accessed on 7 May 2022).
16. Buchhorn, M.; Smets, B.; Bertels, L.; Roo, B.D.; Lesiv, M.; Tsendbazar, N.-E.; Herold, M.; Fritz, S. Copernicus Global Land Service: Land Cover 100m: Collection 3: Epoch 2019: Globe 2020. Available online: https://zenodo.org/record/3939050 (accessed on 7 May 2022).
17. Vermote, E.; Wolfe, R. MODIS/Terra Surface Reflectance Daily L2G Global 1 km and 500 m SIN Grid V061 2021. Available online: https://lpdaac.usgs.gov/products/mod09gav061/ (accessed on 7 May 2022).
18. Huang, Y.; Liu, H.; Yu, B.; Wu, J.; Kang, E.L.; Xu, M.; Wang, S.; Klein, A.; Chen, Y. Improving MODIS Snow Products with a HMRF-Based Spatio-Temporal Modeling Technique in the Upper Rio Grande Basin. *Remote Sens. Environ.* **2018**, *204*, 568–582. [CrossRef]
19. Otsu, N. A Threshold Selection Method from Gray-Level Histograms. *IEEE Trans. Syst. Man Cybern.* **1979**, *9*, 62–66. [CrossRef]
20. Li, C.H.; Tam, P.K.S. An Iterative Algorithm for Minimum Cross Entropy Thresholding. *Pattern Recognit. Lett.* **1998**, *19*, 771–776. [CrossRef]
21. Yen, J.-C.; Chang, F.-J.; Chang, S. A New Criterion for Automatic Multilevel Thresholding. *IEEE Trans. Image Process.* **1995**, *4*, 370–378. [CrossRef] [PubMed]

22. Zack, G.W.; Rogers, W.E.; Latt, S.A. Automatic Measurement of Sister Chromatid Exchange Frequency. *J. Histochem. Cytochem.* **1977**, *25*, 741–753. [CrossRef] [PubMed]
23. Glasbey, C.A. An Analysis of Histogram-Based Thresholding Algorithms. *CVGIP Graph. Models Image Process.* **1993**, *55*, 532–537. [CrossRef]
24. Ridler, T.W.; Calvard, S. Picture Thresholding Using an Iterative Selection Method. *IEEE Trans. Syst. Man Cybern.* **1978**, *8*, 630–632. [CrossRef]
25. Sekertekin, A. A Survey on Global Thresholding Methods for Mapping Open Water Body Using Sentinel-2 Satellite Imagery and Normalized Difference Water Index. *Arch. Comput. Methods Eng.* **2021**, *28*, 1335–1347. [CrossRef]
26. Yang, J.; Jiang, L.; Ménard, C.B.; Luojus, K.; Lemmetyinen, J.; Pulliainen, J. Evaluation of Snow Products over the Tibetan Plateau. *Hydrol. Process.* **2015**, *29*, 3247–3260. [CrossRef]
27. Hall, D.K.; Riggs, G.A.; Salomonson, V.V. *MODIS/Terra Snow Cover 5-Min L2 Swath 500 m, Version 5*; NASA National Snow and Ice Data Center Distributed Active Archive Center: Boulder, CO, USA, 2006.
28. Cohen, J. A Coefficient of Agreement for Nominal Scales. *Educ. Psychol. Meas.* **1960**, *20*, 37–46. [CrossRef]
29. Zhang, H.; Zhang, F.; Zhang, G.; Che, T.; Yan, W.; Ye, M.; Ma, N. Ground-Based Evaluation of MODIS Snow Cover Product V6 across China: Implications for the Selection of NDSI Threshold. *Sci. Total Environ.* **2019**, *651*, 2712–2726. [CrossRef]
30. Dietz, A.J.; Kuenzer, C.; Gessner, U.; Dech, S. Remote Sensing of Snow—A Review of Available Methods. *Int. J. Remote Sens.* **2012**, *33*, 4094–4134. [CrossRef]
31. Webster, C.; Jonas, T. Influence of Canopy Shading and Snow Coverage on Effective Albedo in a Snow-Dominated Evergreen Needleleaf Forest. *Remote Sens. Environ.* **2018**, *214*, 48–58. [CrossRef]
32. Buchhorn, M.; Smets, B.; Bertels, L.; Roo, B.D.; Lesiv, M.; Tsendbazar, N.-E.; Li, L.; Tarko, A. *Copernicus Global Land Service: Land Cover 100 m: Version 3 Globe 2015–2019: Product User Manual*; Zenodo: Geneve, Switzerland, September 2020. [CrossRef]
33. Tong, K.P.; Kyba, C.C.M.; Heygster, G.; Kuechly, H.U.; Notholt, J.; Kolláth, Z. Angular Distribution of Upwelling Artificial Light in Europe as Observed by Suomi–NPP Satellite. *J. Quant. Spectrosc. Radiat. Transf.* **2020**, *249*, 107009. [CrossRef]
34. Román, M.O.; Wang, Z.; Sun, Q.; Kalb, V.; Miller, S.D.; Molthan, A.; Schultz, L.; Bell, J.; Stokes, E.C.; Pandey, B.; et al. NASA's Black Marble Nighttime Lights Product Suite. *Remote Sens. Environ.* **2018**, *210*, 113–143. [CrossRef]
35. Riggs, G.A.; Hall, D.K.; Román, M.O. *MODIS Snow Products Collection 6.1 User Guide*; National Snow and Ice Data Center: Boulder, CO, USA, 2019; Volume 66.
36. Levin, N.; Kyba, C.C.M.; Zhang, Q.; Sánchez de Miguel, A.; Román, M.O.; Li, X.; Portnov, B.A.; Molthan, A.L.; Jechow, A.; Miller, S.D.; et al. Remote Sensing of Night Lights: A Review and an Outlook for the Future. *Remote Sens. Environ.* **2020**, *237*, 111443. [CrossRef]

remote sensing

Article

The VIIRS Day/Night Band: A Flicker Meter in Space?

Christopher D. Elvidge [1,*], Mikhail Zhizhin [1,2], David Keith [3], Steven D. Miller [4], Feng Chi Hsu [1], Tilottama Ghosh [1], Sharolyn J. Anderson [5], Christian K. Monrad [6], Morgan Bazilian [7], Jay Taneja [8], Paul C. Sutton [9], John Barentine [10], William S. Kowalik [11], Christopher C. M. Kyba [12], Dee W. Pack [13] and Dorit Hammerling [14]

1 Earth Observation Group, Payne Institute for Public Policy, Colorado School of Mines, Golden, CO 80401, USA; mzhizhin@mines.edu (M.Z.); fengchihsu@mines.edu (F.C.H.); tghosh@mines.edu (T.G.)
2 Russian Space Research Institute, Moscow 117997, Russia
3 WSP Global Inc., Montreal, QC H3H 1P9, Canada; david.keith@wsp.com
4 Cooperative Institute on Research on the Atmosphere, Colorado State University, Fort Collins, CO 80521, USA; steven.miller@colostate.edu
5 National Park Service—Natural Sounds and Night Skies Team, Fort Collins, CO 80525, USA; sharolyn_anderson@nps.gov
6 Monrad Engineering, Inc., Tucson, AZ 85719, USA; chrismonrad@monradengineeringinc.com
7 Payne Institute for Public Policy, Colorado School of Mines, Golden, CO 80401, USA; mbazilian@mines.edu
8 Electrical and Computer Engineering, University of Massachusetts, Amherst, MA 01003, USA; jtaneja@umass.edu
9 Department of Geography, University of Denver, Denver, CO 80210, USA; paul.sutton@du.edu
10 Dark Sky Consulting, LLC, Tucson, AZ 85730, USA; john@darkskyconsulting.com
11 Oregon Chapter of the International Dark-Sky Association, Bend, OR 97709, USA; wsk412@sbcglobal.net
12 GFZ German Research Centre for Geosciences, Telegrafenberg, 14473 Potsdam, Germany; kyba@gfz-potsdam.de
13 Remote Sensing Department, Aerospace Corporation, El Segundo, CA 90245, USA; dee.w.pack@aero.org
14 Department of Applied Mathematics and Statistics, Colorado School of Mines, Golden, CO 80401, USA; hammerling@mines.edu
* Correspondence: celvidge@mines.edu

Abstract: The VIIRS day/night band (DNB) high gain stage (HGS) pixel effective dwell time is in the range of 2–3 milliseconds (ms), which is about one third of the flicker cycle present in lighting powered by alternating current. Thus, if flicker is present, it induces random fluctuations in nightly DNB radiances. This results in increased variance in DNB temporal profiles. A survey of flicker characteristics conducted with high-speed camera data collected on a wide range of individual luminaires found that the flicker is most pronounced in high-intensity discharge (HID) lamps, such as high- and low-pressure sodium and metal halides. Flicker is muted, but detectable, in incandescent luminaires. Modern light-emitting diodes (LEDs) and fluorescent lights are often nearly flicker-free, thanks to high-quality voltage smoothing. DNB pixel footprints are about half a square kilometer and can contain vast numbers of individual luminaires, some of which flicker, while others do not. If many of the flickering lights are drawing from a common AC supplier, the flicker can be synchronized and leave an imprint on the DNB temporal profile. In contrast, multiple power supplies will throw the flickering out of synchronization, resulting in a cacophony with less radiance fluctuation. The examination of DNB temporal profiles for locations before and after the conversion of high-intensity discharge (HID) to LED streetlight conversions shows a reduction in the index of dispersion, calculated by dividing the annual variance by the mean. There are a number of variables that contribute to radiance variations in the VIIRS DNB, including the view angle, cloud optical thickness, atmospheric variability, snow cover, lunar illuminance, and the compilation of temporal profiles using pixels whose footprints are not perfectly aligned. It makes sense to adjust the DNB radiance for as many of these extraneous effects as possible. However, none of these adjustments will reduce the radiance instability introduced by flicker. Because flicker is known to affect organisms, including humans, the development of methods to detect and rate the strength of flickering from space will open up new areas of research on the biologic impacts of artificial lighting. Over time, there is a trend towards the reduction of flicker in outdoor lighting through the replacement of HID with

Citation: Elvidge, C.D.; Zhizhin, M.; Keith, D.; Miller, S.D.; Hsu, F.C.; Ghosh, T.; Anderson, S.J.; Monrad, C.K.; Bazilian, M.; Taneja, J.; et al. The VIIRS Day/Night Band: A Flicker Meter in Space? *Remote Sens.* **2022**, *14*, 1316. https://doi.org/10.3390/rs14061316

Academic Editors: Ran Goldblatt, Steven Louis Rubinyi and Hogeun Park

Received: 17 January 2022
Accepted: 4 March 2022
Published: 9 March 2022

low-flicker LED sources. This study indicates that the effects of LED conversions on the brightness and steadiness of outdoor lighting can be analyzed with VIIRS DNB temporal profiles.

Keywords: VIIRS; DNB; flicker; artificial lighting

1. Introduction

A "flicker meter" is a device capable of recording and characterizing the brightness cycling of lighting powered by alternating current (AC) [1]. In AC, the electrons flow back and forth at a well-regulated frequency. Mass distribution power supplies around the world operate at either 50 or 60 Hertz (cycles per second). AC is frequently converted to direct current (DC) to provide a more stable power supply to devices and appliances, in a process called "rectification". The full-wave rectification (Figure 1) of a 60 Hz source results in lights flickering at 120 Hz, but with half the amplitude of the original alternating current. Further reductions in the amplitude of flicker can be achieved by voltage smoothing with capacitors (common in fluorescent lights) and "constant current" drivers [2] used to power large-wattage LED sources (Figure 1).

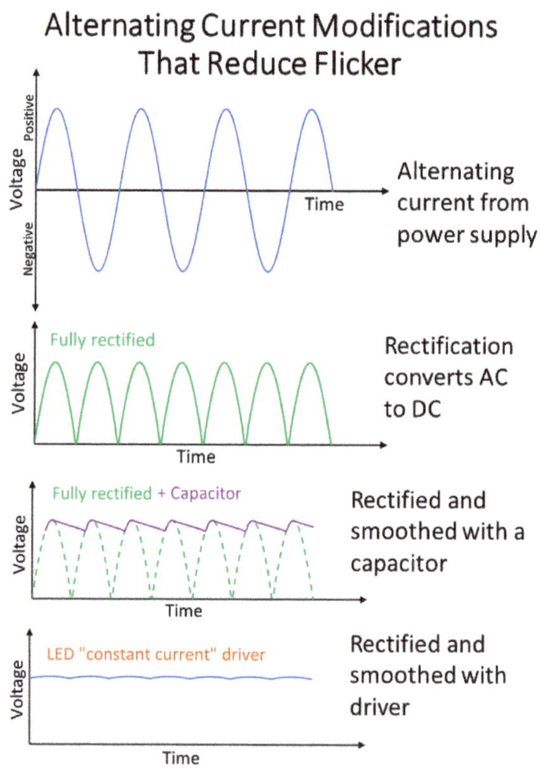

Figure 1. Voltage versus time patterns for alternating current, full-wave rectification, smoothing accomplished with a capacitor, and LED "constant-current".

The "critical flicker frequency" [3] or frame rate of human vision is variable but is placed in the range of 3 to 70 Hz [4]. It is well established that lighting flicker within the range of visual perception can induce neurological symptoms like headaches, loss of attention and visual acuity, irritability, and in some cases, epileptic seizures [5,6]. Most modern luminaires have been engineered to minimize flicker below 50 Hz. However, many light sources flicker in the range of 100 to 120 Hz, which is above the range of conscious perception to the human eye.

The effects of 100–120 Hz flicker on humans remains an active research area. One recent study found that humans are subtly affected by lighting flicker up to 500 Hz [7]. Exposure to flicker at rates above 70 Hz for long periods of time can result in neurological symptoms such as induced photosensitivity [8], chronic migraine and fatigue, and anxiety, depression, and irritability [9]. There is a growing body of evidence that lighting flicker adversely affects other organisms [10]. The satellite detection and characterization of flickering lights would open up a range of new research opportunities in the field of anthropogenic impacts on many organisms, including humans.

To fully resolve the flicker pattern requires temporal sampling that is nearly ten times shorter than the flicker rate being measured [11]. Thus, for 120 Hz flickering, the flicker meter, as one of its design criteria, should measure the brightness of the light at 1200 Hz. This corresponds to a sampling rate of 8.3×10^{-4} seconds (s). According to Table 3.3-6 on page 52 of the VIIRS geolocation Algorithm Theoretical Basis Document (ATBD) [12], the VIIRS DNB pixel dwell time ranges from 2.53×10^{-4} s at nadir to 4.22×10^{-5} s at the edge of scan. If this is true, the DNB pixel dwell time is shorter than the minimum acceptable sampling interval recommended for characterizing 120 Hz flickering.

The standard flicker meter collects many cycles in a single collection session to fully resolve the temporal pattern of the flicker of an individual light. In contrast, the VIIRS DNB collects once or twice per night across many years. At 742 m on a side, DNB pixels may contain hundreds of individual lights of various types. Radiance calibrated VIIRS DNB temporal profiles can be assembled, filtered to remove flawed data, adjusted for a range of radiance contributors such as moonlight, and analyzed for events such as power outages or cycling. Is it also possible to discern differences in AC flicker patterns within VIIRS temporal profiles?

In this paper, we clarify the VIIRS DNB pixel dwell time specifically for the nighttime collection mode, define the flicker characteristics for the major types of lighting, and examine VIIRS DNB temporal profiles for evidence of flicker.

2. Materials and Methods

2.1. Calculation of DNB Pixel Dwell Times for High Gain Stage Data

The VIIRS DNB focal plane assembly contains four detector arrays (Figure 2), which are swept across the earth to construct 16 lines of DNB pixels on each scan [12]. DNB pixel footprints on the earth's surface are maintained at almost 742 m on a side from nadir to edge of scan by the onboard aggregation of different numbers of adjacent detectors in both the track and scan directions. The number of detectors aggregated is largest at nadir and drops gradually, moving toward the edge of scan in fixed patterns, referred to as aggregation zones. Pixels from the first aggregation zone, near nadir, are constructed using 66 detectors in the scan direction and 42 in the track direction (Figure 2). In aggregation zone 32—at the edge of scan—pixels are formed using 11 detectors in the scan direction and 20 detectors in the track direction (Figure 2). All four DNB detector arrays have 672 detectors in the track direction to afford the exact number of detectors required to form the aggregation zone 1 pixels ($16 \times 42 = 672$). Each of the DNB detector arrays are 672 detectors high but vary in width based on the amplification requirements to image clouds from daylit to moonlit conditions. Under daylit conditions, no amplification is required, so the low gain stage (LGS) focal plane array is only one detector wide. The signal amplification is achieved by accumulating charge from multiple samples of the same sub-pixels—a process known as time delay and integration (TDI). The medium gain stage (MGS), designed for cloud imaging in the transition from day to night at dawn and dusk, has three parallel rows of detectors for collecting data with a TDI of three. The highest level of signal amplification comes from the high gain stage (HGS), which is 250 detectors wide. The HGS arrays collect data with a TDI of 250. The HGS is designed to collect at night and is responsible for recording radiances from electric lighting.

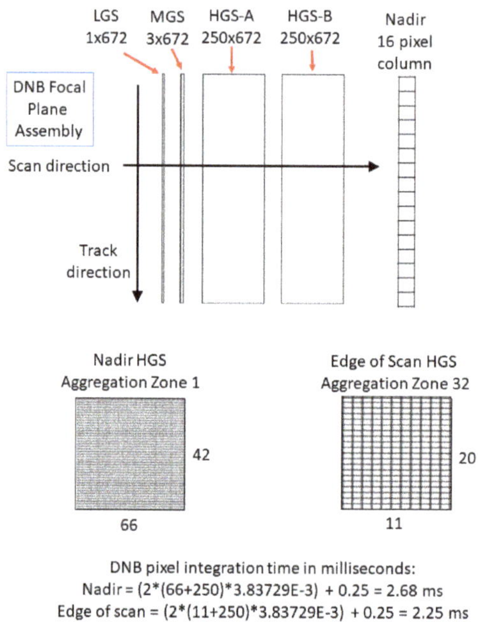

Figure 2. Schematic showing the four DNB detector arrays and the aggregation of charge from variably sized blocks of detectors to maintain near-constant 742 × 742 m ground footprints from nadir to edge of scan. LGS = low gain stage, optimized for daytime collections. MGS = medium gain stage, optimized for collection across the terminator during the transition from day to night. HGS = high gain stage, optimized for nighttime imaging.

There are two HGS detector arrays, referred to as HGS-A and HGS-B. This redundancy is designed to filter out high energy particle detections, which are common when the spacecraft flies through the South Atlantic Anomaly (SAA), which is centered near Rio de Janeiro, Brazil. An onboard algorithm compares the signal from the same ground pixel from HGS-A and HGS-B. If there is a large difference between the two, the brighter pixel is dropped, and the pixel's recorded radiance only comes from one of the two HGS detector arrays. If the difference is modest, the two HGS signals are averaged.

The detector array sampling rate of the DNB is 260.6 KHz [12], corresponding to 3.83729×10^{-3} milliseconds (ms) per detector. The DNB integration table in the ATBD [12] calculates DNB pixel collection times by multiplying the subpixels in the scan direction by 3.83729×10^{-3} s. These numbers are correct for the LGS, which has no TDI, and accumulating all 66 along-scan detectors (aggregation zone 1) yields 2.53×10^{-4} s. To add in the TDI to the DNB pixel HGS collection time, an oversample factor of 250 is applied to each detector in the along-scan direction (Figure 3). Thus, at nadir, the HGS-A pixel collection time is (66 + 250) times 3.83729×10^{-3} ms = 1.213 ms. To this, we add the gap time between HGS-A and HGS-B, which is estimated to be 0.25 ms, and then add the time required to collect HGS-B, which is the same (1.213 ms) as HGS-A. Thus, the total integration time for a HGS DNB pixel at nadir is 2.67 ms, while for aggregation zone 32, it is 2.29 ms.

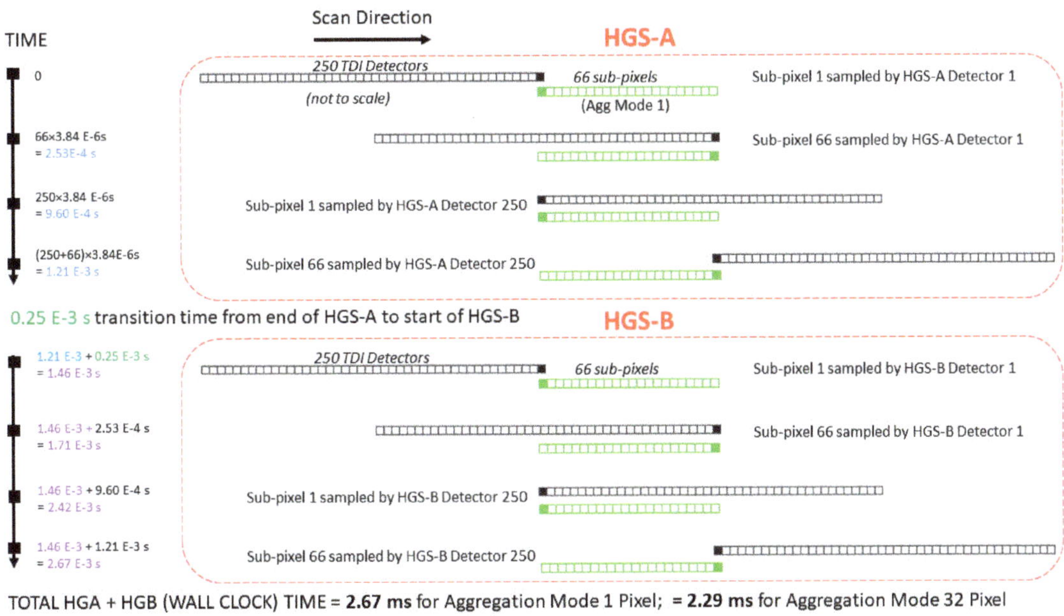

Figure 3. Timeline for the VIIRS DNB HGS pixel aggregation.

It is important to note that the four DNB detector arrays collect photons and accumulate charge continuously during the scan across the earth, but that the zone of active pixel formation gradually tapers from nadir to edge of scan (Figure 4) to maintain the 742 m DNB pixel size. At nadir, all 672 detectors are read out to form pixels. At the edge of scan, the number of detectors actively involved in pixel formation drops to 320. Similar tapering occurs on the LGS and MGS arrays. At any one instant, the HGS detector arrays collect data for a number for DNB pixels. This number can be calculated as 250 divided by the number of sub-pixels in the scan direction. At nadir, this is 250/66 = 2.79 pixels and at the edge of scan, 250/11 = 22.72 pixels. Onboard, the DNB evaluates the pixel voltages from HGS, MGS, and LGS to record the highest unsaturated voltage. Pixels with bright surface lighting will have a usable signal in the HGS but remain undetectable in the MGS and LGS. Conversely, daylit clouds will have a usable voltage from the LGS, while the HGS signal will be saturated. This arrangement makes it possible for the DNB to instantly adjust to abrupt changes in brightness, such as transits across the dawn–dusk lines.

As an interesting side note, we expected that the HGS noise filtering would eliminate the South Atlantic Anomaly high-energy particle detections commonly found in the low-light imaging data collected by the VIIRS predecessor: the U.S. Air Force Defense Meteorological Satellite Program (DMSP) Operational Linescan System (OLS). The SAA-affected pixels have high radiance levels in unexpected places, such as the Atlantic Ocean between Africa and South America. Their numbers appear unchanged between OLS and VIIRS. Several years ago, we conducted an experiment where HGS-A and HGS-B were brought down intact from the satellite along with the normal DNB data stream using one of the VIIRS calibration modes. The HGS-A and HGS-B could not be brought down simultaneously, so they were brought down in alternating granules collected over the Atlantic Ocean south of the equator—a zone famous for SAA detections. We found that SAA hits in both the HGS-A and HGS-B, but these were absent from the normal DNB data. However, the normal DNB data from those granules had a different set of SAA detection pixels. Our conclusion was that the HGS noise filtering is working as designed. Therefore, the SAA detections present in the DNB data stream must be coming from the MGS or LGS, which lack the noise filtering.

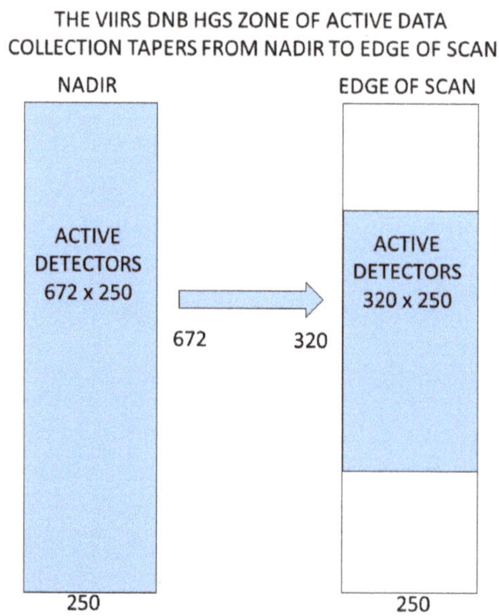

Figure 4. The active zone of DNB pixel formation tapers from 672 detectors in the track direction at nadir to 320 detectors at the edge of scan.

2.2. *Collection of High-Speed Camera Data of Individual Luminaires*

High-speed camera (HSC) data and spectral signatures were collected for the set of individual luminaires listed in Table 1. Most of the data collections were made with a Sony Cyber-Shot DSC-RX100 V 20.1 operated at 1000 frames per second. A second camera was used to collect data from a hovering DJI Mavic Air 2S drone in Moscow using the built-in camera running at 120 frames per second. We also experimented with the collection of flicker data on the panoramic views of city lights with a night vision photomultiplier tube (PMT) adapter (generation 3+ auto-gated white phosphor AGM PVS-14) replacing the fore optics of the Sony camera. Both the original Sony camera and PMT-equipped Sony camera were tested for the presence of flicker or other aberrations with high-speed camera data collected on the moon as a flicker-free light source in the night sky.

The MP4 files collected of the luminaires are available via the links listed in the supplemental material. Temporal profiles were produced from rectangular subsets of the video frames centered on the light sources. A brightness threshold was applied to each frame to minimize the inclusion of background into the temporal profiles, which are also available in the supplemental material.

Luminaire spectra were collected with a BWTek BRC112E spectrometer covering the 400 to 1000 nanometer range. The spectrometer was mounted to record the luminaires' reflectance from a Spectralon panel. Like fingerprints, the spectra were used to confirm the lighting types based on previously published spectral signatures of lights [13].

Table 1. Individual luminaires measured.

Luminaire Type	Designation	Sample	Manufacturer	Model	Wattage
Metal Halide	Ceramic Metal Halide	C0009	Sylvania	MetalARC Mp 100/U/MED	100
	Metal Halide	C0012	EMCO Lighting	ERA20-3H SO-89193	175
	Metal Halide	C0015	Philips	MP400/BU/PS Kr85	400
	Metal Halide RAB	C0016	RAB	RAB LMH250PS	250
High Pressure Sodium	HPS	C0017	Sylvania	LU150/55/ECO	175
	HPS	C0041	General Electric	Lucalox	100
Low Pressure Sodium	LPS	C0007	Osram	N068 80X Great Britain	18
Fluorescent	4 foot tube pair	C0006	Philips	Alto F40T/C50 Sumpreme	40
	Fluorescent ceiling	C0035	Philips	Alto-II F32T8/Tl841 800 series	32
	Compact Fluorescent	C0028	BFLG	13	13
	Compact Fluorescent	C0031	Greenlight	13W/ELS-M/1	13
	Compact Fluorescent	C0032	TCP	ESN11	11
Incandescent	Incandescent	C0014			
	Incandescent	C0027	Sylvania	Capsylite FL-tugsten-halogen	45
	Incandescent	C0034	Soft White		60
LED	LED	C0018	PLT Solutions	PLT-11557	300
	LED 4000 K	C0020	LeoTEK	Beta	100
	LED 3000 K	C0022	LeoTEK	Beta	100
	LED	C0026	Lithonia	DSXPG	51
	iphone	C0036	Apple	iPhone Xr MT472LL/A	?

2.3. Collection of Flicker Data for Mixtures of Lights

To investigate the expression of flicker that might be present in VIIRS DNB data collected on pixels having large numbers of individual luminaires, high-speed camera data were collected for a set of scenes containing mixtures of light sources. We speculated that the magnitude of flicker would decline when large numbers of individual luminaires are viewed together. The first collection in the multi-light series was made on a wide range of luminaires from Table 1 placed together on an office table. This included LED streetlights, metal halides, high-pressure sodium, low-pressure sodium, incandescent, and fluorescent lights (Figure 5).

Figure 5. High-speed camera data were collected on a mixture of luminaires placed together on a table.

Next, we collected high-speed camera data (1000 frames per second) from the ground for three outdoor scenes containing large numbers of individual luminaires. This included: (1) Broomfield, Colorado's civic center (Figure 6), (2) the central zone of Golden, Colorado (Figure 7), collected from a viewpoint located on Lookout Mountain Road, and (3) Boulder, Colorado, viewed at a distance from a viewpoint along Highway 36 (Figure 8). The panoramic collections of Golden and Boulder cover large portions of each city as viewed from an oblique angle. The resulting temporal profiles from Golden and Boulder cover much larger areas than the individual DNB pixel footprint. The examination of the MP4 movies of these collections reveals the presence of a cacophony of sources ranging from lights having prominent flicker to lights having no visually evident flicker.

Figure 6. High-speed camera data were collected on a mixture of lights present at the Civic Center in Broomfield, Colorado. Also included in the scene was the moon—in the upper left-hand corner—which made it possible to analyze the flicker from the camera itself.

Figure 7. Oblique view of the mixture of lights presents in the central Golden, Colorado area collected with the high-speed camera. The data collection was from a viewpoint along Lookout Mountain Road west of Golden.

Figure 8. Panoramic view of Boulder, Colorado, from a viewpoint along Highway 36. High-speed camera data were collected covering the full sweep of lights visible in Boulder from the viewpoint.

To further investigate the remote sensing of flicker from the DNB, digital camera data were collected from a hovering drone carrying a slow-motion camera pointed in the nadir direction towards the earth. The experiment took place in the suburbs of Moscow, Russia (55.6321 north, 37.8622 east), at 01:00 local time on 29 November 2021. The vehicle was a DJI Air 2S drone with a built-in 4K video camera. The data were collected while the drone hovered 315 m above the earth's surface, resulting in a footprint of 600 × 600 m—slightly smaller than the DNB pixel footprint. The nighttime movie of the lights was recorded with FHD resolution 1920 × 1080, shutter speed 1/120 s, f/2.8 aperture, and sensitivity ISO 3200. The data were collected at 120 frames per second. The scene includes numerous flickering high-pressure sodium streetlights and parking lot lights, plus strings of white lights (Figure 9), with some clearly flickering and others with no obvious flickering.

Figure 9. View of lights collected at 120 frames per second from a hovering drone in Moscow, Russia.

2.4. Processing of the High-Speed Camera Data

The digital data from the high-speed camera MP4 movie frames were filtered to remove background pixels, which helped to concentrate the signal coming from the lights. The brightness of the remaining pixels in each frame were averaged and frame step times added. For the Sony camera, the frame steps were in millisecond increments since the camera ran at 1000 frames per second. The time step for the drone camera collection was 8.333 ms, corresponding to 120 frames per second. The result was a set of temporal profiles that were further analyzed to detect the presence and relative intensity of flicker.

An autocorrelation function (ACF) was used for the detection of flicker in the HSC data and a determination of the cycling rate in Hz. The ACF rates the similarity of the time series with itself for a series of evenly spaced temporal or lag increments [14]. For instance, if the data series is broken up into a series of consecutive five ms segments, the ACF calculates the degree of similarity found between the segments. This autocorrelation analysis is performed for all possible lag increments within the time series. The types of detectable temporal phenomena from the ACF analysis are dictated by the temporal increment of the data. For instance, annual cycling cannot be analyzed with annual nighttime lights data but can be analyzed with either nightly radiances [15] or average monthly radiances [16]. The presence of an AC flicker cycling in the high-speed camera data is indicated when an ACF peak exceeds the 95% or 99% confidence level threshold calculated via a z-statistic [17]. For an ACF sample size of 100 points, the 95% confidence threshold is 0.196, and the 99% confidence threshold is 0.258. A flicker rate of 120 Hz is indicated by the presence of a primary ACF peak at 8 ms, exceeding the 95% confidence thresholds (Figure 10).

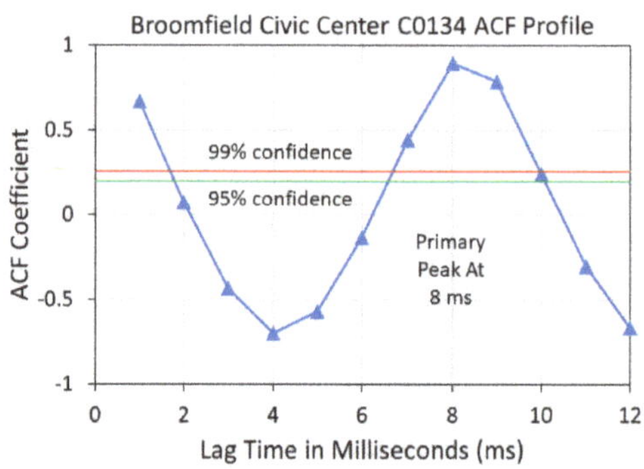

Figure 10. An autocorrelation function (ACF) is generated for each high-speed camera temporal profile to detect the presence of millisecond-scale cycling. The primary peak is defined as the highest local max lag value greater than 1 ms and exceeding 95% confidence. The Sony camera collected at 1000 frames per second. An ACF lag of 8 ms indicates the presence of 120 Hz flickering.

Calculations were made of the percent flicker and flicker index based on the methods described in the Lighting Handbook 10th Edition [18]; examples are shown in Figure 11. Note that in order to apply the flicker index to the randomized and actual DNB profiles, the definition had to be modified. In addition, the index of dispersion is calculated as the variance divided by the mean [19].

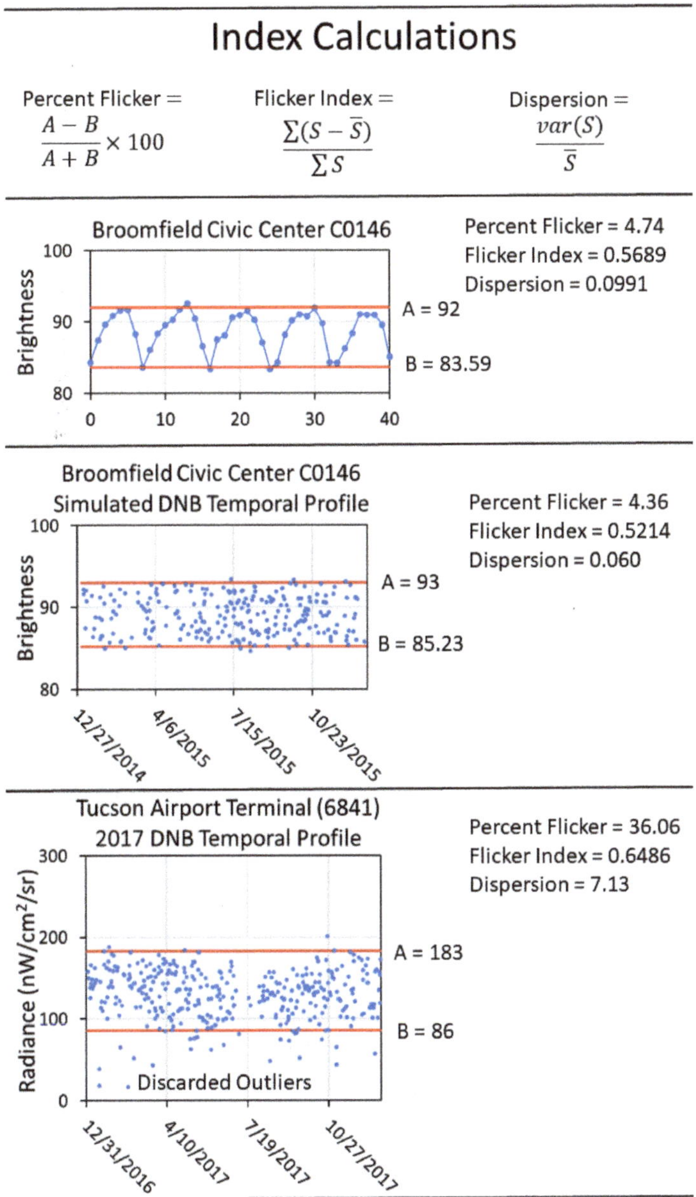

Figure 11. Calculation of percent flicker, flicker index, and index of dispersion.

Percent flicker is designed to quantify the amplitude of the flicker and is defined as:

$$\text{Percent Flicker} = \frac{\text{A} - \text{B}}{\text{A} + \text{B}} \times 100$$

For the high-speed camera temporal profiles, A is the average brightness of the upper peak, while B is the average brightness of the lower peak. For the simulated and actual DNB temporal profiles, A and B are the upper and lower limits of the core data range. Outlier removal and detrending may be required to obtain results that can be compared across samples.

The flicker index is designed to quantify the symmetry of the flicker and is defined as:

$$\text{Flicker Index} = \frac{\sum(S - \overline{S})}{\sum S}$$

where the S is the sample brightness, and $\overline{S}$ is the mean of sample brightness.

The dispersion index is defined as:

$$\text{Dispersion Index} = \frac{\text{var}(S)}{\overline{S}}$$

where $\text{var}(S)$ is the variance of sample brightness.

As with the percent flicker, outlier removal and trend removal may be advisable prior to the flicker index and dispersion calculations to yield consistent results.

2.5. Simulation of DNB Temporal Profiles from the High-Speed Camera Data

Because the VIIRS DNB temporal repeat cycle ranges from hours to days, the flicker patterns in lighting are sampled randomly. It is impossible to discern the lighting flicker rates with the ACF methodology using nightly time series of VIIRS DNB temporal profiles. However, we still want to test the flicker indices to evaluate their utility in analyzing the flicker effects present in DNB temporal profiles. Each DNB scanline has 4064 pixels and the HGS dwell time varies as a function of the sample position. The SNPP orbit processes, with a 16 day repeat cycle. That means the satellite repeats its orbital tracks every 16 days, with only minor deviations. The repeat cycle, combined with the spacecraft's precision orbit control, results in a series of narrowly defined "permissible" sample positions for any location on the ground. To mimic the variable temporal sampling of the HGS, we followed the sampling pattern established for a particular mid-latitude location, as shown in Figure 12. Any location would suffice due to the even mixture of sample positions over time. The Figure 12 template happens to be from Houston, Texas, and has 1946 cloud-free DNB pixels from 2012 to 2020, binned into 20 vertical sample position columns. In simulating the DNB temporal profile from each high-speed camera temporal profile, exactly 1946 samples are drawn randomly from the brightness profile using the sample positions from Figure 12. The randomly drawn sample positions point to the appropriate DNB temporal aggregation time via a look up table. The simulated DNB temporal profiles have the same brightness units and range as the original 1000 frame per second profiles but are randomized. The A and B inflection points used to calculate the percent flicker are taken to be the upper and lower brightness levels of the data cloud (Figure 11).

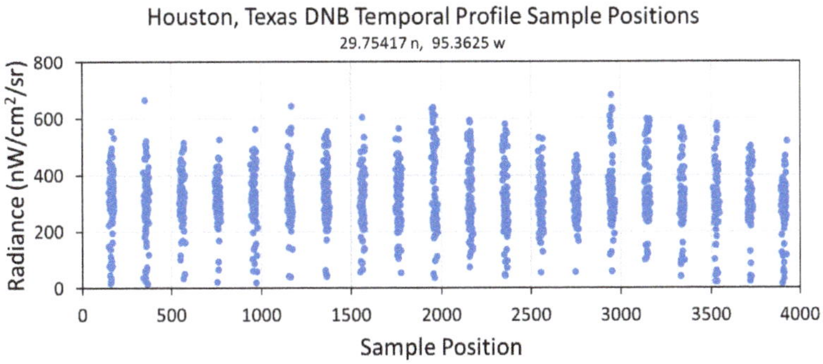

Figure 12. Sample position columns for the VIIRS DNB temporal profile from a 15-arcsecond grid cell in Houston, Texas. These columns define the sampling pattern for simulating the DNB temporal profiles from the high-speed camera data for each of the tested luminaires.

2.6. Examination of VIIRS DNB Temporal Profiles for Flicker Effects

To explore the detection of flicker in actual VIIRS data, a nightly DNB temporal profile grid was constructed for Tucson, Arizona. We generated the profiles at 15-arcsecond spacing, with 100 cells east–west and 72 cells north–south centered on Tucson (Figure 13). The DNB profiles were processed to remove sunlit and cloudy data. A satellite zenith angle normalization was applied to remove view angle effects [15]. A lunar illuminance adjustment was applied to subtract the reflected moonlight [15]. Cloudy pixels were filtered out based on the VIIRS cloud mask [16] generated by NOAA [20]. The annual mean, variance, and dispersion were then calculated for each 15-arcsecond grid cell.

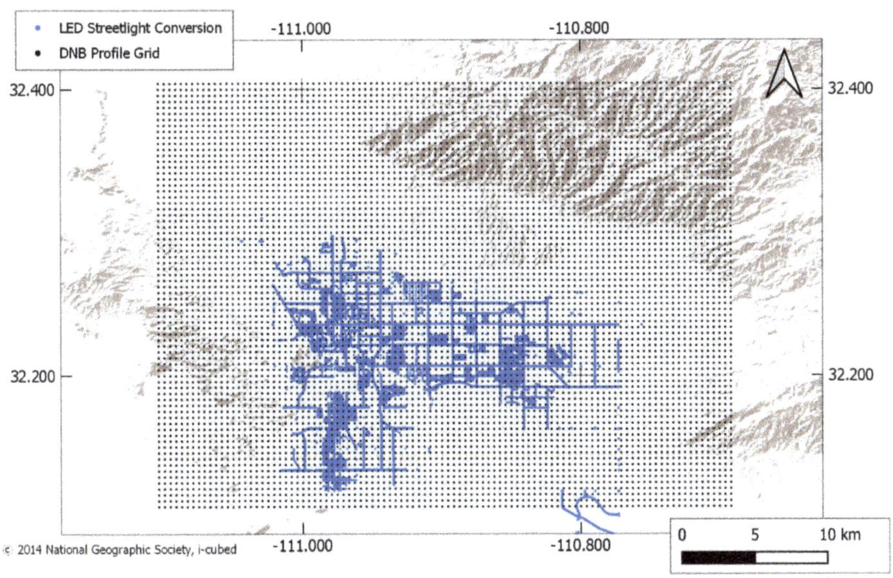

Figure 13. Outline of the VIIRS DNB temporal profile grid constructed for Tucson, Arizona. The locations of LED streetlight conversions are marked blue.

For decades, Tucson has taken actions to reduce light pollution to limit interference with several nearby astronomical observatories, such as Kitt Peak National Observatory. In 2016, the City of Tucson began replacing high-pressure sodium streetlights with LED luminaires [21–23]. The motivation for the streetlight conversions was two-fold. There was a financial incentive, which was to reduce the power consumption and replacement cycle for individual luminaires. The other motivator was that the new system was designed to allow individual streetlights to be dimmed, thereby reducing light pollution. All of the city's nearly 20,000 streetlights were converted. It should be noted that other lighting present in the grid cells was unaffected by the city's LED conversion program. Following the conversion, the majority of the new LED streetlights have been operated at 60% capacity from midnight to dawn [21].

The grid cells are divided into three zones to examine the effects of the streetlight conversions on the index values. First, the grid cells are divided into those with and without detectable lighting (Figure 14). Background areas, devoid of detectable lighting, are defined based on annual average radiance values under one nanowatt/cm^2/sr. The lit grid cells are further divided into sets with and without City of Tucson LED streetlight conversions (Figure 15).

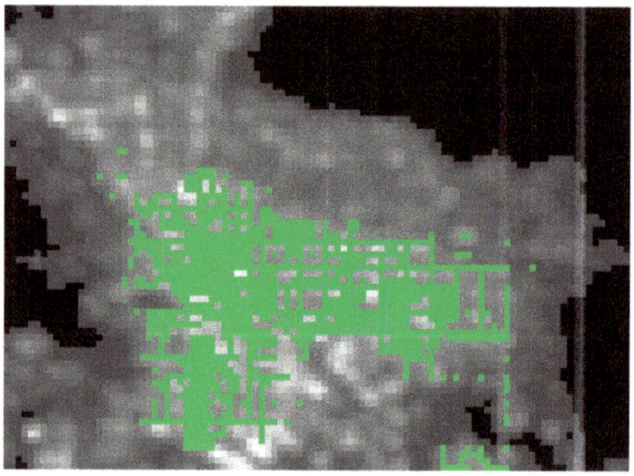

Figure 14. Average radiance image from Tucson for the combined years 2015 and 2017. Background areas, devoid of VIIRS detected lighting, are black. Grid cells with detectable lighting and no LED streetlight conversions are gray, and lit grid cells with LED streetlight conversions are green.

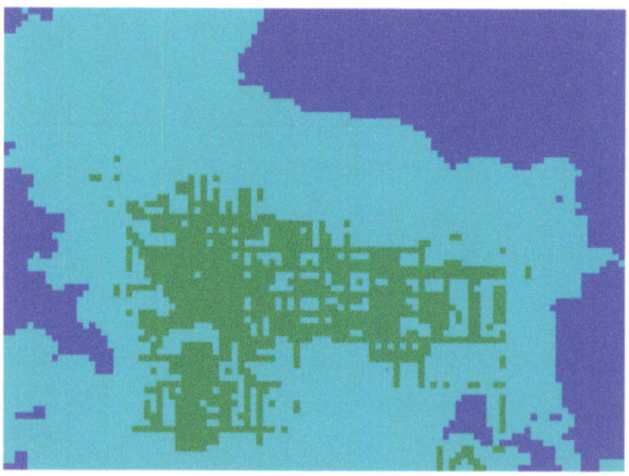

Figure 15. The DNB temporal profiles are divided into three categories: background (blue), lit and zero streetlight conversions (cyan), and lit with streetlight conversions (green). There are 1955 grid cells in the background, 4121 grid cells lit with no-LED conversions, and 1124 grid cells with LED streetlight conversions.

A set of five Tucson grid cells were selected for more detailed examination and testing. This included one grid cell from the background. The other four grid cells were selected because they show abrupt and dramatic changes in radiance levels, with two from the city's LED conversion zone and two from areas outside the city's LED conversion zone. For these five grid cells, the annual percent flicker, flicker index, and index of dispersion values were calculated.

3. Results

3.1. Re-Calculation of DNB Pixel Collection Times for HGS

The re-calculated SNPP DNB HGS pixel integration times are listed for each aggregation zone in Table 2 and shown graphically in Figure 16. The integration time is longest at nadir (2.68 ms) and shortest at the edge of scan (2.25 ms). The DNB pixel HGS-A and HGS-B integration times at nadir and edge of scan, shown overlaying a high-speed camera (1000 frames per second) temporal profile from a high-pressure sodium luminaire, are shown in Figure 17. The DNB HGS integration times are about one-third to one-fourth of the typical lighting flicker rates of 100 to 120 Hz.

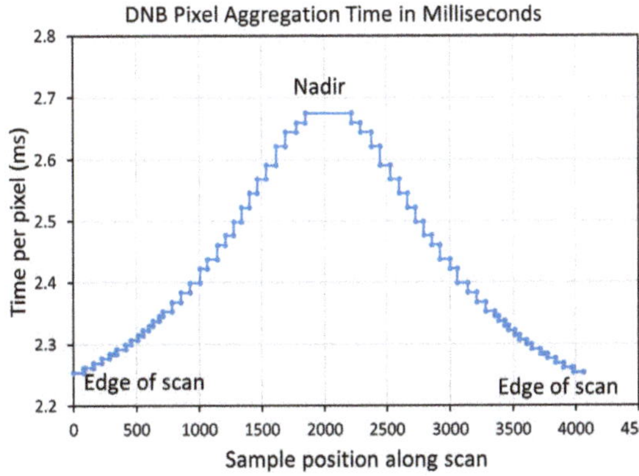

Figure 16. SNPP DNB pixel integration times by aggregation zone.

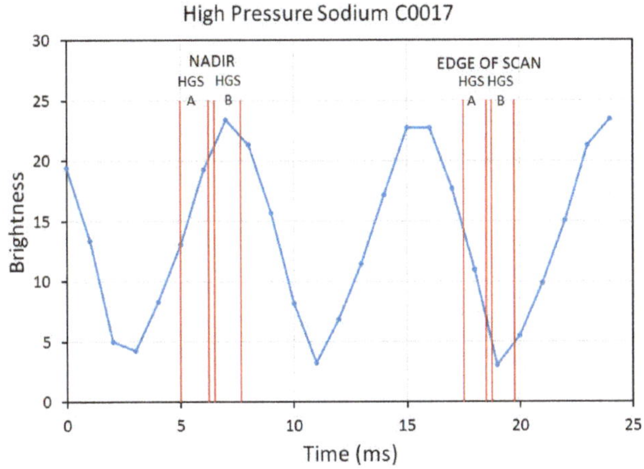

Figure 17. SNPP VIIRS DNB pixel integration times for HGS-A and HGS-B randomly placed over a high-pressure sodium temporal profile collected with the Sony high-speed camera. Each HGS sample consists of an A and a B side, separated by a 0.25 ms gap.

Table 2. SNPP VIIRS DNB pixel integration times by aggregation zone.

Aggregation Mode from Nadir	Number of Sub-pixels per Pixel Track Direction	Scan Direction	Number of Pixels per Mode	Time per Pixel (s) HGS-A	Time on HGS A-B Gap (s)	Time per Pixel (s) HGS-B	Total Time per DNB Pixel (s)
1	42	66	184	1.21E-03	2.50E-04	1.21E-03	2.68E-03
2	42	64	72	1.20E-03	2.50E-04	1.20E-03	2.66E-03
3	41	62	88	1.20E-03	2.50E-04	1.20E-03	2.64E-03
4	40	59	72	1.19E-03	2.50E-04	1.19E-03	2.62E-03
5	39	55	80	1.17E-03	2.50E-04	1.17E-03	2.59E-03
6	38	52	72	1.16E-03	2.50E-04	1.16E-03	2.57E-03
7	37	49	64	1.15E-03	2.50E-04	1.15E-03	2.54E-03
8	36	46	64	1.14E-03	2.50E-04	1.14E-03	2.52E-03
9	35	43	64	1.12E-03	2.50E-04	1.12E-03	2.50E-03
10	34	40	64	1.11E-03	2.50E-04	1.11E-03	2.48E-03
11	33	38	64	1.11E-03	2.50E-04	1.11E-03	2.46E-03
12	32	35	80	1.09E-03	2.50E-04	1.09E-03	2.44E-03
13	31	33	56	1.09E-03	2.50E-04	1.09E-03	2.42E-03
14	30	30	80	1.07E-03	2.50E-04	1.07E-03	2.40E-03
15	29	28	72	1.07E-03	2.50E-04	1.07E-03	2.38E-03
16	28	26	72	1.06E-03	2.50E-04	1.06E-03	2.37E-03
17	27	24	72	1.05E-03	2.50E-04	1.05E-03	2.35E-03
18	27	23	32	1.05E-03	2.50E-04	1.05E-03	2.35E-03
19	26	22	48	1.04E-03	2.50E-04	1.04E-03	2.34E-03
20	26	21	32	1.04E-03	2.50E-04	1.04E-03	2.33E-03
21	25	20	48	1.04E-03	2.50E-04	1.04E-03	2.32E-03
22	25	19	40	1.03E-03	2.50E-04	1.03E-03	2.31E-03
23	24	18	56	1.03E-03	2.50E-04	1.03E-03	2.31E-03
24	24	17	40	1.02E-03	2.50E-04	1.02E-03	2.30E-03
25	23	16	72	1.02E-03	2.50E-04	1.02E-03	2.29E-03
26	23	15	24	1.02E-03	2.50E-04	1.02E-03	2.28E-03
27	22	15	32	1.02E-03	2.50E-04	1.02E-03	2.28E-03
28	22	14	64	1.01E-03	2.50E-04	1.01E-03	2.28E-03
29	21	13	64	1.01E-03	2.50E-04	1.01E-03	2.27E-03
30	21	12	64	1.01E-03	2.50E-04	1.01E-03	2.26E-03
31	20	12	16	1.01E-03	2.50E-04	1.01E-03	2.26E-03
32	20	11	80	1.00E-03	2.50E-04	1.00E-03	2.25E-03
Total			**2032**				7.71E-02

3.2. *Results from High-Speed Camera Temporal Profiles for Individual Sources*

Figure 18 shows representative examples of the high-speed camera temporal profiles from the 23 individual luminaires, including high-pressure sodium, metal halide, fluorescent, incandescent, and LED. The 120 Hz flicker pattern can be seen visually, though the clarity of the pattern is diminished in the LED streetlight.

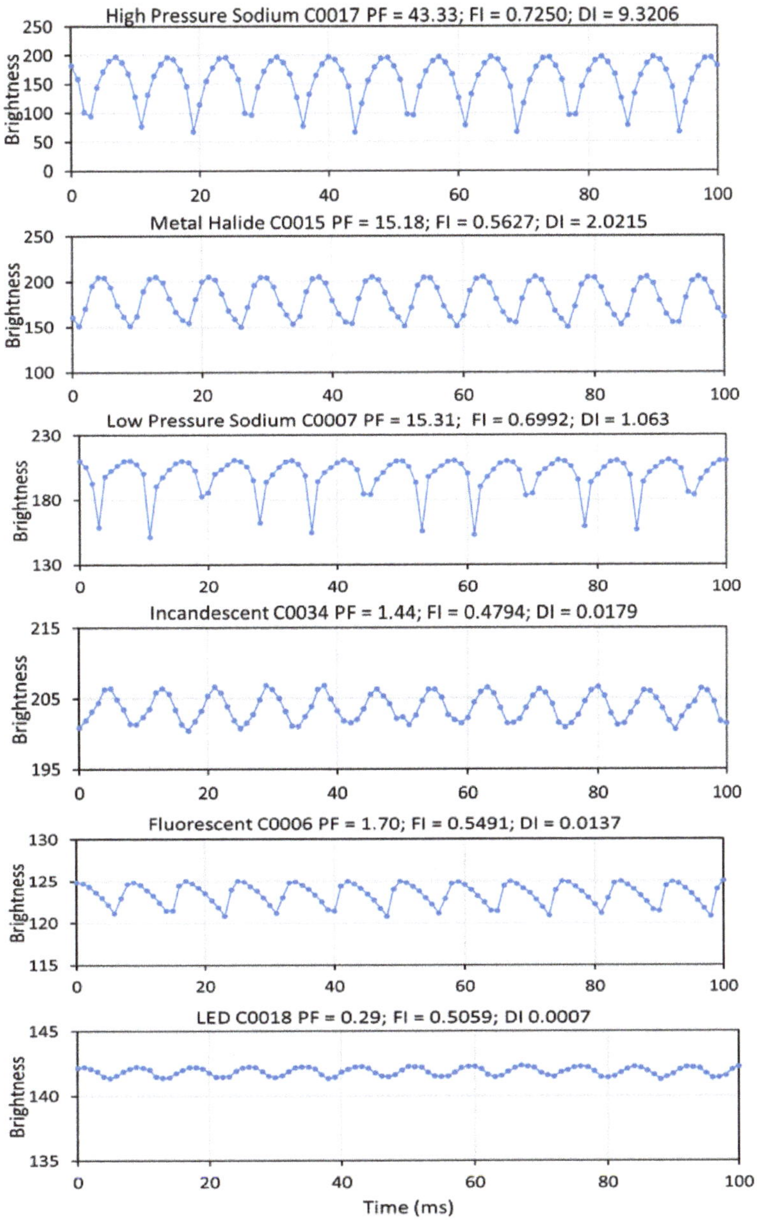

Figure 18. High-speed camera temporal profiles from a variety of individual luminaires. The brightness axis scaling has been adjusted to highlight the flicker pattern, which is weakest in the LED. The title line on each profile records the luminaire type, sample number, percent flicker (PF), flicker index (FI), and dispersion index (DI).

The temporal profiles were resampled to simulate DNB temporal profiles, a process that obliterates the cycling pattern evident in the millisecond temporal profiles (Figure 19). However, histograms from the DNB temporal profiles of individual luminaires continue to exhibit spikes near the low and high ends of brightness range, corresponding to the flicker inflection points. The histogram spike pattern is characteristic of "harmonic oscillators" [24]. The histogram spikes occur due to the increased temporal dwell at nearly the same brightness level when the sample falls across either the upper or lower flicker inflection points.

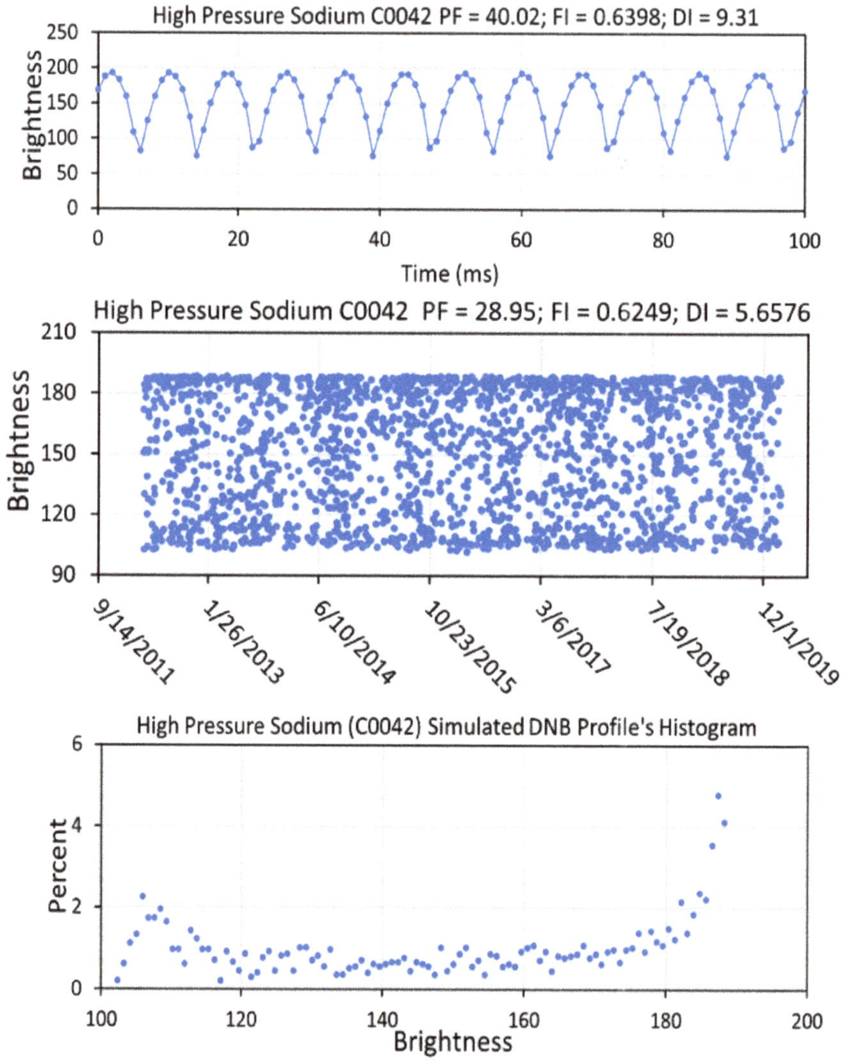

Figure 19. Simulated DNB temporal profile and associated histogram from the resampling of a metal halide high-speed camera temporal profile. Note that the brightness range of the simulated DNB temporal profile is slightly less than the original 1000 frames per second temporal profile, an expression of the temporal aggregation from the longer DNB pixel aggregation. The histogram shows spikes at brightness levels of 10.5 and 19, a vestigial expression of the flickering upper and lower sine wave peaks.

The flicker index and ACF results from the individual luminaires are summarized in Table 3. The table is sorted according to the percent flicker in descending order. The highest percent flicker values are found with the high-intensity discharge lamps, with the two high-pressure sodium sources having the highest percent flicker, followed by the metal halides and low-pressure sodium. The LED streetlights and one of the fluorescent tubes have the lowest percent flicker.

Table 3. Flicker results for individual luminaires. SDNB = simulated DNB.

Luminaires	Number	% Flicker	Flicker Index	Dispersion	Flicker Hz	SDNB % Flicker	SDNB Flicker Index	SDNB Dispersion
Compact Fluorescent	C0028	68.12	0.8216	17.1248	120	38.59	0.7303	8.0688
High Pressure Sodium	C0017	43.33	0.7250	9.3200	120	28.69	0.6356	5.4270
High Pressure Sodium	C0042	40.02	0.6398	9.3078	120	28.95	0.6249	5.6576
Metal halide	C0012	21.71	0.5656	3.6492	120	18.49	0.5816	2.3106
Low Pressure Sodium	C0007	15.31	0.6992	1.0630	120	8.25	0.6051	0.4863
Metal Halide	C0015	15.18	0.5627	2.0215	120	12.40	0.5550	1.3042
Ceramic metal Halide	C0009	14.40	0.5754	1.7200	120	10.73	0.6237	1.0180
Metal Halide RAB	C0016	12.51	0.5956	1.7166	120	9.96	0.5474	0.8887
Incandescent	C0027	5.47	0.5108	0.2264	60	5.26	0.5464	0.2031
Fluorescent	C0006	1.70	0.5491	0.0137	120	1.23	0.5152	0.0080
Incandescent	C0034	1.44	0.4794	0.0179	120	1.23	0.5604	0.0128
Incandescent dual	C0033	1.30	0.4986	0.0135	120	1.14	0.5661	0.0093
Incandescent	C0014	1.06	0.5379	0.0118	120	1.16	0.5075	0.0077
iphone flashlight	C0036	0.77	0.4574	0.0047	0	0.53	0.6268	0.0016
Compact Fluorescent	C0031	0.65	0.5561	0.0031	120	0.47	0.6230	0.0018
Compact Fluorescent	C0032	0.63	0.3580	0.0028	120	0.59	0.5852	0.0013
LED	C0018	0.29	0.5059	0.0007	120	0.30	0.5049	0.0005
LED 4000 K	C0020	0.19	0.4897	0.0003	120	0.22	0.4930	0.0002
Compact Fluorescent	C0011	0.14	0.5450	0.0003	120	0.17	0.5378	0.0003
LED	C0026	0.12	0.5250	0.0002	120	0.12	0.5201	0.0001
LED 3000 K	C0022	0.10	0.4360	0.0001	120	0.11	0.5117	0.0001
Fluorescent Ceiling	C0035	0.08	0.5547	0.00004	120	0.09	0.6592	0.00004

Figure 20 shows representative examples of ACF profiles from the high-speed camera data and a grid cell in Tucson found to have weekly brightness cycling. The ACF analysis found 120 Hz flicker in 19 of 20 individual luminaires (Table 3). The ACF analysis found no detectable flicker from the iPhone flashlight (sample C0036).

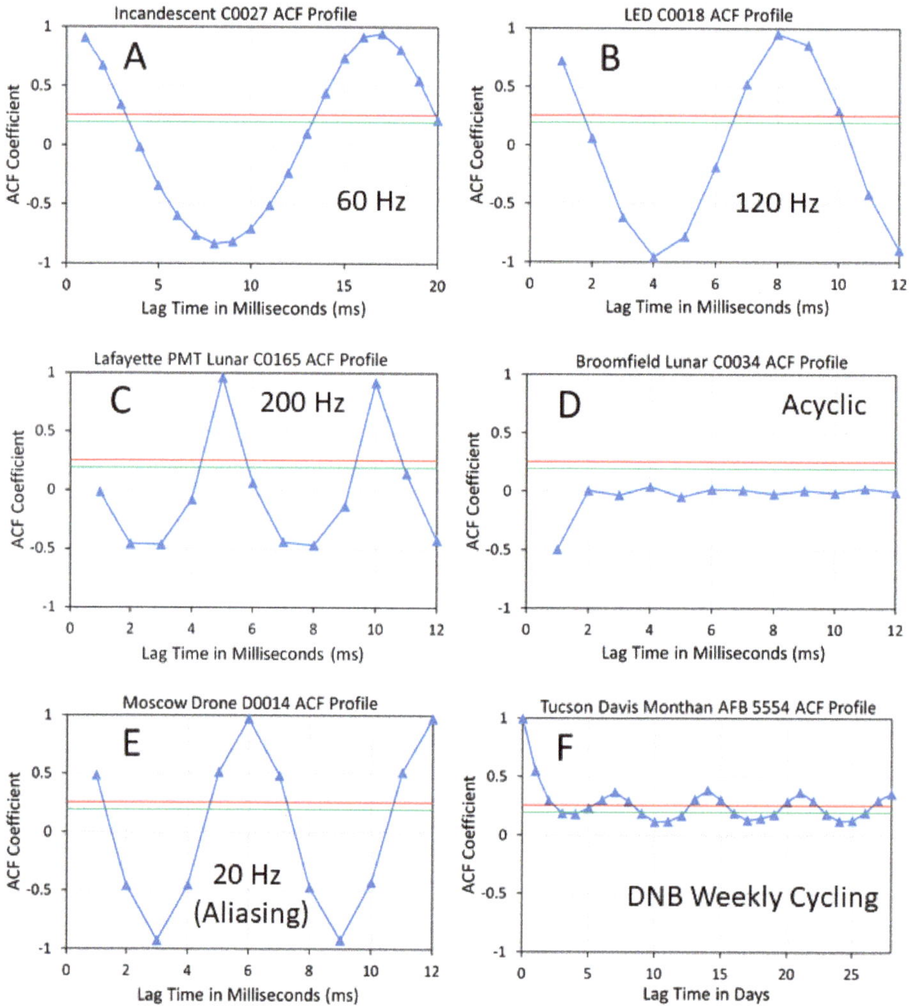

Figure 20. Examples of autocorrelation function (ACF) profiles from the study. (**A**) High-speed camera (HSC) data from a Sylvania incandescent light (C0027), (**B**) HSC data from an LED streetlight (C0018), (**C**) PMT-equipped HSC data from a lunar view, (**D**) HSC data from a lunar view, and (**E**) DJI Air 2S drone camera data of lights in Moscow, Russia, collected at 120 frames per second. The primary peak has a lag of six frames, indicating 20 Hz flickering. The actual flicker rate is 100 Hz. The discrepancy is due to aliasing. (**F**) VIIRS DNB nightly temporal profile of Davis Monthan Air Force Base, Tucson, Arizona. The seven-day lag indicates the presence of weekly cycling in the brightness of the lights.

Scattergrams were generated to examine the relationships between the three indices and the impact of DNB resampling. Figure 21 shows the percent flicker versus flicker index, with labeling of the lighting types. The luminaires with high percent flicker also have relatively high flicker index values, though it is not a well-defined linear relationship. In contrast, the index of dispersion is highly correlated to percent flicker (Figure 22).

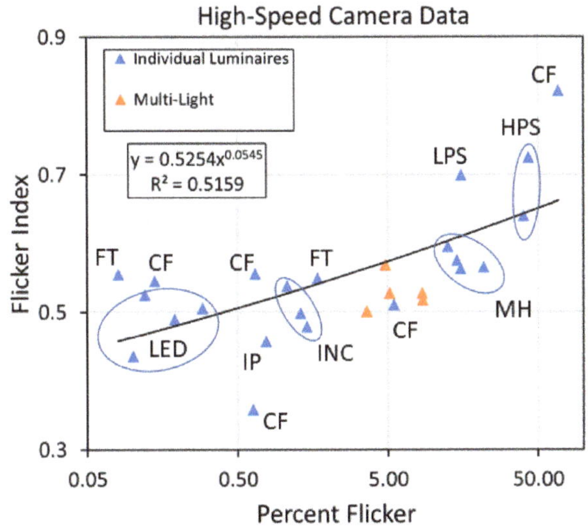

Figure 21. Percent flicker versus flicker index for individual luminaires, multi-light collections, and Tucson DNB profiles. The individual luminaires are clustered into types: HPS = high-pressure sodium, MH = metal halide, LPS = low-pressure sodium, INC = incandescent, CF = compact fluorescent, IP = iPhone flashlight, LED = LED streetlights, FT = fluorescent tubes.

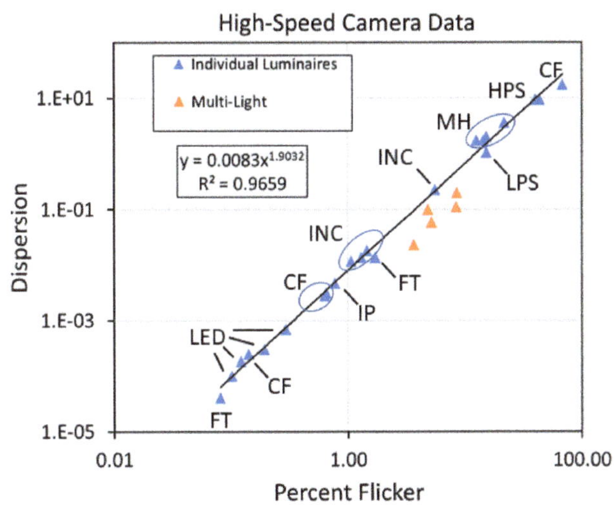

Figure 22. Percent flicker versus the index of dispersion for the HSC collections made of individual luminaires, multi-light collections, and DNB temporal profiles from Tucson.

The scattergram of percent flicker from the individual luminaires versus DNB simulated temporal profiles shows a highly linear relationship (Figure 23). There is a slight decline in the percent flicker values from the DNB-simulated temporal profiles, presumably due to the longer integration time of the DNB versus the Sony camera running at 1000 frames per second. The flicker index values from HSC temporal profiles versus their DNB-simulated equivalents are shown in Figure 24. Note that the DNB simulation results in compression in the flicker index range. Figure 25 shows that the index of dispersion values for HSC and DNB simulated temporal profiles are highly correlated.

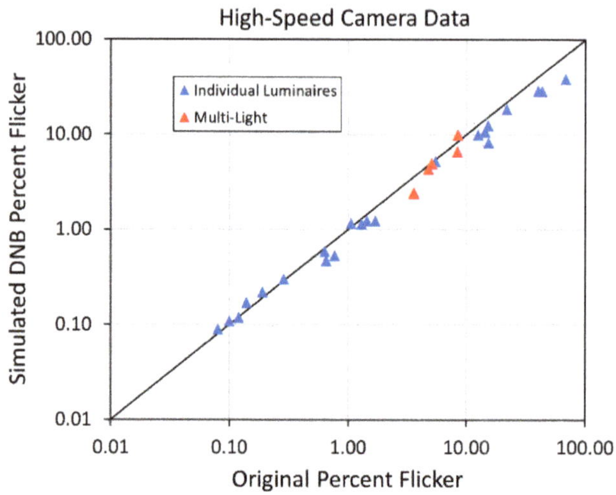

Figure 23. Percent flicker from the high-speed camera collections versus the DNB-simulated temporal profiles. The two are highly corelated, with the data cloud forming a diagonal. There is a slight loss in flicker index values for the DNB-simulated sets, presumably due to the longer integration time of the DNB compared to the digital camera data.

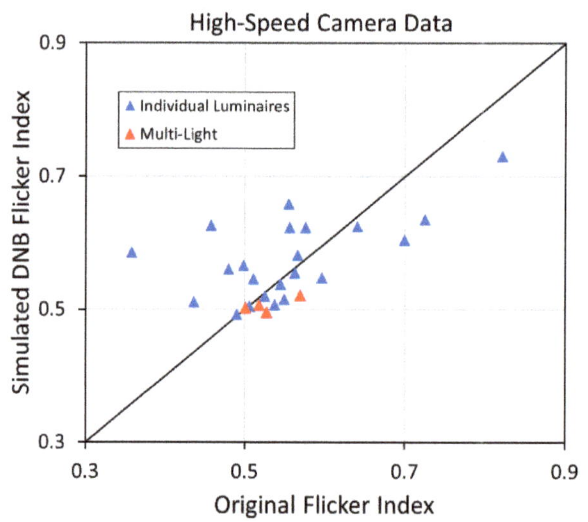

Figure 24. Flicker index values from the high-speed camera collections versus the DNB-simulated temporal profiles. The two data sets are not well correlated. Specifically, the flicker index data range from the simulated DNB temporal profiles is compressed, indicating a loss in the index's ability to distinguish different types of lighting.

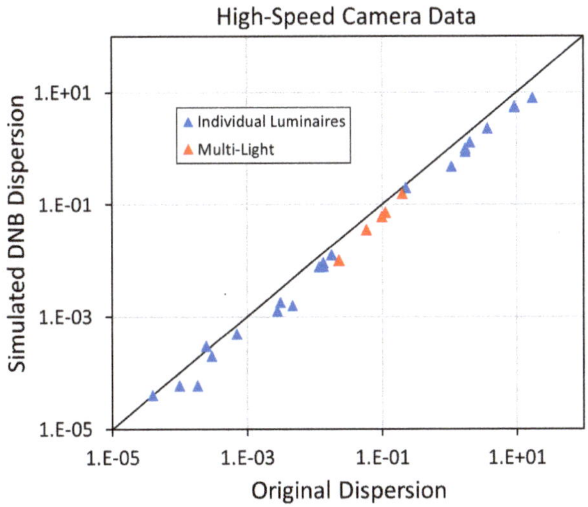

Figure 25. Index of dispersion values from the high-speed camera collections versus the DNB simulated temporal profiles. The two are highly corelated, with the data cloud forming a diagonal. The DNB-simulated profiles produced slightly lower dispersion values, presumably due to the longer integration time of the DNB compared to the digital camera data.

3.3. Results from Lunar Views

The lunar view temporal profiles from the Sony and PMT cameras are shown in Figure 26. These can be considered HSC collections of individual sources since all other sources are masked out. The temporal profiles from the Sony camera show a low level of background noise with no sign of flickering. The PMT's lunar profile shows prominent AC fluctuations. This led to the exclusion of the PMT collections from other parts of this study.

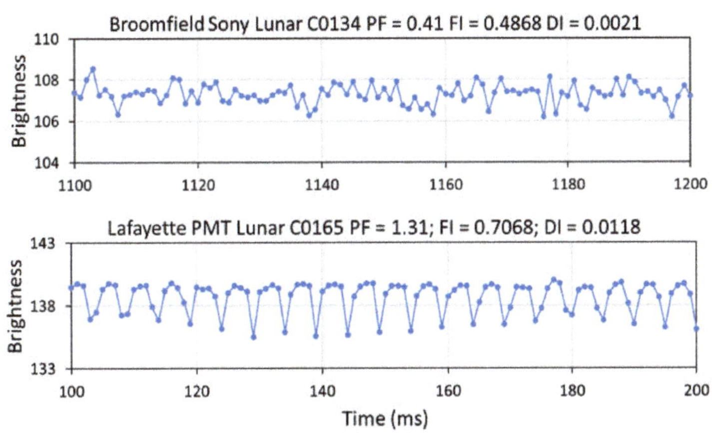

Figure 26. High-speed camera collections of the moon at night from the Sony camera (**top**) and the Sony camera equipped with a photo-multiplier tube (**bottom**). Note the prominent electronic flicker present in the PMT collection. This led to the exclusion of the PMT from further flicker study collections.

3.4. Results from High-Speed Camera Temporal Profiles for Multi-Light Collections

Figure 27 shows the high-speed camera temporal profiles from the multi-light collections from the table-top collection, oblique view of Golden, horizontal view of the Broomfield Civic Center, panoramic view of Boulder, and overhead view of outdoor lighting collected from a hovering drone in Moscow. Flicker is evident in each of the profiles, though it is more weakly expressed in the Golden and Boulder samples where the sample view covered lights present in multiple VIIRS DNB pixel footprints. The multi-light temporal profiles from Colorado feature 120 Hz flicker. Note that the horizontal axis scaling is expanded from 100 ms to 500 ms for the Moscow drone profile where the camera frame rate is 120 per second. The index results are reported in Table 4. All five of the multi-light temporal profiles had flicker present in the range of 3 to 5%. The ACF analysis found the flicker rate from Colorado to be 120 Hz, and the Moscow sample's to be 100 Hz.

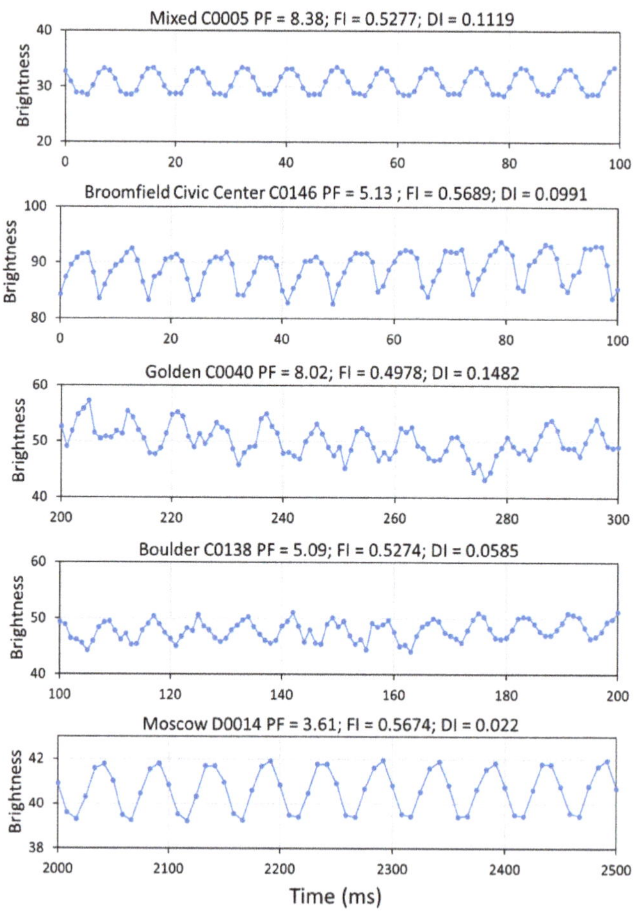

Figure 27. High-speed camera temporal profiles from five multi-light collections. Each of the samples shows visual evidence of AC flicker. The pattern is less clear in the Golden and Boulder examples, where the field-of-view covered multiple DNB pixels. Note that the temporal scale for the Colorado 1000 frames per second collection is standardized to 100 ms. The Moscow drone data were collected at 120 frames per second and a horizontal axis range of 500 ms to expose the flicker pattern. PF = percent flicker, FI = flicker index, DI = dispersion index.

Table 4. Multi-light camera collections. SDNB = simulated VIIRS DNB.

Observed Scene	Number	% Flicker	Flicker Index	Dispersion	Flicker Hz	DNB % Flicker	DNB Flicker Index	DNB Dispersion
Table Multilights	C0005	8.38	0.5277	0.1119	120	6.67	0.4953	0.072
Golden Panorama	C0040	8.43	0.5178	0.1972	120	9.92	0.5069	0.1581
Broomfield Civic Center	C0146	4.79	0.5689	0.0991	120	4.36	0.5214	0.06
Boulder Panorama	C0138	5.09	0.5274	0.0585	120	4.99	0.4956	0.0349
Moscow Drone	D0014	3.61	0.5008	0.0228	100	2.45	0.5026	0.0101

3.5. Examination of DNB Temporal Profiles of Tucson, Arizona

The VIIRS DNB profiles for five grid cells from the Tucson metropolitan area are shown in Figure 28. Percent flicker, flicker index, and index of dispersion calculations were made for each of the five temporal profiles to examine the value of the indices in detecting changes in flicker. For each site, one year is reported as a reference and a second year is reported as a subject. The results are shown in Table 5.

Table 5. Flicker index calculations for five Tucson grid cells.

Location	Cell Number	Reference Year	Ref. % Flicker	Ref. Flicker Index	Ref. Dispersion	2017 % Flicker	2017 Flicker Index	2017 Dispersion	LED Conversions
S. Kino Parkway and 22nd	4738	2015	21.21	0.5797	1.88	23.29	0.5148	0.91	61
Automall	2730	2015	63.77	0.677	13.92	59.09	0.6725	9.01	29
HomeGoods Distribution Center	6746	2014	99.99	0.6175	4.15	41.63	0.5644	10.87	0
Airport terminal	6841	2013	22.59	0.657	15.17	36.06	0.6457	7.13	0
Mt. Lemmon	100	2015	99.97	0.6426	0.422	99.96	0.6554	0.404	0

The South Kino Parkway grid cell covers a lit roadway and a commercial zone with streets and buildings. It had 61 LED streetlight conversions. The temporal profile started out in 2012 with a mean radiance of 46 nW and a variance of 141, more than three times the mean. There was an upward radiance trend from 2012 to 2015, leading to a 2015 mean of 62 nW and a variance of 117. In 2016, there was a rapid decline in both radiance and variance, likely in response to the reduction in streetlight voltages after midnight and the near absence of flicker in the LED streetlights. The radiant emission pattern remained stable from 2017 to 2021.

The Tucson Automall grid cell is centered on a large, brightly lit parking lot featuring a large number of metal halide and high-pressure sodium luminaires. Multiple car dealerships are located together and are reported to begin turning off portions of their lighting after the 8 p.m. showroom closures. The grid cell had 29 LED streetlight conversions. The temporal profile is distinctive relative to the others in Figure 28 for its high level of variance across all years, fluctuating from a high of 1796 in 2013 to a low of 568 in 2018. As with the Kino Parkway grid cell, there is a decline in the mean radiance in 2016 during the

LED conversion period. The mean radiance drops from 101 nW in 2015 to 72 nW in 2018. Following this, the mean radiance then slowly rebounds to 77 nW in 2021.

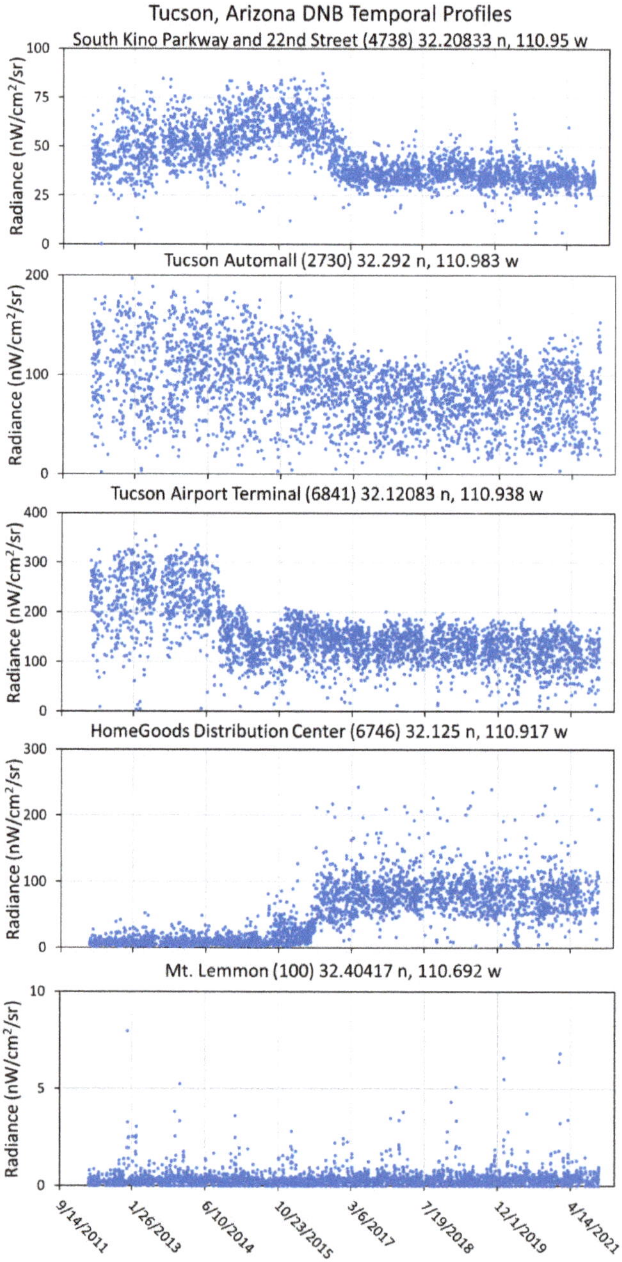

Figure 28. VIIRS DNB temporal profiles for five grid cells from Tucson, Arizona.

The Tucson International Airport (TUS) terminal grid cell contains the terminal building, parking lots, and a portion of the runway. It is one of the brightest locations found in Tucson, with zero LED streetlight conversions. The mean brightness started out with a mean of 220 nW and variance of 3636 in 2012. The mean increased slightly until mid-2014, at which point there was a sharp dip in radiance which continued until mid-2015, a period of renovation with the conversion of many metal halide and high-pressure sodium outdoor lights to highly-shielded LEDs. Once the construction was complete, there was a slight radiance recovery in late 2015, and the emissions then stabilized to means near 120 nW and variances near 900 from 2017 to 2021.

The HomeGoods Distribution Center is an 850,000 square foot facility that held its grand opening on 14 October 2016 [25]. The DNB temporal profile shows a relatively modest and stable level of lighting present from 2012 to July 2015, with mean radiances near 10 nW and variances ranging from 35 to 50. The brightness dropped slightly in August of 2015 and then bumped up with the radiances frequently rising to the 20–30 nW range for the next year—the facilities construction phase. The brightness increased again at the end of June 2016, prior to the grand opening, and remained in a largely stable state, with mean radiances in the 80 to 90 nW range and variances in the 880 to 1000 range through 2021. There was a month-long radiance drop in April 2020, likely the result of a site-wide shut down during the COVID-19 pandemic.

The fifth individual grid cell being examined is representative of the unlit background, taken from the core of Mt. Lemmon, part of the Coronado National Forest. The DNB temporal profile features low mean radiances, in the range of 0.3 to 0.5 nW, indicating the absence of detectable lighting. Variances were also low, under 1 in each of the years. There was a series of radiance spikes, up to 8 nW on a winter night with high lunar illuminance, likely the result of moonlight reflecting off snow.

3.6. Examination of Annual Mean, Variance, and Dispersion across Tucson

The grid cells from the Tucson DNB temporal profile grid were divided into three zones: background (no lighting detected), lit and no-LED conversions, and lit with LED conversions. In this section, we examine the behavior of the lighting in the three zones prior to and following the completion of most of the LED conversions using the annual mean, variance, and index of dispersion from 2015 and 2017.

Figure 29 shows the scattergrams of the annual mean radiances for the three zones. There was a slight increase in the mean radiance for 71.6% of the grid cells in the background zone. The mean radiances were largely unchanged in the lit grid cells with no LED streetlight conversions. In contrast, the mean radiance declined for 97% of the grid cells with LED conversion from 2015 to 2017.

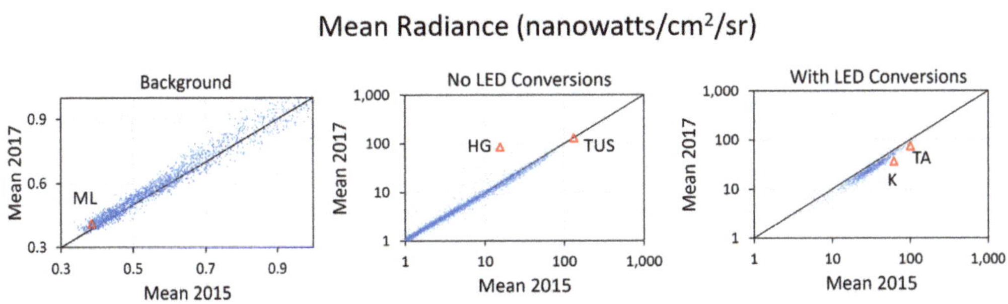

Figure 29. Mean radiance scattergrams for the three zones for 2015 versus 2017. The diagonal lines indicate equal values in the two years. There was a minor increase in the mean radiance in 2017 for the background grid cells. The mean radiance stayed nearly constant for lit grid cells lacking LED streetlight conversions, tracking the diagonal. Mean radiance declined in 97% of the grid cells with LED streetlight conversions. Red triangles correspond to the five Tucson DNB profiles shown in Figure 28: ML = Mount Lemmon; HG = HomeGoods Distribution Center; TUS = Tuscon International Airport; K = South Kino Parkway; TA = Tucson Automall.

Figure 30 shows the scattergrams of the annual variance for the three zones for 2015 versus 2017. The grid cells in the background zone changed very little in variance from 2015 to 2017, with the data cluster centered on the diagonal. There was a discernable decline in variance for a small portion of the background grid cells. For the lit zone with no-LED conversions, 74.6% of the lit grid cells had a decline in variance. Amongst the grid cells with LED streetlight conversions, 92.6% had a decline in variance in 2017 relative to 2015.

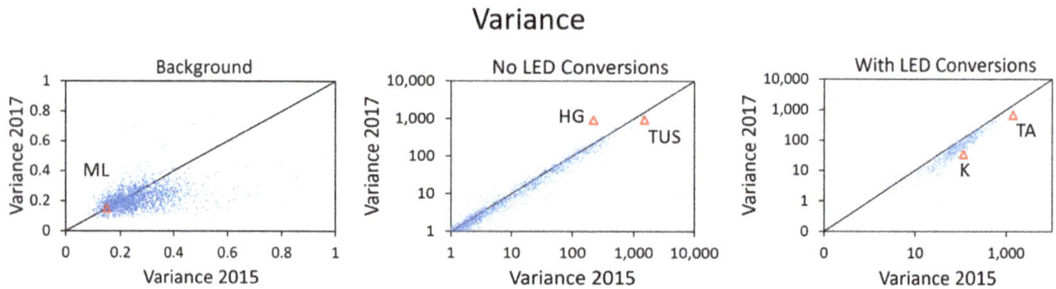

Figure 30. Variance scattergrams from the three zones for 2015 versus 2017. The diagonal lines indicate equal values in the two years. The grid cells in the background zone were largely stable in variance, with only a small portion of the grid cells dropping slightly in variance in 2017 relative to 2015. For the lit grid cells with no LED streetlight conversion, 74.6% had a slight decline in variance in 2017. In contrast, 92.6% of the grid cells with LED streetlight conversions had a decline in variance in 2017. Red triangles correspond to the five Tucson DNB profiles shown in Figure 28: ML = Mount Lemmon; HG = HomeGoods Distribution Center; TUS = Tuscon International Airport; K = South Kino Parkway; TA = Tucson Automall.

Figure 31 shows the scattergrams of the annual index of dispersion for the three zones for 2015 versus 2017. There was a slight decrease in the dispersion for the background zone. The dispersion is largely unchanged in the lit zone with no LED streetlight conversions. There is a decline in the dispersion in 2017 compared to 2015 for 79.3% of the grid cells with LED streetlight conversions.

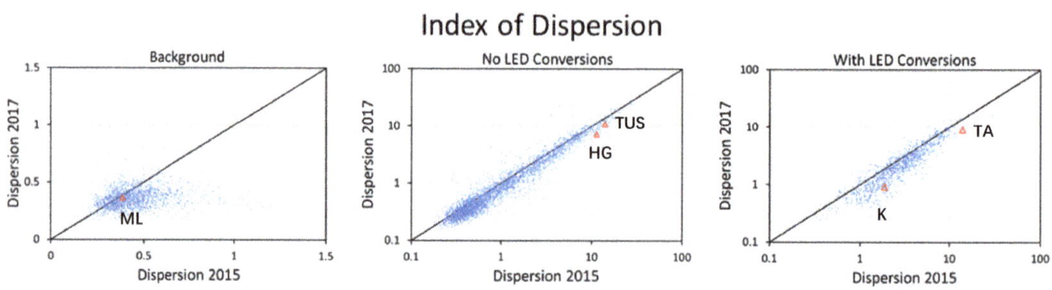

Figure 31. Index of dispersion scattergrams for the three zones for 2015 versus 2017. The diagonal lines indicate equal values in the two years. The grid cells in the background zone were largely stable in dispersion, with only a small portion of the grid cells dropping slightly in variance in 2017 relative to 2015. For the lit grid cells with no LED streetlight conversion, 74.8% had a decline in dispersion in 2017. In contrast, 79.3% of the grid cells with LED streetlight conversions had a decline in dispersion in 2017. Red triangles correspond to the five Tucson DNB profiles shown in Figure 28: ML = Mount Lemmon; HG = HomeGoods Distribution Center; TUS = Tuscon International Airport; K = South Kino Parkway; TA = Tucson Automall.

4. Discussion

The recalculation of the VIIRS DNB HGS pixel aggregation time finds that the pixel collection times vary from 2.25 to 2.68 ms. This is substantially shorter than the 8.3 ms cycle for the lighting flicker associated with a 60 Hz alternating current. Of the lighting types measured, flicker is most prominent in high-intensity discharge (HID) luminaires such as high-pressure sodium and metal halide, which are widely used in outdoor lighting. Compared to HID sources, flicker is reduced in incandescent and fluorescent lights. Fluorescent light fixtures frequently include capacitors, which reduce the voltage fluctuations that cause flicker. The thermal inertia of incandescent filaments serves as an inbuilt capacitor, reducing flicker to a greater degree than HID sources. Over the years, LED manufacturers have invested in the development of "constant-current" drivers [26] that have largely eliminated flicker in today's LED streetlights.

Evidence for the effects of flicker can be seen in VIIRS DNB temporal profiles from Tucson sites where LED conversion projects replaced HID streetlights with low-flicker LED sources. Both the South Kino Parkway and Tucson Airport temporal profiles start out with high radiance levels and high variance. The conversion period is indicated by noticeable drops in radiance levels. Upon completion of the LED conversions, the radiance levels stabilize at a reduced level and both variance and dispersion are reduced. The radiance drop is likely the result of voltage reductions after midnight, plus the improved shielding of the new LED streetlights. In the grid cells where the variance reductions exceed that radiance reduction, there is a decline in dispersion. This occurred in 79.3% of the Tucson grid cells with LED streetlight conversions. This decline in the dispersion indicates that the flicker was reduced when the high-flicker HID lights were replaced with low-flicker LED lights.

Three indices were tested for their ability to detect changes in the flicker levels in VIIRS DNB temporal profiles, including percent flicker, flicker index, and the index of dispersion. Percent flicker and the flicker index are standard measurements developed to detect and quantify flicker by the lighting engineering community. All three indices performed well for the individual luminaires listed in Table 3. For the VIIRS DNB profiles, the index testing contrasted the values from a reference year and a subject year, selected to straddle an obvious change in the lighting behavior. The DNB profile index results (Table 4) indicate that the percent flicker and flicker index did not perform as well as the index of dispersion. The grid cell from Figure 16 with the highest percent flicker is from Mt. Lemmon—a background area devoid of lighting. The percent flicker remained nearly constant for the South Kino Parkway grid cell, which had 61 LED streetlight conversions, resulting in a substantial reduction in variance. Similar issues were found for the flicker index, which jumped by two orders of magnitude for the Mt. Lemmon grid cell and remained largely unchanged for the South Kino Parkway and Tucson Airport grid cells, which had large numbers of HID-to-LED conversions. Our assessment is that the index of dispersion is better than the percent flicker and flicker index for the detection and analysis of flicker in VIIRS DNB temporal profiles.

5. Conclusions

Back to the original question, is the VIIRS DNB a flicker meter in space? The answer is "yes" and "no". The DNB HGS pixel integration time is more than three times shorter than the 100 to 120 Hz flicker found in lighting. However, VIIRS is incapable of measuring the flicker rates in Hz due to the fact that the DNB's repeat cycle is measured in hours rather than milliseconds. The VIIRS DNB differs from conventional flicker meters in that it is unable to collect coherent flicker cycles from spaces. Instead, the DNB integrates radiances from all of the surface lighting present in pixel footprints and collects for 2–3 ms nightly with hour- to day-long gaps between observations. This implies that if flicker is present at the pixel footprint level, the recorded VIIRS radiance will fluctuate unpredictably over time, making it a source of instantaneous dispersion. This instability is expressed as an enhanced level of variance in long-term DNB temporal profiles. Because the VIIRS pixel footprints are about half a square kilometer, there can be many individual lights present and contributing to the radiance observed by the DNB. If the flickering lights are not synchronized, the observable radiance pattern is an erratic cacophony, with multiple offset flicker patterns tending to cancel each other out. An AC flicker pattern can emerge from the cacophony, though it will be a weak signal compared to that found in individual sources.

Several recent studies [15,27] have enumerated the factors that introduce variance into VIIRS DNB temporal profiles, such as the view angle, atmospheric variability, cloud-effects, cycling of lunar illuminance, snow cover, and even deciduous tree phenology. To this, we should add flicker and the variability of flicker within DNB pixel footprints that are assembled in the construction of DNB temporal profiles. The distinction between flicker and these other variables is that flicker is an instantaneous source of dispersion, and the other sources operate with longer time constants. Together, these variables produce a cloud of radiance dispersion in long-term DNB temporal profiles. The stability of DNB temporal profiles can be improved by adjusting the radiance levels for individual pixels designed to remove these influences. The purpose of cloud-screening is to reduce the radiance dispersion induced by cloud obscuration, a variability that is unrelated to actual surface lighting. Satellite zenith angle effects can be analyzed and removed via normalization [15]. It is also possible to isolate and remove fluctuations associated with lunar, weekly and annual cycling from DNB temporal profiles. However, no amount of adjustment for the environmental and geometric effects on DNB radiances can remove the effects of lighting flicker. Thus, flicker, if present at the pixel level, imprints itself indelibly on the DNB temporal profiles and sets an implicit limitation on the uncertainty of applications that utilize the DNB observations of artificial lights.

It is possible to collect data on flicker levels with high-speed cameras. A frame rate of 1000 per second makes it possible to observe flicker patterns in good detail. It appears feasible to collect quantitative flicker information for footprints approximating the DNB with high-speed cameras from hovering drones. One caution is that it is important to test the camera for the presence of electronic flicker by viewing the moon or daylit surface devoid of AC flicker.

The data from a high-speed camera collected on a wide range of luminaires indicate that high-intensity discharge (HID) luminaries, such as high- and low-pressure sodium and metal halides, are particularly prone to high levels of flicker. In contrast, modern LED luminaires are virtually devoid of flicker. This fact should be considered when designing ground-based DNB calibration sources. LED sources are vastly better than HID sources for DNB calibration sources. This fact casts doubt on the value of DNB calibration efforts where HID sources are utilized [28,29].

We examined the DNB temporal profiles from grid cells known to have had HID-to-LED conversions in Tucson, Arizona. The DNB profiles reveal the conversion date range, and the reduction in flicker effects that occurs when temporally unstable HID luminaires are replaced with steady LED sources. Three annual indices were tested for detecting and characterizing flicker in DNB temporal profiles, including percent flicker, flicker index, and the index of dispersion. The preliminary results indicate that the index of dispersion is the best of the three. The flicker index values calculated from the DNB temporal profiles have a compressed range and show little change between the pre- and post-LED streetlight conversion radiance profiles. Percent flicker operates suitably in lit grid cells but produces anomalously high values for background grid cells.

It should be noted that the flicker effects on VIIRS DNB data cannot be detected or analyzed using the current VIIRS monthly or annual nighttime light products, which only report the average cloud-free radiance [30,31]. The analyses from Tucson were based on nightly DNB temporal profiles. Our results indicate that the information value of monthly and annual VIIRS DNB products can be improved by producing the full suite of statistical moments (mean, variance, skew, and kurtosis). The mean and variance can be further processed to calculate the index of dispersion.

Over time, outdoor lighting flicker is being reduced as high-pressure sodium and metal halide sources are replaced with constant-current LED streetlights [32]. The results from Tucson indicate that the loss of flicker and dimming of the upwelling radiances associated with LED conversions can be detected and quantified with VIIRS DNB temporal profiles. While the transition to LED is primarily based on economics, there may be human health impacts. Several studies have noted the adverse health impacts of white LEDs, where the blue component is enhanced [33,34]. In contrast, the reduction in flicker associated with LED conversions may have beneficial effects on humans, a possibility that should be further investigated.

Our results shed some light on the sensor requirements for the observation of flicker rates and amplitudes. Flicker can be detected and analyzed with high-speed camera data. Such data can be collected on the ground or from the air. The collection platform should be stabilized with a tripod on the ground or by hovering in the air. Airborne surveys of flicker will be the easiest way to investigate the spatial patterns of flicker synchronization associated with multiple lights operating from a common AC power supply. It may be possible to map boundaries between power grid entities based on the discontinuities in flickering. From satellites, it may be possible to observe the flicker rates and amplitudes with high-speed, low-detection-limit cameras capable of staring at lights present on the earth's surface. This could be done from geostationary orbit or with pointable sensors in lower orbit.

In conclusion, VIIRS DNB-measured radiances from artificial light sources are affected by several phenomena, including flicker. The flicker effects can be concentrated by filtering out the pixels contaminated by cloud cover and adjusting the DNB radiances for effects such as lunar reflectance, view angle, and atmospheric variability. In addition, outlier and trend removal may be required to achieve consistent results. After the extraneous effects are removed or compensated for, DNB temporal profiles can reveal details of lighting development and conversions from high-flicker to low-flicker sources.

Author Contributions: Conceptualization, data collection, analysis, figure preparation, and writing: C.D.E.; data collection and analysis: M.Z.; data collection and provision of luminaires: D.K.; data collection, manuscript review, and editing: S.D.M., S.J.A., P.C.S. and D.H.; data collection and DNB temporal profile generation: F.C.H.; details on Tucson study sites: J.B., C.K.M. and C.C.M.K.; conceptualization and administrative support: M.B.; conceptualization: J.T.; manuscript review and editing: J.B., W.S.K., D.W.P., C.C.M.K., T.G. and D.H. All authors have read and agreed to the published version of the manuscript.

Funding: This research was funded by the Rockefeller Foundation, grant number 2018 POW 004 (Prof. Jay Taneja, principal investigator) and the NOAA Joint Polar Satellite System (JPSS) proving ground program project on VIIRS detection of COVID impacts on lighting (Prof. Steve Miller, principal investigator). Dr. Kyba is funded by Helmholtz Association Initiative and Networking Fund (CS-003).

Data Availability Statement: The VIIRS DNB temporal profile grid of Tucson is available at: https://eogdata.mines.edu/wwwdata/hidden/dnb_profiles_deliver_licorr/usa_tucson2/ (accessed on 6 March 2022).

Acknowledgments: The VIIRS data used in the research are the result of many years of development and satellite operations sponsored jointly by NASA and NOAA. Funding for the study came from NOAA's Joint Polar Satellite System (JPSS) proving ground study on COVID effects on lighting (Steven Miller, principal investigator) and the Rockefeller Foundation's e-guide project, grant 2018 POW 004 (Jay Taneja, principal investigator). Jeff Puschell of Raytheon Corporation assisted in clarifying the DNB pixel aggregation timeline. We are thankful to Eng. Artem Khuramshin (Hewlett Packard, Russia) for providing and operation of the drone during the slow-motion filming of streetlights in Moscow.

Conflicts of Interest: The authors declare no conflict of interest.

References

1. Mirra, C. *Flicker Measurement and Evaluation*; Union Internationale d'Electrothermie, Group de Travail Perturbations: Paris, France, 1986.
2. Fan, Y.H.; Wu, C.J.; Fan, C.C.; Chih, K.W.; Liao, L.D. A Novel Time Based Current Compensator for LED Back Light Modules. *Key Eng. Mater.* **2007**, *364–366*, 377–382. [CrossRef]
3. Bartley, S.H. The neural determination of critical flicker frequency. *J. Exp. Psychol.* **1937**, *21*, 678. [CrossRef]
4. Wilkins, A.; Veitch, J.; Lehman, B. LED lighting flicker and potential health concerns: IEEE standard PAR1789 update. In Proceedings of the 2010 IEEE Energy Conversion Congress and Exposition, ECCE, Atlanta, GA, USA, 12–16 September 2010; pp. 171–178. [CrossRef]
5. Wilkins, A.J. Health and efficiency in lighting practice. *Energy* **1993**, *18*, 123–129. [CrossRef]
6. Price, L.L.A. Can the Adverse Health Effects of Flicker from LEDs and Other Artificial Lighting Be Prevented? *LEUKOS* **2017**, *13*, 191–200. [CrossRef]
7. Davis, J.; Hsieh, Y.-H.; Lee, H.-C. Humans perceive flicker artifacts at 500 Hz. *Sci. Rep.* **2015**, *5*, 7861. [CrossRef]
8. IEEE Recommended Practices for Modulating Current in High-Brightness LEDs for Mitigating Health Risks to Viewers. IEEE-SA Standards Board, 26 March 2015. Available online: http://www.bio-licht.org/02_resources/info_ieee_2015_standards-1789.pdf (accessed on 5 March 2022).
9. Takahashi, T.; Tsukahara, Y.; Nomura, M.; Matsuko, H. Pokemon Seizures. *Neurol. J. Southeast Asia* **1999**, *4*, 1–11. Available online: http://www.neurology-asia.org/articles/19991_001.pdf (accessed on 5 March 2022).
10. Inger, R.; Bennie, J.; Davies, T.W.; Gaston, K.J. Potential biological and ecological effects of flickering artificial light. *PLoS ONE* **2014**, *9*, e98631. [CrossRef]
11. Schakel, M.; Banerjee, K.; Bergen, T.; Blattner, P.; Bouroussis, C.; Dekker, P.; Klej, A.; Li, C.; Ootake, H.; Reiners, T.; et al. *Guidance on the Measurement of Temporal Light Modulation of Light Sources and Lighting Systems*; Technical Note 12; International Commission on Illumination: Vienna, Austria, 2021. [CrossRef]
12. Baker, N.; Joint Polar Satellite System (JPSS); Visible Infrared Imaging Radiometer Suite (VIIRS); Sensor Data Records (SDR); Algorithm Theoretical Basis Document (ATBD). Goddard Space Flight Centre, Greenbelt Maryland, Ground Project Code, 474,474-00053. 22 April 2011. Available online: https://www.star.nesdis.noaa.gov/jpss/documents/ATBD/D0001-M01-S01-004_JPSS_ATBD_VIIRS-Geolocation_A.pdf (accessed on 7 March 2022).
13. Elvidge, C.D.; Keith, D.M.; Tuttle, B.T.; Baugh, K.E. Spectral identification of lighting type and character. *Sensors* **2010**, *10*, 3961–3988. [CrossRef]
14. Box, G.E.P.; Jenkins, G.M.; Reinsel, G.C. Autocorrelation Function and Spectrum of Stationary Processes. In *Time Series Analysis*; Wiley Online Library: Hoboken, NJ, USA, 2008; ISBN 9781118619193.
15. Elvidge, C.D.; Hsu, F.-C.; Zhizhin, M.; Ghosh, T.; Taneja, J.; Bazilian, M. Indicators of electric power instability from satellite observed nighttime lights. *Remote Sens.* **2020**, *12*, 3194. [CrossRef]
16. Hsu, F.; Zhizhin, M.; Ghosh, T.; Elvidge, C.; Taneja, J. The Annual Cycling of Nighttime Lights in India. *Remote Sens.* **2021**, *13*, 1199. [CrossRef]
17. Ord, J.K.; Getis, A. Testing for local spatial autocorrelation in the presence of global autocorrelation. *J. Reg. Sci.* **2001**, *41*, 411–432. [CrossRef]
18. The Illuminating Engineering Society of North America. *The Lighting Handbook: Reference and Application*, 10th ed.; IESNA: New York, NY, USA, 2011; 1328p.
19. Selby, B. The index of dispersion as a test statistic. *Biometrika* **1965**, *52*, 627–629. [CrossRef]
20. Kopp, T.J.; Thomas, W.; Heidinger, A.K.; Botambekov, D.; Frey, R.A.; Hutchison, K.D.; Iisager, B.D.; Brueske, K.; Reed, B. The VIIRS Cloud Mask: Progress in the First Year of S-NPP toward a Common Cloud Detection Scheme. *J. Geophys. Res. Atmos.* **2014**, *119*, 2441–2456. [CrossRef]
21. Barentine, J.C.; Walker, C.E.; Kocifaj, M.; Kundracik, F.; Juan, A.; Kanemoto, J.; Monrad, C.K. Skyglow changes over Tucson, Arizona, resulting from a municipal LED street lighting conversion. *J. Quant. Spectrosc. Radiat. Transf.* **2018**, *212*, 10–23. [CrossRef]
22. Barentine, J.C.; Kundracik, F.; Kocifaj, M.; Sanders, J.C.; Esquerdo, G.A.; Dalton, A.M.; Foott, B.; Grauer, A.; Tucker, S.; Kyba, C.C.M. Recovering the city street lighting fraction from skyglow measurements in a large-scale municipal dimming experiment. *J. Quant. Spectrosc. Radiat. Transf.* **2020**, *253*, 107120. [CrossRef]
23. Kyba, C.C.M.; Ruby, A.; Kuechly, H.U.; Kinzey, B.; Miller, N.; Sanders, J.; Barentine, J.; Kleinodt, R.; Espey, B. Direct measurement of the contribution of street lighting to satellite observations of nighttime light emissions from urban areas. *Light. Res. Technol.* **2021**, *53*, 189–211. [CrossRef]
24. Dick, R. Chapter 6—Harmonic Oscillators and Coherent States. In *Advanced Quantum Mechanics*; Springer: Berlin/Heidelberg, Germany, 2012.
25. City of Tucson. HomeGoods Distribution Center Holds Grand Opening. 16 October 2016. Available online: https://www.tucsonaz.gov/ward-5/news/homegoods-distribution-center-holds-grand-opening (accessed on 28 November 2021).
26. Dazhong, S.H.Z. Constant Current Driver for Semiconductor Lighting. *Res. Prog. SSE Solid State Electron.* **2006**, *2*.
27. Wang, Z.; Román, M.O.; Kalb, V.L.; Miller, S.D.; Zhang, J.; Shrestha, R.M. Quantifying uncertainties in nighttime light retrievals from Suomi-NPP and NOAA-20 VIIRS Day/Night Band data. *Remote Sens. Environ.* **2021**, *263*, 112557. [CrossRef]

28. Ryan, R.E.; Pagnutti, M.; Burch, K.; Leigh, L.; Ruggles, T.; Cao, C.; Aaron, D.; Blonski, S.; Helder, D. The Terra Vega active light source: A first step in a new approach to perform nighttime absolute radiometric calibrations and early results calibrating the VIIRS DNB. *Remote Sens.* **2019**, *11*, 710. [CrossRef]
29. Cao, C.; Bai, Y. Quantitative analysis of VIIRS DNB nightlight point source for light power estimation and stability monitoring. *Remote Sens.* **2014**, *6*, 11915–11935. [CrossRef]
30. Elvidge, C.D.; Baugh, K.; Zhizhin, M.; Hsu, F.C.; Ghosh, T. VIIRS night-time lights. *Int. J. Remote Sens.* **2017**, *38*, 5860–5879. [CrossRef]
31. Elvidge, C.D.; Zhizhin, M.; Ghosh, T.; Hsu, F.-C.; Taneja, J. Annual time series of global VIIRS nighttime lights derived from monthly averages: 2012 to 2019. *Remote Sens.* **2021**, *13*, 922. [CrossRef]
32. Weinold, M. A long overdue end to flicker: The 2020 EU lighting efficiency regulations. *Camb. J. Sci. Policy* **2020**, *1*, e0628620899.
33. Bullough, J.D. The blue-light hazard: A review. *J. Illum. Eng. Soc.* **2000**, *29*, 6–14. [CrossRef]
34. Bullough, J.D.; Bierman, A.; Rea, M.S. Evaluating the blue-light hazard from solid state lighting. *Int. J. Occup. Saf. Ergon.* **2019**, *25*, 311–320. [CrossRef]

remote sensing MDPI

Technical Note

Evaluating Road Lighting Quality Using High-Resolution JL1-3B Nighttime Light Remote Sensing Data: A Case Study in Nanjing, China

Nuo Xu, Yongming Xu *, Yifei Yan, Zixuan Guo, Baizhi Wang and Xiang Zhou

School of Remote Sensing and Geomatics Engineering, Nanjing University of Information Science and Technology, Nanjing 210044, China
* Correspondence: xym30@nuist.edu.cn

Abstract: A good lighting environment for roads at night is essential for traffic safety. Accurate and timely knowledge of road lighting quality is meaningful for the planning and management of urban road lighting systems. Traditional field observations and mobile observations have limitations for road lightning quality evaluation at a large scale. This study explored the potential of 0.92 m resolution JL1-3B nighttime light remote sensing images to evaluate road lighting quality in Nanjing, China. Combined with synchronous field measurements and JL1-3B data, multiple regression and random forest regression with several independent variable combinations were developed and compared to determine the optimal model for surface illuminance estimation. Cross validation results showed that the random forest model with Hue, saturability, ln(Intensity), ln(Red), ln(Green) and ln(Blue) as the input independent variables had the best performance (R^2 = 0.75 and RMSE = 9.79 lux). Then, this model was used to map the surface illuminance. The spatial scopes of roads were extracted from Google Earth images, and the illuminance within roads was derived to calculate the average, standard deviation and coefficient of variation to indicate the overall brightness level and brightness uniformity of the roads. This study provides a quantitative and effective reference for road lighting evaluation.

Keywords: nighttime light remote sensing; road lighting quality measurements; JL1-3B; Nanjing

Citation: Xu, N.; Xu, Y.; Yan, Y.; Guo, Z.; Wang, B.; Zhou, X. Evaluating Road Lighting Quality Using High-Resolution JL1-3B Nighttime Light Remote Sensing Data: A Case Study in Nanjing, China. *Remote Sens.* **2022**, *14*, 4497. https://doi.org/10.3390/rs14184497

Academic Editors: Ran Goldblatt, Steven Louis Rubinyi and Hogeun Park

Received: 8 July 2022
Accepted: 6 September 2022
Published: 9 September 2022

1. Introduction

In night driving, road night lighting plays an important role in traffic safety. The International Commission on Illumination (CIE) reported that road lighting reduced nighttime accidents by 13~75% over 15 countries [1]. Christopher found that the reported crashes where roads were lit decreased 28.95% in total crashes and 39.21% in injury night crashes [2]. Elvik et al. indicated that road lighting reduced the nighttime crash rate by 23% in Belgium, Britain and Sweden [3]. William found that improving overall uniformity up to approximately 0.4 lowers the night-to-day crash ratio for highways in New Zealand [4]. Good road lighting quality provides a good lit environment, thus reducing traffic crashes at night. Consequently, evaluating road lighting quality is meaningful to improve the nighttime road lit environment and gain urban traffic safety [5–9].

Road lighting quality reflects the photometric performance of road lights aiming at satisfying drivers' visual needs at night, which includes parameters such as average illuminance, overall uniformity, etc., according to CIE [10] and European Committee for Standardization [11] documents.

Field observation is the most popular way to evaluate road lighting quality. Generally, illuminance meters or imaging photometers are employed to measure road brightness at typical places to assess lighting quality. Liu et al. utilized different illuminance meters and handheld luminance meters to measure the illuminance in different orientations and the brightness of roads in Dalian, China [12]. Guo et al. measured the luminance of roads in Espoo, Finland using an imaging photometer LMK Mobile Advanced (IPLMA), which converted photos into luminance values, and calculated the average luminance by the software LMK 2000 [13]. Jägerbrand also applied the IPLMA to obtain road lighting parameters in Sweden [14]. Ekrias et al. used an imaging luminance photometer ProMetric 1400 to measure the luminance of roads in Finland and calculated the average luminance and uniformities [15]. However, static road lighting measurements cannot provide lighting information at large scales. Additionally, observation positions must be placed within roads, which may affect the normal running of traffic flow [16–18].

Some researchers have carried out mobile measurements with imaging luminance devices or photometers boarded on vehicles. Greffier et al. used the High Dynamic Range (HDR) Imaging Luminance Measuring Device (ILMD) mounted on a car to measure road luminance [19]. Zhou et al. developed a mobile road lighting measurement system with a photometer mounted on a vehicle, which was able to record the illuminance and position data simultaneously and successfully applied in Florida, America [20]. However, mobile measurements are disturbed by many factors, such as the head and rear lights of cars, relative positions between observation vehicles and street lamps, and the vibration of observation angles.

Nighttime lighting remote sensing offers a unique way to monitor spatial-continuous nighttime nocturnal lighting at a large scale. Nighttime light remote sensing data, such as DMSP/OLS, NPP/VIIRS and Luojia1-01, are widely used in mapping urbanization processes [21], estimating GDP, investigating poverty and monitoring disasters [22]. However, due to the coarse resolutions of these nighttime satellite remote sensing data (DSMP/OLS: 2.7 km; NPP/VIIRS: 0.75 km; Luojia1-01: 130 m), they cannot map the detailed light environment within roads and therefore cannot be applied for road lighting quality evaluation, thus few studies have been carried out on road lighting environments based on nighttime light data However, there still are some studies. Cheng et al. used JL1-3B nighttime light data to extract and classify the street lights by a local maximum algorithm, achieving an accuracy of above 89% [23]. Zheng et al. used the multispectral feature of JL1-3B data to discriminate light source types using ISODATA algorithm, with an overall accuracy of 83.86% [24]. An unmanned aerial vehicle (UAV) is able to provide high-resolution nighttime lighting images on a relatively large scale. Rabaza et al. used a digital camera onboard the UAV to capture orthoimages of the lit roads in Deifontes, Spain, which was calibrated by a known luminance relationship. Then, the average luminance or illuminance was calculated [25]. UAVs have the outstanding advantages of convenience, low cost and extremely high resolution. However, it is also limited by security, privacy factors, and the short flight distance makes it unsuitable for large areas. The newly launched JL1-3B satellite provides nighttime images with a high spatial resolution of 0.92 m and multispectral data [24,26]. The features of JL1-3B show the potential of evaluating road lighting quality at large scales.

This study aimed to evaluate the road lighting quality in Nanjing, China, using high-resolution JL1-3B nighttime light remote sensing data combined with the in situ measured illuminance on typical roads. Several machine learning models were developed and compared to produce a fine resolution illuminance map with good accuracy. Then, the remotely sensed illuminance within roads was extracted to calculate indicators to evaluate road lighting quality.

2. Study Area and Datasets

2.1. Study Area

Nanjing, the capital city of Jiangsu Province, China, is located in eastern China. It covers a total area of 6587.02 km^2, with latitudes ranging from 31°14′ to 32°37′N and longitudes ranging from 118°22′ to 119°14′E. In recent decades, Nanjing has experienced significant economic and population growth and, accordingly, a rapid increase in the number of motor vehicles. In 2020, the total number of motor vehicles reached 2,799,469, and the total length of roads reached 9796 km [27]. It is difficult to evaluate the lighting quality of such a vast road network by timely in situ observations. Therefore, exploring an effective method for road lighting assessment at a large scale is very meaningful for Nanjing city.

2.2. Data Collection

2.2.1. Remote Sensing Data

JL1-3B, known as "Jilin-1 03B", which was developed by Changguang Satellite Technology Co., Ltd. (Changchun, China) and launched on 9 January 2017. This satellite is on a sun-synchronous orbit with an altitude of 535 km, which provides three imaging models: video imaging, push-broom imaging and night light imaging [23]. Night light imaging models provide radiometrically and geometrically calibrated high spatial resolution (0.92 m) images with multiple bands (red band: 580–723 nm, green band: 489–585 nm, blue band: 437–512 nm). Regions within an off-nadir angle of ±45° can be accessed by the area array camera boarded on the payload. Each tile of the image covers 11 km × 4.5 km of the ground area. Compared with DMSP-OPS, NPP-VIIRS and LJ1-01 nighttime light remote sensing satellites, the high spatial resolution of JL1-3B is capable of depicting the spatially detailed lit environment within road lights. Compared with the high-resolution nighttime light remote sensing satellite EROS-B (0.7 m), multispectral data are provided, allowing for more detailed studies relative to road lighting [23,24,26,28]. The JL1-3B data can be commercially ordered from the website of Changguang Satellite Technology Co., Ltd. (http://www.jl1.cn/Search.aspx?txtSearch=JL1-3B (accessed on 5 October 2020)), of which price is 180 CNY/km^2.

The JL1-3B image in this study was collected at 21:08:37 (Beijing Time) on 23 October 2020, which covered the northern part of the main urban area (Figure 1). The off-nadir angle was 0.02 degrees, avoiding the shielding effect of buildings on the road lit environment. The image was primarily geometrically and radiometrically corrected (Figure 2).

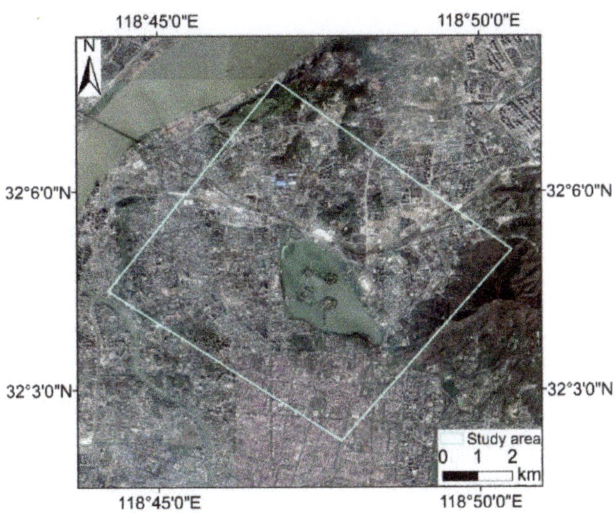

Figure 1. JL1-3B image range of study area without geometric correction (base map: Google Earth).

2.2.2. Field Measurements

Field observations were carried out from 20:30–21:30 on 23 October 2020, to coincide with the JL1-3B overpass time. A TES-1399R illuminance meter was employed to measure the in situ illuminance. Its measurement range is from 0.01 lux to 999,900 lux, and the accuracy is ±3% rdg (calibrated to standard incandescent lamp, 2856K). Five groups carried out observations on the foot with TES-1399R simultaneously on different representative routes, which covered different lighting conditions. At each sample point, the TES-1399R was facing-upward horizontally, placed at a height of 1.5 m to measure the downward illuminance from the street lamps. The locations of these sample points were also recorded by Global Navigation Satellite System (GNSS) instruments. To avoid the influence of the shading effect of street trees, only the sites that were not covered by trees were selected for observation. A total of 214 measurements were collected (Figure 2).

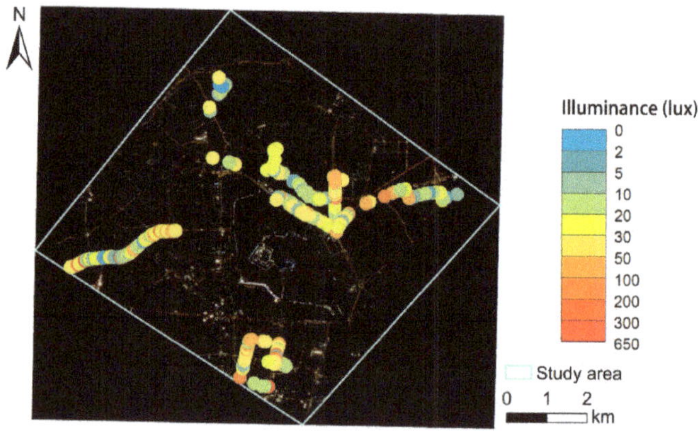

Figure 2. Distribution of measured points (base map: raw JL1-3B image).

3. Methodology

3.1. Workflow

The flowchart of this study is shown in Figure 3, which includes three main steps: (1) preprocessing the JL1-3B data; (2) developing multiple models and selecting the best of them to map illuminance from the JL1-3B image; and (3) evaluating road lighting quality based on the illuminance map.

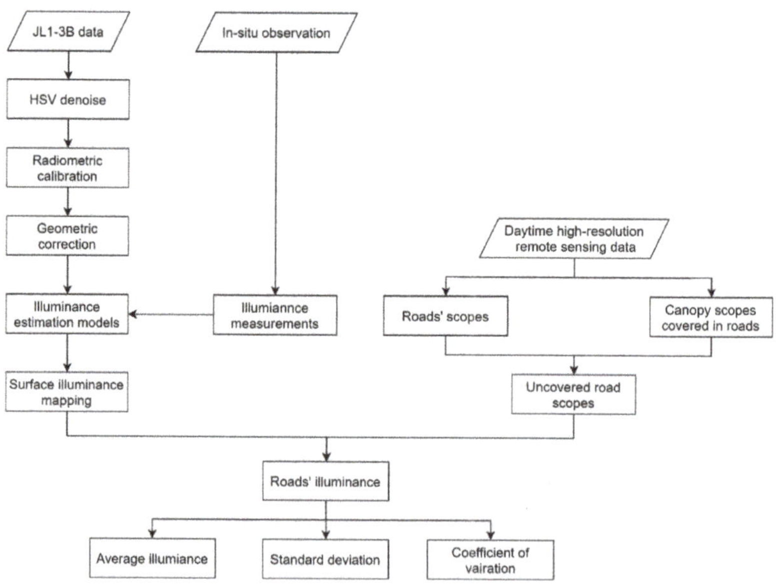

Figure 3. Flow chart.

3.2. JL1-3B Data Preprocessing

The noise was removed using the threshold method in the HSV color space. Noises consist of a patch of pixels with high values of the R, G or B band, which randomly cross the image. Considering that they are characterized by high saturation, the image was transformed from RGB to HSV space to identify noisy pixels by the following rules:

$$\{|H-120| < 10 \; or \; |H-240| < 10 \; or \; |H-360| < 10\} \; and \; \{S > 0.8\} \; and \; \{V > 0.55\}, \quad (1)$$

where H is the hue, S is the saturation and V is the value of the HSV space.

The noisy pixels were identified and then filled using Delaunay triangulation with triangles calculated from the surrounding valid DN values. Figure 4b shows the denoised image. After the denoising process, the monochromatic noisy pixels were effectively removed. Note that the denoising process should be performed before geometric registration because resampling during the registration will mix the monochromatic noisy values and normal values, making it difficult to distinguish noisy pixels.

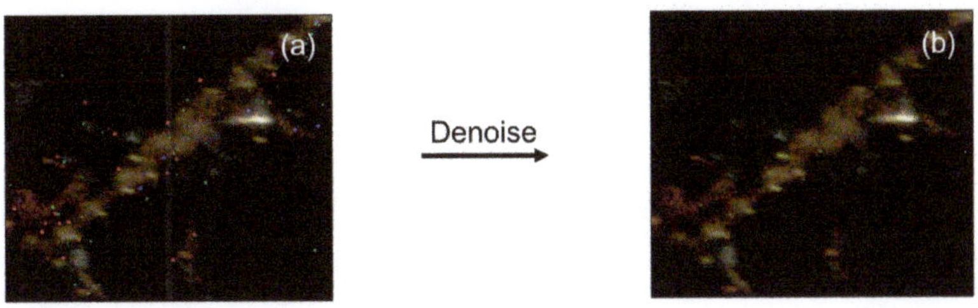

Figure 4. Comparison before (**a**) and after denoising (**b**).

The denoised JL1-3B image was calibrated in radiation according to Equation (2):

$$L_i = \frac{D_i - b_i}{a_i}, \quad (2)$$

where L_i is the radiance ($W \cdot m^{-2} \cdot sr^{-1}$) of band i, D_i is the DN value of band i, and a_i and b_i are the calibrated coefficients of band i (Table 1).

Table 1. Radiance calibration coefficients of JL1-3B.

Wavelength	a	b
Band 1 (Red)	9681	−4.73
Band 2 (Green)	5455	−3.703
Band 3 (Blue)	2997	−4.471

Although the JL1-3B L1A image has been systematically geometrically corrected, it still has obvious geometric deviations. Taking a high-resolution Google Earth image as the reference image, 26 ground control points were selected to register the JL1-3B image. It should be noted that the JL1-3B sensor is an array imaging sensor, and the JL1-3B image of the study area was mosaicked from multiple images. Therefore, the geometric distortion of the whole image varies from different source images. Under this consideration, the Delaunay triangulation method instead of the polynomial method was employed for image wrapping. Figure 5 shows the registered JL1-3B image.

During field observations, the surrounding environment is complex, and the measured values may be affected by a variety of factors, such as the headlights of passing vehicles, landscape lighting and window lights from nearby buildings. These interference factor may affect the consistency between surface observed illuminance and space-born observed illuminance. For example, the in-situ observations that were illuminated by vehicle headlights had relative high measurement values, but the corresponding pixel values in the

JL1-3B image were not high because the filed observation and the satellite overpass were not perfectly synchronous. Therefore, all the sample points were manually checked to remove problematic sample points to improve the estimation accuracy of illuminance. Finally, 168 sample points remained.

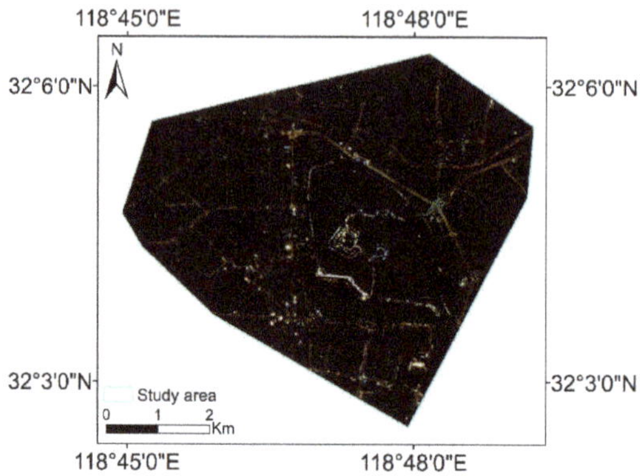

Figure 5. The JL1-3B image after preprocessing.

3.3. Illuminance Mapping

The emitted light from street lamps is reflected by the road surface and then propagates through the atmosphere to reach the satellite sensor. The road surface has similar reflection characteristics, and the atmospheric conditions are relatively uniform at a small scale. In this way, a close relationship between the in situ observed illuminance and the remotely sensed radiance should exist. Figure 6a gives the scatter plots between the observed illuminance and the radiance of the three JL1-3B bands. The illuminance has good relationships with the radiance of the R, G and B bands, with correlation coefficients of 0.638, 0.648, and 0.649, respectively. It can also be noted that there are many unique values in band B, which can be attributed to the relatively strong scattering intensity of the atmosphere for the blue band [29,30]. Figure 6b gives the scatter plots between the observed illuminance and the log values of the radiance of the three bands. The correlations between them were higher than those between the observed illuminance and the radiance values. In order to provide more relevant variables to the illuminance estimation model, HIS color space was introduced under the consideration that it is more robust to changes in illuminance [31]. Thus, the RGB color space was transferred to HIS color space, and the correlation coefficients between illuminance and H, I, S were −0.34, 0.67 and −0.37, respectively. After their logarithmic transformation, log(I) showed the highest correlation coefficient (0.77) with illuminance, and the Pearson coefficients of log(H) and log(S) were both −0.34.

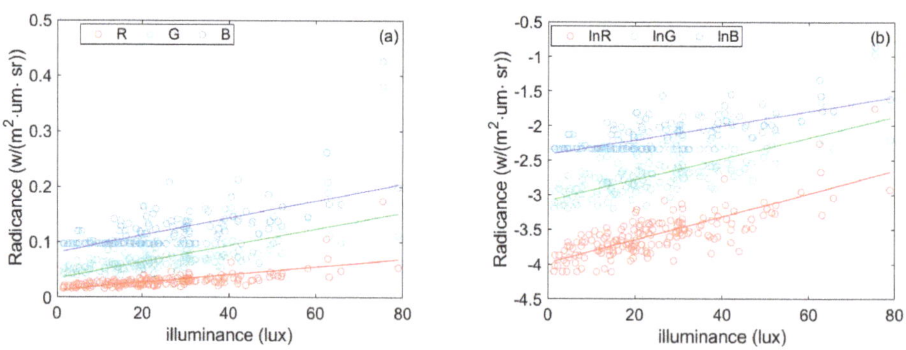

Figure 6. (**a**) RGB scatter plot; and (**b**) lnRGB scatter plot.

Relationships between illuminance and different color components varies [32]. Based on the correlation analysis above, R, G, B, I and their log values show high correlations with observed illuminance. Considering the auxiliary information that the variable Hue and Saturation can also provide, variables in RGB color space and HIS color space were integrated. Then, six combinations of these independent variables were employed (Table 2) to determine the optimal variables for illuminance estimation.

Table 2. Combinations of independent variables.

Combination	Independent Variables
1	R, G, B
2	lnR, lnG, lnB
3	H, S, I
4	H, S, lnI
5	H, S, I, R, G, B
6	H, S, lnI, lnR, lnG, lnB

To properly portray the quantitative relationships between variables in Table 2 and illuminance, multiple linear regression (MLR) and random forest (RF) algorithms were used to develop models for illuminance mapping based on the abovementioned variable combinations. MLR is widely used for its simplicity and interpretability [33,34]. Random forest regression (RF) is an ensemble algorithm combined with tree predictors such that random inputs and features are selected in the process of forming trees [35]. The predicted result is the average value of the overall regression trees, which can effectively reduce the bias of a single tree [36].

Taking the observed illuminance as the dependent variable and the six variable combinations as the independent variables, multiple linear regression and RF models were fitted. The predictive accuracy was evaluated using 10-fold cross-validation. First, the dataset was randomly divided into 10 uniform subsets. Then, one subset (17 samples) was used as the test set, and the remaining 9 subsets (151 samples) were used as the training set to fit the model. This process was repeated 10 times using each subset as the test set once. *RMSE* (Equation (3)) and R^2 (Equation (4)) were calculated to indicate the model performance.

$$RMSE = \sqrt{\frac{\sum_{\{i=1\}}^{n} (\hat{y}_i - y_i)^2}{n}}, \tag{3}$$

$$R^2 = 1 - \frac{\sum_{i=1}^{n} (\hat{y}_i - y_i)^2}{\sum_{i=1}^{n} (y_i - \overline{y})^2}, \tag{4}$$

where $\hat{y}_i$ is the predicted illuminance (lux), y_i is the actual observed illuminance (lux), $\overline{y}$ is the average illuminance (lux) of the whole road, and n is the number of samples.

The number of weak classifiers (n) and the number of variables selected randomly for each tree split (s) are two important parameters of RF. The parameter n influences the fitting effectiveness [37], and the parameter s affects the final results, both of which were tuned to optimize the model

By comparing the accuracies of the models, the best model and variable combination were chosen. Then, this model was applied to the corresponding JL1-3B spatial variables to produce a high-resolution surface illuminance map of the study area.

3.4. Road Lighting Quality Evaluation

The spatial scopes of the different classes of roads where distribution of illuminance was obvious and continuous were extracted from Google Earth images by visual interpretation. Note that different sections of one road may vary in different classes and lighting conditions; thus, sections of roads were considered as the evaluating objects. During the process of extracting roads, some rules were obeyed. (1) The intersection areas with buffers of approximately 20 m were not extracted to avoid the influence of mixed-source lighting; (2) the street tree canopies were also excluded in order to avoid the shaded pixels within the roads (Figure 7); and (3) due to the relatively low accuracy of nighttime lighting data geometrical correction, roads with relatively low degrees of overlap between the JL1-3B preprocessed image and Google Earth image (base map in the preprocess of geometric correction) were excluded.

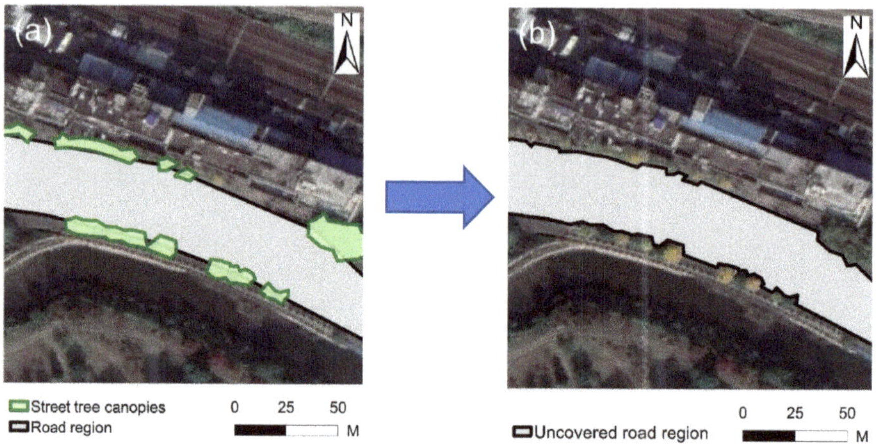

Figure 7. Extraction process illustration of uncovered road region: (**a**) extracted road region and street tree canopies; (**b**) uncovered road region.

Based on the remotely sensed illuminance map and road scopes, the mean value, standard deviation and coefficient of variation (C_v) of the illuminance within each road were calculated. The average illuminance indicates the overall brightness level of the road, and the standard deviation and C_v reflect the brightness uniformity of the road.

$$\overline{E} = \frac{\sum_{i=1}^{n} E_i}{n}, \tag{5}$$

$$\sigma = \sqrt{\frac{\sum_{i=1}^{n} (E_i - \overline{E})^2}{n}}, \tag{6}$$

$$C_v = \frac{\sigma}{\overline{E}}, \tag{7}$$

where $\overline{E}$ is the average illuminance of the road or road section (lux), E_i is the illuminance of pixel i in the road or road section (lux), and n is the number of pixels in the road or road section. σ is the standard deviation of the illuminance of the road or road section (lux), and C_v is the coefficient of variation of the illuminance of the road or road section.

The standard for lightning design of urban roads in China [38] specifies the standards of overall brightness and uniformity. The overall brightness is identified by average illuminance and the threshold values of different classes of road were shown in Table 3. For expressways and main roads, the average illuminance is required to be greater than 20 lux. For secondary roads, the average illuminance should be more than 10 lux, and for branches, the average illuminance should be greater than 8 lux. The uniformity is identified by the ratio of the minimum illuminance to average illuminance. However, this factor is not suitable for remote sensing. The minimum value is not stable, which is easily affected by image noise, shades and other factors. Additionally, it cannot fully utilize the advantages of remote sensing that can map spatial continuous lit environments. In this study, we employed standard deviation and C_v of illuminance to indicate uniformity, and took the mean values of C_v as the reference values for road lighting uniformity.

Table 3. Average illuminance standard for road grades [18,38].

Road Grade	Average Illuminance Standard
Expressway/Main road	≥20 lux
Secondary road	≥10 lux
Branch	≥8 lux

4. Results

4.1. Surface Illuminance Map

Cross-validation results with goodness of fit (R^2) and root mean squared error (RMSE) of MLR and RF models based on different variable combinations are shown in Table 4, and the optimal parameters of RF for each variable combination are listed in Table 5. In terms of R^2, the performance of the RF models was generally better than that of MLR, for which R^2 for all variable combinations was 0.74 or 0.75. Especially for variable combination 1, the R^2 rose from 0.47 (MLR model) to 0.74 (RF model), indicating that the nonlinear relationship between illuminance and the R, G, and B values was more appropriate than the linear relationship. With regard to the performance of MLR models, variable combination 6 was the optimal with the highest R^2 (0.64) and lowest RMSE (9.50 lux), and variable combination 5 was slightly worse than it. From the perspective of variable combinations' color spaces, variable combinations 5 and 6 outperformed variable combinations 3 and 4 as well as combinations 1 and 2 in both the MLR and RF models, suggesting that the models combining RGB and HIS color spaces were superior to the single-color space. Based on the above analysis, RF models with variable combinations 5 and 6 were relatively better choices for estimating the illuminance. Considering that the RMSE of variable combination 5 was only 0.03 smaller than that of variable combination 6, and this study preferred R^2 as the more important metric for models, variable combination 6 applying the RF model with the highest R^2 (0.75) and the relatively low RMSE (9.79 lux) was selected as the optimal model for further estimation.

Table 4. Validation results of the models with different variable combinations.

Models		MLR		RF	
Number	Combinations	R^2	RMSE	R^2	RMSE
1	R, G, B	0.47	11.36	0.74	9.72
2	lnR, lnG, lnB	0.62	9.55	0.74	9.72
3	H, S, I	0.50	10.98	0.75	10.03
4	H, S, lnI	0.62	9.62	0.75	10.03
5	H, S, I, R, G, B	0.63	9.52	0.74	9.76
6	H, S, lnI, lnR, lnG, lnB	0.64	9.50	0.75	9.79

Table 5. Optimal parameters for RF models with different variable combinations.

Combination No.	Number of Classifiers (n)	Number of Variables (s)
1	150	3
2	150	3
3	110	3
4	110	3
5	110	6
6	120	6

Figure 8 shows the scatter plot between the observed and estimated illuminance from the random forest model based on variable combination 6. Most samples were clustered near the 1:1 line, indicating the good fitness of the model. In addition, there was no obvious overestimation or underestimation.

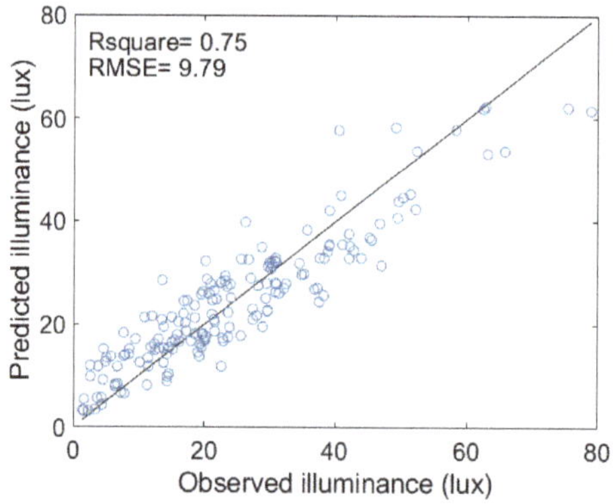

Figure 8. Scatter plot for the optimal model.

The developed random forest model with variable combination 6 was applied to the corresponding spatial independent variables, and the surface illuminance over the study area was mapped (Figure 9).

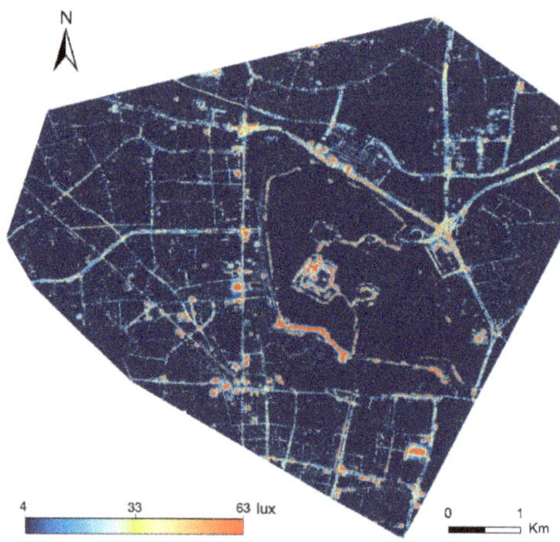

Figure 9. Illuminance map.

4.2. Road Lighting Quality Evalutions

The main 50 roads in the study area were extracted manually, including 16 expressways/main roads, 12 secondary roads and 22 branches. Based on the remotely sensed surface illuminance map and road spatial scopes, the illuminance values within each road were extracted to calculate the average, standard deviation and C_v to assess lighting quality. The results are shown in Table S1. Additionally, the boxplots of the illuminance within each road are shown in Figure S1. Table 6 shows the statistical results of the lighting quality the expressways/main roads, secondary roads and branches. The mean of average illuminance of expressways/main roads, secondary roads and branches were 20.64 lux, 22.35 lux, and 19.46 lux, respectively. Generally, the overall average illuminance of all the three road classes met the average illuminance standard shown in Table 3. However, the minimum values of average illuminance of expressways/main roads, secondary roads and branches were 9.90 lux (Heyan Road), 12.08 lux (Hunan Road), and 8.58 lux (Gaomenlou

Road), respectively, indicating that there were some expressways/main roads and secondary roads that did not meet the average illuminance standard. Overall, 87.5% of the expressways/main roads, all the secondary roads and all the branches met the standard of average illuminance.

Table 6. The statistics of road lighting quality indicators for each road grade.

Grade	Road Lighting Quality Indicators	Maximum	Minimum	Mean
Expressway/Main road	Average illuminance(lux)	27.74	9.90	20.64
	Std (lux)	14.12	8.28	10.89
	C_v	0.98	0.42	0.56
Secondary road	Average illuminance (lux)	36.75	12.08	22.35
	Std (lux)	13.83	8.29	10.65
	C_v	0.90	0.33	0.52
Branch	Average illuminance (lux)	36.13	8.58	19.46
	Std (lux)	12.79	5.30	9.51
	C_v	0.89	0.28	0.54

From the aspects of standard deviation, which represents the absolute uniformity, the mean value of standard deviation of expressways/main roads, secondary roads and branches were 10.89 lux, 10.65 lux, and 9.51 lux, respectively. Though the three road classes showed similar overall standard deviation, their C_v values were quite different because their average illuminances are different. In fact, absolute uniformity (standard deviation) was not suitable for the indicator that was used to compare illuminance variation in roads as it ignores the impact of average illuminance, thus, C_v was considered as the better indicator to evaluate the uniformity of roads. The mean values of C_v of expressways/main roads, secondary roads and branches were 0.56, 0.52 and 0.54, respectively. However, the maximum C_v values of these three road classes were 0.98 (Heyan Road), 0.90 (Hunan Road) and 0.89 (Gaomenlou Road), suggesting that there were some roads that suffered poor uniformity of illuminance. Given that the existing standard road lighting uniformity is not proper for remotely sensed illuminance, we set C_v of the three road classes (0.56, 0.52, and 0.54, respectively) as the reference values of lighting uniformity. The roads with C_v higher than the reference values were considered that had poor road lighting uniformity. Overall, 75% of the expressways/main roads, 66.7% of the secondary roads and 50% of the branches showed good uniformity.

Four typical roads were selected for detailed analysis, including Jianning Road, Hubei Road, Hongshan Road and Longpan Road (Table 7). Jianning Road showed both poor overall brightness and uniformity; Hubei Road showed good overall brightness, but poor uniformity; Hongshan Road had poor overall brightness, but good uniformity; Longpan Road had both good overall brightness and uniformity.

Table 7. Values of typical roads' lighting quality evaluation metrics.

Road Name	Average Illuminance (lux)	Std (lux)	C_v	Grade
Jianning Road	16.65	9.83	0.59	Main road
Hubei Road	23.03	12.79	0.56	Branch
Hongshan Road	19.27	8.51	0.44	Expressway
Longpan Road	27.74	13.44	0.48	Expressway

Figure 10 shows the illuminance maps (local segments) and histograms (all the roads) of these four roads, which presents more details at pixel scale than the table and boxplots. For Jianning Road (Figure 10a), the illuminance map shows that most pixels within the road had low illuminance values indicating the poor overall brightness, and unilluminated sections of road were interspersed, resulting in the awful uniformity. The histogram of this road illustrates that most pixels were under the standard for the main roads (20 lux), and the distribution was obviously skewed to a low illuminance value. For Hubei Road

(Figure 10b), most areas within the road were obviously de-lighted. However, the middle sections were much brighter than the sides. The histogram indicates that most pixels ranged from 15 to 35 lux and there were not relatively fewer extreme illuminated pixels. For Hongshan Road (Figure 10c), the road regions were approximately evenly and fully illuminated, but the overall brightness was relatively low. Most pixels were concentrated between 10–20 lux from the histogram of this road, and few pixels had extreme illuminance values. The illuminance map for Longpan Road (Figure 10d) showed a generally high and even distributed lit environment, and the histogram also indicated this point. The illuminance map and histogram can not only evaluate the lighting quality of the whole road, but can reflect more spatial and distribution details.

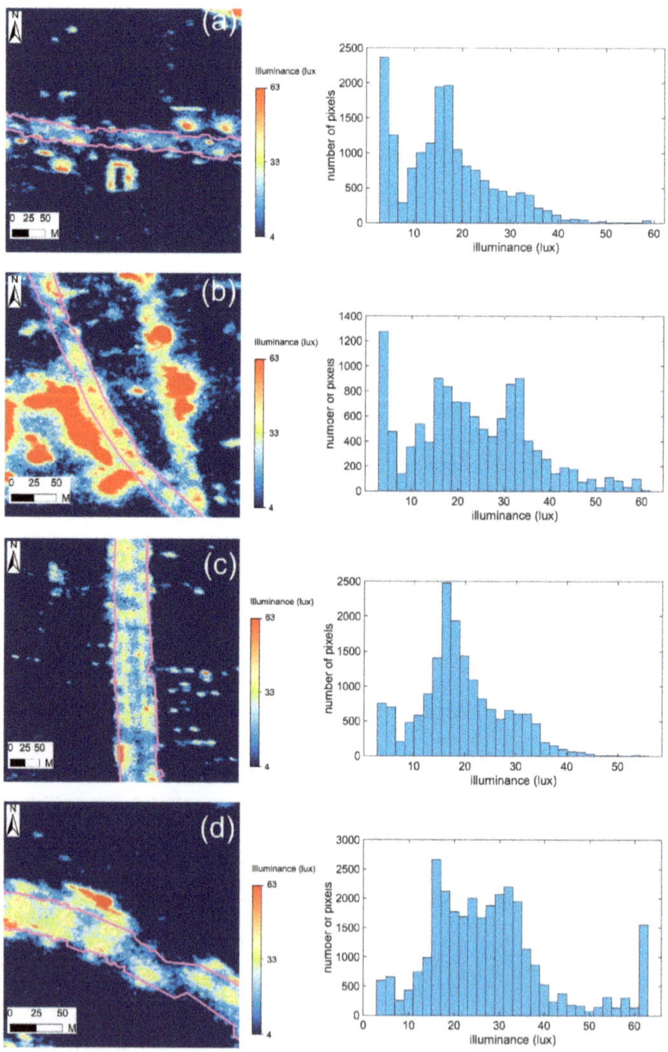

Figure 10. The illuminance maps and histograms of 4 typical roads. The purple box shows the roads' scopes: (**a**) Jianning Road; (**b**) Hubei Road; (**c**) Hongshan Road; and (**d**) Longpan Road. To illustrate the illuminance more clearly, illuminance maps show local sections of the road. Histograms represent the illuminance of the whole road.

5. Discussion

Traditional ground-based measurements are limited by the measuring scopes and usually disturb the normal traffic flow. Nighttime remote sensing can observe spatial continuous lit environments at large scales, providing potential for road lighting quality evaluation. However, due to the relative coarse resolutions, most nighttime remote sensing data cannot be used to map illuminance within roads. The new JL1-3B data has a high spatial resolution of 0.92 m that can capture detailed lighting environment within roads, and it also has three channels that can well characterize the light colors [23,24]. Therefore, JL1-3B is a proper data that can be used for effectively quantifying road lighting quality. This paper explores the method to evaluating road lighting quality by the estimated illuminance derived from JL1-3B data and in situ observations. The high-resolution illuminance map contains large amounts of illuminance data and can effectively depict the lighting environment within roads at pixel scale. Furthermore, with the revisiting period of 9 days, the proposed method using JL1-3B data is able to measure the road lighting quality periodically, providing timely information about large-scale road lighting condition for concerned government departments. Compared with traditional road lighting quality measurements, the application of nighttime light remote sensing data in this study is superior in terms of safety, rapidity, measuring scopes, measuring frequency and information content. In addition to the radiance values of the three bands of JL1-3B, other color components, such as HIS, were also introduced for estimating illuminance. Six combinations of variables were compared to determine the optimal variables combination. The results (Table 4) showed the optimal model involved in all color components, indicating that considering more color components may improve the performance of the estimation model. Furthermore, the random forest model outperformed the multiple regression model. This may be attributed to the fact that different street lamps emit light of different colors, and the tree-based random forest method that can handle different conditions under different branches is more suitable for this complex scenario.

There were also some limitations in this study. During the process of data collection, some deficiencies like discrepancies between the ground and satellite observations and low positioning accuracy of the GNSS system, have negative impacts on the outcomes of the methods. Satellite imaging was almost instantaneous, but in situ observation lasted approximately 1 h. During observation, the lit environment may change, resulting in inconsistencies between field observations and remote sensing data. Although we carefully checked them and removed some obvious problematic samples, there may still be some uncertainties caused by the inconsistency. To overcome this problem, the planned sampling section can be divided into stable and unstable illumination zones. For stable zones, the lighting is steady, and therefore, the illuminance cannot be measured strictly synchronously with the satellite overpass. For unstable zones, the lighting is changeable; therefore, the illuminance should be recorded strictly synchronously with the satellite overpass. Assigning more observation teams also helps to reduce the inconsistency because they can collect abundant samples in less time. Moreover, there are also a lot of pixels that were affected by these factors in the remotely sensed illuminance map. How to identify and remove these problematic pixels is still a difficult problem to overcome in the future. During this in situ observation, normal handheld GNSS instruments were used to record sample coordinates. However, the positioning accuracy of these instruments is generally within 10 m. Given the obvious spatial difference in the light environment, this accuracy level is not sufficient. RTK GNSS instruments with centimeter-level accuracy can effectively improve the spatial consistency between field observations and remote sensing data. There is another important point to note, which is the observation angle of the satellite. Though JL1-3B can image within the off-nadir angle of $\pm 45^{\circ}$, the satellite zenith angle should not be high enough to eliminate the shading effect of buildings and trees, and also reduce the influence of Rayleigh scattering of the atmosphere.

6. Conclusions

This study proposed a new space-borne method for evaluating road lighting quality based on JL1-3B nighttime light remote sensing data. Firstly, synchronous field observations were carried out to measure illuminance in Nanjing, China. After a series of preprocessing of the JL1-3B image, the in situ observed illuminance and the radiance of JL1-3B were combined to map the high-resolution surface illuminance based on the close relationship between them. Two models (multiple linear regression and random forest) with six independent variable combinations were employed and compared to develop the optimal model for illuminance estimation. Results showed that the random forest model with Hue, Saturability, lnI, lnR, lnG and lnB as the independent variables achieved the best performance (R^2 = 0.75, RMSE = 9.79 lux). Additionally, the optimal model was applied to the JL1-3B preprocessed image to derive the surface illuminance map. The average, standard deviation and C_v of the illuminance within roads were calculated to assess their lighting quality.

This study is a preliminary study to develop a technical framework to evaluate road lighting quality using JL1-3B nighttime light remote sensing data. JL1-3B can be ordered from Changguang Co., Ltd., Changchun, China and the instruments for ground-based observation were inexpensive TES-1399R illuminance meters. The devices and methods used in this study have the advantages of low cost, simplicity and reliability, providing a reference for road lighting quality evolution in other regions.

Supplementary Materials: The following supporting information can be downloaded at: https://www.mdpi.com/article/10.3390/rs14184497/s1, Figure S1: The boxplot of the extracted roads; Table S1: Road lighting quality evaluation metrics of the extracted roads.

Author Contributions: Conceptualization, Y.X.; methodology, Y.X. and N.X.; formal analysis, N.X., Y.X. and B.W.; investigation, Y.X., Y.Y., X.Z., Z.G. and B.W.; writing—original draft preparation, N.X. and Y.X.; writing—review and editing, N.X. and Y.X.; supervision, Y.X.; project administration, Y.X. All authors have read and agreed to the published version of the manuscript.

Funding: This research was funded by the National Natural Science Foundation of China (41871028, 42171101) and the Science and Technology Research Plan in Key areas of Xinjiang Production and Construction Corps (2022AB016).

Conflicts of Interest: The authors declare no conflict of interest.

References

1. CIE Central Bureau. *Road Transport Lighting for Developing Countries*; CIE Central Bureau: Vienna, Austria, 2007.
2. Monsere, M.C.; Fischer, E.L. Safety effects of reducing freeway illumination for energy conservation. *Accid. Anal. Prev.* **2008**, *40*, 1773–1780. [CrossRef] [PubMed]
3. Wanvik, P.O. Effects of Road Lighting on Motorways. *Traffic Inj. Prev.* **2009**, *10*, 279–289. [CrossRef] [PubMed]
4. Frith, W.; Tate, F. The safety impact of road lighting on roads with speed limits greater than 70 km/h. *Road Transp. Res. A J. Aust. New Zealand Res. Pract.* **2016**, *25*, 62–72.
5. Xu, Y.; Sun, C.; Ye, Z.; Zhao, Q. *The Influence of Road Lighting on Safety at Crossings*; American Society of Civil Engineers: Reston, VA, USA, 2017; pp. 5056–5066.
6. Dodds, G.; Glenn, J.J.; Robinson, R.; Mullin, F. *Assessment of Road-Lighting Parameters Using Camera Based Systems*; IET: Lucknow, India, 1999; p. 5.
7. Fisher, A.J. Road lighting as an accident counter-measure. *Aust. Road Res.* **1977**, *7*, 3–16.
8. Armas, J.; Laugis, J. *Road Safety by Improved Road Lighting: Road lighting Measurements and Analysis*; Tallinn University of Technology: Tallinn, Estonia, 2007.
9. Beyer, R.F.; Ker, K. *Street Lighting for Preventing Road Traffic Injuries*; Cochrane Database of Systematic Reviews: London, UK, 2009.
10. International Commission on Illumination (CIE). *Lighting of Roads for Motor and Pedestrian Traffic*, 2nd ed.; International Commission on Illumination (CIE): Vienna, Austria, 2010.
11. European Committee for Standardization (CEN). *Road Lighting—Part 2: Performance Requirements*; European Committee for Standardization (CEN): Brussels, Belgium, 2015.
12. Liu, M.; Zhang, B.; Luo, T.; Liu, Y.; Portnov, B.A.; Trop, T.; Jiao, W.; Liu, H.; Li, Y.; Liu, Q. Evaluating Street Lighting Quality in Residential Areas by Combining Remote Sensing Tools and a Survey on Pedestrians' Perceptions of Safety and Visual Comfort. *Remote Sens.* **2022**, *14*, 826. [CrossRef]

13. Guo, L.; Eloholma, M.; Halonen, L. Luminance Monitoring and Optimization of Luminance Metering in Intelligent Road Lighting Control Systems. *Lighting Eng.* **2007**, *9*, 24–40.
14. Jägerbrand, A.K. LED (Light-Emitting Diode) Road Lighting in Practice: An Evaluation of Compliance with Regulations and Improvements for Further Energy Savings. *Energies* **2016**, *9*, 357. [CrossRef]
15. Ekrias, A.; Eloholma, M.; Halonen, L.; Song, X.J.; Zhang, X.; Wen, Y. Road lighting and headlights: Luminance measurements and automobile lighting simulations. *Build. Environ.* **2008**, *43*, 530–536. [CrossRef]
16. CEN. *European Standard, Road Lighting—Part 3: Calculation of Performance (EN 13201-3:2015)*; CEN: Brussels, Belgium, 2015.
17. CIE. *Calculation and Measurement of Luminance and Illuminance in Road Lighting*; CIE: Paris, France, 1990.
18. SAC. *Measurement Methods for Lighting (GB/T 5700-2008)*; China National Standard Press: Beijing, China, 2008; pp. 5–6.
19. Greffier, F.; Muzet, V.; Boucher, V.; Fournela, F.; Dronneau, R. Use of an imaging luminance measuring device to evaluate road lighting performance at different angles of observation. In Proceedings of the 29th Quadrennial Session of the CIE, Washington, DC, USA, 14–22 June 2019; International Commission on Illumination: Vienna, Austria, 2019.
20. Zhou, H.; Pirinccioglu, F.; Hsu, P. A new roadway lighting measurement system. *Transp. Res. Part C Emerg. Technol.* **2009**, *17*, 274–284. [CrossRef]
21. Elvidge, C.D.; Hsu, F.C.; Baugh, K.E.; Ghosh, T. National trends in satellite-observed lighting. *Glob. Urban Monit. Assess. Through Earth Obs.* **2014**, *23*, 97–118.
22. Levin, N.; Kyba, C.C.; Zhang, Q.; de Miguel, A.S.; Román, M.O.; Li, X.; Portnov, B.A.; Molthan, A.L.; Jechow, A.; Miller, S.D. Remote sensing of night lights: A review and an outlook for the future. *Remote Sens. Environ.* **2020**, *237*, 111443. [CrossRef]
23. Cheng, B.; Chen, Z.; Yu, B.; Li, Q.; Wang, C.; Li, B.; Wu, B.; Li, Y.; Wu, J. Automated Extraction of Street Lights from JL1-3B Nighttime Light Data and Assessment of Their Solar Energy Potential. *IEEE J. Sel. Top. Appl. Earth Obs. Remote Sens.* **2020**, *13*, 675–684. [CrossRef]
24. Zheng, Q.; Weng, Q.; Huang, L.; Wang, K.; Deng, J.; Jiang, R.; Ye, Z.; Gan, M. A new source of multi-spectral high spatial resolution night-time light imagery—JL1-3B. *Remote Sens. Environ.* **2018**, *215*, 300–312. [CrossRef]
25. Rabaza, O.; Molero-Mesa, E.; Aznar-Dols, F.; Gómez-Lorente, D. Experimental Study of the Levels of Street Lighting Using Aerial Imagery and Energy Efficiency Calculation. *Sustainability* **2018**, *10*, 4365. [CrossRef]
26. Wang, F.; Zhou, K.; Wang, M.; Wang, Q. The Impact Analysis of Land Features to JL1-3B Nighttime Light Data at Parcel Level: Illustrated by the Case of Changchun, China. *Sensors* **2020**, *20*, 5447. [CrossRef]
27. National Bureau of Statistics of China. *Nanjing Statistical Yearbook*; National Bureau of Statistics of China: Beijing, China, 2021.
28. Du, M.; Wang, L.; Zou, S.; Shi, C. Modeling the Census Tract Level Housing Vacancy Rate with the Jilin1-03 Satellite and Other Geospatial Data. *Remote Sens.* **2018**, *10*, 1920. [CrossRef]
29. Ribas, S.J.; Torra, J.; Paricio, S.; Canal-Domingo, R. How clouds are amplifying (or not) the effects of ALAN. *Int. J. Sustain. Lighting* **2016**, *18*, 32–39. [CrossRef]
30. Kyba, C.; Tong, K.P.; Bennie, J.; Birriel, I.; Birriel, J.J.; Cool, A.; Danielsen, A.; Davies, T.W.; Outer, P.N.D.; Edwards, W.; et al. Worldwide variations in artificial skyglow. *Sci. Rep.* **2015**, *5*, 1–7. [CrossRef]
31. Oh, C.; Kim, B.; Sohn, K. Automatic illumination invariant road detection with stereo vision. In Proceedings of the 2012 7th IEEE Conference on Industrial Electronics and Applications (ICIEA), Singapore, 18–20 July 2012; pp. 889–893.
32. Xu, Y.; Knudby, A.; Côté-Lussier, C. Mapping ambient light at night using field observations and high-resolution remote sensing imagery for studies of urban environments. *Build. Environ.* **2018**, *145*, 104–114. [CrossRef]
33. Pan, J.; Hu, Y. Spatial Identification of Multi-dimensional Poverty in Rural China: A Perspective of Nighttime-Light Remote Sensing Data. *J. Indian Soc. Remote Sens.* **2018**, *46*, 1093–1111. [CrossRef]
34. Li, X.; Chen, X.; Zhao, Y.; Xu, J.; Chen, F.; Li, H. Automatic intercalibration of night-time light imagery using robust regression. *Remote Sens. Lett.* **2013**, *4*, 45–54. [CrossRef]
35. Breiman, L. Random forests. *Mach. Learn.* **2001**, *45*, 5–32. [CrossRef]
36. Cootes, T.F.; Ionita, M.C.; Lindner, C.; Sauer, P. *Robust and Accurate Shape Model Fitting Using Random Forest Regression Voting*; Springer: Berlin, Germany, 2012.
37. Segal, M.R. *Machine Learning Benchmarks and Random Forest Regression*; ResearchGate: Berlin, Germany, 2004.
38. *CJJ 45-2015*; Standard for Lighting Design of Urban Road. China Architecture & Building Press: Beijing, China, 2015. (In Chinese)

MDPI
St. Alban-Anlage 66
4052 Basel
Switzerland
Tel. +41 61 683 77 34
Fax +41 61 302 89 18
www.mdpi.com

Remote Sensing Editorial Office
E-mail: remotesensing@mdpi.com
www.mdpi.com/journal/remotesensing

www.ingramcontent.com/pod-product-compliance
Lightning Source LLC
Chambersburg PA
CBHW061606220926
51850CB00005B/433
9783036571782